Stochastic Processes in Science, Engineering and Finance

Stochastic Processes in Science, Engineering and Finance

Frank Beichelt
University of the Witwatersrand
Johannesburg, Republic of South Africa

Chapman & Hall/CRC
Taylor & Francis Group
Boca Raton London New York

Published in 2006 by
Chapman & Hall/CRC
Taylor & Francis Group
6000 Broken Sound Parkway NW, Suite 300
Boca Raton, FL 33487-2742

Printed in the United States of America on acid-free paper
10 9 8 7 6 5 4 3 2 1

International Standard Book Number-10: 1-58488-493-2 (Hardcover)
International Standard Book Number-13: 978-1-58488-493-4 (Hardcover)
Library of Congress Card Number 2005054991

Library of Congress Cataloging-in-Publication Data

Beichelt, Frank, 1942-
Stochastic processes in science, engineering, and finance / Frank Beichelt.
p. cm.
Includes bibliographical references and index.
ISBN 1-58488-493-2 (alk. paper)
1. Stochastic processes. I. Title.

QA274.B399 2006
519.2'3--dc22 2005054991

Meiner Mutter,

Gerda Johanna Beichelt,
geb. *Venus*,

zum 90. Geburtstag

PREFACE

The book is a self-contained introduction into stochastic processes with special emphasis on their applications in science, engineering, finance, computer science and operations research. It provides theoretical foundations for modeling time-dependent random phenomena in these areas and illustrates their application through the analysis of numerous, practically relevant examples. As a nonmeasure theoretic text, the material is presented in a comprehensible, application-oriented way. Its study only assumes a mathematical maturity which students of applied sciences acquire during their undergraduate studies in mathematics and probability theory. However, readers with this background are not advised to completely ignore the introductory chapter 1. Although this chapter mainly summarizes standard knowledge on random variables, it also deals with subjects as mixed probability distributions, moment generating functions, nonparametric classes of probability distributions, inequalities, convergence criteria, and limit theorems, which undergraduate courses usually not cover. For readers who need to refresh their knowledge of probability theory studying chapter 1 is imperative anyway. The study of stochastic processes as of any other mathematically based science requires less routine effort, but more creative work on one's own. Therefore, numerous exercises have been added to enable readers to assess to which extent they have grasped the subject. Solutions to most of the exercises can be found in an appendix or exercises are given together with their solutions. To make the book attractive to theoretically interested readers as well, some important proofs and challenging examples and exercises have been included. Exercises marked with a star belong to this category. The chapters are organized in such a way that reading a chapter only requires knowledge of some of the previous ones. The book has been developed as a course text for undergraduates. But parts of it may also serve as a basis for preparing senior undergraduate and graduate level courses for students of applied fields.

Generally, this book does not deal with data analysis aspects of stochastic processes. It can be anticipated that readers will use a software package for tackling statistical and numerical problems. However, studying the text will enable readers to work more creatively with such software and develop their own one. On the whole, the author hopes the book will fulfil its main purpose which is to enable readers to apply stochastic modeling in their own fields.

This book has its origin in the author's text *Stochastische Prozesse für Ingenieure* [6] (English translation [7]). This is still visible, but substantial changes in contents and presentation have led to a new work. Helpful comments are welcome and should be sent to the author: Beichelt@stats.wits.ac.za.

Johannesburg, November 2005 *Frank Beichelt*

Acknowledgment

I sincerely appreciate the constructive, helpful cooperation with Theresa Delforn, Dr. Sunil Nair, Helena Redshaw and Sylvia Wood from CRC Press.

I am grateful to my colleague Professor David Lazar Block, Director of the Cosmic Dust Laboratory at the University of the Witwatersrand Johannesburg, for contributing his fitting and beautiful cover photo.

Many thanks to my honours student Ms. Angela Rademeyer, who helped by her probing and persistent questions to identify and eliminate many a weak spot in a previous version of the text.

The author is deeply indebted to the *National Research Foundation of South Africa* (NRF) for their support of the development of this book within his research project.

Frank E. Beichelt

Short Biography of Frank E. Beichelt

Frank E. Beichelt is a professor of operations research in the School of Statistics and Actuarial Science at the University of the Witwatersrand Johannesburg, Republic of South Africa. Previously he was a full professor of mathematics and head of the Department of Mathematics at the University of Applied Sciences Mittweida, Germany, an associate professor for reliability and maintenance theory at the University for Transportation and Communication 'Friedrich List' Dresden, and a deputy head engineer for science and technology in the German lignite mining industry. He holds an MSc in mathematics from the Friedrich-Schiller-University Jena, a Dr. rer. nat. in mathematics from the Mining Academy Freiberg, and a senior doctorate in engineering (Dr. sc. techn.) from the University of Transportation and Communication 'Friedrich List' Dresden. He is author and coauthor of 10 books and about 120 publications in journals and conference proceedings in the field of stochastic modeling.

SYMBOLS AND ABBREVIATIONS

□ ■ ●	symbols after an example, a theorem, a definition
$f(t) \equiv c$	$f(t) = c$ <u>for all</u> $t \in \mathbf{T}$
$f * g$	convolution of two functions f and g
$f^{*(n)}$	nth convolution power of f
$\hat{f}(s),\ L\{f\}$	Laplace transform of a function f
$o(x)$	Landau order symbol
δ_{ij}	Kronecker symbol
x_+	$\max(0, x)$

Probability Theory

$X,\ Y,\ Z$	random variables
$E(X),\ Var(X)$	mean (expected) value of X, variance of X
$f_X(x),\ F_X(x)$	probability density function, (cumulative probability) distribution function of X
$F_Y(y\|x),\ f_Y(y\|x)$	conditional distribution function, density of Y given $X = x$
$X_t, F_t(x)$	residual lifetime of a system of age t, distribution function of X_t
R_t	residual lifetime of a system which is operating at time t
$E(Y\|x)$	conditional mean value of Y given $X = x$
$\lambda(x),\ \Lambda(x)$	failure rate, integrated failure rate (hazard function)
$N(\mu, \sigma^2)$	normally distributed random variable (normal distribution) with mean value μ and variance σ^2
$\varphi(x),\ \Phi(x)$	probability density function, distribution function of a standard normal random variable
$f_{\mathbf{X}}(x_1, x_2, \ldots, x_n)$	joint probability density function of $\mathbf{X} = (X_1, X_2, \ldots, X_n)$
$F_{\mathbf{X}}(x_1, x_2, \ldots, x_n)$	joint distribution function of $\mathbf{X} = (X_1, X_2, \ldots, X_n)$
$Cov(X, Y),\ \rho(X, Y)$	covariance, correlation between X and Y
$M(z)$	z-transform (moment generating function) of a discrete random variable or its probability distribution

Stochastic Processes

$\{X(t), t \in \mathbf{T}\},\ \{X_t, t \in \mathbf{T}\}$	continuous-time, discrete-time stochastic process with parameter space $\mathbf{T}$
$\mathbf{Z}$	state space of a stochastic process
$f_t(x),\ F_t(x)$	probability density, distribution function of $X(t)$
$f_{t_1, t_2, \ldots, t_n}(x_1, x_2, \ldots, x_n),\ F_{t_1, t_2, \ldots, t_n}(x_1, x_2, \ldots, x_n)$	joint probability density, joint distribution function of $(X(t_1), X(t_2), \ldots, X(t_n))$

$m(t)$	trend function of a stochastic process
$C(s,t)$	covariance function of a stochastic process
$C(\tau)$	covariance function of a stationary stochastic process
$C(t),\ \{C(t), t \ge 0\}$	compound random variable, compound stochastic process
$\rho(s,t)$	correlation function of a stochastic process
$\{T_1, T_2, ...\}$	random point process
$\{Y_1, Y_2, ...\}$	sequence of interarrival times, renewal process
N	integer-valued random variable, discrete stopping time
$\{N(t), t \ge 0\}$	(random) counting process
$N(s, t)$	increment of a counting process in $(s, t]$
$H(t),\ H_1(t)$	renewal function of an ordinary, delayed renewal processs
$A(t)$	forward recurrence time, point avaliability
$B(t)$	backward recurrence time
$R(t),\ \{R(t), t \ge 0\}$	risk reserve, -process; repair cost rate, -process
R_t	residual lifetime of a system operating at time t
$A,\ A(t)$	stationary (long-run) availability, point availability
$p_{ij},\ p_{ij}(t)$	transition probabilities of a homogeneous, discrete-, continuous-time Markov chain
$q_{ij},\ q_i$	conditional, unconditional transition rates (transition intensities) of a homogeneous, continuous-time Markov chain
$\{\pi_i; i \in \mathbf{Z}\}$	stationary state distribution of a homogeneous Markov chain
$\lambda_j,\ \mu_j$	birth, death rates
λ, μ, ρ	arrival rate, service rate, traffic intensity λ/μ (in queueing models)
μ_i	mean sojourn time of a semi-Markov process in state i
μ	drift parameter of a Brownian motion process with drift
W	waiting time in a queueing system
L	lifetime, cycle length, queue length, continuous stopping time
$L(x)$	first-passage time with regard to level x
$L(a,b)$	first-passage time with regard to level $\min(a, b)$
$\{B(t), t \ge 0\}$	Brownian motion (process)
σ^2	$\sigma^2 = Var(B(1))$ (volatility)
$\{S(t), t \ge 0\}$	standardized Brownian motion ($\sigma = 1$)
$\{\bar{B}(t), 0 \le t \le 1\}$	Brownian bridge
$\{D(t), t \ge 0\}$	Brownian motion with drift
$\{D_u(t), t \ge 0\}$	shifted Brownian motion with drift
$M(t)$	absolute maximum of the Brownian motion (with drift) in $[0, t]$
M	absolute maximum of the Brownian motion (with drift) in $[0, \infty)$
$\{U(t), t \ge 0\}$	Ornstein-Uhlenbeck process, integrated Brownian motion process

TABLE OF CONTENTS

2 BASICS OF STOCHASTIC PROCESSES

3 RANDOM POINT PROCESSES

4 MARKOV CHAINS IN DISCRETE TIME

5 MARKOV CHAINS IN CONTINUOS TIME

6 MARTINGALES

7 BROWNIAN MOTION

CHAPTER 1

Probability Theory

1.1 RANDOM EVENTS AND THEIR PROBABILITIES

Probability theory comprises mathematically based theories and methods for investigating random phenomena. Formally, random phenomena occur in connection with random experiments. A *random experiment* is characterized by two properties:

1. Repetitions of the experiment, even if carried out under identical conditions, generally have different outcomes.

2. The possible outcomes of the experiment are known.

Thus, the outcomes of a random experiment cannot be predicted with certainty. However, if random experiments are repeated sufficiently frequently under identical conditions, *stochastic* or *statistical regularities* can be found. Examples of random experiments are:

1) Counting the number of vehicles arriving at a filling station a day.

2) Counting the number of shooting stars during a fixed time interval. The possible outcomes are, as in the previous random experiment, nonnegative integers.

3) Recording the daily maximum wind velocity at a fixed location.

4) Recording the lifespans of technical systems or organisms.

5) Recording the daily maximum fluctuation of share prices. The possible outcomes are, as in the random experiments 3 and 4, nonnegative numbers.

6) The total profit sombody makes with his financial investments a year. This 'profit' can be negative, i.e. any real number can be the outcome.

As the examples show, in this context the term 'experiment' has a more abstract meaning than in the customary sense.

Random Events A possible outcome a of a random experiment is called an *elementary* or a *simple event.* The set of all elementary events is called *space of elementary events* or *sample space.* Here and in what follows, the sample space is denoted as $\mathbf{M}$. A sample space $\mathbf{M}$ is *discrete* if it is a finite or a countably infinite set.

A *random event* (briefly: *event*) A is a subset of $\mathbf{M}$. An event A is said to have *occurred* if the outcome a of the random experiment is an element of A: $a \in A$.

Let A and B be two events. Then the set-theoretic operations *intersection* '$\cap$' and *union* '$\cup$' can be interpreted in the following way:

$A \cap B$ is the event that both A and B occur and $A \cup B$ is the event that A or B (or both) occur.

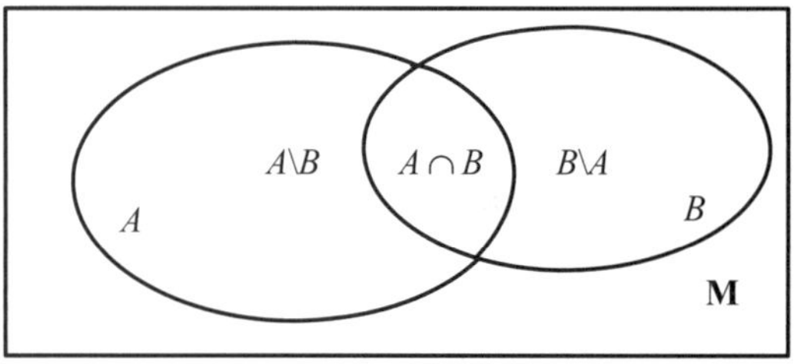

Figure 1.1 Venn Diagram

If $A \subseteq B$ (A is a subset of B), then the occurrence of A implies the occurrence of B.

$A \backslash B$ is the set of all those elementary events which are elements of A, but not of B. Thus, $A \backslash B$ is the event that A occurs, but not B. Note that $A \backslash B = A \backslash (A \cap B)$.

The event $\overline{A} = \mathbf{M} \backslash A$ is the *complement of* A. If A occurs, then $\overline{A}$ cannot occur and vice versa.

Rules of de Morgan Let $A_1, A_2, ..., A_n$ be a sequence of random events. Then

$$\overline{\bigcup_{i=1}^n A_i} = \bigcap_{i=1}^n \overline{A}_i, \quad \overline{\bigcap_{i=1}^n A_i} = \bigcup_{i=1}^n \overline{A}_i. \tag{1.1}$$

In particular, if $n = 2$, $A_1 = A$ and $A_2 = B$, the rules of de Morgan simplify to

$$\overline{A \cup B} = \overline{A} \cap \overline{B}, \quad \overline{A \cap B} = \overline{A} \cup \overline{B}. \tag{1.2}$$

The empty set $\emptyset$ is the *impossible event,* since, for not containing an elementary event, it can never occur. By definition, **M** contains all elementary events so that it must always occur. Hence **M** is called the *certain event.* Two events A and B are called disjoint or (*mutually*) *exclusive* if their joint occurrence is impossible, i.e. if $A \cap B = \emptyset$. In this case the occurrence of A implies that B does not occur and vice versa. In particular, A and $\overline{A}$ are disjoint events (Figure 1.1).

Probability Let $\boldsymbol{M}$ be the set of all those random events A which can occur when carrying out the random experiment, including **M** and $\emptyset$. Further, let $P = P(A)$ be a function on $\boldsymbol{M}$ with properties

I) $P(\emptyset) = 0, \quad P(\mathbf{M}) = 1,$

II) for any event A, $0 \le P(A) \le 1$,

III) for any sequence of disjoint (mutually exclusive) random events $A_1, A_2, ...$, i.e. $A_i \cap A_j = \emptyset$ for $i \ne j$,

$$P\left(\bigcup_{i=1}^\infty A_i \right) = \sum_{i=1}^\infty P(A_i). \tag{1.3}$$

The number $P(A)$ is the *probability* of event A. $P(A)$ characterizes the degree of certainty of the occurrence of A. This interpretation of the probability is justified by the following implications from properties I) to III).

Specifically, the independence of the A_i implies for $k=n$ a direct generalization of formula (1.9):

$$P(A_1 \cap A_2 \cap \cdots \cap A_n) = P(A_1)\,P(A_2)\cdots P(A_n). \tag{1.10}$$

Example 1.1 In a set of traffic lights, the colour 'red' (as well as green and yellow) is indicated by two bulbs which operate independently of each other. Colour 'red' is clearly visible if at least one bulb is operating. What is the probability that in the time interval [0, 200 *hours*] colour 'red' is visible if it is known that a bulb survives this interval with probability 0.95 ? To answer this question, let

A = 'bulb 1 does not fail in [0, 200]', B = 'bulb 2 does not fail in [0, 200]'.

The event of interest is

$C = A \cup B$ = 'red light is clearly visible in [0, 200]'.

Since A and B are independent,

$$P(C) = P(A \cup B) = P(A) + P(B) - P(A \cap B)$$

$$= P(A) + P(B) - P(A)\,P(B) = 0.95 + 0.95 - (0.95)^2 = 0.9975\,.$$

Another possibility of solving this problem is to apply the rule of de Morgan (1.2):

$$P(\overline{C}) = P(\overline{A \cup B}) = P(\overline{A} \cap \overline{B}) = P(\overline{A})\,P(\overline{B})$$

$$= (1-0.95)(1-0.95) = 0.0025.$$

Hence, $P(C) = 1 - P(\overline{C}) = 0.9975.$ □

Example 1.2 1% of the population in a country are HIV-positive. A test procedure for diagnosing whether a person is HIV-positive indicates with probability 0.98 that the person is HIV-positive if it is HIV-positive, and with probability 0.96 that this person is not HIV-positve if it is not HIV-positive. What is the probability that a test person is HIV- positive if the test indicates that?

To solve the problem, random events A and B are introduced:

A = 'The test indicates that a person is HIV-positive.'

B = 'A test person is HIV-positive.'

Then,

$$P(B) = 0.01, \quad P(\overline{B}) = 0.99$$

$$P(A|B) = 0.98, \quad P(\overline{A}|B) = 0.02,$$

$$P(\overline{A}|\overline{B}) = 0.96, \quad P(A|\overline{B}) = 0.04.$$

Since $\{B, \overline{B}\}$ is an exhaustive set of events with $B \cap \overline{B} = \emptyset$, the total probability rule (1.7) is applicable to determining $P(A)$:

$$P(A) = P(A|B)\,P(B) + P(A|\overline{B})\,P(\overline{B})$$

$$= 0.98 \cdot 0.01 + 0.04 \cdot 0.99 = 0.0494\,.$$

Bayes' theorem (1.8) yields the desired probability $P(B|A)$:

$$P(B|A) = \frac{P(A|B)\,P(B)}{P(A)} = \frac{0.98 \cdot 0.01}{0.0494} = 0.1984\,.$$

Although the initial parameters of the test look acceptable, this result is quite unsatisfactory: In view of $P(\overline{B}|A) = 0.8016$, about 80% HIV-negative test persons will be shocked to learn that the test procedure indicates they are HIV-positive. In such a situation the test has to be repeated several times.

The probability that a person is not HIV-positive if the test procedure indicates this is

$$P(\overline{B}|\overline{A}) = \frac{P(\overline{A}|\overline{B})\,P(\overline{B})}{P(\overline{A})} = \frac{0.96 \cdot 0.99}{1 - 0.0494} = 0.99979\,.$$

This result is, of course, an excellent feature of the test. □

1.2 RANDOM VARIABLES

1.2.1 Basic Concepts

All the outcomes of the random experiments 1 to 6 at page 1 are real numbers. But when considering the random experiment 'tossing a die', the set of outcomes is 'head' and 'tail'. With such outcomes, no quantitative analysis of the random experiment is possible. Hence it makes sense to assign, for instance, number 1 to 'head' and number 0 to 'tail'. Or consider a problem in quality control. The possible outcomes when testing a unit be 'faulty' and 'operating'. The random experiment consists in checking the quality of the units in a sample of size n. The simple events of this random experiment are n-dimensional vectors with elements 'faulty' and 'operating'. Usually, one is not primarily interested in these sequences, but in the total number of faulty units in a sample. Thus, when the outcomes of a random experiment are not real numbers or if the outcomes are not of immediate interest, then it makes sense to assign real numbers to the outcomes. This leads to the concept of a random variable:

Given a random experiment with sample space $\mathbf{M}$, a *random variable* X is a real function on $\mathbf{M}$: $X = X(a),\ a \in \mathbf{M}$.

Thus, a random variable associates a number with each outcome of a random experiment. The set of all possible *values* or *realizations* which X can assume is called the *range* of X and is denoted as $\mathbf{R} = \{X(a),\ a \in \mathbf{M}\}$. The range of a random variable is not its most important characteristic, for, in assigning values to simple events, frequently arbitrariness prevails. (When flipping a coin, a '-1' ('+1) may be assigned to head (tail)). Different units of measurement are another source of arbitrariness. By introducing a random variable X, one passes from the sample space $\mathbf{M}$ of a random experiment to the range $\mathbf{R}$ of X, which is simply another sample space for otherwise the same random experiment. Thus, a random variable can be interpreted as the outcome of a random experiment, the simple events of which are real numbers. The advantage of introducing random variables X is that they do not depend on the physical nature of the underlying random experiment. All that needs to be known is the values X can assume and the probabilistic law which controls their occurrence. This 'probabilistic law' is called *probability distribution of* X and will be denoted as $\boldsymbol{P}_X$. For this nonmeasure theoretic textbook, the following explanation is sufficient:

> The probability distribution $\boldsymbol{P}_X$ of a random variable X contains all the information necessary for calculating the *interval probabilities* $P(X \in (a,b])$, $a \le b$.

A *discrete random variable* has a finite or a countably infinite range, i.e. the set of its possible values can be written as a finite or an infinite sequence (examples 1 and 2). Let X be a discrete random variable with range $\mathbf{R} = \{x_0, x_1, x_2, ...\}$. Further, let p_i be the probability of the random event that X assumes value x_i :

$$p_i = P(X = x_i), \quad i = 0, 1, 2, ...$$

The set $\{p_0, p_1, p_2, ...\}$ can be identified with the probability distribution $\boldsymbol{P}_X$ of X, since for any interval $(a, b]$ the interval probabilities are given by

$$P(X \in (a,b]) = P(a < X \le b) = \sum_{x_i \in (a,b]} p_i .$$

Since X must assume one of its values, the probability distribution of any discrete random variable satisfies the *normalizing condition*

$$\sum_{i=0}^{\infty} p_i = 1 .$$

On the other hand, any sequence of nonnegative numbers $\{p_0, p_1, p_2, ...\}$ satisfying the normalizing condition can be considered the probability distribution of a discrete random variable.

The range of a *continuous random variable* X is a finite or an infinite interval. In this case, the *probability distribution* of X can be most simply characterized by its (*cumulative*) *distribution function*:

$$F(x) = P(X \le x), \quad x \in \mathbf{R}_X. \tag{1.11}$$

Thus, $F(x)$ is the probability of the random event that X assumes a value which is less than or equal to x. Any distribution function $F(x)$ has properties

$$1)\ F(-\infty) = 0, \quad F(+\infty) = 1 \qquad 2)\ F(x) \text{ is nondecreasing in } x. \tag{1.12}$$

On the other hand, every function $F(x)$ which is continuous from the right and satisfies properties (1.12) is the distribution function of a certain random variable X (Figure 1.2). For $a < b$, the interval probabilities are given by

$$P(X \in (a,b]) = P(a < X \le b) = F(b) - F(a). \tag{1.13}$$

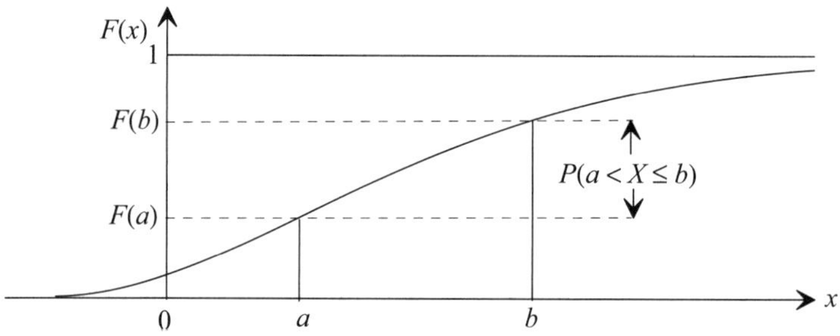

Figure 1.2 Qualitative graph of the distribution function of a continuous random variable

The definition (1.11) of a distribution function applies to discrete random variables X as well. Let $\{x_0, x_1, x_2, ...\}$ be the range of X with $x_i < x_{i+1}$ for $i = 0, 1, ...$ Then,

$$F(x) = P(X \le x) = \begin{cases} 0 & \text{for } x < x_0 \\ \Sigma_{i=0}^{k} p_i & \text{for } x_k \le x < x_{k+1}, \quad k = 0, 1, 2, \cdots \end{cases}. \tag{1.14}$$

If the range of X is finite and x_n is the largest possible value of X, then (1.14) has to be supplemented by $F(x) = 1$ for $x_n \le x$. Thus, the distribution function $F(x)$ of a discrete random variable X is a piecewise constant function with jumps of size p_i at $x = x_i - 0$. Therefore (Figure 1.3),

$$p_i = F(x_i) - F(x_i - 0); \quad i = 0, 1, 2, ...$$

Given $\{p_0, p_1, ...\}$, the distribution function of X can be constructed and, vice versa, given the distribution function of X, the probabilities $p_i = P(X = x_i)$ can be obtained. Hence, the probability distribution of any random variable X can be identified with its distribution function.

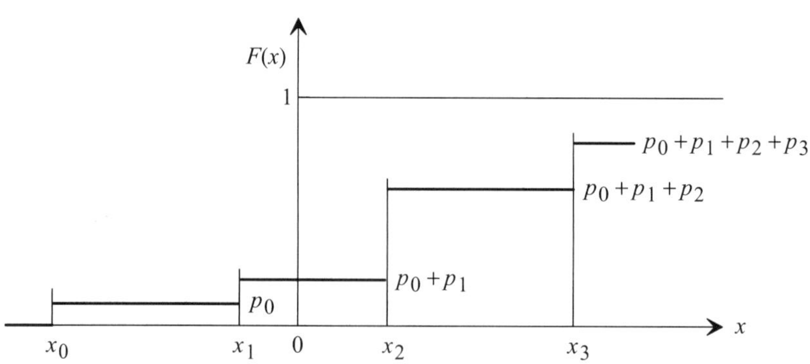

Figure 1.3 Qualitative graph of the distribution function of a discrete random variable

1.2.2 Discrete Random Variables

1.2.2.1 Numerical Parameters

The probability distribution and the range of a random variable X contain all the information on X. However, to get quick information on essential features of a random variable, it is desirable to condense as much as possible of this information to some numerical parameters.

The *mean value* (*mean, expected value*) $E(X)$ of X is defined as

$$E(X) = \Sigma_{i=0}^{\infty} x_i p_i$$

given that

$$\Sigma_{i=0}^{\infty} |x_i| p_i < \infty.$$

Thus, the mean value of a discrete random variable X is a 'weighted mean' of all its possible values x_i. The weights of the x_i are their respective probabilities.

Another motivation of this definition (see section 1.9.2): The arithmetic mean of n values of X, obtained from n independent repetitions of the underlying random experiment, tends to $E(X)$ as n tends to infinity.

If X is nonnegative with range $\{0, 1, 2, ...\}$, then its mean value can be written in the form

$$E(X) = \Sigma_{i=1}^{\infty} P(X \geq i) = \Sigma_{i=1}^{\infty} \Sigma_{k=i}^{\infty} p_k . \tag{1.15}$$

If $y = h(x)$ is a real function, then the mean value of the random variable $Y = h(X)$ can be obtained from the probability distribution of X:

$$E(X) = \Sigma_{i=0}^{\infty} h(x_i) p_i . \tag{1.16}$$

In particular, the mean value of

$$h(X) = (x - E(X))^2$$

is called *variance* of X:

$$Var(X) = \Sigma_{i=0}^{\infty} (x_i - E(X))^2 p_i .$$

Hence, $Var(X)$ is the mean squared deviation of X from its mean value $E(X)$:

$$Var(X) = E((X - E(X)^2).$$

Frequently a shorter notation is used:

$$\mu = E(X) \text{ and } \sigma^2 = Var(X).$$

The *standard deviation* of X is defined as

$$\sigma = \sqrt{Var(X)} ,$$

and the *coefficient of variation* of X is

$$V(X) = \sigma / |\mu| .$$

Variance, standard deviation, and coefficient of variation are measures for the *variability* of X. The coefficient of variation is most informative in this regard for taking into account not only the deviation of X from its mean value, but relates this deviation to the average absolute size of the values of X.

The nth *moment* μ_n of X is the mean value of X^n:

$$\mu_n = E(X^n) = \sum_{i=0}^{\infty} x_i^n p_i .$$

1.2.2.2 Important Discrete Probability Distributions

Uniform Distribution A random variable X with range $\mathbf{R} = \{x_1, x_2, ..., x_n\}$ has a *discrete uniform distribution* if

$$p_i = P(X = x_i) = \frac{1}{n}; \quad i = 1, 2, ..., n .$$

Thus, each possible value has the same probability. Mean value and variance are

$$E(X) = \frac{1}{n} \sum_{i=1}^{n} x_i, \quad Var(X) = \frac{1}{n} \sum_{i=1}^{n} \left(x_i - E(X)\right)^2 .$$

Thus, $E(X)$ is the arithmetic mean of all values which X can assume. In particular, if $x_i = i$, then

$$E(X) = \frac{n(n+1)}{2}, \quad Var(X) = \frac{(n-1)(n+1)}{12} .$$

For instance, if X is the outcome of 'rolling a die', then $\mathbf{R} = \{1, 2, ..., 6\}$ and $p_i = 1/6$.

Geometric Distribution A random variable X with range $\mathbf{R} = \{1, 2, \cdots\}$ has a *geometric distribution* with parameter p, $0 < p < 1$, if

$$p_i = P(X = i) = p\,(1-p)^{i-1}; \quad i = 1, 2, ...$$

Mean value and variance are

$$E(X) = 1/p, \quad Var(X) = (1-p)/p^2 .$$

For instance, if X is the random integer indicating how frequently one has to toss a die to get for the first time a '6', then X has a geometric distribution with $p = 1/6$.

Generally, X denotes the number of independent trials (independent random experiments) one has to carry out to have for the first time a 'success' if the random event 'success' in one trial has probability p.

Sometimes the geometric distribution is defined with range $\mathbf{R} = \{0, 1, \cdots\}$ and

$$p_i = P(X = i) = p\,(1-p)^i; \quad i = 0, 1, ...$$

In this case, mean value and variance are

$$E(X) = \frac{1-p}{p}, \quad Var(X) = \frac{1-p}{p^2} .$$

Poisson Distribution A random variable X with range $\mathbf{R} = \{0, 1, \cdots\}$ has a *Poisson distribution* with parameter λ if

$$p_i = P(X = i) = \frac{\lambda^i}{i!} e^{-\lambda}; \quad i = 0, 1, ...; \quad \lambda > 0.$$

The parameter λ is equal to mean value and variance of X:

$$E(X) = \lambda, \quad Var(X) = \lambda.$$

Bernoulli Distribution A random variable X with range $\mathbf{R} = \{0, 1\}$ has a *Bernoulli distribution* or a (0,1)-*distribution* with parameter p, $0 < p < 1$, if

$$p_0 = P(X = 0) = 1 - p, \quad p_1 = P(X = 1) = p.$$

Mean value and variance are

$$E(X) = p \quad \text{and} \quad Var(X) = p(1 - p).$$

Since X can only assume two values, it is called a *binary random variable.* In case $\mathbf{R} = \{0, 1\}$, X is a $(0, 1)$-variable.

Binomial Distribution A random variable X with range $\mathbf{R} = \{0, 1, ..., n\}$ has a *binomial distribution with parameters p* and n if

$$p_i = P(X = i) = \binom{n}{i} p^i (1 - p)^{n-i}; \quad i = 0, 1, 2, ..., n; \quad 0 \le p \le 1.$$

Frequently the following notation is used:

$$p_i = b(i, n, p) = \binom{n}{i} p^i (1 - p)^{n-i}.$$

Mean value and variance are

$$E(X) = np, \quad Var(X) = np(1 - p).$$

The binomial distribution occurs in the following situation: A random experiment, the outcome of which is a (0,1)-variable, is independently repeated n times. Such a series of experiments is called a *Bernoulli trial* of length n. The outcome X_i of experiment i can be considered the *indicator variable* of a random event A with probability $p = P(A)$:

$$X_i = \begin{cases} 1 & \text{if } A \text{ occurs} \\ 0 & \text{if } \bar{A} \text{ occurs} \end{cases}; \quad i = 1, 2, ..., n.$$

If the occurrence of event A is interpreted as 'success', then the sum

$$X = \Sigma_{i=1}^{n} X_i$$

is equal to the number of successes in a Bernoulli trial of length n. Moreover, X has a binomial distribution with parameters n and p.

Note that the number of experiments which have to be performed in a Bernoulli trial till the first occurrence of event A has a geometric distribution with parameter p and range $\{1, 2, ...\}$.

Negative Binomial Distribution A random variable X with range $\{0, 1, ...\}$ has a *negative binomial distribution* with parameters p and r, $0 < p < 1$, $r > 0$, if

$$P(X = i) = \binom{r+i-1}{i} p^i (1-p)^r; \quad i = 0, 1, ...$$

Equivalently,

$$P(X = i) = \binom{-r}{i} (-p)^i (1-p)^r; \quad i = 0, 1, ...$$

Mean value and variance are

$$E(X) = \frac{p\,r}{1-p}, \qquad Var(X) = \frac{p\,r}{(1-p)^2}.$$

Note that the number of non-successes (event $\overline{A}$) in a Bernoulli trial till the occurrence of the rth success has a negative binomial distribution, $r = 1, 2, ...$ (see geometric distribution).

Hypergeometric Distribution A random variable X with range

$$\mathbf{R} = \{0, 1, ..., \min(n, M)\}$$

has a *hypergeometric distribution* with parameters M, N, and n, $M \le N$, $n \le N$, if

$$p_m = P(X = m) = \frac{\binom{M}{m}\binom{N-M}{n-m}}{\binom{N}{n}}; \quad m = 0, 1, ..., \min(n, M).$$

As an application, consider the lottery '5 out of 45'. In this case, $M = n = 5$, $N = 45$ and p_m is the probability that a gambler has hit exactly m winning numbers with one coupon. More importantly, as example 1.4 indicates, the hypergeometric distribution plays a key role in statistical quality control.

Approximations In view of the binomial coefficients involved in the definition of the binomial and hypergeometric distribution, the following approximations are useful for numerical analysis:

Poisson Approximation to the Binomial Distribution If n is sufficiently large and p is sufficiently small, then

$$\binom{n}{i} p^i (1-p)^{n-i} \approx \frac{\lambda^i}{i!} e^{-\lambda}; \quad \lambda = np, \quad i = 0, 1, ..., n.$$

Binomial Approximation to the Hypergeometric Distribution If N is sufficiently large compared to n, then

$$\frac{\binom{M}{m}\binom{N-M}{n-m}}{\binom{N}{n}} \approx \binom{n}{m} p^m (1-p)^{n-m}, \qquad p = \frac{M}{N}.$$

Poisson Approximation to the Hypergeometric Distribution If n is sufficiently large and $p = M/N$ is sufficiently small, then

$$\frac{\binom{M}{m}\binom{N-M}{n-m}}{\binom{N}{n}} \approx \frac{\lambda^m}{m!}e^{-\lambda}, \quad \text{where } \lambda = np.$$

Example 1.3 On average, only 0.01% of trout eggs will develop into adult fishes. What is the probability p_a that at least three adult fishes arise from 40,000 eggs?

Let X be the random number of eggs out of 40,000 which develop into adult fishes. It is assumed that the eggs develop independently of each other. Then X has a binomial distribution with parameters $n = 40,000$ and $p = 0.0001$. Thus,

$$p_i = P(X = i) = \binom{40,000}{i}(0.0001)^i(0.9999)^{40,000-i},$$

where $i = 1, 2, ..., 40,000$. Since n is large and p is small, the Poisson distribution with parameter $\lambda = np = 4$ can be used to approximately calculating the p_i:

$$p_i \approx \frac{4^i}{i!}e^{-4}; \quad i = 0, 1, ...$$

The desired probability is

$$p_a = 1 - p_0 - p_1 - p_2 \approx 1 - 0.0183 - 0.0733 - 0.1465 = 0.7619.$$ □

Example 1.4 A delivery of 10,000 transistors contains 200 defective ones. According to agreement, the customer accepts a percentage of 2% defective transistors. A sample of size $n = 100$ is taken. The customer will reject the delivery if there are no more than 4 defective transistors in the sample. The probability of rejection p_r is the *producer's risk*, since the delivery is in line with the agreement.

To determine p_r, the hypergeometric distribution with $N = 10,000$, $M = 200$, and $n = 100$ has to be applied. Let X be the random number of defective transistors in the sample. Then the producer's risk is

$$p_r = 1 - p_0 - p_1 - p_2 - p_3 - p_4$$

with

$$p_m = P(X = m) = \frac{\binom{200}{m}\binom{9800}{100-m}}{\binom{10,000}{100}}.$$

Since N is large enough compared to n, the binomial approximation with $p = 0.02$ can be applied:

$$p_m \approx \binom{100}{m}(0.02)^m(0.98)^{100-m}; \quad m = 0, 1, 2, 3, 4.$$

Thus, the delivery is rejected with probability $p_r \approx 0.051$. For the sake of comparison: The Poisson approximation with $\lambda = np = 2$ yields $p_r \approx 0.055$. □

1.2.3 Continuous Random Variables

1.2.3.1 Probability Density and Numerical Parameters

As mentioned before, the range of a continuous random variable is a noncountably infinite set. This property of a continuous random variable results from its definition:

> A random variable is called *continuous* if its distribution function $F(x)$ has a first derivative.

Equivalently, a random variable is called *continuous* if there exists a function $f(x)$ so that

$$F(x) = \int_{-\infty}^{x} f(u)\, du.$$

The function

$$f(x) = F'(x) = dF(x)/dx, \quad x \in \mathbf{R}_X$$

is called the *probability density function* of X (briefly: *probability density* or simply *density*). Sometimes the term *probability mass function* is used. A density has property (Figure 1.4)

$$\int_{-\infty}^{+\infty} f(x)\, dx = F(\infty) = 1.$$

Conversely, every nonnegative function $f(x)$ satisfying this condition is the probability density of a certain random variable X. As with its distribution function, the probability distribution $\boldsymbol{P}_X$ of a continuous random variable X can be identified with its probability density. The range of X coincides with the set of all those x for which its density is positive: $\mathbf{R} = \{x, f(x) > 0\}$ (Figure 1.4).

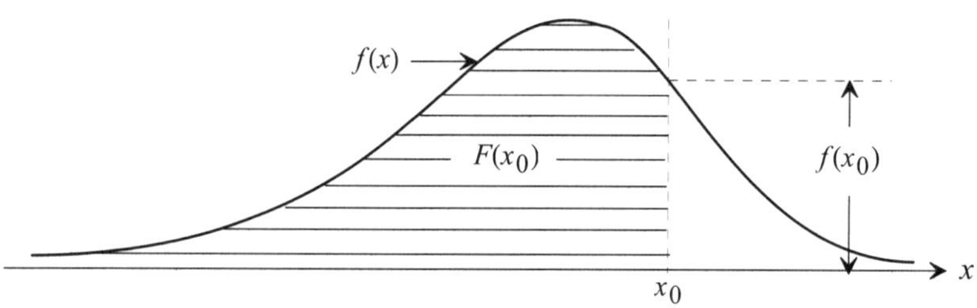

Figure 1.4 Distribution function and density

The *mean value* of X (*mean, expected value*) is defined as

$$E(X) = \int_{-\infty}^{+\infty} x\, f(x)\, dx$$

given that

$$\int_{-\infty}^{+\infty} |x|\, f(x)\, dx < \infty.$$

In terms of its distribution function, the mean value of X is given by

$$E(X) = \int_0^{\infty}[1 - F(x)]\,dx - \int_{-\infty}^{0} F(x)\,dx.$$

In particular, for nonnegative random variables, the analogue to (1.15) is

$$E(X) = \int_0^{\infty}[1 - F(x)]\,dx. \tag{1.17}$$

If $h(x)$ is a real function and X any continuous random variable with density $f(x)$, then the mean value of the random variable $Y = h(X)$ can directly be obtained from the density of X:

$$E(Y) = \int_{-\infty}^{+\infty} h(x)\, f(x)\,dx. \tag{1.18}$$

In particular, the mean value of $h(X) = (X - E(X))^2$ is the *variance* of X:

$$Var(X) = \int_{-\infty}^{+\infty}(x - E(X))^2 f(x)\,dx.$$

Hence, the variance of a random variable is its mean squared deviation from its mean value. *Standard deviation* and *coefficient of variation* are defined and motivated as with discrete random variables.

The nth *moment* of X is

$$\mu_n = E(X^n) = \int_{-\infty}^{+\infty} x^n f(x)\,dx; \quad n = 0, 1, \ldots$$

The following relationship between variance, second moment and mean value is also valid for discrete random variables:

$$Var(X) = E(X^2) - (E(X))^2 = \mu_2 - \mu^2. \tag{1.19}$$

For a continuous random variable X, the *interval probability* (1.13) can be written as follows:

$$P(a < X \le b) = F(b) - F(a) = \int_a^b f(x)\,dx.$$

The α*–percentile* x_α (also denoted as α*–quantile* q_α) of a random variable X is defined as

$$F(x_\alpha) = \alpha.$$

This implies that in a long series of random experiments with outcome X, about α% of the observed values of X will be equal to or less than x_α. The 0.5-percentile is called the *median* of X or of its probability distribution. Thus, in a long series of random experiments with outcome X, about 50% of the observed values will be to the left and to the right of $x_{0.5}$ each.

A probability distribution is *symmetric* with symmetry center a if $f(x)$ satisfies

$$f(a - x) = f(a + x) \quad \text{for all } x.$$

For symmetric distributions, symmetry center, mean value, and median coincide:

$$a = E(X) = x_{0.5}.$$

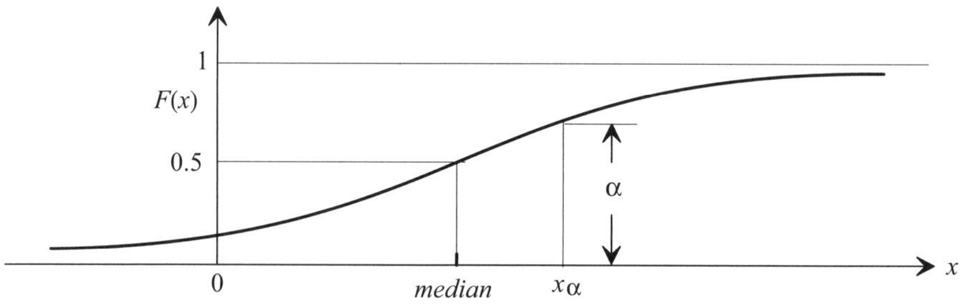

Figure 1.5 Illustration of the percentiles

A *mode m* of a random variable is an x-value at which $f(x)$ assumes a relative maximum. A density $f(x)$ is called *unimodal* if it has only one maximum.

Standardization A random variable Z (discrete or continuous) with

$$E(Z) = 0 \text{ and } Var(Z) = 1$$

is called a *standardized random variable.* For any random variable X with finite mean value μ and variance σ, the random variable

$$Z = \frac{X - \mu}{\sigma}$$

is a standardized random variable. Z is called the *standardization* of X.

1.2.3.2 Important Continuous Probability Distributions

In this section some important probability distributions of continuous random variables X will be listed. If the distribution function is not explicitely given, it can only be represented as an integral over the density.

Uniform Distribution A random variable X has a uniform distribution over the interval $[c, d]$ with $c < d$ if it has distribution function and density

$$F(x) = \begin{cases} 0, & x < c \\ \frac{x-c}{d-c}, & c \le x \le d \\ 1, & d < x \end{cases}, \qquad f(x) = \begin{cases} \frac{1}{d-c}, & c \le x \le d \\ 0, & x \notin [c, d] \end{cases}, \qquad c < d.$$

Thus, for any subinterval $[a, b]$ of $[c, d]$, the corresponding interval probability is

$$P(a < X \le b) = \frac{b-a}{d-c}.$$

This probability depends only on the length of the interval $[a, b]$, but not on its position within the interval $[c, d]$, i.e. all subintervals of $[c, d]$ of the same length have the same chance that X takes on a value out of it.

Mean value and variance of X are

$$E(X) = \frac{c+d}{2}, \quad Var(X) = \frac{1}{12}(d-c)^2.$$

Pareto Distribution A random variable X has a *Pareto distribution* over the interval $[d, \infty)$ if it has distribution function and density

$$F(x) = 1 - \left(\frac{d}{x}\right)^c, \quad f(x) = \frac{c}{d}\left(\frac{d}{x}\right)^{c+1}, \quad x \geq d \geq 0.$$

Mean value and variance are

$$E(X) = \frac{c\,d}{c-1}, \quad c > 1,$$

$$Var(X) = \frac{c\,d^2}{(c-1)^2\,(c-2)}, \quad c > 2.$$

Exponential Distribution A random variable X has an *exponential distribution* with parameter λ if it has distribution function and density

$$F(x) = 1 - e^{-\lambda x}, \quad f(x) = \lambda e^{-\lambda x}, \; x \geq 0, \; \lambda > 0.$$

Mean value and variance are

$$E(X) = 1/\lambda, \quad Var(X) = 1/\lambda^2.$$

In view of their simple structure and convenient properties, the exponential distribution is quite popular in all sorts of applications. Frequently, the parameter λ is denoted as $1/\mu$.

Erlang Distribution A random variable X has an *Erlang distribution* with parameters λ and n if it has distribution function and density

$$F(x) = 1 - e^{-\lambda x} \sum_{i=0}^{n-1} \frac{(\lambda x)^i}{i\,!},$$

$$f(x) = \lambda \frac{(\lambda x)^{n-1}}{(n-1)\,!} e^{-\lambda x}; \quad x \geq 0, \; \lambda > 0, \; n = 1, 2, \ldots$$

Mean value and variance are

$$E(X) = n/\lambda, \quad Var(X) = n/\lambda^2.$$

The exponential distribution is a special case of the Erlang distribution $(n = 1)$.

Gamma Distribution A random variable X has a *Gamma distribution* with parameters α and β if it has density

$$f(x) = \frac{\beta^\alpha}{\Gamma(\alpha)} x^{\alpha-1} e^{-\beta x}, \quad x > 0, \; \alpha > 0, \; \beta > 0,$$

where the *Gamma function* $\Gamma(z)$ is defined by

$$\Gamma(z) = \int_0^\infty x^{z-1} e^{-x}\, dx, \quad z > 0.$$

Mean value and variance are

$$E(X) = \alpha/\beta, \quad Var(X) = \alpha/\beta^2.$$

Special cases: Exponential distribution for $\alpha = 1$ and $\beta = \lambda$, Erlang distribution for $\alpha = n$ and $\beta = \lambda$.

Beta Distribution A random variable X has a *Beta distribution* in the interval $[0, 1]$ with parameters α and β if it has density

$$f(x) = \frac{1}{B(\alpha, \beta)} x^{\alpha-1}(1-x)^{\beta-1}, \quad 0 \le x \le 1,\ \alpha > 0,\ \beta > 0.$$

Mean value and variance are

$$E(X) = \frac{\alpha}{\alpha+\beta}, \quad Var(X) = \frac{\alpha\beta}{(\alpha+\beta)^2(\alpha+\beta+1)}.$$

The *Beta function* $B(x,y)$ is defined by

$$B(x,y) = \frac{\Gamma(x)\,\Gamma(y)}{\Gamma(x+y)}; \quad x > 0,\ y > 0.$$

Weibull Distribution A random variable X has a *Weibull distribution* with scale parameter θ and form parameter β if it has distribution function and density (Figure 1.6)

$$F(x) = 1 - e^{(x/\theta)^\beta}, \quad f(x) = \frac{\beta}{\theta}\left(\frac{x}{\theta}\right)^{\beta-1} e^{(x/\theta)^\beta}; \quad x > 0,\ \beta > 0,\ \theta > 0.$$

Mean value and variance are

$$E(X) = \theta\,\Gamma\left(\tfrac{1}{\beta} + 1\right), \quad Var(X) = \theta^2\left[\Gamma\left(\tfrac{2}{\beta} + 1\right) - \left(\Gamma\left(\tfrac{1}{\beta} + 1\right)\right)^2\right].$$

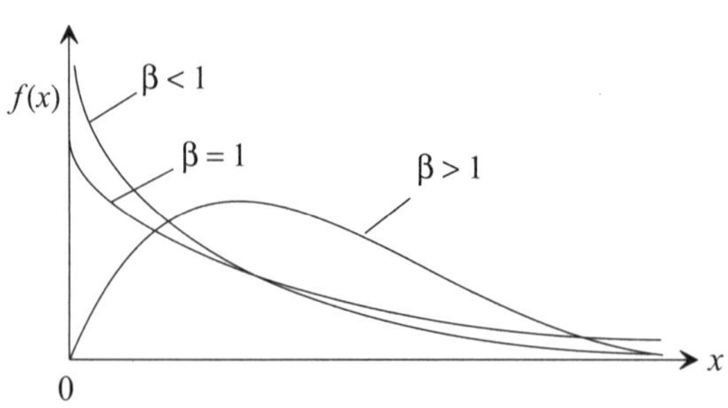

Figure 1.6 Densities of the Weibull distribution

Special cases: Exponential distribution for $\theta = 1/\lambda$ and $\beta = 1$, Rayleigh distribution for $\beta = 2$.

The Weibull distribution was found by the German mining engineers *E. Rosin* and *E. Rammler* in the late twenties of the past century when investigating the distribution of the size of stone, coal and other particles after a grinding process (see, for example, [68]). In the forties of the past century, the Swedish engineer *W. Weibull* came across this distribution type when investigating mechanical wear.

Normal Distribution A random variable X has a *normal* (or *Gaussian*) *distribution* with parameters μ and σ^2 if it has density (Figure 1.7)

$$f(x) = \frac{1}{\sqrt{2\pi}\,\sigma} e^{-\frac{1}{2}\left(\frac{x-\mu}{\sigma}\right)^2}, \quad -\infty < x < +\infty, \quad -\infty < \mu < +\infty, \ \sigma > 0.$$

As the notation of the parameters indicates, mean value and variance are

$$E(x) = \mu, \quad Var(X) = \sigma^2.$$

A normally distributed random variable (or, generally, the normal distribution) with parameters μ and σ^2 is denoted as $N(\mu, \sigma^2)$. Different from most other probability distributions, the standardization of a normally distributed random variable also has a normal distribution. Therefore, if $X = N(\mu, \sigma^2)$, then

$$N(0, 1) = \frac{X - \mu}{\sigma}.$$

The density of the standardized normal distribution is denoted as $\varphi(x)$:

$$\varphi(x) = \frac{1}{\sqrt{2\pi}} e^{-x^2/2}, \quad -\infty < x < +\infty.$$

The corresponding distribution function $\Phi(x)$ can only be represented as an integral, but the percentiles of this distribution are widely tabulated.

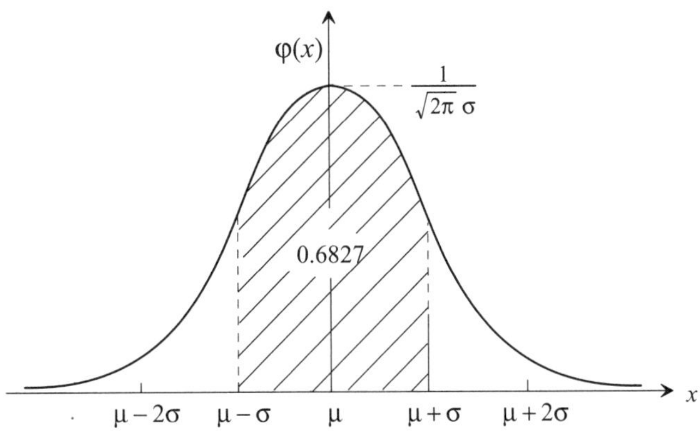

Figure 1.7 Density of the normal distribution (Gaussian bell curve)

Since $\varphi(x)$ is symmetric with symmetry center 0,

$$\Phi(x) = 1 - \Phi(-x).$$

Hence there is the following relationship between the α- and the $(1-\alpha)$- percentiles of the standardized normal distribution:

$$-x_\alpha = x_{1-\alpha}, \quad 0 < \alpha < 1/2.$$

This is the reason for introducing the following notation (analogously for other distributions with symmetry center 0):

$$z_\alpha = x_{1-\alpha}, \quad 0 < \alpha < 1/2.$$

Hence,

$$P(-z_{\alpha/2} \le N(0,1) \le z_{\alpha/2}) = \Phi(z_{\alpha/2}) - \Phi(-z_{\alpha/2}) = 1 - \alpha .$$

Generally, if $X = N(\mu, \sigma^2)$, the interval probabilities (1.13) can be calculated by using the standardized normal distribution:

$$P(a \le X \le b) = \Phi\left(\frac{b-\mu}{\sigma}\right) - \Phi\left(\frac{a-\mu}{\sigma}\right).$$

Logarithmic Normal Distribution A random variable X has a *logarithmic normal distribution* with parameters μ and σ if it has density

$$f(y) = \frac{1}{\sqrt{2\pi}\,\sigma y} \exp\left\{-\frac{1}{2}\left(\frac{\ln y - \mu}{\sigma}\right)^2\right\}; \quad y > 0,\ \sigma > 0,\ -\infty < \mu < \infty$$

Thus, X has a logarithmic normal distribution with parameters μ and σ if it has structure $X = e^Y$, where $Y = N(\mu, \sigma^2)$. Equivalently, X has a logarithmic normal distribution if $Y = \ln X$ has a normal distribution. Therefore, the distribution function of X is

$$F(y) = \Phi\left(\frac{\ln y - \mu}{\sigma}\right), \quad x > 0.$$

Mean value and variance are

$$E(X) = e^{\mu + \sigma^2/2}, \quad Var(X) = e^{2\mu + \sigma^2}\left(e^{\sigma^2} - 1\right).$$

Cauchy Distribution A random variable X has a *Cauchy distribution* with parameters λ and μ if it has density

$$f(x) = \frac{\lambda}{\pi\,[\lambda^2 + (x-\mu)^2]}, \quad -\infty < x < \infty,\ \lambda > 0,\ -\infty < \mu < \infty .$$

Mean value and variance do not exist.

Inverse Gaussian Distribution A random variable X has an *inverse Gaussian distribution* with parameters α and β if it has density

$$f(x) = \sqrt{\frac{\alpha}{2\pi x^3}}\,\exp\left(-\frac{\alpha(x-\beta)^2}{2\beta^2 x}\right), \quad x > 0,\ \alpha > 0,\ \beta > 0.$$

The corresponding distribution function is

$$F(x) = \Phi\left(\frac{x-\beta}{\beta\sqrt{\alpha x}}\right) + e^{-2\alpha/\beta}\,\Phi\left(-\frac{x+\beta}{\beta\sqrt{\alpha x}}\right), \quad x > 0.$$

Mean value and variance are

$$E(X) = \beta, \quad Var(X) = \beta^3/\alpha.$$

Logistic Distribution A random variable X has a *logistic distribution* with parameters μ and σ if it has density

$$f(x) = \frac{\pi\exp\left(-\frac{\pi}{\sqrt{3}}\frac{x-\mu}{\sigma}\right)}{\sqrt{3}\,\sigma\left[1+\exp\left(-\frac{\pi}{\sqrt{3}}\frac{x-\mu}{\sigma}\right)\right]^2}, \quad -\infty < x < \infty,\ \sigma > 0,\ -\infty < \mu < \infty.$$

Mean value and variance are

$$E(X) = \mu, \quad Var(X) = \sigma^2.$$

Example 1.5 A company needs wooden shafts of a length of 600 *mm*. It accepts deviations of maximal $\pm 6\,mm$. The producer delivers shafts of random length X which has an $N(200, \sigma^2)$-distribution.

1) What percentage is rejected by the company if $\sigma = 3\,mm$? The probability that a shaft will be rejected is

$$\begin{aligned} P(|X-600| > 6) &= 1 - P(|X-600| \le 6) = 1 - P(594 \le X \le 606) \\ &= 1 - \Phi\left(\frac{606-600}{3}\right) - \Phi\left(\frac{594-600}{3}\right) \\ &= 1 - [\Phi(2) - \Phi(-2)] = 2\,[1-\Phi(2)] \\ &= 2\cdot[1-0.97725] \\ &= 0.0455. \end{aligned}$$

Thus, 4.55 % of the shafts are rejected.

2) What is the value of σ if the company rejects on average 10% of the shafts?

By making use of the previous derivation with $\sigma = 3$ replaced by σ,

$$P(|X-600|>6)=1-[\Phi(6/\sigma)-\Phi(-6/\sigma)]=2\,[1-\Phi(6/\sigma)].$$

This probability must be equal to 0.1. Hence, the parameter σ has to be determined from $\Phi(6/\sigma)=0.95$, or, equivalently, from

$$6/\sigma=z_{0.95}=1.64,$$

since the 0.95-percentile of the standardized normal distribution is $z_{0.95}=1.64$. Thus, $\sigma=3.658$. □

Example 1.6 In a certain geographical area of Southern Africa, mean value and variance of the lifetime of the African wild dog have been determined as

$$\mu=8.86230 \text{ [years] and } \sigma^2=21.45964.$$

1) Assuming that the lifetime of an African wild dog has a Weibull distribution, the parameters θ and β of this distribution satisfy

$$E(X)=\theta\,\Gamma(1+1/\beta)=8.86230,$$
$$Var(X)=\theta^2\left[\Gamma(1+2/\beta)-(\Gamma(1+1/\beta))^2\right]=21.45964.$$

Combining these equations yields an equation in β :

$$\frac{\Gamma(1+2/\beta)}{[\Gamma(1+1/\beta)]^2}=1.27323.$$

The solution is $\beta=2$ (Rayleigh-distribution). Hence, $\theta=10$.

2) What is the probability that an African wild dog will survive 10 years on condition that it has survived 5 years? According to (1.4), the probability of interest is

$$P(X>10|X>5))=\frac{P(X>10)}{P(X>5)}=\frac{e^{-(10/10)^2}}{e^{-(5/10)^2}}$$
$$=e^{-0.75}=0.47237.$$

Note that the (unconditional) probability for an African wild dog to reach an age of at least 10 years is $e^{-(10/10)^2}=e^{-1}=0.36788$. □

1.2.4 Mixtures of Random Variables

The probability distribution $\boldsymbol{P}_X$ of any random variable X depends on one or more numerical parameters. To emphasize the dependency on a special parameter θ, in this section the notation $\boldsymbol{P}_{X,\theta}$ instead of $\boldsymbol{P}_X$ is used. Equivalently, in terms of distribution function and density of X (if the latter exists),

$$F_X(x)=F_X(x,\theta),\quad f_X(x)=f_X(x,\theta).$$

Mixtures of random variables or their probability distributions arise from the assumption that the parameter θ is a realization of a random parameter Θ, and all the probability distributions being elements of the set $\{\boldsymbol{P}_{X,\theta}, \theta \in \mathbf{R}_{\Theta}\}$ are mixed.

1. Discrete Random Variable Θ with range $\mathbf{R}_{\Theta} = \{\theta_0, \theta_1, ...\}$ Let the random parameter Θ have probability distribution

$$\boldsymbol{P}_{\Theta} = \{q_0, q_1, ...\} \text{ with } q_n = P(\Theta = \theta_n\}; \quad n = 0, 1, ...$$

Then the *mixture of probability distributions* of type $\boldsymbol{P}_{X,\theta}$ is defined as

$$G(x) = \sum_{n=0}^{\infty} F_X(x, \theta_n)\, q_n .$$

2. Continuous Random Variable Θ with range $\mathbf{R}_{\Theta} \subseteq (-\infty, +\infty)$ Let the random parameter Θ have probability density $f_{\Theta}(\theta)$, $\theta \in \mathbf{R}_{\Theta}$. Then the *mixture of probability distributions* of type $\boldsymbol{P}_{X,\theta}$ is defined as

$$G(x) = \int_{\mathbf{R}_{\Theta}} F_X(x, \theta)\, f_{\Theta}(\theta)\, d\theta .$$

Thus, if Θ is discrete, then $G(x)$ is the weighted sum of the $F_X(x, \theta_n)$ with *weights* q_n given by the probability distribution of Θ. If Θ is continuous, $G(x)$ is the weighted integral of $F_X(x, \theta)$ with *weight function* $f_{\Theta}(x, \theta)$. In either case, the function $G(x)$ satisfies properties (1.12). Hence, $G(x)$ is the distribution function of a *mixed random variable* Y and the probability distribution of Y is the *weighted mixture of probability distributions* of type $\boldsymbol{P}_{X,\theta}$.

If X is continuous, the respective densities of Y are

$$g(x) = \sum_{n=0}^{\infty} f_X(x, \theta_n)\, q_n \quad \text{and} \quad g(x) = \int_{\mathbf{R}_{\Theta}} f_X(x, \theta) f_{\Theta}(\theta)\, d\theta .$$

In either case, by (1.16) and (1.18), $G(x)$ is the mean value of the random variable $F_X(x, \Theta)$, and $g(x)$ is the mean value of the random variable $f_X(x, \Theta)$:

$$G(x) = E(F_X(x, \Theta)), \quad g(x) = E(f_X(x, \Theta)).$$

If X is discrete with probability distribution

$$\boldsymbol{P}_{X,\theta} = \{p_i(\theta) = P(X = x_i; \theta); \quad i = 0, 1, ...\},$$

then the probability distribution of Y, given so far by its distribution function $G(x)$, can equivalently be characterized by its individual probabilities

$$P(Y = x_i) = \sum_{n=0}^{\infty} p_i(\theta_n)\, q_n; \quad i = 0, 1, ... \tag{1.20}$$

if Θ is discrete, and

$$P(Y = x_i) = \int_{\mathbf{R}_{\Theta}} p_i(\theta) f_{\Theta}(\theta)\, d\theta; \quad i = 0, 1, ... \tag{1.21}$$

if Θ is continuous.

The probability distribution of Θ is sometimes called *structure* or *mixing distribution*. Hence, the probability distribution $\boldsymbol{P}_Y$ of the '*mixed random variable*' Y is a *mixture of probability distributions of type* $\boldsymbol{P}_{X,\theta}$ *with regard to a structure distribution* $\boldsymbol{P}_\Theta$.

The mixture of probability distributions provides a method for producing types of probability distributions, which are specifically tailored to serve the needs of certain applications.

Example 1.7 (***mixture of exponential distributions***) Let X have an exponential distribution with parameter λ:

$$F_X(x,\lambda) = P(X \le x) = 1 - e^{-\lambda x}, \quad x \ge 0.$$

This distribution is to be mixed with regard to a structure distribution $\boldsymbol{P}_L$, where L is exponentially distributed with density

$$f_L(\lambda) = \mu\, e^{-\mu\lambda}.$$

Mixing yields the distribution function

$$G(x) = \int_0^{+\infty} F_X(x,\lambda) f_L(\lambda)\, d\lambda = \int_0^{+\infty} (1 - e^{-\lambda x}) \mu e^{-\mu\lambda}\, d\lambda$$

$$= 1 - \mu/(x+\mu).$$

Hence, mixing exponential distributions with regard to an exponential structure distribution gives distribution function and density

$$G(x) = \frac{x}{x+\mu}, \quad g(x) = \frac{\mu}{(x+\mu)^2}, \quad x \ge 0.$$

This is a Pareto distribution. □

Example 1.8 (***mixture of binomial distributions***) Let X have a binomial distribution with parameters n and p:

$$P(X=i) = \binom{n}{i} p^i (1-p)^{n-i}, \quad i = 0,1,2,...,n.$$

The parameter n is considered to be a value of a Poisson with parameter λ distributed random variable N:

$$P(N=n) = \frac{\lambda^n}{n!} e^{-\lambda}; \quad n = 0,1,... \ (\lambda \text{ fixed}).$$

Then, from (1.20), using

$$\binom{n}{i} = 0 \text{ for } n < i,$$

the mixture of binomial distributions $\boldsymbol{P}_{X,n}$, $n = 0,1,...$ with regard to the structure distribution $\boldsymbol{P}_N$ is obtained as follows:

$$P(Y=i) = \sum_{n=0}^{\infty} \binom{n}{i} p^i (1-p)^{n-i} \frac{\lambda^n}{n!} e^{-\lambda}$$

$$= \sum_{n=i}^{\infty} \binom{n}{i} p^i (1-p)^{n-i} \frac{\lambda^n}{n!} e^{-\lambda}$$

$$= \frac{(\lambda p)^i}{i!} e^{-\lambda} \sum_{k=0}^{\infty} \frac{[\lambda(1-p)]^k}{k!}$$

$$= \frac{(\lambda p)^i}{i!} e^{-\lambda} e^{\lambda(1-p)}.$$

Thus,

$$P(Y=i) = \frac{(\lambda p)^i}{i!} e^{-\lambda p}; \quad i = 0, 1, ...$$

This is a Poisson distribution with parameter λp. □

Mixed Poisson Distribution Let X have a Poisson distribution with parameter λ:

$$\boldsymbol{P}_{X,\lambda} = \{P(X=i) = \frac{\lambda^i}{i!} e^{-\lambda};\ i = 0, 1, ...;\ \lambda > 0\}.$$

Then a random variable Y with range $\{0, 1, ...\}$ is said to have *mixed Poisson distribution* if its probability distribution is a mixture of the Poisson distributions $\boldsymbol{P}_{X,\lambda}$ with regard to any structure distribution. For instance, if the structure distribution is given by the density $f_L(\lambda)$ of a positive random variable L (i.e. the parameter λ of the Poisson distribution is a realization of L), the distribution of Y is given by

$$P(Y=i) = \int_0^{\infty} \frac{\lambda^i}{i!} e^{-\lambda} f_L(\lambda)\, d\lambda, \quad i = 0, 1, ...$$

A mixed Poisson distributed random variable Y has the following properties:

(1) $E(Y) = E(L)$

(2) $Var(Y) = E(L) + Var(L)$

(3) $P(Y > n) = \int_0^{\infty} \frac{\lambda^n}{n!} e^{-\lambda} \overline{F}_L(\lambda))\, d\lambda,$

where $F_L(\lambda) = P(L \le \lambda)$ is the distribution function of L and $\overline{F}_L(\lambda) = 1 - F_L(\lambda)$.

Example 1.9 (*mixed Poisson distribution, gamma structure distribution*) Let the random structure variable L have a gamma distribution with density

$$f_L(\lambda) = \frac{\beta^\alpha}{\Gamma(\alpha)} \lambda^{\alpha-1} e^{-\beta\lambda}, \quad \lambda > 0,\ \alpha > 0,\ \beta > 0.$$

The corresponding mixed Poisson distribution is obtained as follows:

$$P(Y=i) = \int_0^\infty \frac{\lambda^i}{i!} e^{-\lambda} \frac{\beta^\alpha}{\Gamma(\alpha)} \lambda^{\alpha-1} e^{-\beta\lambda}\, d\lambda$$

$$= \frac{1}{i!} \frac{\beta^\alpha}{\Gamma(\alpha)} \int_0^\infty \lambda^{i+\alpha-1} e^{-\lambda(\beta+1)}\, d\lambda$$

$$= \frac{1}{i!} \frac{\beta^\alpha}{\Gamma(\alpha)} \frac{1}{(\beta+1)^{i+\alpha}} \int_0^\infty x^{i+\alpha-1} e^{-x}\, dx$$

$$= \frac{1}{i!} \frac{\Gamma(i+\alpha)}{\Gamma(\alpha)} \frac{\beta^\alpha}{(\beta+1)^{i+\alpha}}$$

$$= \binom{i-1+\alpha}{i} \left(\frac{1}{\beta+1}\right)^i \left(\frac{\beta}{\beta+1}\right)^\alpha; \quad \alpha > 0,\ \beta > 0, \quad i = 0, 1, \ldots$$

This is a negative binomial distribution with parameters $r = \alpha$ and $p = 1/(\beta+1)$. In deriving this result, the following property of the gamma function has been used:

$$\Gamma(i+\alpha) = (i-1+\alpha)\,\Gamma(i-1+\alpha); \ i = 1, 2, \ldots$$ □

1.2.5 Functions of a Random Variable

Let X be a continuous random variable and $y = h(x)$ a real function. This chapter deals with the probability distribution of the random variable $Y = h(X)$.

Theorem 1.1 Let X and Y be linearly dependent: $Y = \alpha X + \beta$. Then,

$$F_Y(y) = F_X\left(\frac{y-\beta}{\alpha}\right) \quad \text{for } \alpha > 0,$$

$$F_Y(y) = 1 - F_X\left(\frac{y-\beta}{\alpha}\right) \quad \text{for } \alpha < 0,$$

$$f_Y(y) = \left|\frac{1}{\alpha}\right| f_X\left(\frac{y-\beta}{\alpha}\right) \quad \text{for } \alpha \neq 0,$$

$$E(Y) = \alpha E(X) + \beta, \quad Var(Y) = \alpha^2 Var(X).$$

Proof The distribution function of Y is obtained as follows:

$$F_Y(y) = P(Y \le y) = P(\alpha X + \beta \le y) = P\left(X \le \frac{y-\beta}{\alpha}\right) = F_X\left(\frac{y-\beta}{\alpha}\right) \quad \text{for } \alpha > 0.$$

$$F_Y(y) = P(Y \le y) = P(\alpha X + \beta \le y) = P\left(X > \frac{y-\beta}{\alpha}\right) = 1 - F_X\left(\frac{y-\beta}{\alpha}\right) \quad \text{for } \alpha < 0.$$

The corresponding density $f_Y(y)$ is obtained by differentiation of $F_Y(y)$.

For $\alpha > 0$, the variance of Y is

$$Var(Y) = \int (y - E(Y))^2 f_Y(y)dy = \int (y - \alpha E(X) - \beta)^2 \frac{1}{\alpha} f_X\left(\frac{y-\beta}{\alpha}\right) dy.$$

Substituting $x = (y - \beta)/\alpha$ yields

$$Var(Y) = \int (\alpha x - \alpha E(X))^2 \frac{1}{\alpha} f_X(x)\, \alpha\, dx = \alpha^2 Var(X).$$

(The integrals involved refer to the ranges of X and Y.) The case $\alpha < 0$ is done analogously. ■

If $X = N(\mu, \sigma^2)$, then the standardization of X, namely

$$Z = \frac{X - \mu}{\sigma} = \frac{1}{\sigma} X - \frac{\mu}{\sigma},$$

also has a normal distribution. More generally, every linear transform $Y = \alpha X + \beta$ of X has a normal distribution. Usually, $Y = \alpha X + \beta$ has not the same distribution type as X. For instance, if X has distribution function

$$F_X(x) = 1 - e^{-\lambda x},\ x \geq 0,$$

then the distribution function of $Y = \alpha X + \beta$ is

$$F_Y(y) = F_X\left(\frac{y-\beta}{\alpha}\right) = 1 - e^{-\lambda \frac{y-\beta}{\alpha}},\ y \geq \beta,\ \alpha > 0.$$

This distribution function characterizes the class of *shifted exponential distributions*. As a consequence, the standardization of an exponentially distributed random variable does not have an exponential distribution.

Strictly Monotone Function $y = h(x)$ Let $y = h(x)$ be a strictly monotone function with inverse function $x = h^{-1}(y)$.

If $y = h(x)$ is strictly increasing, then, for any random variable X, the distribution function of $Y = h(X)$ is

$$F_Y(y) = P(h(X) \leq y) = P(X \leq h^{-1}(y)) = F_X(h^{-1}(y)).$$

If $y = h(x)$ is strictly decreasing, then, for any random variable X,

$$F_Y(y) = P(h(X) \leq y) = P(X > h^{-1}(y)).$$

Hence,

$$F_Y(y) = 1 - F_X(h^{-1}(y)).$$

By differentiation, applying the chain rule, the density of Y is in either case seen to be

$$f_Y(y) = f_X(h^{-1}(y)) \left| \frac{dh^{-1}(y)}{dy} \right| = f_X(x(y)) \left| \frac{dx(y)}{dy} \right|.$$

Note that the formulas given are only valid for y being element of the range of Y. Outside of this range, the distribution function of Y is 0 or 1 and the density of Y is 0.

Example 1.10 A solid of mass m moves along a straight line with a random velocity X, which is uniformly distributed over the interval $[0, V]$. The random kinetic energy of the solid is

$$Y = \frac{1}{2} m X^2 .$$

In view of $y = h(x) = \frac{1}{2} m x^2$, it follows that

$$x = h^{-1}(y) = \sqrt{2y/m} \quad \text{and} \quad \frac{dx}{dy} = \sqrt{1/(2my)}\,, \quad 0 < y < \frac{1}{2} m V^2 .$$

Since

$$f_X(x) = 1/V, \quad 0 \le x \le V,$$

the density of Y is

$$f_Y(y) = \frac{1}{V}\sqrt{\frac{1}{2my}}\,, \quad 0 \le y \le \frac{1}{2} m V^2 .$$

The mean kinetic energy of the solid is

$$E(Y) = \int_0^{mV^2/2} y \frac{1}{V}\sqrt{1/(2my)}\; dy = \frac{1}{V}\sqrt{1/2m} \int_0^{mV^2/2} y^{1/2} dy$$

$$= \frac{2}{3V}\sqrt{1/2m}\left[y^{3/2}\right]_0^{mV^2/2} = \frac{1}{6} m V^2 .$$

It is more convenient to determine $E(Y)$ by means of (1.18):

$$E(Y) = \int_0^V \frac{1}{2} m x^2 \frac{1}{V} dx = \frac{1}{2} m \frac{1}{V} \int_0^V x^2 dx = \frac{1}{6} m V^2 . \qquad \square$$

1.3 TRANSFORMATION OF PROBABILITY DISTRIBUTIONS

The probability distributions or at least moments of random variables can frequently be obtained from special functions, so called (*probability- or moment-*) *generating functions* of random variables or, equivalently, of their probability distributions. This is of importance, since it is in many applications of stochastic methods easier to determine the generating function of a random variable instead of its probability distribution. Examples will be considered in the following chapters. The method of determining the probability distribution or moments of a random variable from its generating function is theoretically justified, since to every probability distribution belongs exactly one generating function of a given type and vice versa. Formally, going over from a probability distribution to its generating function is a *transformation* of this distribution. This section deals with the *z-transformation* for discrete nonnegative random variables and with the *Laplace transformation* for continuous random variables.

1.3.1 z-Transformation

The discrete random variable X has range $\{0, 1, ...\}$ and probability distribution

$$\{p_0, p_1, ...\} \text{ with } p_i = P(X = i);\; i = 0, 1, ...$$

The *z-transform* of X, or, equivalently, of its probability distribution, is defined as

$$M(z) = \Sigma_{i=0}^{\infty}\, p_i z^i,$$

where z is a complex number. For our purposes it is sufficient to assume that z is a real number. If misunderstandings are possible, the notation $M_X(z)$ is used instead of $M(z)$. From (1.16), $M(z)$ is the mean value of the random variable $Y = z^X$:

$$M(z) = E(z^X). \tag{1.22}$$

$M(z)$ converges absolutely for $|z| \le 1$:

$$|M(z)| \le \Sigma_{i=0}^{\infty}\, p_i \left|z^i\right| \le \Sigma_{i=0}^{\infty}\, p_i = 1.$$

Therefore, $M(z)$ can be differentiated (as well as integrated) term by term:

$$M'(z) = \Sigma_{i=0}^{\infty}\, i p_i z^{i-1}.$$

Letting $z = 1$ yields

$$M'(1) = \Sigma_{i=0}^{\infty}\, i p_i = E(X).$$

Taking the second derivative of $M(z)$ gives

$$M''(z) = \Sigma_{i=0}^{\infty}(i-1)\, i p_i z^{i-2}.$$

Letting $z = 1$ yields

$$M''(1) = \Sigma_{i=0}^{\infty}\,(i-1)\, i p_i = \Sigma_{i=0}^{\infty}\, i^2 p_i - \Sigma_{i=0}^{\infty}\, i p_i\;.$$

Therefore, $M''(1) = E(X^2) - E(X)$. Thus, the first two moments of X are

$$E(X) = M'(1), \quad E(X^2) = M''(1) + M'(1).$$

Continuing in this way, all moments of X can be generated by derivatives of $M(z)$. Hence, the z-transform is indeed a moment generating function. In view of (1.19),

$$E(X) = M'(1), \qquad Var(X) = M''(1) + M'(1) - \left[M'(1)\right]^2. \tag{1.23}$$

On the other hand, by expanding a given z-transform $M(z)$ into a power series in z, the resulting coefficients of z^i are the probabilities $p_i = P(X = i)$. Hence, $M(z)$ is also called a *probability generating function.*

Poisson Distribution X has a Poisson distribution with parameter λ:

$$p_i = P(X = i) = \frac{\lambda^i}{i!}e^{-\lambda}; \quad i = 0, 1, ...$$

Then,

$$M(z) = \sum_{i=0}^{\infty} \frac{\lambda^i}{i!} e^{-\lambda} z^i = e^{-\lambda} \sum_{i=0}^{\infty} \frac{(\lambda z)^i}{i!} = e^{-\lambda} e^{+\lambda z}.$$

Hence,

$$M(z) = e^{\lambda(z-1)}.$$

The first two derivatives are

$$M'(z) = \lambda e^{\lambda(z-1)}, \quad M''(z) = \lambda^2 e^{\lambda(z-1)}.$$

Letting $z = 1$ yields

$$M'(1) = \lambda, \quad M''(1) = \lambda^2.$$

Thus, mean value, second moment and variance of X are

$$E(X) = \lambda, \quad Var(X) = \lambda, \quad E(X^2) = \lambda(\lambda + 1).$$

Binomial Distribution X has a binomial distribution with parameters n and p:

$$p_i = P(X = i) = \binom{n}{i} p^i (1-p)^{n-i}; \quad i = 0, 1, ..., n.$$

Then,

$$M(z) = \Sigma_{i=0}^{n} \binom{n}{i} p^i (1-p)^{n-i} z^i = \Sigma_{i=0}^{n} \binom{n}{i} (p z)^i (1-p)^{n-i}.$$

This is a binomial series so that

$$M(z) = [pz + (1-p)]^n.$$

By differentiation,

$$M'(z) = np[pz + (1-p)]^{n-1},$$

$$M''(z) = (n-1)\, np^2 [pz + (1-p)]^{n-2}.$$

Hence,

$$M'(1) = np \quad \text{and} \quad M''(1) = (n-1)\, np^2$$

so that

$$E(X) = np, \quad E(X^2) = (n-1)np^2 + np, \quad Var(X) = np(1-p).$$

Convolution Let $\{p_0, p_1, ...\}$ and $\{q_0, q_1, ...\}$ be the respective probability distribution of the discrete random variables X and Y with joint range $\{0,1,...\}$ and let a sequence $\{r_0, r_1, ...\}$ be defined as follows:

$$r_n = \Sigma_{i=0}^{n} p_i q_{n-i} = p_0 q_n + p_1 q_{n-1} + \cdots + p_n q_0, \quad n = 0, 1, ... \tag{1.24}$$

The sequence $\{r_0, r_1, ...\}$ is called the *convolution* of the probability distributions $\{p_0, p_1, ...\}$ and $\{q_0, q_1, ...\}$. The convolution is the probability distribution of a certain random variable, since

$$\Sigma_{n=0}^{\infty} r_n = 1, \quad r_n \geq 0.$$

For deriving the z-transform of the convolution, the following formula is needed:

$$\sum_{n=0}^{\infty}\sum_{i=0}^{n} a_{in} = \sum_{i=0}^{\infty}\sum_{n=i}^{\infty} a_{in}\,. \tag{1.25}$$

If Z denotes that random variable whose probability distribution is the convolution $\{r_0, r_1, ..., \}$, then its z-transform is

$$\begin{aligned} M_Z(z) &= \sum_{n=0}^{\infty} r_n z^n = \sum_{n=0}^{\infty}\sum_{i=0}^{n} p_i q_{n-i} z^n \\ &= \sum_{i=0}^{\infty} p_i z^i \left(\sum_{n=i}^{\infty} q_{n-i} z^{n-i}\right) \\ &= \left(\sum_{i=0}^{\infty} p_i z^i\right)\left(\sum_{k=0}^{\infty} q_k z^k\right). \end{aligned}$$

Thus, the z-transform of Z is the product of the z-transforms of X and Y:

$$M_Z(z) = M_X(z) \cdot M_Y(z)\,. \tag{1.26}$$

1.3.2 Laplace Transformation

Let $f(x)$ be any real-valued function on $[0, +\infty)$ with properties

1) $f(x)$ is piecewise continuous,

2) there exist real constants a and s_0 such that $f(x) \le a\, e^{s_0 x}$ for all $x \ge 0$.

The *Laplace transform* $\hat{f}(s)$ of $f(x)$ is defined as the parameter integral

$$\hat{f}(s) = \int_0^{\infty} e^{-sx} f(x)\, dx\,,$$

where the parameter s is any complex number satisfying $Re(s) > s_0$.

Notation If $z = x + iy$ is any complex number (i.e. $i = \sqrt{-1}$ and x, y are real numbers, then $R(z)$ denotes the *real part* of z: $R(z) = x$.

Assumptions 1 and 2 make sure that $\hat{f}(s)$ exists. With regard to the applications considered in this book, s can be assumed to be real. In this case, under the assumptions 1 ans 2, $\hat{f}(s)$ exists in the half-plane given by $\{s, s > s_0\}$.

Specifically, if $f(x)$ is the probability density of a nonnegative random variable X, then $\hat{f}(s)$ has a simple interpretation:

$$\hat{f}(s) = E(e^{-sX})\,. \tag{1.27}$$

This relationship is identical to (1.22) if there z is written in the form $z = e^{-s}$.

The n fold derivative of $\hat{f}(s)$ with respect to s is

$$\frac{d^n \hat{f}(s)}{ds^n} = (-1)^n \int_0^{\infty} x^n e^{-sx} f(x)\, dx\,.$$

Hence, if $f(x)$ is the density of a random variable X, then its moments of all orders can be obtained from $\hat{f}(s)$:

$$E(X^n) = (-1)^n \left. \frac{d^n \hat{f}(s)}{ds^n} \right|_{s=0} \quad ; \quad n = 0, 1, ... \tag{1.28}$$

Thus, the Laplace transform is a *moment generating function*. However, the Laplace transform is also a *probability* (*density*) *generating function*, since via a (complex) inversion formula the density of X can be obtained from its Laplace transform.

In what follows, it is more convenient to use the notation

$$\hat{f}(s) = L\{f\}.$$

Partial integration in $\hat{f}(s)$ yields ($s > s_0 \geq 0$)

$$L\left\{\int_0^x f(u)\, du\right\} = \frac{1}{s}\hat{f}(s) \tag{1.29}$$

and

$$L\left\{\frac{df(x)}{dx}\right\} = L\{f'(x)\} = s\,\hat{f}(s) - f(0). \tag{1.30}$$

More generally, if $f^{(n)}(x)$ denotes the nth derivative of $f(x)$ with respect to x, then

$$\hat{f}^{(n)}(s) = s^n \hat{f}(s) - s^{n-1} f(0) - s^{n-2} f'(0) - \cdots - s^1 f^{(n-2)}(0) - f^{(n-1)}(0).$$

Let f_1 and f_2 be any two functions satisfying assumptions 1) and 2). Then,

$$L\{f_1 + f_2\} = L\{f_1\} + L\{f_2\} = \hat{f}_1(s) + \hat{f}_2(s). \tag{1.31}$$

Convolution The *convolution* $f_1 * f_2$ of two functions f_1 and f_2, which are defined on the interval $[0, +\infty)$, is given by

$$(f_1 * f_2)(x) = \int_0^x f_2(x-u) f_1(u)\, du.$$

The following formula is the 'continuous' analogue to (1.26):

$$L\{f_1 * f_2\} = L\{f_1\} L\{f_2\} = \hat{f}_1(s)\,\hat{f}_2(s). \tag{1.32}$$

A proof of this relationship is easily established:

$$\begin{aligned} L\{f_1 * f_2\} &= \int_0^\infty e^{-sx} \int_0^x f_2(x-u) f_1(u)\, du\, dx \\ &= \int_0^\infty e^{-su} f_1(u) \int_u^\infty e^{-s(x-u)} f_2(x-u)\, dx\, du \\ &= \int_0^\infty e^{-su} f_1(u) \int_0^\infty e^{-sy} f_2(y)\, dy\, du \\ &= \hat{f}_1(s)\,\hat{f}_2(s). \end{aligned}$$

Verbally, formula (1.32) means that the Laplace transform of the convolution of two functions is equal to the product of the Laplace transforms of these functions.

In proving (1.32), *Dirichlet's formula* had been applied:

$$\int_0^z \int_0^y f(x,y)\,dx\,dy = \int_0^z \int_x^z f(x,y)\,dy\,dx. \tag{1.33}$$

Obviously, formula (1.33) is the 'continuous analogue' to formula (1.25):

Retransformation The Laplace transform $\hat{f}(s)$ is called the *image* of $f(x)$ and $f(x)$ is the *pre-image* of $\hat{f}(s)$. Finding the pre-image of a given Laplace transform (*retransformation*) can be a difficult task. Properties (1.31) and (1.32) of the Laplace transformation suggest that Laplace transforms should be decomposed as far as possible into terms and factors (for instance, decomposing a fraction into partial fractions), because the retransformations of the arising less complex terms and factors are usually easier done than the retransformation of the original image. Retransformation is facilitated by *contingency tables.* These tables contain important functions and their Laplace transforms. As already mentioned, there exists an explicit formula for obtaining the pre-image of a given Laplace transform. Its application requires knowledge of complex calculus.

Example 1.11 Let X have an exponential distribution with parameter λ:

$$f(x) = \lambda e^{-\lambda x}, \quad x \ge 0.$$

The Laplace transform of $f(x)$ is

$$\hat{f}(s) = \int_0^\infty e^{-sx}\lambda e^{-\lambda x}\,dx = \lambda \int_0^\infty e^{-(s+\lambda)x}\,dx = \frac{\lambda}{s+\lambda}.$$

It exists for $s > -\lambda$. The nth derivative of $\hat{f}(s)$ is

$$\frac{d^n \hat{f}(s)}{ds^n} = (-1)^n \frac{\lambda\, n!}{(s+\lambda)^{n+1}}.$$

Thus, the nth moment is

$$E(X^n) = \frac{n!}{\lambda^n}; \quad n = 0, 1, \ldots$$

□

Example 1.12 The definition of the Laplace transform can be extended to functions defined on the whole real axis $(-\infty, +\infty)$. For instance, consider the density of an $N(\mu, \sigma^2)$-distribution:

$$f(x) = \frac{1}{\sqrt{2\pi}\,\sigma} e^{-\frac{(x-\mu)^2}{2\sigma^2}}; \quad x \in (-\infty, +\infty).$$

Its Laplace transform is defined as

$$\hat{f}(s) = \frac{1}{\sqrt{2\pi}\,\sigma} \int_{-\infty}^{+\infty} e^{-sx} e^{-\frac{(x-\mu)^2}{2\sigma^2}}\,dx.$$

Obviously, this improper parameter integral exists for all s. Substituting $u = (x - \mu)/\sigma$ yields

$$\hat{f}(s) = \frac{1}{\sqrt{2\pi}} e^{-\mu s} \int_{-\infty}^{+\infty} e^{-\sigma s u} e^{-u^2/2} du$$

$$= \frac{1}{\sqrt{2\pi}} e^{-\mu s + \frac{1}{2}\sigma^2 s^2} \int_{-\infty}^{+\infty} e^{-\frac{1}{2}(u+\sigma s)^2} du.$$

The last integral is equal to $\sqrt{2\pi}$. Hence,

$$\hat{f}(s) = e^{-\mu s + \frac{1}{2}\sigma^2 s^2}.$$

□

For probability densities $f(x)$, two important variants of the Laplace transform are the *moment generating function* and the *characteristic function.*

a) Moment Generating Function Let X be a random variable with density $f(x)$ and t a real parameter. Then the parameter integral

$$M(t) = E\left(e^{tX}\right) = \int_{-\infty}^{+\infty} e^{tx} f(x)\, dx$$

is called the *moment generating function* of X. $M(t)$ arises from the Laplace transform of $f(x)$ by letting $s = -t$. (The terminology is a bit confusing, since, as mentioned before, the Laplace transform is moment generating as well.)

b) Characteristic Function Let X be a random variable with density $f(x)$, t a real parameter and $i = \sqrt{-1}$. Then the parameter integral

$$\psi(t) = E\left(e^{itX}\right) = \int_{-\infty}^{+\infty} e^{itx} f(x)\, dx$$

is called the *characteristic function* of X. Obviously, $\psi(t)$ is the Fourier transform of $f(x)$. The characteristic function $\psi(t)$ is obtained from the Laplace transform by letting $s = -it$.

Characteristic functions belong to the most important mathematical tools for solving probability theoretic problems, e.g. for proving limit theorems and for characterizing and analyzing stochastic processes.

One of their main advantages to the Laplace transform and to the moment generating function is that they always exist:

$$|\psi(t)| = \left| \int_{-\infty}^{+\infty} e^{itx} f(x)\, dx \right| \le \int_{-\infty}^{+\infty} \left| e^{itx} \right| f(x)\, dx = \int_{-\infty}^{+\infty} f(x)\, dx = 1.$$

The characteristic function has quite analogous properties to the Laplace transform (if the latter exists) with regard to its relationship to the probability distribution of sums of independent random variables.

1.4 CLASSES OF PROBABILITTY DISTRIBUTIONS BASED ON AGING BEHAVIOUR

This section is restricted to the class of nonnegative random variables. Lifetimes of technical systems and organisms are likely to be the most prominent members of this class. Hence, a terminology is used tailored to this application. The lifetime of a system is the time span from its starting up time point (birth) to its failure (death), where 'failure' is assumed to be an instantaneous event. In the engineering context, a failure of a system need not be equivalent to the end of its useful life. If X is a lifetime with distribution function $F(\cdot)$, then $F(x)$ is called *failure probability* and $\overline{F}(x) = 1 - F(x)$ is called *survival probability* with regard to the interval $[0, x]$, because $F(x)$ and $\overline{F}(x)$ are the respective probabilities that the system does or does not fail in $[0, x]$.

Residual Lifetime Let $F_t(x)$ be the distribution function of the *residual lifetime* X_t of a system, which has already worked for t time units without failing:

$$F_t(x) = P(X_t \le x) = P(X - t \le x \,|\, X > t).$$

According to (1.6),

$$F_t(x) = \frac{P(X - t \le x \cap X > t)}{P(X > t)} = \frac{P(t < X \le t + x)}{P(X > t)}.$$

Formula (1.13) yields the desired result:

$$F_t(x) = \frac{F(t+x) - F(t)}{\overline{F}(t)}; \quad x \ge 0, \ t \ge 0. \tag{1.34}$$

The corresponding *conditional survival probability* $\overline{F}_t(x) = 1 - F_t(x)$ is given by

$$\overline{F}_t(x) = \frac{\overline{F}(t+x)}{\overline{F}(t)}; \quad x \ge 0, \ t \ge 0. \tag{1.35}$$

Hence, using (1.17), the mean residual lifetime $\mu(t) = E(X_t)$ of a system is

$$\mu(t) = \frac{1}{\overline{F}(t)} \int_t^\infty \overline{F}(x)\, dx. \tag{1.36}$$

Example 1.13 (***uniform distribution***) The random variable X has uniform distribution over $[0, T]$. Then its density and distribution function are

$$f(x) = \begin{cases} 1/T & \text{for} \quad 0 \le x \le T, \\ 0, & \text{elsewhere}, \end{cases} \qquad F(x) = \begin{cases} 0 & \text{for} \quad x < 0, \\ x/T & \text{for} \quad 0 \le x \le T, \\ 1 & \text{for} \quad T < x. \end{cases}$$

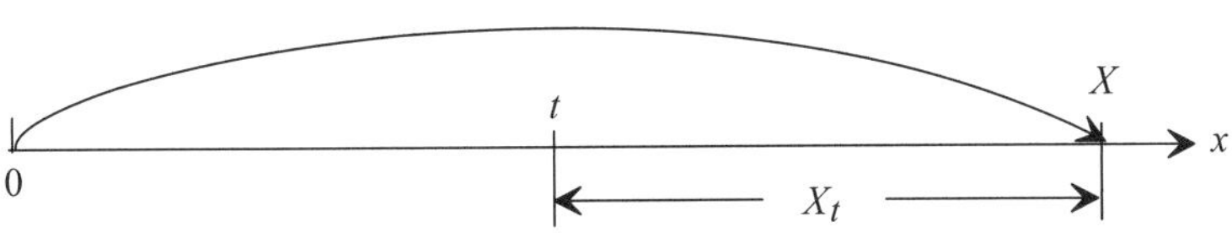

Figure 1.8 Illustration of the residual lifetime

The conditional failure probability is

$$F_t(x) = \frac{x}{T-t}; \quad 0 \le t < T, \ 0 \le x \le T-t.$$

Thus, X_t is uniformly distributed over the interval $[0, T-t]$, and the conditional failure probability is increasing with increasing t, $t < T$. □

Example 1.14 (*exponential distribution*) Let X have an exponential distribution with parameter λ, i.e. its density and distribution function are

$$f(x) = \lambda e^{-\lambda x}, \quad F(x) = 1 - e^{-\lambda x}, \quad x \ge 0.$$

Given t, the corresponding conditional failure probability is for $x \ge 0$ and $t \ge 0$

$$F_t(x) = \frac{(1 - e^{-\lambda (t+x)}) - (1 - e^{-\lambda t})}{e^{-\lambda t}} = 1 - e^{-\lambda x} = F(x). \tag{1.37}$$

Thus, the residual lifetime of the system has the same distribution function as the lifetime of a new system, namely an exponential distribution with parameter λ. The exponential distribution is the only continuous probability distribution, which has this so-called *memoryless property* or *lack of memory property*. Consequently, the age of an operating system with exponential lifetime has no influence on its future failure behaviour. Or, equivalently, if the system has not failed in the interval $[0, t]$, then, with respect to its failure behaviour in $[t, \infty)$, it is at time t *as good as new*. Complex systems and electronic hardware often have this property if they have survived the 'early failure time period'.

The fundamental relationship $F_t(x) = F(x)$ is equivalent to

$$\overline{F}(t+x) = \overline{F}(t)\,\overline{F}(x). \tag{1.38}$$

It can be shown that the distribution function of the exponential distribution is the only one which satisfies the functional equation (1.38). □

The engineering (biological) background of the conditional failure probability motivates the following definition.

Definition 1.1 A system is *aging* (*rejuvenating*) in the interval $[t_1, t_2]$, $t_1 < t_2$, if for an arbitrary but fixed x, the conditional failure probability $F_t(x)$ is increasing (decreasing) for increasing t, $t_1 \le t \le t_2$. ●

In case of technical systems, periods of rejuvenation may be due to maintenance actions and, in case of human beings, due to successful medical treatments or adopting a healthier lifestyle. Note that here and in what follows the terms 'increasing' and 'decreasing' have the meaning of 'nondecreasing' and 'nonincreasing', respectively.

Provided the existence of the density $f(x) = F'(x)$, another approach to modeling the aging behaviour of a system is based on the concept of its failure rate. To derive this concept, the conditional system failure probability $F_t(\Delta t)$ of a system in $[t, t+\Delta t]$ is considered relative to the length Δt of this interval. This is a conditional failure probability per unit time, i.e. a 'failure probability rate':

$$\frac{1}{\Delta t} F_t(\Delta t) = \frac{F(t+\Delta t) - F(t)}{\Delta t} \cdot \frac{1}{\overline{F}(t)} .$$

For $\Delta t \to 0$, the first ratio on the right hand side tends to $f(t)$. Hence,

$$\lim_{\Delta t \to 0} \frac{1}{\Delta t} F_t(\Delta t) = f(t) \big/ \overline{F}(t) .$$

This limit is called *failure rate* or *hazard function* and denoted as $\lambda(t)$:

$$\lambda(t) = f(t) \big/ \overline{F}(t) . \tag{1.39}$$

(In demograpy and in actuarial science, $\lambda(t)$ is called *force of mortality*.) $\lambda(t)$ gives information on both the instantaneous tendency of a system to fail and its 'state of wear' at age t. Integration on both sides of (1.39) from $t = 0$ to $t = x$ yields

$$F(x) = 1 - e^{-\int_0^x \lambda(t)\, dt}, \quad x \geq 0.$$

If introducing the *integrated failure rate*

$$\Lambda(x) = \int_0^x \lambda(t)\, dt,$$

$F(x)$, $F_t(x)$ and the corresponding survival probabilities can be written as follows:

$$F(x) = 1 - e^{-\Lambda(x)}, \quad \overline{F}(x) = e^{-\Lambda(x)},$$

$$F_t(x) = 1 - e^{-[\Lambda(t+x) - \Lambda(t)]}, \tag{1.40}$$

$$\overline{F}_t(x) = e^{-[\Lambda(t+x) - \Lambda(t)]}; \quad x \geq 0, \ t \geq 0.$$

This representation of $\overline{F}_t(x)$ implies an important property of the failure rate:

A system ages in $[t_1, t_2]$, $t_1 < t_2$, if its failure rate $\lambda(t)$ is increasing in this interval.

For many applications, the following property of $\lambda(t)$ is crucial:

$$P(X - t \leq \Delta t | X > t) = \lambda(t)\, \Delta t + o(\Delta t),$$

where $o(x)$ is the *Landau order symbol* with respect to $x \to 0$, i.e. any function of x satisfying

$$\lim_{x \to 0} \frac{o(x)}{x} = 0. \tag{1.41}$$

Thus, for Δt being sufficiently small, $\lambda(t)\, \Delta t$ is approximately the probability that the system fails in $(t, t + \Delta t]$ if it has survived interval $[0, t]$. This property of the failure rate can be used for its statistical estimation: At time $t = 0$ a specified number of independently operating, identical systems start working. Then the failure rate of these systems in the interval $[t, t+\Delta t]$ is approximately equal to the number of systems, which fail in $[t, t+\Delta t]$, divided by the product of Δt and the number of systems which are still operating at time t.

For instance, if X has a *Weibull distribution* with parameters β and θ, then

$$\lambda(x) = (\beta/\theta)\,(x/\theta)^{\beta-1}, \quad x > 0.$$

Consequently, the failure rate is increasing in $[0, \infty)$ if $\beta > 1$, and it is decreasing in $[0, \infty)$ if $\beta < 1$. If $\beta = 1$, the failure rate is identically constant: $\lambda(t) \equiv \lambda = 1/\theta$.

Based on the behaviour of the conditional failure probability of a system, several nonparametric classes of probability distributions have been proposed and investigated during the past 50 years. Originally, they were defined with regard to applications in reliability engineering. Nowadays these classes also play an important role in fields as demography and actuarial science. The most obvious classes are *IFR* (*increasing failure rate*) and *DFR* (decreasing failure rate).

IFR- (DFR-) Distribution $F(x)$ is an IFR- (DFR-) distribution (briefly: $F(x)$ is IFR (DFR)) if $F_t(x)$ is increasing (decreasing) in t for fixed, but arbitrary x.

If the density $f(x) = F'(x)$ exists, then, from (1.40):

> $F(x)$ is IFR (DFR) if and only if the corresponding failure rate $\lambda(t)$ is increasing (decreasing) in t.

Another characterization of IFR and DFR is based on the Laplace transform $\hat{f}(s)$ of the density $f(x) = F'(x)$. For $n = 1, 2, \ldots,$ let

$$a_{-1}(s) \equiv 1, \; a_0(s) = \frac{1}{s}\left[1 - \hat{f}(s)\right], \; a_n(s) = \frac{(-1)^n}{n!}\frac{d^n a_0(s)}{ds^n}. \tag{1.42}$$

Then $F(x)$ is IFR (DFR) if and only if

$$a_n^2(s) \underset{(\le)}{\ge} a_{n-1}(s)\, a_{n+1}(s); \quad n = 0, 1, \ldots$$

(*Vinogradov* [85]). If $f(x)$ does not exist, then this statement remains valid if $\hat{f}(s)$ is the Laplace-Stieltjes transform of $F(x)$.

The example of the Weibull distribution shows that, within one and the same parametric class of probability distributions, different distribution functions may belong to different nonparametric classes of probability distributions:

If $\beta > 1$, then $F(x)$ is IFR, if $\beta < 1$, then $F(x)$ is DFR, if $\beta = 1$ (exponential distribution), then $F(x)$ is both IFR and DFR.

The IFR- (DFR-) class is equivalent to the aging (rejuvenation) concept proposed in definition 1.1. The following nonparametric classes present modifications and more general concepts of aging and rejuvenation than the ones given by definition 1.1.

IFRA- (DFRA-) Distribution The failure rate (force of mortality) of human beings (as well as of other organisms), is usually not (strictly) increasing. In short time periods, for instance, after having overcome a serious illness or another life-threatening situation, the failure rate is likely to decrease. But the average failure rate will definitely increase. Analogously, technical systems, which operate under different, time-dependent stress levels (temperature, pressure, speed) will not have a (strictly) increasing failure rate. Hence, the definition of the classes IFRA (increasing failure rate average) and DFRA (decreasing failure rate average) makes sense:

$F(x)$ is an IFRA- (DFRA-) distribution if the function

$$-\frac{1}{t}\ln \bar{F}(t)$$

is increasing (decreasing) in t.

This definition is motivated by the fact that, assuming the existence of the probability density $f(x) = F'(x)$, according to (1.39), the average failure rate over the interval $[0, t]$ is

$$\bar{\lambda}(t) = \frac{1}{t}\int_0^t \lambda(x)\,dx = -\frac{1}{t}\ln \bar{F}(t).$$

Another, equivalent characterization of IFRA (DFRA) is: $F(x)$ is IFRA (DFRA) if

$$\bar{F}(ax) \underset{(\le)}{\ge} \left[\bar{F}(x)\right]^a, \quad a > 1,\ x \ge 0.$$

NBU- (NWU-) Distribution Since

$$F_t(x) = F(x)$$

is equivalent to $\bar{F}(t+x) = \bar{F}(t)\bar{F}(x)$, a new system has a smaller failure probability than a used system of age t if and only if

$$\bar{F}(t+x) \le \bar{F}(t)\bar{F}(x).$$

This motivates the concepts of NBU (new better than used) and NWU (new worse than used):

$F(t)$ is an NBU- (NWU-) distribution if

$$\bar{F}(t+x) \underset{(\ge)}{\le} \bar{F}(t)\bar{F}(x) \tag{1.43}$$

for all $x \ge 0,\ t \ge 0$.

(Note that the equation $F_t(x) \equiv F(x)$ means that a 'used' system has the same lifetime distribution as a new one.) As the classes IFR and DFR (as well as other classes), NBU and NWU can be characterized by properties of Laplace transforms of its probability densities (*Vinogradov* [85]): With the notation (1.42),

$F(x)$ is NBU (NWU) if and only if

$$a_n(s)\,a_m(s) \underset{(\le)}{\ge} a_{n+m+1}(s) \quad \text{for all } m = 0, 1, \ldots;\ n = 0, 1, \ldots, \text{ and } s \ge 0.$$

NBUE- (NWUE-) Distribution According to (1.17) and (1.36), the mean life of a new system μ and the mean residual lifetime $\mu(t)$ of a system, which is still operating at age t (used system) are given by

$$\mu = \int_0^\infty \overline{F}(x)\,dx, \qquad \mu(t) = \frac{1}{\overline{F}(t)} \int_t^\infty \overline{F}(x)\,dx. \tag{1.44}$$

When comparing μ and $\mu(t)$, one arrives at the classes NBUE (*new better than used in expectation*) and NWUE (*new worse than used in expectation*):

$F(x)$ is an NBUE- (NWUE-) distribution if

$$\frac{1}{\mu} \int_t^\infty \overline{F}(x)\,dx \underset{(\geq)}{\leq} \overline{F}(t) \quad \text{for all } t \geq 0.$$

The survival function on the left-hand side of this inequality plays an important role in renewal theory (section 3.3). There it is denoted as

$$\overline{F}_S(t) = \frac{1}{\mu} \int_t^\infty \overline{F}(x)\,dx.$$

The corresponding distribution function is

$$F_S(t) = 1 - \overline{F}_S(t) = \frac{1}{\mu} \int_0^t \overline{F}(x)\,dx. \tag{1.45}$$

Hence, $F(x)$ is an NBUE- (NWUE-) distribution if and only if

$$F_S(x) \underset{(\geq)}{\leq} F(x) \quad \text{for all } x \geq 0.$$

Note that, if $F(x)$ is IFR (DFR), then $F_S(x)$ is IFR (DFR), too.

2-NBU- (2-NWU-) Distribution $F(x)$ is a 2-NBU- (2-NWU-) *distribution* if the corresponding distribution function $F_S(x)$, defined by (1.45), satisfies

$$\overline{F}_S(t+x) \underset{(\geq)}{\leq} \overline{F}_S(t)\,\overline{F}_S(x).$$

Obviously, this is equivalent to $F_S(x)$ being NBU (NWU).

NBUL- (NWUL-) Distribution When applying the Laplace transform with s as a real, nonnegative number to both sides of the defining relation (1.43) one obtains for NBU

$$\int_0^\infty e^{-sx}\,\overline{F}(t+x)dx \leq \overline{F}(t) \int_0^\infty e^{-sx}\,\overline{F}(x)dx,$$

and for NWU,

$$\int_0^\infty e^{-sx}\,\overline{F}(t+x)dx \geq \overline{F}(t) \int_0^\infty e^{-sx}\,\overline{F}(x)dx.$$

This leads to the following definition:

$F(x)$ is an NBUL- (NWUL-) distribution (*new better* (*worse*) *than used in Laplace ordering*) if

$$\frac{\int_t^\infty e^{-s\,u}\,\overline{F}(u)\,du}{\int_0^\infty e^{-s\,u}\,\overline{F}(u)\,du} \underset{(\geq)}{\leq} e^{-s\,t}\,\overline{F}(t), \quad s, t \geq 0.$$

Equivalently, $F(x)$ is NBUL (NWUL) if

$$\int_0^\infty e^{-sx}\,\overline{F}_t(x)dx \underset{(\geq)}{\leq} \int_0^\infty e^{-sx}\,\overline{F}(x)\,dx; \quad s, t \geq 0.$$

IMRL- (DMRL-) Distribution The monotonicity behaviour of the mean residual lifetime $\mu(t)$ motivates another class of nonparametric probability distributions:

$F(x)$ is an IMRL- (DMRL-) distribution if

$$\mu(t_2) \underset{(\leq)}{\geq} \mu(t_1) \text{ for } 0 \leq t_1 \leq t_2 .$$

Implications between some classes of nonparametric distribution classes are:

$$IFR \Rightarrow IFRA \Rightarrow NBU \Rightarrow NBUE$$
$$DFR \Rightarrow DFRA \Rightarrow NWU \Rightarrow NWUE$$

Knowledge of the nonparametric class a distribution function belongs to and knowledge of some of its numerical parameters allow the construction of lower and/or upper bounds on this otherwise unknown distribution function. The first and most important results along this line can be found in Barlow and Proschan ([3, 4]).

1) Let $F(x) = P(X \leq x)$ be IFR and $\mu_n = E(X^n)$ the nth moment of X. Then,

$$\overline{F}(x) \geq \begin{cases} \exp[-x\,(n!/\mu_n)^{1/n}] & \text{for } x \leq \mu_n^{1/n} \\ 0 & \text{for } x > \mu_n^{1/n} \end{cases} .$$

In particular, for $n = 1$, with $\mu = \mu_1 = E(X)$,

$$\overline{F}(x) \geq \begin{cases} e^{-x/\mu} & \text{for } x \leq \mu \\ 0 & \text{for } x > \mu \end{cases} . \tag{1.46}$$

2) The lower bound (1.46) can be improved (Solov'ev [77]):

$$\overline{F}(x) \geq \begin{cases} e^{-\alpha x/\mu} & \text{for } x \leq \beta\mu \\ 0 & \text{for } x > \beta\mu \end{cases} ,$$

where

$$\beta = \frac{\mu_2}{\mu^2} + \left(\frac{\mu_2}{\mu^2} - 1\right)\alpha\left(\ln\frac{1}{1-\alpha}\right)^{-1} .$$

The parameter α satisfies $0 < \alpha < 1$ and is solution of the equation

$$\frac{\mu_2}{\mu^2} - 1 = \frac{2\alpha - \alpha^2 + 2(1-\alpha)\ln(1-\alpha)}{\alpha^2} .$$

3) If $F(x)$ is DFR, then

$$\bar{F}(x) \le \begin{cases} e^{-x/\mu} & \text{for } x \le \mu \\ \mu(ex)^{-1} & \text{for } x > \mu \end{cases}$$

and (Brown [14])

$$\bar{F}(x) \ge e^{-(\gamma + x/\mu)}, \quad x \ge 0,$$

where

$$\gamma = \frac{\mu_2}{2\mu^2} - 1 .$$

It can be shown that

$$\gamma \underset{(\ge)}{\le} 0 \text{ if } F(x) \text{ is IFR (DFR).}$$

The constant γ also occurs in the estimates given under 4) and 5).

4) If $F(x)$ is IFR, then (Solov'ev [77])

$$\sup_x \left| \bar{F}(x) - e^{-x/\mu} \right| \le 1 - \sqrt{2\gamma - 1} ,$$

5) If $F(x)$ is DFR, then (Brown [14]),

$$\sup_x \left| \bar{F}(x) - e^{-x/\mu} \right| \le 1 - e^{-\gamma},$$

$$\sup_x \left| \bar{F}(x) - \bar{F}_S(x) \right| \le 1 - e^{-\gamma},$$

where $F_S(x)$ is given by (1.45).

6) If $F(x)$ is IFRA, then

$$\bar{F}(x) \le \begin{cases} 1 & \text{for } x < \mu \\ e^{-rx} & \text{for } x \ge \mu \end{cases},$$

where $r = r(x, \mu)$ is solution of

$$1 - r\mu = e^{-rx}.$$

7) If $F(x)$ is NBUE, then,

$$F(x) \le x/\mu \quad \text{for } x \le \mu .$$

8) If $F(x)$ is NBUE (NWUE), then

$$F_S(x) \underset{(\ge)}{\le} e^{-x/\mu}, \quad x \ge 0 .$$

Other results on nonparametric classes of distributions will be needed in subsequent chapters and presented in connection with specific applications.

1.5 ORDER RELATIONS BETWEEN RANDOM VARIABLES

Most classes of nonparametric probability distributions introduced in the previous section can be embedded into the more general framework of order relations between random variables. These 'stochastic orders' have proved a powerful tool for the approximate analysis of complex stochastic models, which elude a mathematically rig- orous treatment, in particular in queueing-, inventory-, and reliability theory, and recently in actuarial science. The breakthrough in theory and application of stochastic orders came with the publication of the English edition of the monograph Stoyan [79], see [80]. The present state of art of theory and applications can be found in the monograph Müller and Stoyan [62].

In this section, the nonnegative random variables X and Y are assumed to have distribution (survival) functions $F(x)$ and $G(x)$ ($\overline{F}(x)$ and $\overline{G}(x)$).

Usual Stochastic Order X is smaller than Y with regard to the *usual stochastic order* if

$$\overline{F}(x) \le \overline{G}(x) \quad \text{for all } x. \tag{1.47}$$

Thus, X assumes large values with lower probability than Y. This order relation between two random variables had been for many years the only one to be known. For that reason it was simply called the *stochastic order*. Mann and Whitney [58] were probably the first ones who introduced and used this concept.

Notation: $X \underset{st}{\le} Y$

With regard to the previous section: $F(x)$ is IFR (DFR) if and only if

$$X_{t_2} \underset{st}{\le} X_{t_1} \ \left(X_{t_2} \underset{st}{\ge} X_{t_1}\right) \ \text{for } 0 \le t_1 \le t_2,$$

where X_t is the residual lifetime of a system operating at time t.

$F(x)$ is NBU (NWU) if and only if

$$X_t \underset{st}{\le} X \ \left(X_t \underset{st}{\ge} X\right).$$

Let the random variable X_S have the distribution function $F_S(x)$ given by (1.45), and $X_{S,t}$ be the corresponding residual lifetime. Then,

$F(x)$ is 2-NBU (2-NWU) if and only if

$$X_{S,t} \underset{st}{\le} X_S \ \left(X_{S,t} \underset{st}{\ge} X_S\right).$$

Properties of the usual stochastic order (mean values are assumed to exist):

1) If $X \underset{st}{\le} Y$, then $E(X) \le E(Y)$.

2) If $X \underset{st}{\le} Y$, then $E(h(X)) \le E(h(Y))$ for all increasing functions $h(\cdot)$ and vice versa.

3) If $X \underset{st}{\le} Y$ and $E(X) = E(Y)$, then X and Y have the same distribution functions.

Hazard Rate Order This stochastic order is closely related to the distribution function of the residual lifetime. Let the residual lifetimes of systems with respective lifetimes X and Y be denoted as X_t and Y_t. If the usual stochastic order

$$X_t \underset{st}{\leq} Y_t$$

is required to hold for all $t \geq 0$, then, according to (1.35), this is equivalent to

$$\frac{\overline{F}(t+x)}{\overline{F}(t)} \leq \frac{\overline{G}(t+x)}{\overline{G}(t)} \quad \text{for all } t \geq 0,$$

or

$$\frac{\overline{F}(t+x)}{\overline{G}(t+x)} \leq \frac{\overline{F}(t)}{\overline{G}(t)}\,; \quad t \geq 0,\ x \geq 0.$$

This relationship motivates the following order relation:

X is smaller than Y with respect to the *hazard rate order* (*failure rate order*) if the ratio $\overline{F}(t)/\overline{G}(t)$ is decreasing with increasing t.

Notation: $X \underset{hr}{\leq} Y$

Properties of the hazard rate order:

1) If X and Y have continuous densities so that the respective failure rates $\lambda_X(t)$ and $\lambda_Y(t)$ exist, then $X \underset{hr}{\leq} Y$ if and only if $\lambda_X(t) \geq \lambda_Y(t)$ for $t \geq 0$.

2) Let $X \underset{hr}{\leq} Y$ and $h(\cdot)$ be an increasing real function. Then, $h(X) \underset{hr}{\leq} h(Y)$.

3) If $X \underset{hr}{\leq} Y$, then $X \underset{st}{\leq} Y$.

Convex Orders The usual stochastic order and the hazard rate order refer to the absolute sizes of the random variables to be compared. However, for many applications it is useful to include the variability aspect. If random variables X and Y have about the same mean, usually the one with the smallest variability is preferred. This aspect is taken into account by *convex orders*.

(a) X is said to be smaller than Y *in convex order* if for all real-valued convex functions $h(\cdot)$ with property that $E(h(X))$ and $E(h(Y))$ exist,

$$E(h(X)) \leq E(h(Y)).$$

Notation: $X \underset{cx}{\leq} Y$

(b) X is said to be smaller than Y *in increasing convex order* if for all real-valued increasing convex functions $h(\cdot)$ with property that $E(h(X))$ and $E(h(Y))$ exist,

$$E(h(X)) \leq E(h(Y)).$$

Notation: $X \underset{icx}{\leq} Y$

(c) X is said to be smaller than Y *in increasing concave order* if for all real-valued concave functions $h(\cdot)$ with property that $E(h(X))$ and $E(h(Y))$ exist,

$$E(h(X)) \le E(h(Y)). \tag{1.48}$$

Notation: $X \underset{icv}{\le} Y$

Before stochasticians started to thoroughly investigate these orders, some of them had already been known in applied sciences for a couple of years. In actuarial science, 'increasing convex order' had been known as 'stop-loss order', whereas in decision theory 'increasing concave order' had been called 'second order stochastic dominance'.

Properties of convex orders:

1) $X \underset{icx}{\le} Y$ if and only if $-Y \underset{icv}{\le} -X$.

Hence, only one of these stochastic orders needs to be investigated.

2) $X \underset{cx}{\le} Y$ holds if and only if

$$X \underset{icx}{\le} Y \text{ and } E(X) = E(Y).$$

3) If $X \underset{cx}{\le} Y$, then

$$E(X^n) \le E(Y^n) \text{ and } E((X-E(X))^n) \le E((Y-E(Y))^n) \quad \text{for } n = 2, 4, \ldots$$

Specifically,

$$\text{if } X \underset{cx}{\le} Y, \text{ then } Var(X) \le Var(Y).$$

4) Let $(c-x)_+ = \max(0, c-x)$. Then

$$X \underset{icx}{\le} Y$$

holds if and only if for all x

$$E((X-x)_+) \le E((Y-x)_+). \tag{1.49}$$

Thus, for defining $X \underset{icx}{\le} Y$, condition (1.48) needs to be checked only for a simple class of convex functions, namely the so-called *wedge functions*

$$h(x) = (c-x)_+ \,.$$

Note that the function

$$\pi_X(x) = E((X-x)_+) = \int_x^\infty \overline{F}(u)\,du \tag{1.50}$$

is convex and decreasing in x. In actuarial science, this function is called the *stop-loss-transform*, since the net premium of a stop-loss reinsurance contract has this structure.

1.6 MULTIDIMENSIONAL RANDOM VARIABLES

1.6.1 Basic Concepts

Let $(X_1, X_2, ..., X_n)$ be an n-dimensional vector, the components of which are random variables. Then $(X_1, X_2, ..., X_n)$ is called a *random vector, a multidimensional random variable* or, more precisely, an *n-dimensional random vector* or an *n-dimensional random variable.* Its *joint distribution function* or simply the joint distribution function of the random variables $X_1, X_2, ..., X_n$ is defined by

$$F(x_1, x_2, ..., x_n) = P(X_1 \le x_1, X_2 \le x_2, ..., X_n \le x_n). \tag{1.51}$$

This function characterizes the *probability distribution* of $(X_1, X_2, ..., X_n)$. The distribution functions of the X_i, denoted as

$$F_{X_i}(x) = P(X_i \le x_i),$$

can be obtained from the joint distribution function:

$$F_{X_i}(x_i) = F(\infty, ..., \infty, x_i, \infty, ..., \infty); \quad i = 1, 2, ..., n. \tag{1.52}$$

The one-dimensional distribution functions

$$F_{X_1}(x),\ F_{X_2}(x),\ ...,\ F_{X_n}(x)$$

are the *marginal distributions* of $(X_1, X_2, ..., X_n)$. The marginal distributions of a random vector cannot fully characterize its probability distribution, since they do not contain information on the statistical dependency between the X_i. Only if the random variables X_i are independent, joint distribution and the set of the marginal distributions contain the same amount of information on $X_1, X_2, ..., X_n$.

Independence The random variables $X_1, X_2, ..., X_n$ are said to be *independent* if for all vectors $(x_1, x_2, ..., x_n)$

$$F(x_1, x_2, ..., x_n) = F_{X_1}(x_1) F_{X_2}(x_2) \cdots F_{X_n}(x_n). \tag{1.53}$$

In this case, the distribution functions of the X_i fully determine the joint distribution function.

Identical Distribution The random variables $X_1, X_2, ..., X_n$ are called *identically distributed* if they have the same distribution function (probability distribution):

$$F(x) = F_{X_i}(x); \quad i = 1, 2, ..., n.$$

For independent, identically distributed (iid) random variables,

$$F(x_1, x_2, ..., x_n) = F(x_1) F(x_2) \cdots F_n(x_n).$$

Thus, the joint distribution function of a random vector with independent components is equal to the product of its marginal distribution functions.

1.6.2 Two-Dimensional Random Vectors

1.6.2.1 Discrete Components

Consider a random vector (X, Y), the components X and Y of which are discrete random variables with respective ranges $\{x_0, x_1, ...\}$ and $\{y_0, y_1, ...\}$ and probability distributions

$$\{p_i = P(X = x_i;\ i = 0, 1, ...\} \text{ and } \{q_j = P(Y = x_j;\ j = 0, 1, ...\}.$$

Furthermore, let

$$r_{ij} = P(X = x_i \cap Y = y_j).$$

The set of probabilities $\{r_{ij};\ i,j = 0, 1, ...\}$ is the *joint* or *two-dimensional probability distribution* of the random vector (X, Y). From the definition of the r_{ij},

$$p_i = \Sigma_{j=0}^{\infty} r_{ij}, \quad q_j = \Sigma_{i=0}^{\infty} r_{ij}. \tag{1.54}$$

In accordance with the terminology introduced in section 1.6.1, the probability distributions $\{p_i,\ i = 0, 1, ...\}$ and $\{q_i,\ i = 0, 1, ...\}$ constitute the *marginal distribution* of (X, Y). By (1.6), the conditional probabilities of $X = x_i$ given $Y = y_j$ and $Y = y_j$ given $X = x_i$ are

$$P(X = x_i | Y = y_j) = \frac{r_{ij}}{q_j}, \quad P(Y = y_j | X = x_i) = \frac{r_{ij}}{p_i}.$$

The sets

$$\left\{\frac{r_{ij}}{q_j};\ i = 0, 1, ...\right\} \text{ and } \left\{\frac{r_{ij}}{p_i};\ j = 0, 1, ...\right\}$$

are the *conditional probability distributions of X given* $Y = y_j$ and *of Y given* $X = x_i$, respectively. The corresponding conditional mean values are

$$E(X|Y = y_j) = \sum_{i=0}^{\infty} x_i \frac{r_{ij}}{q_j}, \quad E(Y|X = x_i) = \sum_{j=0}^{\infty} y_j \frac{r_{ij}}{p_i}.$$

The conditional mean value $E(X|Y)$ of X given Y is a random variable, since the condition is random. Its range is

$$\{E(X|Y = y_0),\ E(X|Y = y_1), ...\}.$$

The mean value of $E(X|Y)$ is

$$E(E(X|Y)) = \sum_{j=0}^{\infty} E(X|Y = y_j) P(Y = y_j) = \sum_{j=0}^{\infty}\sum_{i=0}^{\infty} x_i \frac{r_{ij}}{q_j} q_j$$

$$= \sum_{i=0}^{\infty} x_i \sum_{j=0}^{\infty} r_{ij} = \sum_{i=0}^{\infty} x_i p_i = E(X).$$

Because the roles of X and Y can be changed,

$$E(E(X|Y)) = E(X) \text{ and } E(E(Y|X)) = E(Y). \tag{1.55}$$

From (1.53): X and Y are independent if and only if the random events "$X = x_i$" and "$Y = y_j$" are independent for all $i, j = 0, 1, 2, \ldots$ Hence, if X and Y are independent,

$$r_{ij} = p_i q_j; \quad i,j = 0, 1, \ldots$$

1.6.2.2 Continuous Components

Let X and Y be continuous random variables with respective distribution functions and densities $F_X(x)$, $F_Y(y)$, $f_X(x)$, $f_Y(y)$. The joint distribution function of (X, Y),

$$F_{X,Y}(x,y) = P(X \le x, Y \le y),$$

has the following properties:

1) $F_{X,Y}(-\infty,-\infty) = 0, \quad F_{X,Y}(+\infty,+\infty) = 1$

2) $0 \le F_{X,Y}(x,y) \le 1$

3) $F_{X,Y}(x,+\infty) = F_X(x), \quad F_{X,Y}(+\infty,y) = F_Y(y)$ (1.56)

4) For $x_1 \le x_2$ and $y_1 \le y_2$

$$F_{X,Y}(x_1,y_1) \le F_{X,Y}(x_2,y_1) \le F_{X,Y}(x_2,y_2)$$
$$F_{X,Y}(x_1,y_1) \le F_{X,Y}(x_1,y_2) \le F_{X,Y}(x_2,y_2)$$

Conversely, any function of two variables which has these properties is the joint distribution function of a random vector (X, Y). (Properties 1 to 4 also hold for random vectors with discrete components.) The probability distributions of X and Y are called the *marginal distributions* of the two-dimensional random variable (X, Y).

Assuming its existence, the partial derivative of $F_{X,Y}(x,y)$ with respect to x and y,

$$f_{X,Y}(x,y) = \frac{\partial F_{X,Y}(x,y)}{\partial x\, \partial y},$$

is called the *joint probability density* of (X, Y). Equivalently, the joint density can be defined as a function $f(x,y)$ satisfying

$$F_{X,Y}(x,y) = \int_{-\infty}^{x} \int_{-\infty}^{y} f_{X,Y}(u,v)\, du\, dv \tag{1.57}$$

for all x, y. Every joint (probability) density has properties

$$f_{X,Y}(x,y) \ge 0, \quad \int_{-\infty}^{+\infty} \int_{-\infty}^{+\infty} f_{X,Y}(x,y)\, dx\, dy = 1.$$

Conversely, any function of two variables x and y satisfying these two conditions can be considered to be the joint density of a random vector (X, Y). Combining (1.56) and (1.57), one obtains the marginal densities of (X, Y):

$$f_X(x) = \int_{-\infty}^{+\infty} f_{X,Y}(x,y)\, dy, \quad f_Y(y) = \int_{-\infty}^{+\infty} f_{X,Y}(x,y)\, dx. \tag{1.58}$$

Thus, the respective marginal densities of (X, Y) are simply the densities of X and Y.

If X and Y are independent, then, according to (1.52),

$$F_{X,Y}(x,y) = F_X(x)\,F_Y(y).$$

Hence, in terms of the densities, if X and Y are independent, then the joint density of the random vector (X, Y) is the product of its marginal densities:

$$f_{X,Y}(x,y) = f_X(x) f_Y(y).$$

The conditional distribution function of Y given $X = x$ and the corresponding conditional density of Y given $X = x$ are denoted as

$$F_Y(y|x) = P(Y \le y \,|\, X = x)$$

$$f_Y(y|x) = dF_Y(y|x)/dy.$$

For continuous random variables, condition $X = x$ has probability 0 so that formula (1.4) cannot directly be applied to deriving $F_Y(y|x)$. Hence, consider for $\Delta x > 0$

$$P(Y \le y | x \le X \le x + \Delta x) = \frac{P(Y \le y \cap x \le X \le x + \Delta x)}{P(x \le X \le x + \Delta x)}$$

$$= \frac{\int_{-\infty}^{y} \frac{1}{\Delta x}\left(\int_{x}^{x+\Delta x} f_{X,Y}(u,v)\,du\right) dv}{\frac{1}{\Delta x}\left[F_X(x+\Delta x) - F_X(x)\right]}.$$

If $\Delta x \to 0$, then, assuming $f_X(x) > 0$,

$$F_Y(y|x) = \frac{1}{f_X(x)} \int_{-\infty}^{y} f_{X,Y}(x,v)\,dv.$$

Differentiation yields the desired conditional density:

$$f_Y(y|x) = \frac{f_{X,Y}(x,y)}{f_X(x)}. \tag{1.59}$$

By changing the roles of X and Y, the *conditional density of X given Y* is

$$f_X(x|y) = \frac{f_{X,Y}(x,y)}{f_Y(y)}.$$

The *conditional mean value of Y given $X = x$* is

$$E(Y|x) = \int_{-\infty}^{+\infty} y f_Y(y|x)\,dy.$$

The *conditional mean value of Y given X* is

$$E(Y|X) = \int_{-\infty}^{+\infty} y f_Y(y|X)\,dy.$$

$E(Y|X)$ is a random variable with properties

$$E(XY) = E(XE(Y|X))$$

$$E(E(Y|X)) = E(Y)$$

$$E(Y_1 + Y_2|X) = E(Y_1|X) + E(Y_2|X). \tag{1.60}$$

If X and Y are independent, then

$$E(X|Y=y) = E(X|Y) = E(X) \tag{1.61}$$

$$E(XY) = E(X)E(Y). \tag{1.62}$$

The *covariance* $Cov(X, Y)$ between random variables X and Y is defined as

$$Cov(X, Y) = E\{[x - E(X)][Y - E(Y)]\}. \tag{1.63}$$

This representation of the covariance is equivalent to

$$Cov(X, Y) = E(XY) - E(X)E(Y). \tag{1.64}$$

In particular, $Cov(X, X)$ is the variance of X:

$$Cov(X, X) = Var(X) = E((X - E(X))^2).$$

From (1.62), if X and Y are independent, then covariance between these two random variables is 0: $Cov(X, Y) = 0$. But if $Cov(X, Y) = 0$, then X and Y are not necessarily independent. The covariance can assume any value between $-\infty$ and $+\infty$. Nevertheless, it serves as a parameter giving information on the strength of the stochastic relationship between two random variables X and Y.

The *correlation coefficient* between X and Y is defined as

$$\rho(X, Y) = \frac{Cov(X,Y)}{\sqrt{Var(X)}\ \sqrt{Var(Y)}}. \tag{1.65}$$

The correlation coefficient has the following properties:

1) If X and Y are independent, then $\rho(X, Y) = 0$.

2) $\rho(X, Y) = \pm 1$ if and only if there exist constants a and b so that $Y = aX + b$.

3) For any random variables X and Y, $-1 \le \rho(X, Y) \le 1$.

The correlation coefficient is, therefore, a measure of the strength of the linear stochastic relationship between random variables.

X and Y are said to be *uncorrelated* if $\rho(X, Y) = 0$. Otherwise they are called *positively* or *negatively correlated* depending on the sign of $\rho(X, Y)$. Obviously, X and Y are uncorrelated if and only if

$$E(XY) = E(X)E(Y).$$

Thus, if X and Y are independent, then they are uncorrelated. But if X and Y are uncorrelated, they need not be independent. To show this, two examples are given. Example 1.15 takes into account that the definitions of covariance and correlation coefficient and properties derived from them also refer to discrete random variables.

Table 1.1 Joint distribution for example 1.15

Y \ X	-1	0	+1
-2	1/16	1/16	1/16
-1	2/16	1/16	2/16
+1	2/16	1/16	2/16
+2	1/16	1/16	1/16

Example 1.15 Let X and Y be discrete random variables with ranges

$$\mathbf{R}_X = \{-2, -1, 1, 2\} \text{ and } \mathbf{R}_Y = \{-1, 0, 1\}.$$

Table 1.1 shows the joint distribution of (X, Y). Accordingly, the mean values of X and Y are:

$$E(X) = \frac{3}{16} \cdot (-2) + \frac{5}{16} \cdot (-1) + \frac{5}{16} \cdot 1 + \frac{3}{16} \cdot 2 = 0$$

$$E(Y) = \frac{6}{16} \cdot (-1) + \frac{4}{16} \cdot 0 + \frac{6}{16} \cdot 1 = 0$$

The mean value of the product XY is

$$\begin{aligned} E(XY) = & \frac{1}{16} \cdot (-2)(-1) + \frac{1}{8} \cdot (-1)(-1) + \frac{1}{8} \cdot 1 \cdot (-1) + \frac{1}{16} \cdot 2 \cdot (-1) \\ & + \frac{1}{16} \cdot (-2) \cdot 0 + \frac{1}{16} \cdot (-1) \cdot 0 + \frac{1}{16} \cdot 1 \cdot 0 + \frac{1}{16} \cdot 2 \cdot 0 \\ & + \frac{1}{16} \cdot (-2) \cdot 1 + \frac{1}{8} \cdot (-1) \cdot 1 + \frac{1}{8} \cdot 1 \cdot 1 + \frac{1}{16} \cdot 2 \cdot 1 = 0. \end{aligned}$$

Hence, $E(XY) = E(X)\,E(Y) = 0$ so that X and Y are uncorrelated.

On the other hand,

$$P(X = 2,\ Y = -1) = \frac{1}{16} \neq P(X = 2) \cdot P(Y = -1) = \frac{3}{16} \cdot \frac{6}{16} = \frac{18}{256} = \frac{9}{128}.$$

Thus, X and Y are uncorrelated, but not independent. □

Example 1.16 Let the random vector (X, Y) have the joint probability density

$$f_{X,Y}(x, y) = \frac{x^2+y^2}{4\pi} \exp\left\{-\left(\frac{x^2+y^2}{2}\right)\right\}, \quad -\infty < x, y < \infty.$$

According to (1.58), the marginal density $f_X(x)$ is obtained as follows:

$$f_X(x) = \int_{-\infty}^{+\infty} \frac{x^2+y^2}{4\pi} \exp\left\{-\left(\frac{x^2+y^2}{2}\right)\right\} dy$$

$$= \frac{e^{-x^2/2}}{2\sqrt{2\pi}} \left(x^2 \int_{-\infty}^{+\infty} \frac{1}{\sqrt{2\pi}} e^{-y^2/2}\, dy + \int_{-\infty}^{+\infty} y^2 \frac{1}{\sqrt{2\pi}} e^{-y^2/2}\, dy \right).$$

The integrand in the first integral is the density of an $N(0, 1)$-distributed random variable, the second integral is the variance of an $N(0, 1)$-distributed random variable. Hence, both integrals are equal to 1. Thus,

$$f_X(x) = \frac{1}{2\sqrt{2\pi}} (x^2 + 1) e^{-x^2/2}, \quad -\infty < x < +\infty.$$

Since $f_{X,Y}(x, y)$ is symmetric in x and y,

$$f_Y(y) = \frac{1}{2\sqrt{2\pi}} (y^2 + 1) e^{-y^2/2}, \quad -\infty < y < +\infty.$$

Obviously, $f_{X,Y}(x, y) \neq f_X(x) \cdot f_Y(y)$. Hence, X and Y are statistically dependent random variables.

For $f_X(y)$ and $f_Y(y)$ being symmetric with regard to the origin, $E(X) = E(Y) = 0$. On the other side, the mean value of the product XY is

$$E(XY) = \int_{-\infty}^{+\infty} \int_{-\infty}^{+\infty} xy \frac{x^2+y^2}{4\pi} \exp\left\{ -\left(\frac{x^2+y^2}{2} \right) \right\} dx\, dy$$

$$= \frac{1}{4\pi} \left(\int_{-\infty}^{+\infty} x^3 e^{-x^2/2}\, dx \right) \cdot \left(\int_{-\infty}^{+\infty} y^3 e^{-y^2/2}\, dy \right).$$

The integrands in the integrals of the second line are asymmetric with regard to the origin. Thus, both integrals are equal to 0. Hence,

$$E(XY) = E(X)\, E(Y) = 0.$$

This proves that X and Y are uncorrelated, but not independent. □

The following example shows that the correlation coefficient may give absolutely wrong information on the degree of the statistical dependency between two random variables other than the linear one.

Example 1.17 Let $Y = \sin X$ with X uniformly distributed over the interval $[0, \pi]$:

$$f_X(x) = 1/\pi, \quad 0 \le x \le \pi.$$

The mean values of interest are

$$E(X) = \pi/2, \quad E(Y) = \tfrac{1}{\pi} \int_0^{\pi} \sin x\, dx = 2/\pi, \quad E(XY) = \tfrac{1}{\pi} \int_0^{\pi} x \sin x\, dx = 1.$$

Thus, the covariance between X and Y is 0:

$$Cov(X, Y) = 1 - \tfrac{\pi}{2} \cdot \tfrac{2}{\pi} = 0.$$

Hence, $\rho(X, Y) = 0$. Despite the functional relationship between the random variables X and Y, they are uncorrelated. □

Bivariate Normal Distribution The random vector (X, Y) has a *bivariate normal* or a *bivariate Gaussian distribution* with parameters

$$\mu_x, \mu_y, \sigma_x, \sigma_y \text{ and } \rho, \quad -\infty < \mu_x, \mu_y < \infty, \ \sigma_x > 0, \sigma_y > 0, \ -1 < \rho < 1$$

if it has joint density

$$f_{X,Y}(x,y) = \tag{1.66}$$

$$= \frac{1}{2\pi\sigma_x\sigma_y\sqrt{1-\rho^2}} \exp\left\{-\frac{1}{2(1-\rho^2)}\left(\frac{(x-\mu_x)^2}{\sigma_x^2} - 2\rho\frac{(x-\mu_x)(y-\mu_y)}{\sigma_x\sigma_y} + \frac{(y-\mu_y)^2}{\sigma_y^2}\right)\right\}$$

with $-\infty < x, y < +\infty$. By (1.58), the corresponding marginal densities are

$$f_X(x) = \frac{1}{\sqrt{2\pi}\,\sigma_x} \exp\left(-\frac{(x-\mu_x)^2}{2\sigma_x^2}\right), \quad -\infty < x < +\infty$$

$$f_Y(x) = \frac{1}{\sqrt{2\pi}\,\sigma_y} \exp\left(-\frac{(y-\mu_y)^2}{2\sigma_y^2}\right), \quad -\infty < y < +\infty.$$

Thus, if (X, Y) has a bivariate normal distribution with parameters $\mu_x, \sigma_x, \mu_y, \sigma_y$, and ρ, then the random variables X and Y have each a normal distribution with respective parameters μ_x, σ_x and μ_y, σ_y. Since the independence of X and Y is equivalent to $f_{X,Y}(x,y) = f_X(x) f_Y(y)$, X and Y are independent if and only if $\rho = 0$. It can easily be shown that the parameter ρ is equal to the correlation coefficient between X and Y. Therefore:

> If the random vector (X, Y) has a bivariate normal distribution, then X and Y are independent if and only if they are uncorrelated.

The conditional density of Y given $X = x$ is obtained from $f(x,y)$ and (1.59):

$$f_Y(y|x) = \frac{1}{\sqrt{2\pi}\,\sigma_y\sqrt{1-\rho^2}} \exp\left\{-\frac{1}{2\sigma_y^2(1-\rho^2)}\left[y - \rho\frac{\sigma_y}{\sigma_x}(x-\mu_x) - \mu_y\right]^2\right\}. \tag{1.67}$$

Thus, on condition $X = x$, the random variable Y has a normal distribution with parameters

$$E(Y|X=x) = \rho\frac{\sigma_y}{\sigma_x}(x-\mu_x) + \mu_y, \quad Var(Y|X=x) = \sigma_y^2(1-\rho^2). \tag{1.68}$$

Example 1.18 The daily consumptions of tap water X and Y of two neighbouring houses have a joint normal distribution with parameters

$$\mu_x = \mu_y = 16\,[\mathrm{m}^3],\ \sigma_x = \sigma_y = 2\,[\mathrm{m}^3] \text{ and } \rho = 0.5.$$

The conditional probability density of Y on condition $X=x$ has parameters

$$E(Y|x) = \rho\,\frac{\sigma_y}{\sigma_x}(x-\mu_x)+\mu_y = 0.5\cdot\frac{2}{2}(x-16) = \frac{x}{2}+8$$

$$Var(Y|x) = \sigma_y^2(1-\rho^2) = 4\,(1-0.5^2) = 3\,.$$

Hence,

$$f_Y(y|x) = \frac{1}{\sqrt{2\pi}\,\sqrt{3}}\,\exp\left\{-\frac{1}{2}\left(\frac{y-\frac{x}{2}-8}{\sqrt{3}}\right)^2\right\}, \quad -\infty<y<+\infty.$$

This is the density of an $N(8+x/2,\,3)$–distributed random variable. Some conditional interval probabilities are:

$$P(14<Y\le 16|X=10) = \Phi\left(\frac{16-13}{\sqrt{3}}\right)-\Phi\left(\frac{14-13}{\sqrt{3}}\right) = 0.958-0.718 = 0.240$$

$$P(14<Y\le 16|X=14) = \Phi\left(\frac{16-15}{\sqrt{3}}\right)-\Phi\left(\frac{14-15}{\sqrt{3}}\right) = 0.718-0.282 = 0.436.$$

The corresponding unconditional probability is

$$P(14<Y\le 16) = \Phi\left(\frac{16-16}{2}\right)-\Phi\left(\frac{14-16}{2}\right) = 0.500-0.159 = 0.341.$$ □

In what follows, the joint density of a vector (X, Y) is applied to determining the probability distribution of a product and a ratio of two random variables.

Distribution of the Product of two Random Variables Let (X, Y) be a random vector with joint probability density $f_{X,Y}(x,y)$, and

$$Z = XY.$$

The distribution function of Z is given by (Figure 1.9)

$$F_Z(z) = \iint\limits_{\{(x,y);\, xy\le z\}} f_{X,Y}(x,y)\,dx\,dy$$

with $\{(x,y);\, xy\le z\} = \{-\infty<x\le 0,\ z/x\le y<\infty\}\cup\{0\le x<\infty,\ -\infty<y\le z/x\}$. Hence,

$$F_Z(z) = \int_{-\infty}^{0}\int_{z/x}^{+\infty} f_{X,Y}(x,y)\,dy\,dx + \int_{0}^{+\infty}\int_{-\infty}^{z/x} f_{X,Y}(x,y)\,dy\,dx.$$

Differentiation with regard to z yields the probability density of Z:

$$f_Z(z) = \int_{-\infty}^{0}\left(-\frac{1}{x}\right) f_{X,Y}(x,\tfrac{z}{x})\,dx + \int_{0}^{\infty}\frac{1}{x}\, f_{X,Y}(x,\tfrac{z}{x})\,dx.$$

This representation can be simplified:

$$f_Z(z) = \int_{-\infty}^{+\infty}\left|\frac{1}{x}\right| f_{X,Y}(x,\tfrac{z}{x})\,dx, \quad z\in(-\infty,+\infty). \tag{1.69}$$

In case of nonnegative X and Y,

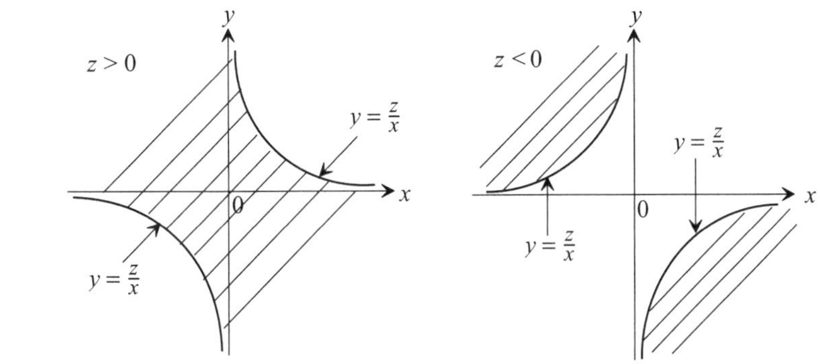

Figure 1.9 Derivation of the distribution function of a product

$$F_Z(z) = \int_0^{+\infty} \int_0^{z/x} f_{X,Y}(x,y)\,dy\,dx, \quad z \ge 0,$$

$$f_Z(z) = \int_0^{+\infty} \frac{1}{x} f_{X,Y}(x, \tfrac{z}{x})\,dx, \quad z \ge 0. \tag{1.70}$$

Distribution of the Ratio of two Random Variables Let (X, Y) be a random vector with joint probability density $f_{X,Y}(x,y)$, and

$$Z = Y/X.$$

The distribution function of Z is given by (Figure 1.10)

$$F_Z(z) = \iint\limits_{\left\{(x,y);\ \frac{y}{x} \le z\right\}} f_{X,Y}(x,y)\,dx\,dy$$

with $\left\{(x,y);\ \frac{y}{x} \le z\right\} = \{-\infty < x \le 0,\ zx \le y < \infty\} \cup \{0 \le x < \infty,\ -\infty < y \le zx\}$. Hence

$$F_Z(z) = \int_{-\infty}^{0} \int_{zx}^{+\infty} f_{X,Y}(x,y)\,dy\,dx + \int_0^{+\infty} \int_{-\infty}^{zx} f_{X,Y}(x,y)\,dy\,dx.$$

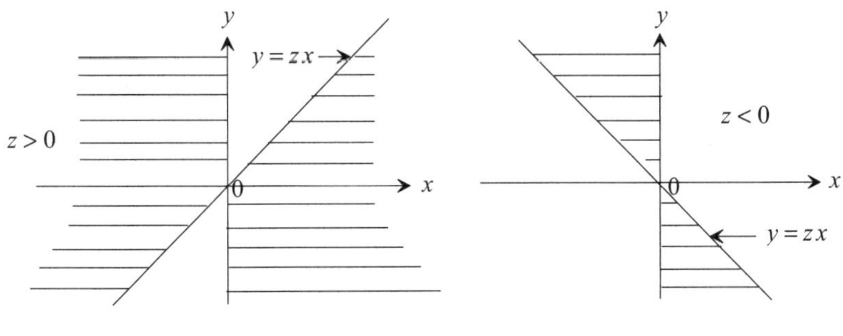

Figure 1.10 Derivation of the distribution function of a ratio

Differentiation with regard to z yields the probability density of Z:

$$f_Z(z) = \int_{-\infty}^{+\infty} |x|\, f_{X,Y}(x, zx)\, dx. \tag{1.71}$$

In case of nonnegative X and Y,

$$F_Z(z) = \int_0^{+\infty} \int_0^{zx} f_{X,Y}(x,y)\, dy\, dx, \quad z \ge 0$$

$$f_Z(z) = \int_0^{+\infty} x\, f_{X,Y}(x, zx)\, dx, \quad z \ge 0. \tag{1.72}$$

Example 1.19 The random vector (X, Y) has the joint density

$$f_{X,Y}(x,y) = \lambda\mu e^{-(\lambda x + \nu y)}, \quad x \ge 0,\ y \ge 0;\ \lambda > 0,\ \nu > 0.$$

The structure of this joint density implies that X and Y are independent and have exponential distributions with parameters λ and ν, respectively. Hence, the density of the ratio $Z = Y/X$ is

$$f_Z(z) = \int_0^{\infty} x\lambda\nu\, e^{-(\lambda+\nu z)x}\, dx, \quad z \ge 0.$$

A slight transformation yields

$$f_Z(z) = \frac{\lambda\nu}{\lambda+\nu z} \int_0^{\infty} x\,(\lambda+\nu z) e^{-(\lambda+\nu z)x} dx, \quad z \ge 0.$$

The integral is the mean value of an exponentially distributed random variable with parameter $\lambda + \nu z$. Hence,

$$f_Z(z) = \frac{\lambda\nu}{(\lambda+\nu z)^2}, \quad F_Z(z) = 1 - \frac{\lambda}{\lambda+\nu z}, \quad z \ge 0.$$

The mean value of Z does not exist. (Try to apply (1.17) to determining $E(Z)$.) □

Example 1.20 A system has the random lifetime (= time to failure) X. After a failure it is replaced with a new system. It takes Y time units to replace a failed system. Thus, within a (lifetime-replacement-) cycle, the random fraction the system is operating, is

$$A = \frac{X}{X+Y}.$$

A is the *availability* of the system in a cycle. Determining the distribution function of A can be reduced to determining the distribution function of $Z=Y/X$ since

$$F_A(t) = P\left(\frac{X}{X+Y} \le t\right) = 1 - P\left(\frac{Y}{X} < \frac{1-t}{t}\right).$$

Hence,

$$F_A(t) = 1 - F_Z\left(\frac{1-t}{t}\right), \quad 0 < t \le 1.$$

Differentiation with respect to t yields the probability density of A:

$$f_A(t) = \frac{1}{t^2} f_Z\left(\frac{1-t}{t}\right), \quad 0 < t \le 1.$$

Specifically, if $f_Z(z)$ is the same as in example 1.19, then

$$f_A(t) = \frac{\lambda\nu}{[(\lambda-\nu)t+\nu]^2}, \quad F_A(t) = \frac{\lambda t}{(\lambda-\nu)t+\nu}, \quad 0 \le t \le 1.$$

For $\lambda \neq \nu$, the mean value of A is

$$E(A) = \frac{\nu}{\nu-\lambda}\left[1 + \frac{\lambda}{\nu-\lambda}\right] \ln\frac{\lambda}{\nu}. \tag{1.73}$$

If $\lambda = \nu$, then A is uniformly distributed over $[0, 1]$. In this case, $E(A) = 1/2$. □

1.6.3 *n*-Dimensional Random Variables

Let $(X_1, X_2, ..., X_n)$ be a random vector with joint distribution function

$$F(x_1, x_2, ..., x_n) = P(X_1 \le x_1, X_2 \le x_2, ..., X_n \le x_n).$$

Provided its existence, the nth mixed partial derivative of the joint distribution function with respect to the $x_1, x_2, \dots, x_n$ is called the *joint* (*probability*) *density* of the random vector $(X_1, X_2, ..., X_n)$:

$$f(x_1, x_2, ..., x_n) = \frac{\partial^n F(x_1, x_2, \dots, x_n)}{\partial x_1 \partial x_2 \cdots \partial x_n}. \tag{1.74}$$

The characteristic properties of two-dimensional distribution functions and probability densities can be extended in a straightforward way to n-dimensional distribution functions and densities. Hence they will not be given here.

The marginal distribution functions are given by (1.52), whereas the marginal densities are

$$f_{X_i}(x_i) = \int_{-\infty}^{+\infty}\int_{-\infty}^{+\infty}\cdots\int_{-\infty}^{+\infty} f(x_1, ..., x_i, ..., x_n)\, dx_1 \cdots dx_{i-1} dx_{i+1} \cdots dx_n. \tag{1.75}$$

If the X_i are independent, then, from (1.53), the joint density of $(X_1, X_2, ..., X_n)$ is equal to the product of the densities of the X_i:

$$f(x_1, x_2, ..., x_n) = f_{X_1}(x_1) f_{X_2}(x_2) \cdots f_{X_n}(x_n). \tag{1.76}$$

The joint distribution function (density) also allows for determining the joint probability distributions of all subsets of $\{X_1, X_2, ..., X_n\}$. For instance, the joint distribution function of the random vector (X_i, X_j), $i < j$, is

$$F_{X_i, X_j}(x_i, x_j) = F(\infty, ..., \infty, x_i, \infty, ..., \infty, x_j, \infty, ..., \infty)$$

and the joint distribution function of $X_1, X_2, ..., X_k$, $k < n$, is

$$F_{X_1, X_2, ..., X_k}(x_1, x_2, ..., x_k) = F(x_1, x_2 ..., x_k, \infty, \infty, ..., \infty). \tag{1.77}$$

The corresponding joint densities are

$$f_{X_i,X_j}(x_i,x_j) =$$

$$\int_{-\infty}^{+\infty}\int_{-\infty}^{+\infty}\cdots\int_{-\infty}^{+\infty} f(x_1,x_2,...,x_n)\,dx_1\cdots dx_{i-1}dx_{i+1}\cdots dx_{j-1}dx_{j+1}\cdots dx_n$$

and

$$f_{X_1,X_2,...,X_k}(x_1,x_2,...,x_k) \tag{1.78}$$

$$= \int_{-\infty}^{+\infty}\int_{-\infty}^{+\infty}\cdots\int_{-\infty}^{+\infty} f(x_1,x_2,...,x_k,x_{k+1}\cdots x_n)\,dx_{k+1}dx_{k+2}\cdots dx_n .$$

Conditional densities can be obtained analogously to the two-dimensional case: For instance, the conditional density of $(X_1,X_2,...,X_n)$ given $X_i = x_i,\ i = 1,2,...,n$, is

$$f(x_1,...,x_{i-1},x_{i+1},...,x_n\Big|x_i\Big) = \frac{f(x_1,x_2,...,x_n)}{f_{X_i}(x_i)} \tag{1.79}$$

and the conditional density of $(X_1,X_2,...,X_n)$ given '$X_1 = x_1,\ X_2 = x_2,...,X = x_k$' is

$$f(x_{k+1},x_{k+2},...,x_n|x_1,x_2,...,x_k) = \frac{f(x_1,x_2,...,x_n)}{f_{X_1,X_2,...,X_k}(x_1,x_2,...,x_k)}, \quad k < n. \tag{1.80}$$

Let $y = h(x_1,x_2,...,x_n)$ be a function of n variables. Then the mean value of the random variable $Y = h(X_1,X_2,...,X_n)$ is defined as

$$E(Y) = \int_{-\infty}^{+\infty}\int_{-\infty}^{+\infty}\cdots\int_{-\infty}^{+\infty} h(x_1,x_2,\ldots,x_n)\,f(x_1,x_2,...,x_n)\,dx_1dx_2\cdots dx_n. \tag{1.81}$$

In particular, the mean value of the product $Y = X_1\,X_2\cdots X_n$ is

$$E(X_1\,X_2\cdots X_n) = \int_{-\infty}^{+\infty}\int_{-\infty}^{+\infty}\cdots\int_{-\infty}^{+\infty} x_1\,x_2\cdots x_n\,f(x_1,x_2,...,x_n)\,dx_1dx_2\cdots dx_n .$$

In view of (1.76), for independent X_i this n-dimensional integral simplifies to

$$E(X_1\,X_2\cdots X_n) = E(X_1)\,E(X_2)\cdots E(X_n). \tag{1.82}$$

The mean value of the product of independent random variables is equal to the product of the mean values of these random variables.

The *conditional mean value of* $Y = h(X_1,X_2,...,X_n)$ *given*

$$'X_1 = x_1, X_2 = x_2,\cdots,X_k = x_k'$$

is

$$E(Y|x_1, x_2, \cdots, x_k) = \tag{1.83}$$

$$= \int_{-\infty}^{+\infty}\int_{-\infty}^{+\infty}\cdots\int_{-\infty}^{+\infty} h(x_1, x_2, ..., x_n) \frac{f(x_1, x_2, ..., x_n)}{f_{X_1, X_2, ..., X_k}(x_1, x_2, ..., x_k)} dx_{k+1} dx_{k+2} \cdots dx_n .$$

Replacing in (1.83) the $x_1, x_2, ..., x_k$ with the random variables $X_1, X_2, ..., X_k$ yields the corresponding random mean value of Y given $X_1, X_2, ..., X_k$:

$$E(Y| X_1, X_2, ..., X_k).$$

The mean value of this random variable (with respect to all $X_1, X_2, ..., X_k$) is

$$E_{X_1, X_2, ..., X_k}(E(Y| X_1, X_2, ..., X_k)) = E(Y). \tag{1.84}$$

The mean value of $E(Y| X_1, X_2, ..., X_k)$ with respect to, for instance, the random variables $X_1, X_2, ..., X_{k-1}$ is again a random variable:

$$E_{X_1, X_2, ..., X_{k-1}}(E(Y| X_1, X_2, ..., X_k)) = E(Y| X_k). \tag{1.85}$$

From this it is obvious how to obtain the conditional mean value

$$E(Y|x_{i_1}, x_{i_2}, \cdots, x_{i_k})$$

and its random analogue

$$E(Y| X_{i_1}, X_{i_2}, \cdots, X_{i_k})$$

with regard to any subsets $\{x_{i_1}, x_{i_2}, \cdots, x_{i_k}\}$ and $\{X_{i_1}, X_{i_2}, \cdots, X_{i_k}\}$ of the respective sets $\{x_1, x_2, ..., x_n\}$ and $\{X_1, X_2, ..., X_n\}$.

The conditional mean values of a sum of random variables have properties

$$E(Y_1 + Y_2 + \cdots + Y_m|x_{i_1}, x_{i_2}, \cdots, x_{i_k}) = \Sigma_{i=1}^m EY_i|x_{i_1}, x_{i_2}, \cdots, x_{i_k}) \tag{1.86}$$

and

$$E(Y_1 + Y_2 + \cdots + Y_m| X_{i_1}, X_{i_2}, \cdots, X_{i_k}) = \Sigma_{i=1}^m EY_i|X_{i_1}, X_{i_2}, \cdots, X_{i_k}). \tag{1.87}$$

Let

$$c_{ij} = Cov(X_i, X_j)$$

be the covariance between X_i and X_j; $i, j = 1, 2, ..., n$. It is useful to unite the c_{ij} in the *covariance matrix* $\mathbf{C}$:

$$\mathbf{C} = ((c_{ij})) \; ; \quad i, j = 1, 2, ..., n .$$

The main diagonal of $\mathbf{C}$ consists of the variances of the X_i :

$$c_{ii} = Var(X_i); \; i = 1, 2, ..., n.$$

n-Dimensional Normal Distribution Let $(X_1, X_2, \cdots, X_n)$ be an n-dimensional random vector with vector of mean values $\boldsymbol{\mu} = (\mu_1, \mu_2, \cdots, \mu_n)$ and covariance matrix $\mathbf{C} = ((c_{ij}))$. Furthermore, let $|\mathbf{C}|$ and $\mathbf{C}^{-1}$ be the positive determinant and the inverse of $\mathbf{C}$, respectively, and $\mathbf{x} = (x_1, x_2, \cdots, x_n)$. Then $(X_1, X_2, \cdots, X_n)$ has an *n-dimensionally normal* (or *Gaussian*) *distribution* if it has joint density

$$f(\mathbf{x}) = \frac{1}{\sqrt{(2\pi)^n |\mathbf{C}|}} \exp\left(-\frac{1}{2}(\mathbf{x}-\boldsymbol{\mu})\,\mathbf{C}^{-1}(\mathbf{x}-\boldsymbol{\mu})^T\right), \tag{1.88}$$

where $(\mathbf{x}-\boldsymbol{\mu})^T$ is the transpose of the vector

$$\mathbf{x}-\boldsymbol{\mu} = (x_1-\mu_1, x_2-\mu_2, \cdots, x_n-\mu_n).$$

By doing the matrix-vector-multiplication in (1.88), $f(\mathbf{x})$ becomes

$$f(x_1, x_2, ..., x_n) = \tag{1.89}$$

$$\frac{1}{\sqrt{(2\pi)^n |\mathbf{C}|}} \exp\left(-\frac{1}{2|\mathbf{C}|} \sum_{i=1}^n \sum_{j=1}^n C_{ij}(x_i-\mu_i)(x_j-\mu_j)\right),$$

where C_{ij} is the cofactor of c_{ij}.

For $n = 2$, $x_1 = x$ and $x_2 = y$, (1.89) becomes the density of the bivariate normal distribution (1.66). Generalizing from the bivariate special case, it can be shown that the random variables X_i have an $N(\mu_i, \sigma_i^2)$–distribution with $\sigma_i^2 = c_{ii}$; $i = 1, 2, ..., n$, if $(X_1, X_2, \cdots, X_n)$ has an n-dimensional normal distribution. If the X_i are uncorrelated, then $\mathbf{C} = ((c_{ij}))$ is a diagonal matrix with $c_{ij} = 0$ for $i \neq j$ so that the product form (1.76) of the joint density and, therefore, the independence of the X_i follows:

$$f_{\mathbf{X}}(x_1, x_2, \cdots, x_n) = \prod_{i=1}^n \left[\frac{1}{\sqrt{2\pi}\,\sigma_i} \exp\left(-\frac{1}{2}\left(\frac{x_i-\mu_i}{\sigma_i}\right)^2\right)\right]. \tag{1.90}$$

Theorem 1.2 If the random vector $(X_1, X_2, ..., X_n)$ has an n-dimensionally normal distribution and the random variables $Y_1, Y_2, ..., Y_m$ are linear combinations of the X_i, i.e. if there exist constants a_{ij} so that

$$Y_i = \sum_{j=1}^{m_i} a_{ij} X_j; \quad i = 1, 2, ..., m,$$

then the random vector $(Y_1, Y_2, ..., Y_m)$ is m-dimensionally normally distributed. ■

Maximum of n Independent Random Variables Let $X_1, X_2, ..., X_n$ be independent random variables and

$$X = \max\{X_1, X_2, ..., X_n\}.$$

Then the random event '$X \leq x$' occurs if and only if

$$'X_1 \leq x,\ X_2 \leq x, ...,\ X_n \leq x'.$$

Thus, the distribution function of the maximum of n independent random variables is

$$F_X(x) = F_{X_1}(x) \cdot F_{X_2}(x) \cdots F_{X_n}(x). \tag{1.91}$$

Minimum of n Independent Random Variables Let $X_1, X_2, ..., X_n$ be independent random variables

$$Y = \min\{X_1, X_2, ..., X_n\}.$$

Then,

$$P(Y > x) = P(X_1 > x, X_2 > x, ..., X_n > x).$$

Hence,

$$\overline{F}_Y(x) = P(Y > x) = \overline{F}_{X_1}(x) \cdot \overline{F}_{X_2}(x) \cdots \overline{F}_{X_n}(x), \tag{1.92}$$

Thus, the distribution function of the minimum of n independent random variables is

$$F_Y(x) = 1 - \overline{F}_{X_1}(x) \cdot \overline{F}_{X_2}(x) \cdots \overline{F}_{X_n}(x). \tag{1.93}$$

Example 1.21 a) A system consists of n subsystems with independent lifetimes $X_1, X_2, ..., X_n$. The system operates if at least one of its subsystems operates (*parallel system*). Hence, its lifetime is

$$X = \max\{X_1, X_2, ..., X_n\}$$

and has distribution function (1.91). In particular, if the lifetimes of the subsystems are identically exponentially distributed with parameter λ,

$$F_X(x) = P(X \le x) = (1 - e^{-\lambda x})^n; \; x \ge 0.$$

By (1.17), the corresponding mean system lifetime is

$$E(X) = \int_0^\infty [1 - (1 - e^{-\lambda x})^n] dx.$$

Substituting $u = 1 - e^{-\lambda x}$ yields

$$E(X) = \frac{1}{\lambda} \int_0^1 \frac{1 - u^n}{1 - u} du = \frac{1}{\lambda} \int_0^1 [1 + u + \cdots + u^{n-1}] du.$$

Hence,

$$E(X) = \frac{1}{\lambda}\left[1 + \frac{1}{2} + \frac{1}{3} + \cdots + \frac{1}{n}\right].$$

b) Under otherwise the same assumptions as in case a), the system fails as soon as the first subsystem fails (*series system*). Thus, its lifetime is

$$Y = \min\{X_1, X_2, ..., X_n\}$$

and has distribution function (1.93). In particular, if the lifetimes of the subsystems are identically exponentially distributed with parameter λ, then

$$F_Y(x) = 1 - e^{-n\lambda x}, \; x \ge 0.$$

The corresponding mean system lifetime is $E(Y) = 1/n\lambda$. □

1.7 SUMS OF RANDOM VARIABLES

1.7.1 Sums of Discrete Random Variables

Mean Value of a Sum The random vector (X, Y) has the two discrete components X and Y and its joint distribution is

$$\{r_{ij} = P(X = x_i \cap Y = y_j\};\ i,j = 0, 1, ...\}.$$

Then the mean value of the sum $Z = X + Y$ is

$$E(Z) = \Sigma_{i=0}^{\infty} \Sigma_{j=0}^{\infty} (x_i + y_j) r_{ij} = \Sigma_{i=0}^{\infty} x_i \Sigma_{j=0}^{\infty} r_{ij} + \Sigma_{i=0}^{\infty} y_j \Sigma_{j=0}^{\infty} r_{ij}.$$

Thus, in view of (1.54),

$$E(X + Y) = E(X) + E(Y). \tag{1.94}$$

By induction, for any discrete random variables $X_1, X_2, ..., X_n$,

$$E(X_1 + X_2 + \cdots + X_n) = E(X_1) + E(X_2) + \cdots + E(X_n). \tag{1.95}$$

Distribution of a Sum Let X and Y be independent random variables with common range $\mathbf{R} = \{0, 1, ...\}$ and probability distributions

$$\{p_i = P(X = i;\ i = 0, 1, ...\} \text{ and } \{q_j = P(Y = j;\ j = 0, 1, ...\}.$$

Then,

$$P(Z = k) = P(X + Y = k) = \Sigma_{i=0}^{k} P(X = i) P(Y = k - i).$$

Letting $r_k = P(Z = k)$ yields for all $k = 0, 1, ...$

$$r_k = p_0 q_k + p_1 q_{k-1} + \cdots + p_k q_0.$$

Thus, according to (1.24), the discrete probability distribution $\{r_k;\ k = 0, 1, ...\}$ is the convolution of the probability distributions of X and Y. Hence, by (1.26),

$$M_Z(z) = M_X(z)\, M_Y(z). \tag{1.96}$$

> The z-transform $M_Z(z)$ of the the sum $Z = X + Y$ of two independent discrete random variables X and Y with common range $\mathbf{R} = \{0, 1, ...\}$ is equal to the product of the z-transforms of X and Y.

By induction, if $Z = X_1 + X_2 + \cdots + X_n$ with independent X_i, then

$$M_Z(z) = M_{X_1}(z)\, M_{X_2}(z) \cdots M_{X_n}(z). \tag{1.97}$$

Example 1.22 Let $Z = X_1 + X_2 + \cdots + X_n$ be a sum of independent random variables, where X_i has a Poisson distribution with parameter $\lambda_i;\ i = 1, 2, ..., n$. The z-transform of X_i is (section 1.3.1)

$$M_{X_i}(z) = e^{\lambda_i (z-1)}.$$

From (1.97),

$$M_Z(z) = e^{(\lambda_1+\lambda_2+\cdots+\lambda_n)(z-1)}.$$

Thus, the sum of independent, Poisson distributed random variables has a Poisson distribution the parameter of which is the sum of the parameters of the Poisson distributions of these random variables. □

1.7.2 Sums of Continuous Random Variables

In this section, X_i; $i = 1, 2, ..., n$; are random variables with respective distribution functions, densities, mean values and variances

$$F_{X_i}(x_i),\ f_{X_i}(x_i),\ E(X_i),\ \text{and}\ Var(X_i);\ i = 1, 2, ..., n.$$

The joint density of the $X_1, X_2, ..., X_n$ is denoted as $f(x_1, x_2, ..., x_n)$. All mean values and variances are assumed to be finite.

Mean Value of a Sum Applying (1.81) with $h(x_1, x_2, ..., x_n) = x_1 + x_2 + \cdots + x_n$ yields the mean value of a sum of n random variables:

$$E\left(\Sigma_{i=0}^{n} X_i\right) = \int_{-\infty}^{+\infty}\int_{-\infty}^{+\infty}\cdots\int_{-\infty}^{+\infty} (x_1 + x_2 + \cdots + x_n) f_{\mathbf{X}}(x_1, x_2, ..., x_n)\, dx_1 dx_2 \cdots dx_n.$$

From (1.75),

$$E\left(\Sigma_{i=0}^{n} X_i\right) = \Sigma_{i=0}^{n} \int_{-\infty}^{+\infty} x_i f_{X_i}(x_i)\, dx_i.$$

Hence,

$$E(X_1 + X_2 + \cdots + X_n) = E(X_1) + E(X_2) + \cdots + E(X_n). \tag{1.98}$$

> The mean value of the sum of (discrete or continuous) random variables is equal to the sum of the mean values of these random variables.

Variance of a Sum The variance of the sum of n random variables is

$$Var\left(\Sigma_{i=0}^{n} X_i\right) = \Sigma_{i=0}^{n} \Sigma_{j=0}^{n} Cov(X_i, X_j). \tag{1.99}$$

Since

$$Cov(X_i, X_i) = Var(X_i) \ \text{and}\ Cov(X_i, X_j) = Cov(X_j, X_i),$$

formula (1.99) can be written in the form

$$Var\left(\Sigma_{i=1}^{n} X_i\right) = \Sigma_{i=1}^{n} Var(X_i) + 2\,\Sigma_{i,j=1;i<j}^{n} Cov(X_i, X_j). \tag{1.100}$$

Thus, for uncorrelated X_i,

$$Var(X_1 + X_2 + \cdots + X_n) = Var(X_1) + Var(X_2) + \cdots + Var(X_n). \tag{1.101}$$

The variance of a sum of uncorrelated random variables is equal to the sum of the variances of these random variables.

Let $\alpha_1, \alpha_2, \cdots, \alpha_n$ be any sequence of finite real numbers. Then,

$$E\left(\sum_{i=1}^n \alpha_i X_i\right) = \sum_{i=1}^n \alpha_i E(X_i) \tag{1.102}$$

$$Var\left(\sum_{i=1}^n \alpha_i X_i\right) = \sum_{i=1}^n \alpha_i^2 Var(X_i) + 2\sum_{i,j=1,\, i<j}^n \alpha_i \alpha_j Cov(X_i, X_j). \tag{1.103}$$

If the X_i are uncorrelated,

$$Var\left(\sum_{i=1}^n \alpha_i X_i\right) = \sum_{i=1}^n \alpha_i^2 Var(X_i). \tag{1.104}$$

For independent, identically distributed random variables with mean μ and variance σ^2, formulas (1.74) and (1.75) simplify to

$$E\left(\sum_{i=1}^n X_i\right) = n\mu, \qquad Var\left(\sum_{i=1}^n X_i\right) = n\sigma^2. \tag{1.105}$$

Note Formulas (1.98) to (1.105) hold for discrete and continuous random variables.

Distribution of a Sum Let X and Y be two independent, continuous random variables with distribution functions $F_X(x)$, $F_Y(y)$ and densities $f_X(x)$, $f_Y(y)$. On condition $Y = y$, the distribution function of the sum $Z = X + Y$ is

$$F_Z(Z \le z \mid Y = y) = P(X + y \le z) = P(X \le z - y) = F_X(z - y)$$

and, on condition $X = x$,

$$F_Z(Z \le z \mid X = x) = P(Y + x \le z) = P(Y \le z - x) = F_Y(z - x).$$

Hence,

$$F_Z(z) = \int_{-\infty}^{+\infty} F_X(z-y) f_Y(y)\, dy = \int_{-\infty}^{+\infty} F_Y(z-x) f_X(x)\, dx. \tag{1.106}$$

By differentiation, the probability density of the sum $Z = X + Y$ is seen to be

$$f_Z(z) = \int_{-\infty}^{+\infty} f_X(z-y) f_Y(y)\, dy = \int_{-\infty}^{+\infty} f_Y(z-x) f_X(x)\, dx. \tag{1.107}$$

The integrals in (1.107) are equivalent definitions of the *convolution of the densities* f_X and f_Y.

Notation $f_Z(z) = (f_X * f_Y)(z) = (f_Y * f_X)(z)$

In terms of the distribution functions, since $dF(x) = f(x)dx$, (1.106) can be written as

$$F_Z(z) = \int_{-\infty}^{+\infty} F_Y(z-x)\, dF_X(x) = \int_{-\infty}^{+\infty} F_X(z-y)\, dF_Y(y). \tag{1.108}$$

The integrals in (1.108) are equivalent definitions of the *convolution of the distribution functions* F_X and F_Y.

Notation $F_Z(z) = F_X * F_Y(z) = F_Y * F_X(z)$

The distribution function (probability density) of the sum of two independent random variables is given by the convolution of their distribution functions (probability densities).

Note With regard to the general definition of the convolution in mathematics (which applies to our definition of the convolution of densities), the convolution of two distribution functions F and G with respective densities f and g is, by (1.106), simply the convolution of F and g or, equivalently, the convolution of G and f.

If X and Y are nonnegative, then (1.106) and (1.107) become

$$F_Z(z) = \int_0^z F_X(z-x)f_Y(x)dx = \int_0^z F_Y(z-y)f_X(y)dy, \quad z \geq 0, \tag{1.109}$$

$$f_Z(z) = \int_0^z f_Y(z-x)f_X(x)\,dx = \int_0^z f_X(z-y)f_Y(y)\,dy, \quad z \geq 0. \tag{1.110}$$

Moreover, if $L(f)$ denotes the Laplace transform of a function f defined on $[0,\infty)$ (its existence provided), then, by (1.32),

$$L(f_Z) = L(f_X * f_Y) = L(F_Y)L(f_X). \tag{1.111}$$

$$L(F_Z) = L(F_X * f_Y) = L(F_X)L(f_Y). \tag{1.112}$$

The Laplace transform of the density of the sum of two nonnegative, independent random variables is equal to the product of their Laplace transforms.

By (1.29), $L(F_Y) = L(f_Y)/s$ so that

$$L(F_Z) = s\,L(F_X)L(F_Y). \tag{1.113}$$

The density of a sum $Z = X_1 + X_2 + \cdots + X_n$ of n independent, continuous random variables X_i is obtained by repeated application of formula (1.107). The resulting function is the *convolution* of the densities $f_{X_1}, f_{X_2}, \ldots, f_{X_n}$ denoted as

$$f_Z(z) = f_{X_1} * f_{X_2} * \cdots * f_{X_n}(z). \tag{1.114}$$

In particular, if the X_i are identically distributed with density f, then f_Z is the *n-fold convolution* of f with itself or, equivalently, the nth *convolution power* $f^{*(n)}(z)$ of f. $f^{*(n)}(z)$ can be recursively obtained as follows:

$$f^{*(i)}(z) = \int_{-\infty}^{+\infty} f^{*(i-1)}(z-x)f(x)dx, \tag{1.115}$$

$i = 2, 3, \ldots, n$; $f^{*(1)}(x) \equiv f(x)$. For nonnegative random variables, this formula simplifies to

$$f^{*(i)}(z) = \int_0^z f^{*(i-1)}(z-x)f(x)\,dx, \quad z \geq 0. \tag{1.116}$$

From (1.111), by induction: The Laplace transform of the density f_Z of the sum of n nonnegative, independent random variables $Z = X_1 + X_2 + \cdots + X_n$ is equal to the product of the Laplace transforms of these random variables:

$$L(f_Z) = L(f_{X_1}) L(f_{X_2}) \cdots L(f_{X_n}). \tag{1.117}$$

The repeated application of (1.108) yields the distribution function of a sum of the n independent random variables $X_1, X_2, ..., X_n$ in the form

$$F_Z(z) = F_{X_1} * F_{X_2} * \cdots * F_{X_n}(z). \tag{1.118}$$

In particular, if the X_i are independent and identically distributed with distribution function F, then $F_Z(z)$ is equal to the nth *convolution power* of F:

$$F_Z(z) = F^{*(n)}(z). \qquad (1.119$$

$F_Z(z)$ can be recursively obtained from

$$F^{*(i)}(z) = \int_{-\infty}^{+\infty} F^{*(i-1)}(z-x)\, dF(x); \tag{1.120}$$

$n = 2, 3, ...;$ $F^{*(0)}(x) \equiv 1,$ $F^{*(1)}(x) \equiv F(x)$. If the X_i are nonnegative, then formula (1.120) becomes

$$F^{*(i)}(z) = \int_0^z F^{*(i-1)}(z-x)\, dF(x). \tag{1.121}$$

Example 1.23 (*Erlang distribution*) Let the random variables X_1 and X_2 be independent and exponentially distributed with parameters λ_1 and λ_2:

$$f_{X_i}(x) = \lambda_i e^{-\lambda_i x}, \quad F_{X_i}(x) = 1 - e^{-\lambda_i x}; \; x \geq 0, \; i = 1, 2.$$

(1.110) yields the density of $Z = X_1 + X_2$:

$$\begin{aligned} f_Z(z) &= \int_0^z \lambda_2\, e^{-\lambda_2 (z-x)}\, \lambda_1 e^{-\lambda_1 x}\, dx \\ &= \lambda_1 \lambda_2 e^{-\lambda_2 z} \int_0^z e^{-(\lambda_1 - \lambda_2)x}\, dx. \end{aligned}$$

If $\lambda_1 = \lambda_2 = \lambda$, then

$$f_Z(z) = \lambda^2 z e^{-\lambda z}, \quad z \geq 0. \tag{1.122}$$

This is the density of an Erlang distribution with parameters $n = 2$ and λ (section 1.3).
If $\lambda_1 \neq \lambda_2$, then

$$f_Z(z) = \frac{\lambda_1 \lambda_2}{\lambda_1 - \lambda_2} \left(e^{-\lambda_2 z} - e^{-\lambda_1 z} \right), \quad z \geq 0.$$

Now let $X_1, X_2, ..., X_n$ be independent, identically distributed exponential random variables with density $f(x) = \lambda e^{-\lambda x}$; $x \geq 0$. The Laplace transform of f is

$$\hat{f}(s) = \lambda / (s + \lambda).$$

Hence, by (1.114), the Laplace transform of the density of $Z = X_1 + X_2 + \cdots + X_n$ is

$$\hat{f}_Z(s) = \left(\frac{\lambda}{s+\lambda}\right)^n.$$

The pre-image of this Laplace transform is

$$f_Z(z) = \lambda \frac{(\lambda z)^{n-1}}{(n-1)!} e^{-\lambda z}, \quad z \geq 0.$$

Hence, Z has an Erlang distribution with parameters n and λ. □

Example 1.24 (*Normal distribution*) The random variables X_i are independent and have a normal distribution with parameters μ_i and σ_i^2; $i = 1, 2$:

$$f_{X_i}(x) = \frac{1}{\sqrt{2\pi}\,\sigma_i} \exp\left(-\frac{1}{2}\frac{(x-\mu_i)^2}{\sigma_i^2}\right); \quad i = 1, 2.$$

According to example 1.12 (page 33), the Laplace transforms of the X_i are

$$\hat{f}_{X_i}(s) = e^{-\mu_i s + \frac{1}{2}\sigma_i^2 s^2}; \quad i = 1, 2.$$

By (1.111), the density of the sum $Z = X_1 + X_2$ has the Laplace transform

$$\hat{f}_Z(s) = \hat{f}_{X_1}(s)\,\hat{f}_{X_2}(s) = e^{-(\mu_1+\mu_2)s + \frac{1}{2}(\sigma_1^2+\sigma_2^2)s^2}.$$

But this is the Laplace transform of an $N(\mu_1 + \mu_2, \sigma_1^2 + \sigma_2^2)$-distributed random variable. Thus, the sum of two independent, normally distributed random variables also has a normal distribution. By induction, if

$$Z = X_1 + X_2 + \cdots + X_n$$

is a sum of independent random variables with $X_i = N(\mu_i, \sigma_i^2)$; $i = 1, 2, ..., n$; then

$$Z = N(\mu_1 + \mu_2 + \cdots + \mu_n,\ \sigma_1^2 + \sigma_2^2 + \cdots + \sigma_n^2). \tag{1.123}$$

As a corollary from this result:
If $X = N(\mu, \sigma^2)$, then, for every $n = 1, 2, ...,$ X can be represented as sum of independent, identically as $N(\mu/n, \sigma^2/n)$-distributed random variables. □

According to theorem 1.2, if $(X_1, X_2, ..., X_n)$ has a joint normal distribution, then the sum $X_1 + X_2 + \cdots + X_n$ has a normal distribution. In particular, if (X, Y) has a bivariate normal distribution with parameters μ_x, μ_y, σ_x, σ_y and ρ, then $X + Y$ has a normal distribution with

$$E(X+Y) = \mu_x + \mu_y, \quad Var(X+Y) = \sigma_x^2 + 2\rho\sigma_x\sigma_y + \sigma_y^2. \tag{1.124}$$

1.7.3 Sums of a Random Number of Random Variables

Frequently, sums of a random number of random variables have to be investigated. For instance, the total claim size an insurance company is confronted with a year is the sum of a random number of random individual claim sizes.

Theorem 1.3 (*Wald's identities*) Let $X_1, X_2, ...$ be a sequence of independent random variables, which are identically distributed as X with $E(X) < \infty$. Let further N be a positive, integer-valued random variable, which is independent of all $X_1, X_2, ...$ Then mean value and variance of the sum $Z = X_1 + X_2 + \cdots + X_N$ are

$$E(Z) = E(X)\, E(N) \tag{1.125}$$

$$Var(Z) = Var(X)\, E(N) + [E(X)]^2\, Var(N). \tag{1.126}$$

Proof By conditioning,

$$E(Z) = \Sigma_{n=1}^{\infty} E(X_1 + X_2 + \cdots + X_N | N = n)\, P(N = n)$$
$$= \Sigma_{n=1}^{\infty} E(X_1 + X_2 + \cdots + X_n)\, P(N = n)$$
$$= E(X)\ \Sigma_{n=1}^{\infty} n\, P(N = n).$$

This proves (1.125).
To verify (1.126), the second moment of Z is determined:

$$E(Z^2) = \Sigma_{n=1}^{\infty} E(Z^2 | N = n)\, P(N = n)$$
$$= \Sigma_{n=1}^{\infty} E([X_1 + X_2 + \cdots + X_n]^2)\, P(N = n).$$

By making use of (1.19),

$$E(Z^2) = \Sigma_{n=1}^{\infty} \{Var(X_1 + X_2 + \cdots + X_n) + [E(X_1 + X_2 + \cdots + X_n)]^2\}\, P(N = n)$$
$$= \Sigma_{n=1}^{\infty} \{n\, Var(X) + n^2\, [E(X)]^2\}\, P(N = n)$$
$$= Var(X)\, E(N) + [E(X)]^2\, E(N^2).$$

Hence,

$$Var(Z) = E(Z^2) - [E(Z)]^2$$
$$= Var(X)\, E(N) + [E(X)]^2\, E(N^2) - [E(X)]^2 [E(N)]^2$$
$$= Var(X)\, E(N) + [E(X)]^2\, Var(N).$$

This is the identity (1.126). ■

Wald's identity (1.125) remains valid if the assumption that N is independent of all $X_1, X_2, ...$ is somewhat weakened. To see this, the concept of a stopping time is introduced.

Definition 1.2 (***stopping time***) A positive, integer-valued random variable N is said to be a *stopping time* for the sequence of independent random variables $X_1, X_2, ...$ if the occurrence of the random event '$N = n$' is completely determined by the sequence $X_1, X_2, ..., X_n$, and, therefore, independent of all $X_{n+1}, X_{n+2}, ..., \; n = 1, 2, ...$. ●

Hint A random event A is said to be *independent of a random variable X* if A is independent of the events '$x < X \le y$' for all x and y with $x < y$.

Sometimes, a stopping time defined in this way is called a *Markov time* and only a finite Markov time is called a stopping time. (Note that a random variable Y is said to be *finite* if $P(Y < \infty) = 1$. In this case, $E(Y) < \infty$.) The notation 'stopping time' can be motivated as follows: The $X_1, X_2, ...$ are observed one after the other. As soon as the event '$N = n$' occurs, the observation is stopped, i.e. the $X_{n+1}, X_{n+2}, ...$ will not be observed.

Theorem 1.4 Under otherwise the same assumptions and notation as in theorem 1.3, let N be a finite stopping time for the sequence $X_1, X_2, ...$. Then

$$E(Z) = E(X)\, E(N). \tag{1.127}$$

Proof Let binary random variables Y_i be defined as follows:

$$Y_i = \begin{cases} 1, & \text{if } N \ge i \\ 0, & \text{if } N < i \end{cases}, \quad i = 1, 2, ...$$

$Y_i = 1$ holds if and only if no stopping has occurred after having observed the $i-1$ random variable $X_1, X_2, ..., X_{i-1}$. Since N is a stopping time, Y_i is independent of $X_i, X_{i+1}, ...$ Since $E(Y_i) = P(N \ge i)$ and $E(X_i\, Y_i) = E(X_i)\, E(Y_i)$.

$$E\left(\Sigma_{i=1}^{N} X_i\right) = E\left(\Sigma_{i=1}^{\infty} X_i\, Y_i\right)$$
$$= \Sigma_{i=1}^{\infty} E(X_i)\, E(Y_i) = E(X)\, \Sigma_{i=1}^{\infty} E(Y_i)$$
$$= E(X)\, \Sigma_{i=1}^{\infty} P(N \ge i).$$

Now formula (1.15) implies (1.127). ■

Example 1.25 a) Let $X_i = 1$ if the ith flipping of a fair coin yields 'head' and $X_i = 0$ otherwise. Then

$$N = \min\{n;\; X_1 + X_2 + \cdots + X_n = 8\} \tag{1.128}$$

is a finite stopping time for the sequence $\{X_1, X_2, ...\}$. From (1.127),

$$E(X_1 + X_2 + \cdots + X_n) = \frac{1}{2} E(N).$$

According to the definition of N, $X_1 + X_2 + \cdots + X_n = 8$. Hence, $E(N) = 16$.

b) Let $X_i = 1$ if the ith flipping of a fair coin yields 'head' and $X_i = -1$ otherwise. Then N given by (1.128) is again a finite stopping time for $\{X_1, X_2, ...\}$. A formal application of Wald's equation (1.127) yields

$$E(X_1 + X_2 + \cdots + X_N) = E(X)\,E(N).$$

The left hand side of this equation is equal to 8. The right hand side contains factor $E(X) = 0$. Therefore, Wald's equation (1.127) is not applicable. □

1.8 INEQUALITIES IN PROBABILITY THEORY

1.8.1 Inequalities for Probabilities

Inequalities in probability theory are useful tools for estimating probabilities and moments of random variables the exact calculation of which is only possible with extremely high effort or is even impossible in view of incomplete information on the underlying probability distribution. All occurring mean values are assumed to exist.

Inequality of Chebyshev For any random variable X with mean value $\mu = E(X)$ and variance $\sigma^2 = Var(X)$,

$$P(|X - \mu| \geq \varepsilon) \leq \sigma^2/\varepsilon^2. \tag{1.129}$$

To proof (1.129), assume for simplicity that X has density $f(x)$. Then, for any $\varepsilon > 0$,

$$\sigma^2 = \int_{-\infty}^{+\infty}(x-\mu)^2 f(x)\,dx \geq \int_{\{x,\,|x-\mu|\geq\varepsilon\}}(x-\mu)^2 f(x)dx$$

$$\geq \int_{\{x,\,|x-\mu|\geq\varepsilon\}} \varepsilon^2 f(x)dx = \varepsilon^2 P(|X-\mu| \geq \varepsilon).$$

This proves the *two-sided Chebychev inequality* (1.129). The following *one-sided Chebychev inequality* is proved analogously:

$$P(X - \mu \geq \varepsilon) \leq \frac{\sigma^2}{\sigma^2 + \varepsilon^2}.$$

Example 1.26 The height X of trees in a forest stand has mean value $\mu = 20\,m$ and standard deviation $\sigma = 2\,m$. To obtain an upper limit of the probability that the height of a tree differs at least $4\,m$ from μ, Chebyshev's inequality is applied:

$$P(|X - 20| \geq 4) \leq 4/16 = 0.250\,.$$

For the sake of comparison, assume that the height of trees in this forest stand has a normal distribution. Then the exact probability that the height of a tree differs at least $4\,m$ from μ is

$$P(|X-20|\geq 4) = P(X-20\geq 4)+P(X-20\leq -4)$$
$$= 2\,\Phi(-2) = 0.046\,.$$

Thus, Chebyshev's inequality gives a rather rough estimate. □

***n*σ-Rules a)** For any random variable X,

$$P(|X-\mu|\leq n\,\sigma)\geq 1-1/n^2\,;\quad n=1,2,...$$

This results from (1.128) by letting there $\varepsilon = n\sigma$.

b) For any random variable X with a bell-shaped density $f(x)$ and mode equal to μ,

$$P(|X-\mu|\leq n\,\sigma)\geq 1-\frac{4}{9n^2};\quad n=1,2,...$$

(Any probability density is called *bell-shaped* if it has exactly one mode.)

Inequalities of Markov Type Let $y=h(x)$ be a nonnegative, strictly increasing function on $[0,\infty)$. Then, for any $\varepsilon>0$, there holds the *general Markov inequality*

$$P(|X|\geq\varepsilon)\leq\frac{E(h(|X|))}{h(\varepsilon)}. \tag{1.130}$$

(1.130) is proved as follows:

$$E(h(|X|)) = \int_{-\infty}^{+\infty} h(|y|)f(y)dy$$
$$\geq \int_{+\varepsilon}^{+\infty} h(|y|)f(y)dy + \int_{-\infty}^{-\varepsilon} h(|y|)f(y)dy$$
$$\geq h(|\varepsilon|)\int_{+\varepsilon}^{+\infty} f(y)dy + h(|\varepsilon|)\int_{-\infty}^{-\varepsilon} f(y)dy$$
$$= h(|\varepsilon|)\,P(|X|\geq\varepsilon).$$

The special case $h(x)=x^a,\ a>0$, yields *Markov's inequality* as such:

$$P(|X|\geq\varepsilon)\leq\frac{E(|X|^a)}{\varepsilon^a}. \tag{1.131}$$

From (1.131) Chebychev's inequality is obtained by letting $a=2$ and replacing X with $X-\mu$.

If $h(x)=e^{bx},\ b>0$, Markov's inequality (1.131 yields an *exponential inequality*:

$$P(|X|\geq\varepsilon)\leq e^{-b\varepsilon}E\left(e^{b|X|}\right). \tag{1.132}$$

Markov's inequality (1.131) and the exponential inequality (1.132) are usually superior to Chebychev's inequality, since, given X and ε, their right hand sides can be minimized with respect to a and b.

1.8.2 Inequalities for Moments

Inequalities of Chebychev Let functions $g(x)$ and $h(x)$ be either both nonincreasing or both nondecreasing. Then,

$$E(g(X))\,E(h(X)) \le E(g(X)\,h(X)). \tag{1.133}$$

If g is nonincreasing and h nondecreasing or vice versa, then

$$E(g(X))\,E(h(X)) \ge E(g(X)\,h(X)). \tag{1.134}$$

As an important special case, let

$$g(x) = x^r \text{ and } h(x) = x^s; \;\; r, s \ge 0.$$

Then

$$E(|X^r|)\,E(|X^s|) \le E(\left|X^{r+s}\right|). \tag{1.135}$$

Inequality of Schwarz

$$[\,E(|X\,Y|)]^2 \le E(|X\,|^2)\;E(|\,Y\,|^2)].$$

Hölder's Inequality Let r and s be positive numbers satisfying

$$\frac{1}{r} + \frac{1}{s} = 1.$$

Then

$$E(|X\,Y|) \le [E(|X\,|^r)]^{1/r}\,[E(|\,Y\,|^s)]^{1/s}.$$

For $r = s = 2$, Hölder's inequality implies the inequality of Schwarz.

Inequality of Minkovski For $r \ge 1$,

$$[E(|X + Y|^r)]^{1/r} \le [E(|X|^r)]^{1/r} + [E(|\,Y|^r)]^{1/r}.$$

Inequality of Jensen Let $h(x)$ be a convex (concave) function. Then, for any X,

$$h(E(X)) \underset{(\ge)}{\le} E(h(X)). \tag{1.136}$$

In particular, if X is nonnegative and $h(x) = x^a$ (convex for $a \ge 1$ and $a \le 0$, concave for $0 \le a \le 1$), $h(x) = e^x$ (convex), and $h(x) = \ln x$ (concave), the respective inequalities of Jensen are

$$[E(X)]^a \le E(X^a) \quad \text{for } a > 1 \text{ or } a < 0,$$

$$[E(X)]^a \ge E(X^a) \quad \text{for } 0 < a < 1,$$

$$\exp(E(X)) \le E(\exp(X)),$$

$$\ln E(X) \ge E(\ln X).$$

1.9 LIMIT THEOREMS

1.9.1 Convergence Criteria for Sequences of Random Variables

Limit theorems in probability theory are based on certain convergence criteria for sequences of random variables, which next have to be introduced.

1) Convergence in Probability A sequence of random variables $\{X_1, X_2, ...\}$ converges *in probability* towards a random variable X if for all $\varepsilon > 0$,

$$\lim_{i\to\infty} P(|X_i - X| > \varepsilon) = 0. \tag{1.137}$$

2) Mean Convergence of *p*th Order A sequence of random variables $\{X_1, X_2, ...\}$ with property

$$E(|X_i|^p) < \infty; \; i = 1, 2, ...$$

converges *in mean of the* pth *order* towards a random variable X if, for all p with $1 \le p < \infty$,

$$\lim_{n\to\infty} E(|X_i - X|^p) = 0 \quad \text{and} \quad E(|X|^p) < \infty. \tag{1.138}$$

Specifically, if $p = 1$, then the sequence $\{X_1, X_2, ...\}$ converges *in mean* towards X. If $p = 2$, then $\{X_1, X_2, ...\}$ converges *in mean square* or *in square mean* towards X.

3) Convergence with Probability 1 A sequence of random variables $\{X_1, X_2, ...\}$ converges *with probability* 1 or *almost sure* towards a random variable X if

$$P(\lim_{i\to\infty} X_i = X) = 1.$$

4) Convergence in Distribution Let the random variables X_i have distribution functions $F_{X_i}(x)$; $i = 1, 2, ...$ Then the sequence $\{X_1, X_2, ...\}$ converges towards a random variable X with distribution function $F_X(x)$ *in distribution* if, for all points of continuity x of $F_X(x)$,

$$\lim_{i\to\infty} F_{X_i}(x) = \lim_{i\to\infty} P(X_i \le x) = P(X \le x) = F_X(x).$$

Implications

a) 3 implies 4, 2 implies 1, and 1 implies 4. Moreover:

b) If $\{X_1, X_2, ...\}$ converges towards a finite constant a in distribution, then $\{X_1, X_2, ...\}$ converges towards a in probability. Hence, if the limit is a finite constant, convergence in distribution and convergence in probability are equivalent.

c) If $\{X_1, X_2, ...\}$ converges towards a random variable X in probability, then there exists a subsequence $\{X_{i_1}, X_{i_2}, ...\}$ of $\{X_1, X_2, ...\}$, which converges towards X with probability 1.

1.9.2 Laws of Large Numbers

There are *weak* and *strong laws of large numbers*. They essentially deal with the convergence behaviour of arithmetic means $\overline{X}_n$ for $n \to \infty$, where

$$\overline{X}_n = \frac{1}{n} \sum_{i=1}^{n} X_i.$$

Theorem 1.5 Let $\{X_1, X_2, ...\}$ be a sequence of independent, identically distributed random variables with finite mean μ and variance σ^2. Then the sequence of arithmetic means $\{\overline{X}_1, \overline{X}_2, ...\}$ converges in probability towards μ:

$$\lim_{n\to\infty} P\left(|\overline{X}_n - \mu| > \varepsilon \right) = 0.$$

Proof In view of $Var(\overline{X}_n) = \sigma^2/n$, Chebyshev's inequality (1.129) yields

$$P\left(|\overline{X}_n - \mu| > \varepsilon \right) \le \frac{\sigma^2}{n\varepsilon^2}.$$

Letting $n \to \infty$ proves the theorem. ■

A generalization of theorem 1.5 is the following one.

Theorem 1.6 Let $\{X_1, X_2, ...\}$ be a sequence of (not necessarily independent) random variables X_i with finite means $\mu_i = E(X_i);\ i = 1, 2, ...$ On condition

$$\lim_{i\to\infty} Var(X_i) = 0,$$

the sequence $\{X_1 - \mu_1, X_2 - \mu_2, ...\}$ converges in probability towards 0. ■

Example 1.27 Let X be the indicator variable of the occurrence of random event A:

$$X = \begin{cases} 1 & \text{if } A \text{ occurs} \\ 0 & \text{otherwise} \end{cases}$$

with

$$p = P(A) = P(X = 1),\ 1 - p = P(X = 0) = P(\overline{A}).$$

Thus, X has a Bernoulli distribution with

$$E(X) = p, \quad Var(X) = p(1-p).$$

To estimate the probability $p = P(A)$, the random experiment with outcomes A and $\overline{A}$ is repeated n times independently of each other. The corresponding sequence of indicator variables be $X_1, X_2, ..., X_n$. The X_i are independent and identically distributed as X. Hence, theorem 1.5 is applicable: With respect to convergence in probability,

$$\lim_{n\to\infty} \overline{X}_n = \lim_{n\to\infty} \frac{1}{n} \Sigma_{i=1}^n X_i = p.$$

Note that $\hat{p}_n(A) = \overline{X}_n$ is the *relative frequency* of the occurrence of random event A in a series of n random experiments (section1.1). Thus, $\hat{p}_n(A)$ is a suitable estimator for the probability of any random event A. □

The following theorem does not need assumptions on variances. Instead, the pairwise independence of the sequence $\{X_1, X_2, ...\}$ is required, i.e. X_i and X_j are independent for $i \neq j$.

Theorem 1.7 Let $\{X_1, X_2, ...\}$ be a sequence of pairwise independent, identically distributed random variables with finite mean μ. Then the corresponding sequence of arithmetic means $\{\overline{X}_1, \overline{X}_2, ...\}$ converges in probability towards μ. ■

Theorems 1.5 to 1.7 are called *weak laws of great numbers*, whereas the following two theorems are *strong laws of great numbers*, since the underlying convergence criterion is *convergence with probability* 1.

Theorem 1.8 Let $\{X_1, X_2, ...\}$ be a sequence of independent, identically distributed random variables with finite mean μ. Then the corresponding sequence of arithmetic means $\{\overline{X}_1, \overline{X}_2, ...\}$ converges with probability 1 towards μ. ■

Theorems 1.5 and 1.8 imply that the sequence of relative frequencies

$$\{\hat{p}_1(A),\ \hat{p}_2(A),\ ...\}$$

converges towards $p = P(A)$ both with respect to convergence in probability and with probability 1. The following theorem abandons the assumption of identically distributed random variables.

Theorem 1.9 Let $\{X_1, X_2, ...\}$ be a sequence of independent random variables with parameters

$$\mu_i = E(X_i) \text{ and } \sigma_i^2 = Var(X_i);\ i = 1, 2, ...$$

On condition that

$$\Sigma_{i=1}^{\infty} (\sigma_i / i)^2 < \infty,$$

the sequence $\{Y_1, Y_2, ...\}$ with

$$Y_n = \overline{X}_n - \frac{1}{n} \Sigma_{i=1}^n \mu_i$$

converges with probability 1 towards 0. ■

1.9.3 Central Limit Theorem

The central limit theorem provides the theoretical base for the dominant role of the normal distribution in probability theory and its applications. Intuitively, it states that a random variable which arises from additive superposition of many random influences, where none of them is dominant, has approximately a normal distribution. There are several variations of the central limit theorem. The simplest is the following one.

Theorem 1.10 (*Lindeberg* and *Lèvy*) Let $Y_n = X_1 + X_2 + \cdots + X_n$ be the sum of n independent, identically distributed random variables X_i with finite mean $E(X_i) = \mu$ and finite variance $Var(X_i) = \sigma^2$, and let Z_n be the standardization of Y_n:

$$Z_n = \frac{Y_n - n\mu}{\sigma\sqrt{n}}.$$

Then,

$$\lim_{n\to\infty} P(Z_n \le x) = \Phi(x) = \frac{1}{\sqrt{2\pi}} \int_{-\infty}^{x} e^{-u^2/2} du. \qquad \blacksquare$$

Corollary Under the conditions of theorem 1.10, Y_n has approximately a normal distribution with mean value $n\mu$ and variance σ^2/n:

$$Y_n \approx N(n\mu, \sigma^2/n). \tag{1.139}$$

Thus, Y_n is *asymptotically normally distributed* as $n \to \infty$. (The fact that Y_n has mean value $n\mu$ and variance $n\sigma^2$ follows from (1.105).)

As a rule of thumb, (1.139) gives satisfactory results if $n \ge 20$. The following theorem shows that the assumptions of theorem 1.10 can be weakened.

Theorem 1.11 (*Lindeberg and Feller*) Let $Y_n = X_1 + X_2 + \cdots + X_n$ be the sum of independent random variables X_i with finite means $\mu_i = E(X_i)$ and finite variances $\sigma_i^2 = Var(X_i)$, and let Z_n be the standardization of Y_n:

$$Z_n = \frac{Y_n - E(Y_n)}{\sqrt{Var(Y_n)}} = \frac{Y_n - \Sigma_{i=1}^n \mu_i}{\sqrt{\Sigma_{i=1}^n \sigma_i^2}}.$$

Then Z_n has the properties

$$\lim_{n\to\infty} P(Z_n \le x) = \Phi(x) = \frac{1}{\sqrt{2\pi}} \int_{-\infty}^{x} e^{-u^2/2} du, \tag{1.140}$$

$$\lim_{n\to\infty} E(Z_n) \to \infty, \tag{1.141}$$

and

$$\lim_{n\to\infty} \max_{i=1,2,\ldots,n} (\sigma_i/E(Z_n)) \to 0 \tag{1.142}$$

if and only if the *Lindeberg condition*

$$\lim_{n\to\infty} \frac{1}{Var(Z_n)} \sum_{i=1}^{n} \int_{|x-\mu_i|>\varepsilon\sqrt{Var(Z_n)}} (x-\mu_i)^2 f_{X_i}(x)dx = 0$$

is fulfilled for all $\varepsilon > 0$. ■

Conditions (1.141) and (1.142) imply that no term X_i in the sum dominates the rest and that, for $n \to \infty$, the contributions of the X_i to the sum uniformly tend to 0. Under the assumptions of theorem 1.10, the X_i a priori have this property.

Example 1.28 On weekdays, a car dealer sells on average one car (of a certain make) per $\mu = 2.4$ days with a standard deviation of $\sigma = 1.6$.

1) What is the probability that the dealer sells at least 35 cars during a quarter (75 weekdays)? Let $X_i;\ i = 1, 2, ..., X_0 = 0$ be the time span between selling the $(i-1)$th and the ith car. Then $Y_n = X_1 + X_2 + \cdots + X_n$ is the time point, at which the nth car is sold (selling times negligibly small). Hence, the probability

$$P(Y_{35} \le 75)$$

has to be determined. If the X_i are assumed to be independent,

$$E(Y_{35}) = 35 \cdot 2.4 = 84 \quad \text{and} \quad Var(Y_{35}) = 35 \cdot 1.6^2 = 89.6.$$

In view of (1.139), Y_{35} has approximately an $N(84, 89.6)$-distribution. Hence,

$$P(Y_{35} \le 75) \approx \Phi\left(\frac{75-84}{9.466}\right) = \Phi(-0.95) = 0.171.$$

2) How many cars $n_{\min}$ has the dealer at least to stock at the beginning of a quarter to make sure that every customer can immediately buy a car with probability 0.95? (It is assumed that this special make of a car is delivered by the manufacturer at no other times.) Obviously, $n = n_{\min}$ is the smallest n with property that

$$P(Y_{n+1} > 75) \ge 0.95.$$

Equivalently, $n_{\min}$ is the smallest n with property

$$P(Y_{n+1} \le 75) \le 0.05 \quad \text{or} \quad \Phi\left(\frac{75 - 2.4\,(n+1)}{1.6\sqrt{n+1}}\right) \le 0.05.$$

Since the 0.05-percentile of an $N(0, 1)$-distribution is $x_{0.05} = -1.64$, the latter inequality is equivalent to

$$\frac{75 - 2.4\,(n+1)}{1.6\sqrt{n+1}} \le -1.64 \quad \text{or} \quad (n - 30.85)^2 \ge 37.7.$$

Hence, $n_{\min} = 37$. □

Normal Approximation to the Binomial Distribution As pointed out in section 1.2, the binomial distribution arises in connection with a Bernoulli trial: Let A be a random event and X its indicator variable:

$$X = \begin{cases} 1 & \text{if } A \text{ occurs} \\ 0 & \text{otherwise} \end{cases}$$

with

$$p = P(X = 1) = P(A) \text{ and } 1 - p = P(X = 0) = P(\overline{A}).$$

A series of n random experiments with respective outcomes $X_1, X_2, ..., X_n$ is carried out, where the X_i are independent and identically distributed as X. Then

$$Y_n = X_1 + X_2 + \cdots + X_n$$

is the number of random experiments with outcome A, whereas $n - Y_n$ is the number of random experiments with outcome $\overline{A}$. The random variable Y_n has a binomial distribution with parameters n and p. Hence, its mean value and variance are

$$E(Y_n) = np, \quad Var(Y_n) = np(1-p).$$

Since the assumptions of theorem 1.10 are fulfilled, Y_n has approximately a normal distribution:

$$Z_n = N(np, np(1-p)).$$

Thus,

$$P(i_1 \le Z_n \le i_2) \approx \Phi\left(\frac{i_2 + \frac{1}{2} - np}{\sqrt{np(1-p)}}\right) - \Phi\left(\frac{i_1 - \frac{1}{2} - np}{\sqrt{np(1-p)}}\right); \quad 0 \le i_1 \le i_2 \le n.$$

In particular,

$$P(Z_n = i) = \binom{n}{i} p^i (1-p)^{n-i}$$

$$\approx \Phi\left(\frac{i + \frac{1}{2} - np}{\sqrt{np(1-p)}}\right) - \Phi\left(\frac{i - \frac{1}{2} - np}{\sqrt{np(1-p)}}\right), \quad 0 \le i \le n.$$

The term $\pm 1/2$ is called a *continuity correction*. It improves the accuracy of this approximation, since a discrete distribution is approximated by a continuous one. These approximation formulas are the better, the larger n is and the nearer p is to 1/2. The 'normal approximation' of the binomial distribution yields satisfactory results if

$$E(Z_n) = np > 35 \quad \text{and} \quad Var(Z_n) = np(1-p) > 10.$$

The approximation of the binomial distribution by the normal distribution is known as the *central limit theorem of Moivre-Laplace*.

Theorem 1.12 (***Moivre-Laplace***) If the random variable X has a binomial distribution with parameters n and p, then, for all x,

$$\lim_{n\to\infty} P\left(\frac{X-np}{\sqrt{np(1-p)}} \le x\right) = \frac{1}{\sqrt{2\pi}} \int_{-\infty}^{x} e^{-u^2/2} du. \qquad \blacksquare$$

Example 1.29 Electronic circuits are subjected to a quality test. It is known that 5% of the production is faulty. What is the probability that the proportion of faulty units in a sample of 1000 circuits is between 4% and 6%?

Let X be the random number of faulty circuits in the sample. Then X has a binomial distribution with parameters

$$n = 1000 \text{ and } p = 0.05.$$

Hence, the desired probability is

$$P(40 \le X \le 60) = \sum_{i=40}^{60} \binom{1000}{i} (0.05)^i (0.95)^{1000-i}.$$

For numerical reasons, it makes sense to apply the normal approximation: Since

$$E(X) = 1000 \cdot 0.05 = 50 > 35 \quad \text{and} \quad Var(X) = 1000 \cdot 0.05 \cdot 0.95 = 47.5 > 10,$$

its application will yield satisfactory results:

$$P(40 \le X \le 60) \approx \Phi\left(\frac{60+0.5-50}{6.892}\right) - \Phi\left(\frac{40-0.5-50}{6.892}\right)$$

$$= \Phi(1.523) - \Phi(-1.523)$$

$$= 0.972. \qquad \square$$

Normal Approximation to the Poisson Distribution Let

$$Y_n = X_1 + X_2 + \cdots + X_n$$

be the sum of n independent, Poisson distributed random variables $X_1, X_2, \cdots, X_n$ with respective parameters $\lambda_1, \lambda_2, ..., \lambda_n$. Then, by example 1.22 (section 1.7.1),

$$M_{Y_n}(z) = e^{(\lambda_1+\lambda_2+\cdots+\lambda_n)z}.$$

Thus, Y_n has a Poisson distribution with parameter $\lambda_1 + \lambda_2 + \cdots + \lambda_n$. As a consequence, every random variable X which has a Poisson distribution with parameter λ can be represented as a sum of n independent random variables, each of which has a Poisson distribution with parameter λ/n. Since the assumptions of theorem 1.10 are fulfilled, it is justified to approximate the Poisson distribution by the normal distribution: If X has a Poisson distribution with parameter λ, then

$$E(X) = \lambda \text{ and } Var(X) = \lambda.$$

Therefore,

$$X \approx N(\lambda, \lambda), \quad F_X(x) \approx \Phi\left(\frac{x-\lambda}{\sqrt{\lambda}}\right).$$

so that, using the continuity correction 1/2 as in the case of the normal approximation to the binomial distribution,

$$P\left(i_1 \le X \le i_2\right) \approx \Phi\left(\frac{i_2 + \frac{1}{2} - \lambda}{\sqrt{\lambda}}\right) - \Phi\left(\frac{i_1 - \frac{1}{2} - \lambda}{\sqrt{\lambda}}\right),$$

$$P(X = i) \approx \Phi\left(\frac{i + \frac{1}{2} - \lambda}{\sqrt{\lambda}}\right) - \Phi\left(\frac{i - \frac{1}{2} - \lambda}{\sqrt{\lambda}}\right).$$

Since the distribution of a nonnegative random variable is approximated by the normal distribution, the assumption

$$E(X) = \lambda > 3\sqrt{Var(X)} = 3\sqrt{\lambda}$$

has to be made. Hence, the normal approximation to the Poisson distribution should only be applied if $\lambda > 9$.

Example 1.30 The number X of traffic accidents in a town a day is known to have a Poisson distribution with parameter $\lambda = E(X) = 12$.

1) What is the probability that there are exactly 10 traffic accidents a day?

$$P(X = 10) = \frac{12^{10}}{10!} e^{-12} = 0.104.$$

The normal approximation yields

$$P(X = 10) \approx \Phi\left(\frac{10 + 0.5 - 12}{\sqrt{12}}\right) - \Phi\left(\frac{10 - 0.5 - 12}{\sqrt{12}}\right)$$

$$= 0.3325 - 0.2330$$

$$= 0.0995.$$

2) What is the probability that there are at least 10 traffic accidents a day?

For computational reasons, it is convenient to apply the normal approximation:

$$P(X \ge 10) = \sum_{i=10}^{\infty} \frac{12^i}{i!} e^{-12} \approx 1 - \Phi\left(\frac{9 + 0.5 - 12}{\sqrt{12}}\right)$$

$$= 0.7673.$$

□

1.10 EXERCISES

Sections 1.1 to section 1.3

1.1) Castings are produced weighing either 1, 5, 10 or 20 kg. Let A, B and C be the events that a casting does not weigh more than 1 or 5 kg, exactly 10 kg, and at least 20 kg, respectively. Characterize verbally the events

$$A \cap B, \ A \cup B, \ A \cap \overline{C}, \text{ and } (\overline{A} \cup \overline{B}) \cap C.$$

1.2) Three persons have been tested for the occurrence of gene g. Based on this random experiment, three random events are introduced as follows:

$$A = \text{'no person has gene } g\text{'}$$

$$B = \text{'at least one person has gene } g\text{'}$$

$$C = \text{'not more than one person has gene } g\text{'}$$

(1) Characterize verbally the random events $A \cap B$, $B \cup C$ and $(A \cup B) \cap \overline{C}$.

(2) By introducing a suitable sample space, determine the sets of elementary events which characterize the random events occurring under (1).

1.3) Let $P(A) = 0.3$; $P(B) = 0.5$ and $P(A \cap B) = 0.2$.

Determine the probabilities $P(A \cup B)$, $P(\overline{A} \cap B)$ and $P(\overline{A} \cup \overline{B})$.

1.4) 200 plates are checked for surface quality (acceptable, non acceptable) and for satisfying given tolerance limits of the diameter (yes, no). The results are:

		surface quality	
		acceptable	*unacceptable*
diameter	*yes*	170	15
	no	8	7

A plate is selected at random from these 200. Let A be the event that its diameter is within the tolerance limits, and let B the event that its surface quality is acceptable.

(1) Determine the probabilities $P(A)$, $P(B)$ and $P(A \cap B)$ from the matrix. By using the rules developed in section 1.1, determine $P(A \cup B)$ and $P(\overline{A} \cup \overline{B})$.

(2) Are A and B independent?

1.5) A company optionally equips its newly developed PC *Ibson* with 2 or 3 hard disk drives and with or without extra software and analyzes the first 1000 orders:

		hard disk drives	
		three	*two*
extra software	*yes*	520	90
	no	70	320

A PC is selected at random from the first 1000 orders. Let A be the event that this PC has three hard disk drives and let B be the event this PC has extra software.

(1) Determine the probabilities

$$P(A),\ P(B),\ \text{and}\ P(A \cap B)$$

from the matrix.

(2) By using the rules developed in section 1.1 determine the probabilities

$$P(A \cup B),\ P(A|B),\ P(B|A),\ P(A \cup B|\overline{B})\ \text{and}\ P(\overline{A}|\overline{B}).$$

1.6) 1000 bits are independently transmitted from a source to a sink. The probability of a faulty transmission of a bit is 0.0005.

What is the probability that the transmission of at least two bits is not successful?

1.7) To construct a circuit a student needs, among others, 12 chips of a certain type. The student knows that 4% of these chips are defective.

How many chips have to be provided so that, with a probability of not less than 0.9, the student has a sufficient number of nondefective chips in order to be able to construct the circuit?

1.8) It costs \$50 to find out whether a spare part required for repairing a failed device is faulty or not. Installing a faulty spare part causes a damage of \$1000.

Is it on average more profitable to use a spare part without checking if

(1) 1% of all spare parts of that type

(2) 3% of all spare parts of that type

(3) 10 % of all spare parts of that type

are faulty ?

1.9) A test procedure for diagnosing faults in circuits indicates no fault with probability 0.99 if the circuit is faultless. It indicates a fault with probability 0.90 if the circuit is faulty. Let the probability that a circuit is faulty be 0.02.

(1) What is the probability that a circuit is faulty if the test procedure indicates a fault?

(2) What is the probability that a circuit is faultless if the test procedure indicates that it is faultless?

1.10) Suppose 2% of cotton fabric rolls and 3% of nylon fabric rolls contain flaws. Of the rolls used by a manufacturer, 70% are cotton and 30% are nylon.

(1) What is the probability that a randomly selected roll used by the manufacturer contains flaws?

(2) Given that a randomly selected roll used by the manufacturer does not contain flaws, what is the probability that it is a nylon fabric roll?

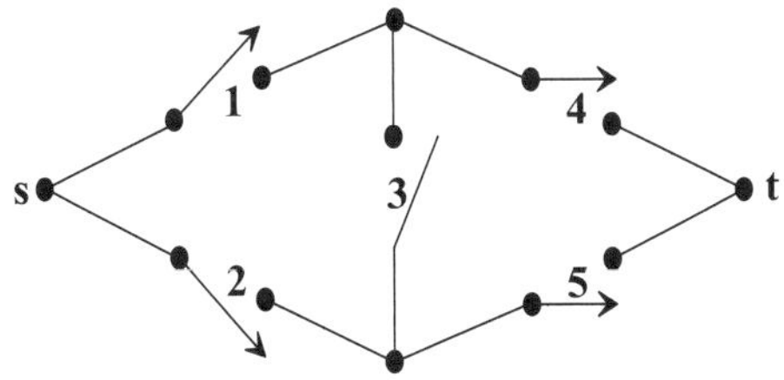

1.11) Transmission of information between computers s and t (see figure) is possible if there is at least one closed path between s and t. The figure indicates the possible interruption of an edge (connection between two nodes of the transmission graph) by a switch. In practice, such an interruption may be caused by a cable break or if the transmission capacity of a channel is exceeded. All 5 switches operate independently. Each one is closed with probability p and open with probability $1-p$. Only switch 3 allows for transmitting information into both directions.

(1) What is the probability $w_{s,t}(p)$ that s can send information to t?

(2) Draw the graph of $w_{s,t}(p)$ as a function of p, $0 \le p \le 1$.

1.12) From a source, symbols 0 and 1 are transmitted independently of each other in proportion $1:4$. Random noise may cause transmission failures: If a 0 was sent, then a 1 will arrive at the sink with probability 0.1. If a 1 was sent, then a 0 will arrive at the sink with probability 0.05. (Figure).

(1) A 1 has arrived. What is the probability that a 1 had been sent?

(2) A 0 has arrived. What is the probability that a 1 had been sent?

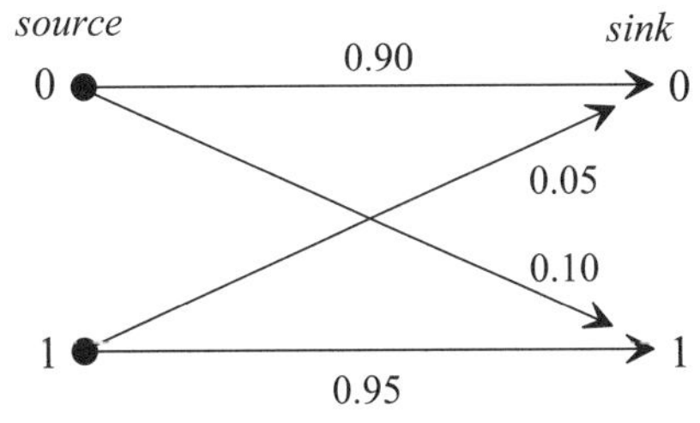

1.13) A biologist measured the weight of 132 eggs of a certain bird species [*gram*]:

i	1	2	3	4	5	6	7	8	9	10
weight x_i	38	41	42	43	44	45	46	47	48	50
number of eggs n_i	4	6	7	10	13	26	33	16	10	7

There are no eggs weighing less than 38 or more than 49. Let X denote the weight of an egg selected randomly from this population.

(1) Determine the probability distribution of X, i.e. $p_i = P(X = x_i);\ i = 1, 2, ..., 10.$

(2) Determine $P(43 \le X \le 48)$ and $P(X > 45)$.

(3) Draw the distribution function of X.

1.14) 120 nails are classified by length:

i		1	2	3	4	5	6	
length x_i (in mm)	< 15.0	15.0	15.1	15.2	15.3	15.4	15.5	> 15.5
number of nails n_i	0	8	26	42	24	15	5	0

Let X denote the length of a nail selected randomly from this population.

(1) Determine the probabilities $p_i = P(X = x_i);\ i = 1, 2, ..., 6.$

(2) Determine the probabilities $P(X \le 15.1)$, $P(X > 15.4)$, and $P(15.0 < X \le 15.5)$.

(3) Draw the distribution function of X.

1.15) Let X be given by exercise 1.13. Determine $E(X)$ and $Var(X)$.

1.16) Let X be given by exercise 1.14. Determine $E(X)$ and $Var(X)$.

1.17) Because it happens that not all airline passengers show up for their reserved seats, an airline would sell 602 tickets for a flight that holds only 600 passengers. The probability that for some reason or other a passenger does not show up is 0.008. The passengers behave independently.

What is the probability that every passenger who shows up will have a seat?

1.18) Water samples are taken from a river once a week. Let X denote the number of samples taken over a period of 20 weeks which are polluted. It is known that on average 10% of the samples are polluted. Assuming independence of the outcomes of the sample analyses, what is the probability that X exceeds its mean by more than one standard deviation?

1.19) From the 300 chickens of a farm, 100 have attracted bird flue. If four chickens are randomly selected from the population of 300, what is the probability that all of them have bird flue?

1.20) Some of the 140 trees in a park are infested with a fungus. A sample of 10 randomly selected trees is taken.

(1) If 25 trees from the 140 are infested, what is the probability that the sample contains at least one infested tree?

(2) If 5 trees from the 140 are infested, what is the probability that the sample contains at least two infested trees?

1.21) Flaws occur at random along the length of a thin copper wire. Suppose that the number of flaws follows a Poisson distribution with a mean of 0.15 flaws per centimetre. What is the probability of more than 2 flaws in a section of length 10 centimetre?

1.22) The number of dust particles which occur on the reflector surface of a telescope has a Poisson distribution with intensity 0.1 per centimetre squared. What is the probability of not more than 2 particles on an area of 10 squared centimetres?

1.23) The random number of crackle sounds produced per hour by an old radio has a Poisson distribution with parameter $\lambda = 12$. What is the probability that there is no crackle sound during the 4 minutes transmission of a listener's favourite hit?

1.24) Show that the following functions are probability density functions for some value of c and determine c:

(1) $f(x) = c\,x^2, \quad 0 \le x \le 4$

(2) $f(x) = c\,(1 + 2x), \quad 0 \le x \le 2$

(3) $f(x) = c\,e^{-x}, \quad 0 \le x < \infty$

These functions are assumed to be identically 0 outside their respective ranges.

1.25) Consider a random variable X with probability density function

$$f(x) = x\,e^{-x^2/2}, \quad x \ge 0.$$

Determine x such that

$$P(X < x) = 0.5, \ P(X \le x) = 0.5, \text{ and } P(X > x) = 0.95.$$

1.26) A road traffic light is switched on every day at 5:00 a.m. It always begins with 'red' and holds this colour 2 minutes. Then it changes to 'green' and holds this colour 4 minutes. This cycle continues till midnight. A car driver arrives at this traffic light at a time point which is uniformly distributed between 9:00 and 9:10 a.m.

(1) What is the probability that the driver has to wait in front of the traffic light?

(2) Determine the same probability on condition that the driver's arrival time point has a uniform distribution over the interval [8:58, 9:08]?

1.27) According to the timetable, a lecture begins at 8:15. The arrival time of professor *Durrick* in the venue is uniformly distributed between 8:13 and 8:20, whereas the arrival time of student *Slugish* is uniformly distributed between 8:05 to 8:30.

What is the probability that *Sluggish* arrives after *Durrick* in the venue?

1.28) Determine $E(X)$ and $Var(X)$ of the three random variables X with probability density functions specified in exercise 1.24.

1.29) The lifetimes of bulbs of a particular type have an exponential distribution with parameter λ [h^{-1}]. Five bulbs of this type are switched on at time $t = 0$. Their lifetimes can be assumed independent.

(1) What is the probability that at time $t = 1/\lambda$ a) all 5, b) at least 3 bulbs are failed?

(2) What is the probability that at least one bulb survives $5/\lambda$ hours?

1.30) The probability density function of the annual energy consumption of an enterprise [in 10^8kwh] is

$$f(x) = 30(x-2)^2\left[1 - 2(x-2) + (x-2)^2\right], \quad 2 \le x \le 3.$$

(1) Determine the distribution function of X.

(2) What is the probability that the annual energy consumption exceeds 2.8?

(3) What is the mean annual energy consumption?

1.31) Assume X is normally distributed with mean 5 and standard deviation 4. Determine the respective values of x which satisfy $P(X > x) = 0.5$, $P(X > x) = 0.95$, $P(x < X < 9) = 0.2$, $P(3 < X < x) = 0.95$, and $P(-x < X < +x) = 0.99$.

1.32) The response time of an average male car driver is normally distributed with mean value 0.5 and standard deviation 0.06 (in seconds).

(1) What is the probability that the response time is greater than 0.6 seconds?

(2) What is the probability that the response time is between 0.5 and 0.55 seconds?

1.33) The tensile strength of a certain brand of polythene sheet can be modeled by a normal distribution with mean 36 psi and standard deviation 4 psi.

(1) Determine the probability that the tensile strength of a sample is at least 28 psi.

(2) If the specifications require the tensile strength to exceed 30 psi, what proportion of the production has to be scrapped?

1.34) The total monthly sick-leave time X of employees of a small company has a normal distribution with mean 100 hours and standard deviation 20 hours. (1) What is the probability that the total monthly sick-leave time is between 50 and 80 hours?

(2) How much time has to be budgeted for sick leave to make sure that the budgeted amount is exceeded with a probability not greater than 0.1?

1.35) Let $X = X_\theta$ have a geometric distribution with

$$p_i = P(X = i) = (1-\theta)\,\theta^i; \; i = 0, 1, \ldots; \; 0 \le \theta \le 1.$$

By mixing the X_θ with regard to a suitable structure distribution density, show that

$$\sum_{i=0}^{\infty} \frac{1}{(i+1)(i+2)} = 1.$$

1.36) A random variable $X = X_\alpha$ have distribution function

$$F_\alpha(x) = e^{-\alpha/x};\ \alpha > 0,\ x > 0$$

(*Frechet distribution*). What distribution type arises when mixing the F_α with regard to the structure distribution density

$$f(\alpha) = \lambda e^{-\lambda\alpha},\ \lambda > 0,\ \alpha > 0\,?$$

Sections 1.4 and 1.5

1.37) The times between the arrivals of taxis at a rank are independent and identically exponentially distributed with parameter $\lambda = 4\,[h^{-1}]$. Assume that an arriving customer does not find an available taxi, the previous one left 3 minutes ago, and no other customers are waiting. What is the probability that the customer has to wait at least 5 minutes for the next free taxi?

1.38) The random variable X has distribution function

$$F(x) = \lambda x/(1+\lambda x),\ \ \lambda > 0,\ x \ge 0.$$

Check whether there is a subinterval of $[0, \infty)$ on which $F(x)$ is DFR.

1.39)* Consider lifetimes X and Y with the respective probability densities

$$f(x) = \begin{cases} 1/4, & 0 \le x \le 4 \\ 0, & \text{otherwise} \end{cases}, \qquad g(x) = \begin{cases} \frac{1}{10}x, & 0 \le x \le 2 \\ \frac{5}{10}x, & 2 \le x \le 3 \\ \frac{3}{10}x, & 3 \le x \le 4 \\ 0, & \text{otherwise} \end{cases}.$$

With the notation introduced in section 1.4, let X_2 and Y_2 be the corresponding residual lifetimes given that $X > 2$ and $Y > 2$, respectively.

(1) Show that $X \underset{st}{\le} Y$. (2) Check whether $X_2 \underset{st}{\le} Y_2$ and interpret the result.

1.40)* Let the random variables A and B have uniform distributions over $[0, a]$ and $[0, b]$, $a < b$, respectively.

(1) Show that $A \underset{st}{\le} B$ and $A \underset{hr}{\le} B$.

(2) Let X be defined by $P(X = 0) = P(X = 1) = 1/2$. Show that if X is independent of A and B then $A + X \underset{hr}{\not\le} B + X$.

(3) Let A_X and B_X be the random variables arising by mixing A and B, respectively, with regard to the distribution of X as structure distribution. Show that

$$A_X \underset{hr}{\not\le} B_X.$$

Sections 1.6 to 1.9

1.41) Every day a car dealer sells X cars of type I and Y cars of type II. The table shows the joint distribution $\{r_{ij} = P(X = i, Y = j);\ \ i, j = 0, 1, 3\}$ of (X, Y) :

	Y	0	1	2
X	0	0.1	0.1	0
	1	0.1	0.3	0.1
	2	0	0.2	0.1

(1) Determine the marginal distributions of (X, Y).

(2) Are X and Y independent?

(3) Determine the conditional mean values $E(X|Y = 1)$ and $E(Y|X = 0)$.

1.42) The random vector (X, Y) has joint density

$$f_{X,Y}(x,y) = x + y, \quad 0 \le x, y \le 1.$$

(1) Are X and Y independent?

(2) Determine the probability density of $Z = XY$.

1.43) The random vector (X, Y) has joint density

$$f_{X,Y}(x,y) = 6x^2 y, \quad 0 \le x, y \le 1.$$

(1) Are X and Y independent? (2) Determine the density of $Z = XY$.

1.44) A supermarket employes 24 shop-assistants. 20 of them achieve an average daily turnover of \$8000, whereas 4 achieve an average daily turnover of \$ 10,000. The corresponding standard deviations are \$2400 and \$3000, respectively. The daily turnovers of all shop-assistants are independent and have a normal distribution. Let Z be the daily total turnover of all shop-assistants.

(1) Determine $E(Z)$ and $Var(Z)$.

(2) What is the probability that the daily total turnover Z is greater than \$ 190,000?

1.45) A helicopter is allowed to carry at most 8 persons provided that their total weight does not exceed 620 kg. The weights of the passengers are independent, identically normally distributed random variables with mean value 76 kg and standard deviation 18 kg.

(1) What are the probabilities of exceeding the permissible load with 7 and 8 passengers, respectively?

(2) What would the maximum total permissible load have to be to ensure that, with probability 0.99, the helicopter will be allowed to fly 8 passengers?

1.46) A freighter has to be loaded with 2000 tons of hard coal. The coal arrives at the harbor by railway carriages each of which holds independently of each other a random load X with $E(X) = 50$ and $Var(X) = 64$.

What is the smallest number $n = n_{min}$ of railway carriages which are necessary to make sure that with a probability of not less than 0.99 the freighter can be loaded with at least 2000 tons of coal?

1.47) In a certain geographical region, the height X of women has a normal distribution with $E(X) = 168$ cm and $Var(X) = 64$ cm, whereas the height Y of men has a normal distribution with $E(Y) = 175$ cm and $Var(Y) = 100$ cm.

Determine the probability that a randomly selected woman is taller than a randomly selected man.

Hint The desired probability has structure $P(X \geq Y) = P(X + (-Y) \geq 0)$.

1.48)* Let X_1 and X_2 be independent and identically distributed with density

$$f(x) = \frac{1}{\pi} \frac{\lambda}{1+x^2}, \quad x \in (-\infty, +\infty).$$

This is a Cauchy distribution with parameters $\lambda = 1$ and $\mu = 0$ (section 1.2.3.2). Verify that $X_1 + X_2$ has a Cauchy distribution with parameters $\lambda = 2$ and $\mu = 0$.

1.49) Let X have a geometric distribution with parameter p, $0 < p < 1$:

$$P(X = i) = p\,(1-p)^{i-1}; \ i = 1, 2, ...$$

(1) Determine the z-transform of X and by means of it $E(X)$ and $Var(X)$.

(2) Let X_1 and X_2 be independent random variables, identically distributed as X. Determine the z-transform of $Z = X_1 + X_2$ and by means of it $E(Z)$ and $Var(Z)$.

Verify the 2nd moment obtained in this way by another method.

1.50) Let $X_1, X_2, ..., X_k$ be independent, binomially distributed random variables with respective parameters $(n_1, p_1), (n_2, p_2), ..., (n_k, p_k)$.

Under which condition has the sum $Z = X_1 + X_2 + \cdots + X_k$ a binomial distribution?

Hint Determine the z-transform of Z.

1.51) (X, Y) has a uniform distribution over the square $[0 \leq x \leq T, 0 \leq y \leq T]$, i.e. its joint density is

$$f_{X,Y}(x, y) = \begin{cases} 1/T^2, & 0 \leq x, y \leq T \\ 0, & \text{otherwise} \end{cases}.$$

(1) Are X and Y independent?

(2) Determine the density of the sum $Z = X + Y$.

1.52) Let X have a Laplace distribution with parameters λ and μ, i.e. X has density

$$f(x) = \frac{\lambda}{2} e^{-\lambda|x-\mu|}, \quad \lambda > 0, \quad -\infty < \mu < +\infty, \quad -\infty < x < +\infty.$$

Determine the Laplace transform of $f(x)$ and, by means of it,

$$E(X), \; E(X^2), \; \text{and } Var(X).$$

1.53) 6% of the citizens in a large town suffer from severe hypertension. Let B_n be the number of people in a sample of n randomly selected citizens from this town which suffer from this desease (Bernoulli trial).

(1) By making use of the Chebychev inequality find a positive integer n_0 with property

$$P\left(\left|\frac{B_n}{n} - 0.06\right| \geq 0.01\right) \leq 0.05 \;\; \text{for all } n \text{ with } n \geq n_0. \tag{i}$$

(2) Find a positive integer n_0 satisfying relationship (i) by making use of the central limit theorem.

CHAPTER 2

Basics of Stochastic Processes

2.1 MOTIVATION AND TERMINOLOGY

A random variable X is the outcome of a random experiment under fixed conditions. A change of these conditions will influence the outcome of the experiment, i.e. the probability distribution of X will change. Varying conditions can be taken into account by considering random variables which depend on a deterministic parameter t: $X = X(t)$. This approach leads to more general random experiments than the ones defined in section 1.1. To illustrate such generalized random experiments, two simple examples will be considered.

Example 2.1 a) At a fixed geographical point, the temperature is measured every day at 12:00. Let x_i be the temperature measured on the ith day of a year. The value of x_i will vary from year to year and, therefore, it can be considered a realization of a random variable X_i. Thus, X_i is the (random) temperature measured on the ith day of a year at 12:00. Apart from random fluctuations of the temperature, the X_i also depend on a deterministic parameter, namely on the time, or, more precisely, on the day of the year. However, if one is only interested in the temperatures X_1, X_2, X_3 on the first 3 days (or any other 3 consecutive days) of the year, then these temperatures are at least approximately identically distributed. Nevertheless, indexing the daily temperatures is necessary, because modeling the obviously existing statistical dependence between the daily temperatures requires knowledge of the joint probability distribution of the random vector (X_1, X_2, X_3). This situation and the problems connected with it motivate the introduction of the generalized random experiment 'daily measurement of the temperature at a given geographical point at 12:00 during a year'. The random outcomes of this generalized random experiment are sequences of random variables $\{X_1, X_2, ..., X_{365}\}$ with the X_i being generally neither independent nor identically distributed. If on the ith day temperature x_i has been measured, then the vector $(x_1, x_2, ..., x_{365})$ can be interpreted as a function $x = x(t)$, defined at discrete time points t, $t \in [1, 2, ..., 365]$: $x(t) = x_i$ for $t = i$. Vector $(x_1, x_2, ..., x_{365})$ is a realization of the random vector $(X_1, X_2, ..., X_{365})$.

b) If a sensor graphically records the temperature over the year, then the outcome of the measurement is a continuous function of time t: $x = x(t)$, $0 \le t \le 1$, where $x(t)$ is realization of the random temperature $X(t)$ at time t at a fixed geographical location. Hence it makes sense to introduce the generalized random experiment 'continuous measurement of the temperature during a year at a given geographical location'. It will be denoted as $\{X(t),\ 0 \le t \le 1\}$.

A complete probabilistic characterization of this generalized random experiment requires knowledge of the joint probability distributions of all possible random vectors

$$(X(t_1), X(t_2), ..., X(t_n)); \quad 0 \le t_1 < t_2 < \cdots < t_n \le 1; \;\; n = 1, 2, ...$$

This knowledge allows for statistically modelling the dependence between the $X(t_i)$ in any sequence of random variables

$$X(t_1), X(t_2), ..., X(t_n).$$

It is quite obvious that, for small time differences $t_{i+1} - t_i$, there is a strong statistical dependence between $X(t_i)$ and $X(t_{i+1})$. However, there is also a dependence between $X(t_i)$ and $X(t_k)$ for large time differences $t_k - t_i$ due to the inertia of weather patterns over an area. □

Example 2.2 The deterministic parameter, which influences the outcome of a random experiment, need not be time. For instance, if at a fixed time point and a fixed observation point the temperature is measured along a vertical of length L to the earth's surface, then one obtains a function $x = x(h)$ with $0 \le h \le L$ which obviously depends on the distance h of the measurement point to the earth's surface. But if the experiment is repeated in the following years under the same conditions (same time, location and measurement procedure), then, in view of the occurrence of nonpredictable influences, different functions $x = x(h)$ will be obtained. Hence, the temperature at distance h is a random variable $X(h)$ and the generalized random experiment 'measuring the temperature along a vertical of length L', denoted as $\{X(h), 0 \le h \le L\}$, has outcomes, which are real functions of h: $x = x(h)$, $0 \le h \le L$.

In this situation, it also makes sense to consider the temperature in dependence of both h and the time point of observation t:

$$x = x(h, t); \;\; 0 \le h \le L, \;\; t \ge 0.$$

Then the observation x depends on a vector of deterministic parameters:

$$x = x(\boldsymbol{\theta}), \;\; \boldsymbol{\theta} = (h, t).$$

In this case, the outcomes of the corresponding generalized random experiment are surfaces in the (h, t, x)-space. However, this book only considers one-dimensional parameter spaces.

An already 'classical' example for illustrating the fact that the parameter need not be time is essentially due to *Cramer* and *Leadbetter* [22]: A machine is required to continuously produce ropes of length $10\,m$ with a given nominal diameter of $5\,mm$. Despite maintaining constant production conditions, minor variations of the rope diameter can technologically not be avoided. Thus, when measuring the actual diameter x of a single rope at a distance d from the origin, one gets a function $x = x(d)$ with $0 \le d \le 10$. This function will randomly vary from rope to rope. This suggests the introduction of the generalized random experiment 'continuous measurement of the rope diameter in dependence on the distance d from the origin'. If $X(d)$ denotes the

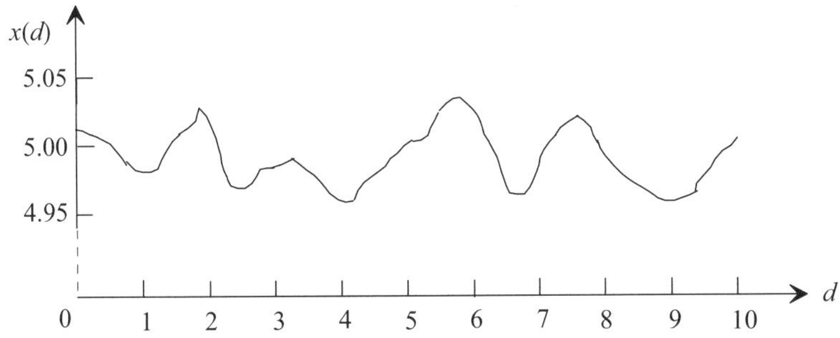

Figure 2.1 Random variation of the diameter of a nylon rope

diameter of a randomly selected rope at a distance d from the origin, then it makes sense to introduce the corresponding generalized random experiment

$$\{X(d),\ 0 \le d \le 10\}$$

with outcomes $x = x(d)$, $0 \le d \le 10$ (Figure 2.1). □

In contrast to the random experiments considered in chapter 1, the outcomes of which are real numbers, the outcomes of the generalized random experiments, dealt with in examples 2.1 and 2.2, are real functions. Hence, in literature, such generalized random experiments are frequently called *random functions*. However, the terminology *stochastic processes* is more common and will be used throughout the book. In order to characterize the concept of a stochastic process more precisely, further notation is required: Let the random variable of interest X depend on a parameter t which assumes values from a set $\mathbf{T}$: $X = X(t),\ t \in \mathbf{T}$. To simplify the terminology and in view of the overwhelming majority of applications, in this book the parameter t is interpreted as time. Thus, $X(t)$ is the random variable X at time t and $\mathbf{T}$ denotes the whole observation time span. Further, let $\mathbf{Z}$ denote the set of all values, the random variables $X(t)$ can assume for all $t \in \mathbf{T}$.

Stochastic process A family of random variables $\{X(t),\ t \in \mathbf{T}\}$ is called a *stochastic process* with parameter space $\mathbf{T}$ and *state space* $\mathbf{Z}$.

If $\mathbf{T}$ is a finite or countably infinite set, then $\{X(t),\ t \in \mathbf{T}\}$ is called a *stochastic process in discrete time* or a *discrete-time stochastic process.* Such processes can be written as a sequence of random variables $\{X_1, X_2, ...\}$ (example 2.1 a). On the other hand, every sequence of random variables can be thought of as a stochastic process in discrete time. If $\mathbf{T}$ is an interval, then $\{X(t),\ t \in \mathbf{T}\}$ is a *stochastic process in continuous time* or a *continuous-time stochastic process.* A stochastic process $\{X(t),\ t \in \mathbf{T}\}$ is said to be *discrete* if its state space $\mathbf{Z}$ is a finite or a countably infinite set, and a stochastic process $\{X(t),\ t \in \mathbf{T}\}$ is said to be *continuous* if $\mathbf{Z}$ is an interval. Thus, there are discrete stochastic processes in discrete time, discrete stochastic processes in continuous time, continuous stochastic processes in discrete time, and continuous stochastic processes in continuous time. Throughout this book the state space **Z** is assumed to be a subset of the real axis.

If the stochastic process $\{X(t), t \in \mathbf{T}\}$ is observed over the whole time period **T**, i.e. the values of $X(t)$ are registered for all $t \in \mathbf{T}$, then one obtains a real function $x = x(t),\ t \in \mathbf{T}$. Such a function is called a *sample path*, a *trajectory* or a *realization* of the stochastic process. In this book the concept *sample path* is used. The sample paths of a stochastic process in discrete time are, therefore, sequences of real numbers, whereas the sample paths of stochastic processes in continuous time can be any functions of time. The sample paths of a discrete stochastic process in continuous time are piecewise constant functions (step functions). The set of all sample paths of a stochastic process with parameter space **T** is, therefore, a subset of all functions over the domain **T**.

In engineering, science and economics there are many time-dependent random phenomena which can be modeled by stochastic processes: In an electrical circuit it is not possible to keep the voltage strictly constant. Random fluctuations of the voltage are for instance caused by *thermal noise.* If $v(t)$ denotes the voltage measured at time point t, then $v = v(t)$ is a sample path of a stochastic process $\{V(t), t \ge 0\}$ where $V(t)$ is the random voltage at time t (Figure 2.2). Producers of radar and satellite supported communication systems have to take into account a phenomenon called *fading.* This is characterized by random fluctuations in the energy of received signals caused by the dispersion of radio waves as a result of inhomogeneities in the atmosphere and by *meteorological* and *industrial noise.* (Both meteorological and industrial noise cause electrical discharges in the atmosphere which occur at random time points with randomly varying intensity.) 'Classic' applications of stochastic processes in economics are modeling the development of share prices, profits, and prices of commodities over time. In operations research, stochastic processes describe the development in time of the 'states' of queueing, inventory and reliability systems. In statistical quality control, they model the fluctuation of quality criteria over time. In medicine, the development in time of 'quality parameters' of health as blood pressure and cholesterol level are typical examples of stochastic processes. One of the first applications of stochastic processes can be found in biology: modeling the development in time of the number of species in a population.

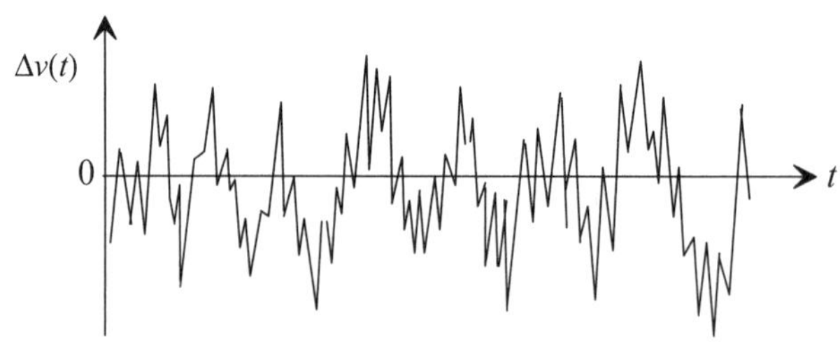

Figure 2.2 Voltage fluctuations caused by random noise

2.2 CHARACTERISTICS AND EXAMPLES

From the mathematical point of view, the heuristic explanation of a stochastic process given needs to be supplemented. Let $F_t(x)$ be the distribution function of $X(t)$:

$$F_t(x) = P(X(t) \le x), \ \ t \in \mathbf{T}.$$

The family of the one-dimensional distribution functions

$$\{F_t(x), t \in \mathbf{T}\}$$

is the *one-dimensional probability distribution* of $\{X(t), t \in \mathbf{T}\}$. In view of the statistical dependence, which generally exists between the $X(t_1), X(t_2), ..., X(t_n)$ for any $t_1, t_2, ..., t_n$, the family of the one-dimensional distribution functions $\{F_t(x), t \in \mathbf{T}\}$ does not completely characterize a stochastic process (see examples 2.1 and 2.2). A stochastic process $\{X(t), t \in \mathbf{T}\}$ is only then completely characterized if, for all $n = 1, 2, ...,$ for all n-tuples $\{t_1, t_2, ..., t_n\}$ with $t_i \in \mathbf{T}$, and for all $\{x_1, x_2, ..., x_n\}$ with $x_i \in \mathbf{Z}$, the joint distribution function of the random vector

$$(X(t_1), X(t_2), ..., X(t_n))$$

is known:

$$F_{t_1, t_2, ..., t_n}(x_1, x_2, ..., x_n) = P(X(t_1) \le x_1, X(t_2) \le x_2, ..., X(t_n) \le x_n). \tag{2.1}$$

The set of all these joint distribution functions defines the *probability distribution* of the stochastic process. For a discrete stochastic process, it is generally simpler to characterize its probability distribution by the probabilities

$$P(X(t_1) \in A_1, X(t_2) \in A_2, \dots, X(t_n) \in A_n)$$

for all $t_1, t_2, ..., t_n$ with $t_i \in \mathbf{T}$ and $A_i \subseteq \mathbf{Z}$; $i = 1, 2, ..., n$; $n = 1, 2, ...$

Trend Function Assuming the existence of $E(X(t))$ for all $t \in \mathbf{T}$, the *trend* or *trend function* of the stochastic process $\{X(t), t \in \mathbf{T}\}$ is the mean value of $X(t)$ as a function of t:

$$m(t) = E(X(t)), \quad t \in \mathbf{T}. \tag{2.2}$$

Thus, the trend function of a stochastic process describes its average development in time. If the densities $f_t(x) = dF_t(x)/dx$ exist, then

$$m(t) = \int_{-\infty}^{+\infty} x\, f_t(x)\, dx, \quad t \in \mathbf{T}.$$

Covariance Function The *covariance function* of a stochastic process $\{X(t), t \in \mathbf{T}\}$ is the covariance between random variables $X(s)$ and $X(t)$ as a function of s and t. Hence, in view of (1.63) and (1.64),

$$C(s,t) = Cov(X(s), X(t)) = E([X(s) - m(s)]\,[X(t) - m(t)])\,; \ \ s, t \in \mathbf{T}, \tag{2.3}$$

or

$$C(s,t) = E(X(s)X(t)) - m(s)m(t); \ \ s, t \in \mathbf{T}. \tag{2.4}$$

In particular,

$$C(t,t) = Var(X(t)). \tag{2.5}$$

The covariance function is a symmetric function of s and t:

$$C(s,t) = C(t,s). \tag{2.6}$$

Since the covariance function $C(s,t)$ is a measure for the degree of the statistical dependence between $X(s)$ and $X(t)$, one expects that

$$\lim_{|t-s|\to\infty} C(s,t) = 0. \tag{2.7}$$

However, example 2.3 shows that this need not be the case.

Correlation Function The *correlation function* of $\{X(t),\ t \in \mathbf{T}\}$ is the correlation coefficient $\rho(s,t) = \rho(X(s), X(t))$ between $X(s)$ and $X(t)$ as a function of s and t. According to (1.65),

$$\rho(s,t) = \frac{Cov\,(X(s), X(t))}{\sqrt{Var(X(s)}\,\sqrt{Var(X(t)}}. \tag{2.8}$$

The covariance function of a stochastic process is also called *autocovariance function* and the correlation function *autocorrelation function.* This is useful when considering covariances and correlations between $X(s)$ and $Y(s)$ with regard to different stochastic processes $\{X(t),\ t \in \mathbf{T}\}$ and $\{Y(t),\ t \in \mathbf{T}\}$.

Example 2.3 (*cosine wave with random amplitude*) Let

$$X(t) = A\,\cos\omega t\,,$$

where A is a nonnegative random variable with $E(A) < \infty$. The process $\{X(t),\ t \geq 0\}$ can be interpreted as the output of an oscillator which is selected from a set of identical ones. (Random deviations of the amplitudes from a nominal value are technologically unavoidable.) The trend function of this process is

$$m(t) = E(A)\,\cos\omega t.$$

By (2.4), its covariance function is

$$\begin{aligned} C(s,t) &= E([A\cos\omega s][A\cos\omega t]) - m(s)m(t) \\ &= [E(A^2) - (E(A))^2](\cos\omega s)(\cos(\omega t)). \end{aligned}$$

Hence,

$$C(s,t) = Var(A)(\cos\omega s)(\cos\omega t).$$

Obviously, the process does not have property (2.7). Since there is a functional relationship between $X(s)$ and $X(t)$ for any s and t, $X(s)$ and $X(t)$ cannot tend to become independent as $|t-s| \to \infty$. Actually, the correlation function between $X(s)$ and $X(t)$ is identically equal to 1: $\rho(s,t) \equiv 1$. For a modification of this process, see example 2.6. □

The stochastic process considered in example 2.3 has a special feature: Once the random variable A has assumed a value a, the process develops in a strictly deterministic way. That means, by only observing a sample path of such a process over an arbitrarily small time interval, one can predict the further development of the sample path with absolute certainty. (The same comment refers to examples 2.6 and 2.7).

More complicated stochastic processes arise when random influences continuously, or at least repeatedly, affect the phenomenon of interest. The following example belongs to this category.

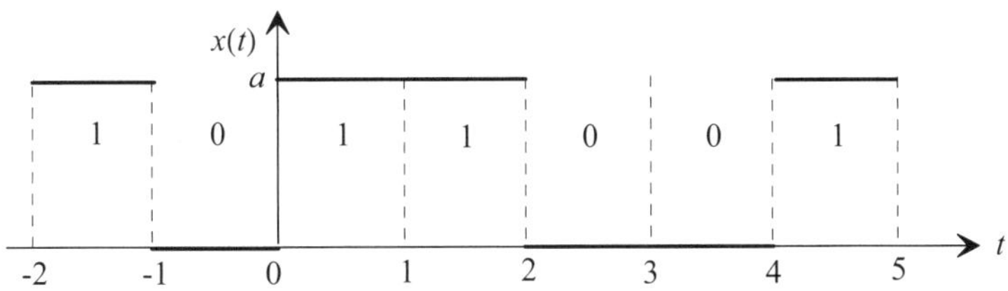

Figure 2.3 Pulse code modulation

Example 2.4 (***pulse code modulation***) A source generates symbols 0 or 1 independently with probabilities p and $1-p$, respectively. The symbol 0 is transmitted by sending nothing during a time interval of length one. The symbol 1 is transmitted by sending a pulse with constant amplitude a during a time unit of length one. The source has started operating in the past. A stochastic signal (sequence of symbols) generated in this way is represented by the stochastic process $\{X(t), t \in (-\infty, +\infty)\}$ with

$$X(t) = \sum_{n=-\infty}^{+\infty} A_n h(t-n), \quad n \le t < n+1, \tag{2.9}$$

where the A_n; $n = 0, \pm 1, \pm 2, ...$; are independent binary random variables defined by

$$A_n = \begin{cases} 0 & \text{with probability} \quad p \\ a & \text{with probability} \quad 1-p \end{cases},$$

and $h(t)$ is given by

$$h(t) = \begin{cases} 1 & \text{for} \quad 0 \le t < 1 \\ 0 & \quad \text{elsewhere} \end{cases}.$$

For any t,

$$X(t) = \begin{cases} 0 & \text{with probability} \quad p \\ a & \text{with probability} \quad 1-p \end{cases}.$$

For example, the section of a sample path $x = x(t)$ plotted in Figure 2.3 is generated by the following partial sequence of a signal:

$$\cdots 1\ 0\ 1\ 1\ 0\ 0\ 1 \cdots$$

Note that the time point $t = 0$ coincides with the beginning of a new transmission period. The process has a constant trend function:

$$m(t) \equiv a \cdot P(X(t) = a) + 0 \cdot P(X(t) = 0) = a(1-p).$$

For $n \le s, t < n+1;\ n = 0, \pm 1, \pm 2, ...,$

$$E(X(s)X(t)) = E(X(s)X(t)|X(s) = a) \cdot P(X(s) = a)$$
$$+ E(X(s)X(t)|X(s) = 0) \cdot P(X(s) = 0)$$
$$= a^2(1-p).$$

If $m \le s < m+1$ and $n \le t < n+1$ with $m \ne n$, then $X(s)$ and $X(t)$ are independent. Hence the covariance function of $\{X(t),\ t \in (-\infty, +\infty)\}$ is

$$C(s,t) = \begin{cases} a^2p(1-p) & \text{for} \quad n \le s, t < n+1;\ \ n = 0, \pm 1, \pm 2, ... \\ 0 & \text{elsewhere} \end{cases}.$$

Although the stochastic process analyzed in this example has a rather simple structure, it is of considerable importance in physics, electrical engineering, and communication. A modification of the pulse code modulation process is considered in example 2.8. As the following example shows, the pulse code modulation is a special shot noise process. □

Example 2.5 (*shot noise process*) At time points T_n, pulses of random intensity A_n are induced. The sequences $\{T_1, T_1, ...\}$ and $\{A_1, A_2, ...\}$ are assumed to be discrete-time stochastic processes with properties

1) With probability 1, $T_1 < T_2 < \cdots$ and $\lim_{n\to\infty} T_n = \infty$,

2) $E(A_n) < \infty;\ n = 1, 2, ...$

In communication theory, the sequence $\{(T_n, A_n);\ \ n = 1, 2, ...\}$ is called a *pulse process*. (In section 3.1, it will be called a *marked point process*.) Let function $h(t)$, the *response* of a system to a pulse, have properties

$$h(t) = 0 \ \text{ for } t < 0 \quad \text{and} \quad \lim_{t\to\infty} h(t) = 0. \tag{2.10}$$

The stochastic process $\{X(t),\ t \in (-\infty, +\infty)\}$ defined by

$$X(t) = \Sigma_{n=1}^{\infty} A_n\, h(t - T_n) \tag{2.11}$$

is called a *shot noise process* or just *shot noise*. It quantifies the additive superposition of the responses of a system to pulses. The factors A_n are sometimes called *amplitudes* of the shot noise process. In many applications, the A_n are independent, identically distributed random variables, or, as in example 2.4, even constant.

If the sequences of the T_n and A_n are doubly infinite,

$$\{T_n;\ n = 0, \pm 1, \pm 2, ...\} \quad \text{and} \quad \{A_n;\ n = 0, \pm 1, \pm 2, ...\},$$

then the shot noise process $\{X(t),\ t \in (-\infty, +\infty)\}$ is defined as

$$X(t) = \sum_{n=-\infty}^{\infty} A_n\, h(t - T_n). \tag{2.12}$$

A well-known physical phenomenon, which can be modeled by a shot noise process, is the fluctuation of the anode current in vacuum tubes ('tube noise'). This fluctuation is caused by random current impulses, which are initiated by emissions of electrons from the anode at random time points (*Schottky effect*).

The term *shot noise* has its origin in the fact that the effect of firing small shot at a metal slab can be modeled by a stochastic process of structure (2.11). More examples of shot noise processes are discussed in chapter 3, where special assumptions on the underlying pulse process are made. □

2.3 CLASSIFICATION OF STOCHASTIC PROCESSES

Stochastic processes are classified with regard to properties which reflect for instance their dependence on time, the statistical dependence of their developments over disjoint time intervals, and the influence of the history or the current state of a stochastic process on its future evolvement. In the context of example 2.1: Has the date any influence on the daily temperature at 12:00? (That need not be the case if the measurement point is near to the equator.) Or, has the sample path of the temperature in January any influence on the temperature curve in February? For reliably predicting tomorrow's temperature at 12:00, is it sufficient to know the present temperature or would knowledge of the temperature curve during the past two days allow a more accurate prediction? What influence has time on trend or covariance function?

Special importance have those stochastic processes, for which the joint distribution functions (2.1) only depend on the distances between t_i and t_{i+1}, i.e. only the relative positions of $t_1, t_2, ..., t_n$ to each other have an impact on the joint distribution of the random variables $X(t_1), X(t_2), ..., X(t_n)$.

Strong Stationarity A stochastic process $\{X(t),\ t \in \mathbf{T}\}$ is said to be *strongly stationary* or *strictly stationary* if for all $n = 1, 2, ...,$ for any h, for all n-tuples

$$(t_1, t_2, ..., t_n) \text{ with } t_i \in \mathbf{T} \text{ and } t_i + h \in \mathbf{T};\ i = 1, 2, ..., n;$$

and for all n-tuples $(x_1, x_2, ..., x_n)$, the joint distribution function of the random vector $(X(t_1), X(t_2), ..., X(t_n))$ has the following property:

$$F_{t_1, t_2, ..., t_n}(x_1, x_2, ..., x_n) = F_{t_1+h, t_2+h, ..., t_n+h}(x_1, x_2, ..., x_n). \tag{2.13}$$

Thus, the probability distribution of a strongly stationary stochastic process is invariant against absolute time shifts. In particular, by letting $n = 1$, property (2.13) implies that the one-dimensional distribution functions $F_t(x)$ do not depend on t. In this case there exists a distribution function $F(x)$ so that

$$F_t(x) \equiv F(x), \; t \in \mathbf{T}. \tag{2.14}$$

Hence, trend- and variance function of $\{X(t), t \in \mathbf{T}\}$ do not depend on t either:

$$m(t) = E(X(t)) \equiv m = \text{constant}, \tag{2.15}$$
$$Var(X(t)) \equiv \text{constant}.$$

The trend function of a strongly stationary process is, therefore, a parallel to the time axis and the fluctuations of its sample paths around the trend function experience no systematic changes with increasing t.

Substituting $n = 2$, $t_1 = 0$, $t_2 = t - s$ and $h = s$ in (2.13) yields for all $s < t$,

$$F_{0,t-s}(x_1, x_2) = F_{s,t}(x_1, x_2),$$

i.e. the joint distribution function of the random vector (X_s, X_t), and, therefore, the mean value of the product $X_s X_t$, depend only on the difference $\tau = t - s$, and not on the absolute values of s and t. Since, according to (2.4),

$$C(s,t) = E[X(s)X(t)] - m^2 \quad \text{for } s, t \in \mathbf{T},$$

$C(s,t)$ must have the same property:

$$C(s,t) = C(s, s + \tau) = C(0, \tau) = C(\tau).$$

Therefore, the covariance function of strongly stationary processes depends only on one variable: For all $s \in \mathbf{T}$,

$$C(\tau) = Cov(X(s), X(s + \tau)). \tag{2.16}$$

Since the covariance function $C(s,t)$ of a stochastic process is symmetric in the variables s and t, the covariance function of a strongly stationary process is symmetric with respect to $\tau = 0$, i.e. $C(\tau) = C(-\tau)$ or, equivalently,

$$C(\tau) = C(|\tau|). \tag{2.17}$$

In practical situations it is generally not possible to determine the probability distributions of all possible random vectors $\{X(t_1), X(t_2), \cdots, X(t_n)\}$ in order to check whether a stochastic process is strongly stationary or not. The user of stochastic processes is, therefore, frequently satisfied with the validity of properties (2.15) and (2.16). Hence, based on these two properties, another concept of stationarity has been introduced. It is, however, only defined for second order processes.

Second Order Process A stochastic process $\{X(t), t \in \mathbf{T}\}$ is said to be a *second order process* if

$$E(X^2(t)) < \infty \quad \text{for all } t \in \mathbf{T}. \tag{2.18}$$

The existence of the second moments of $X(t)$ as required by assumption (2.18) implies the existence of the covariance function $C(s,t)$ for all s and t, and, therefore, the existence of the variances $Var(X(t))$ and mean values $E(X(t))$ for all $t \in \mathbf{T}$ (see inequality of Schwarz, section 1.8.2).

Weak Stationarity A stochastic process $\{X(t), t \in \mathbf{T}\}$ is said to be *weakly stationary* if it is a second order process, which has properties (2.15) and (2.16).

A strongly stationary process is not necessarily weakly stationary, since there are strongly stationary processes, which are not second order processes. But, if a second order process is strongly stationary, then, as shown above, it is also weakly stationary. Weakly stationary processes are also called *wide-sense stationary*, *covariance stationary* or *second-order stationary*. Further important properties of stochastic processes are based on properties of their increments.

Homogeneous Increments The *increment* of a stochastic process $\{X(t), t \in \mathbf{T}\}$ with respect to the interval $[t_1, t_2]$ is the difference $X(t_2) - X(t_1)$.

A stochastic process $\{X(t), t \in \mathbf{T}\}$ is said to have *homogeneous* or *stationary increments* if for arbitrary, but fixed $t_1, t_2 \in \mathbf{T}$ the increment $X(t_2 + \tau) - X(t_1 + \tau)$ has the same probability distribution for all τ with property $t_1 + \tau \in \mathbf{T}$, $t_2 + \tau \in \mathbf{T}$.

An equivalent definition of processes with homogeneous increments is the following one: $\{X(t), t \in \mathbf{T}\}$ has homogeneous increments if the probability distribution of $X(t+\tau) - X(t)$ does not depend on t for any fixed τ; $t, t+\tau \in \mathbf{T}$.

A stochastic process with homogeneous (stationary) increments need not be stationary in any sense.

Independent Increments A stochastic process $\{X(t), t \in \mathbf{T}\}$ has *independent increments* if for all $n = 2, 3, ...$ and for all n-tuples $(t_1, t_2, ..., t_n)$ with $t_i \in \mathbf{T}$ and

$$t_1 < t_2 < t_3 < \cdots < t_n$$

the increments

$$X(t_2) - X(t_1),\ X(t_3) - X(t_2),\ \cdots,\ X(t_n) - X(t_{n-1})$$

are independent random variables.

Gaussian Process A stochastic process $\{X(t), t \in \mathbf{T}\}$ is a *Gaussian process* if the random vectors $(X(t_1), X(t_2), ..., X(t_n))$ have a joint Normal (Gaussian) distribution for all n-tuples $(t_1, t_2, ..., t_n)$ with $t_i \in \mathbf{T}$ and $t_1 < t_2 < \cdots < t_n$; $n = 1, 2, ...$

Gaussian processes have an important property: A Gaussian process is strongly stationary if and only if it is weakly stationary. Important Gaussian processes will be considered later.

Markov Process A stochastic process $\{X(t), t \in \mathbf{T}\}$ has the *Markov(ian) property* if for all $(n+1)$–tuples $(t_1, t_2, ..., t_{n+1})$ with $t_i \in \mathbf{T}$ and $t_1 < t_2 < \cdots < t_{n+1}$, and for any $A_i \subseteq \mathbf{Z}$; $i = 1, 2, ..., n+1$;

$$P(X(t_{n+1}) \in A_{n+1} | X(t_n) \in A_n, X(t_{n-1}) \in A_{n-1}, ..., X(t_1) \in A_1)$$
$$= P(X(t_{n+1}) \in A_{n+1} | X(t_n) \in A_n). \tag{2.19}$$

The Markov property has the following implication: If t_{n+1} is a time point in the future, t_n the present time point and, correspondingly, $t_1, t_2, ..., t_{n-1}$ time points in the past, then the future development of a process having the Markov property does not depend on its evolvement in the past, but only on its present state. Stochastic processes having the Markov property are called *Markov processes.*

A Markov process with finite or countably infinite parameter space **T** is called a *discrete-time Markov process.* Otherwise it is called a *continuous-time Markov process.* Markov processes with finite or countably infinite state spaces **Z** are called *Markov chains.* Thus, a discrete-time Markov chain has both a discrete state space and a discrete parameter space. However, deviations from this notation can be found in literature.

Markov processes play an important role in all sorts of applications, mainly for four reasons: 1) Many practical phenomena can be modeled by Markov processes. 2) The input necessary for their practical application is generally more easily provided than the necessary input for other classes of stochastic processes. 3) Computer algorithms are available for numerical evaluations. 4) Stochastic processes with independent increments always have the Markov property. In this book, the practical importance of Markov processes is illustrated by many examples.

Theorem 2.1 A Markov process is strongly stationary if and only if its one-dimensional probability distributions do not depend on time, i.e. if there exists a distribution function $F(x)$ with

$$F_t(x) = P(X(t) \le x) = F(x) \quad \text{for all } t \in \mathbf{T}. \qquad \blacksquare$$

Hence condition (2.14) is necessary and sufficient for a Markov process to be strongly stationary.

Mean-Square Continuous A second order process $\{X(t), t \in \mathbf{T}\}$ is said to be *mean-square continuous at point* $t = t_0 \in \mathbf{T}$ if

$$\lim_{h \to 0} E([X(t_0 + h) - X(t_0)]^2) = 0. \tag{2.20}$$

The process $\{X(t), t \in \mathbf{T}\}$ is said to be *mean-square continuous in the region* $\mathbf{T}_0$, $\mathbf{T}_0 \subseteq \mathbf{T}$, if it is mean-square continuous at all points $t \in \mathbf{T}_0$.

According to section 1.9.1, the convergence used in (2.20) is called *convergence in mean square.*

There is a simple criterion for a second order stochastic process to be mean-square continuous at t_0: A second order process $\{X(t), t \in \mathbf{T}\}$ is mean-square continuous at t_0 if and only if its covariance function $C(s,t)$ is continuous at $(s,t) = (t_0, t_0)$.

As a corollary from this statement: A weakly stationary process $\{X(t), t \in (-\infty, +\infty)\}$ is mean-square continuous in $(-\infty, +\infty)$ if and only if it is mean-square continuous at time point $t = 0$.

The following two examples make use of two addition formulas from trigonometry:

$$\cos\alpha\,\cos\beta = \frac{1}{2}[\cos(\beta-\alpha)+\cos(\alpha+\beta)],$$
$$\cos(\beta-\alpha) = \cos\alpha\,\cos\beta + \sin\alpha\,\sin\beta.$$

Example 2.6 (*cosine wave with random amplitude and random phase*) In modifying example 2.3, let

$$X(t) = A\,\cos(\omega t + \Phi),$$

where A is a nonnegative random variable with finite mean value and finite variance. The random parameter Φ is assumed to be uniformly distributed over $[0, 2\pi]$ and independent of A. The stochastic process $\{X(t),\ t\in(-\infty,+\infty)\}$ can be thought of the output of an oscillator, selected from a set of oscillators of the same kind and having been turned on at different times. Since

$$E(\cos(\omega t+\Phi)) = \frac{1}{2\pi}\int_0^{2\pi}\cos(\omega t+\varphi)\,d\varphi$$
$$= \frac{1}{2\pi}[\sin(\omega t+\varphi)]_0^{2\pi} = 0,$$

the trend function of this process is identically zero: $m(t)\equiv 0$. From (2.4), its covariance function is

$$C(s,t) = E\{[A\cos(\omega s+\Phi)][A\cos(\omega t+\Phi)]\}$$
$$= E(A^2)\frac{1}{2\pi}\int_0^{2\pi}\cos(\omega s+\varphi)\,\cos(\omega t+\varphi)\,d\varphi$$
$$= E(A^2)\frac{1}{2\pi}\int_0^{2\pi}\frac{1}{2}\{\cos\omega(t-s)+\cos[\omega(s+t)+2\varphi]\}\,d\varphi.$$

The first integrand is a constant with respect to integration. Since the integral of the second term is zero, $C(s,t)$ depends only on the difference $\tau = t-s$:

$$C(\tau) = \frac{1}{2}E(A^2)\,\cos w\tau.$$

Thus, the process is weakly stationary. □

Example 2.7 Let A and B be two uncorrelated random variables satisfying

$$E(A) = E(B) = 0 \ \text{ and } \ Var(A) = Var(B) = \sigma^2 < \infty.$$

The stochastic process $\{X(t),\ t\in(-\infty,+\infty)\}$ be defined by

$$X(t) = A\cos\omega t + B\sin\omega t.$$

Since $Var(X(t)) = \sigma^2 < \infty$ for all t, $\{X(t),\ t\in(-\infty,+\infty)\}$ is a second order process. Its trend function is identically zero: $m(t)\equiv 0$. Thus, from (2.4),

$$C(s,t) = E(X(s)\,X(t)).$$

For A and B being uncorrelated, $E(AB) = E(A)\,E(B)$. Hence,

$$C(s,t) = E(A^2 \cos \omega s \cos \omega t + B^2 \sin \omega s \sin \omega t)$$
$$+E(AB \cos \omega s \sin \omega t + AB \sin \omega s \cos \omega t)$$
$$= \sigma^2 (\cos \omega s \cos \omega t + \sin \omega s \sin \omega t)$$
$$+E(AB)(\cos \omega s \sin \omega t + \sin \omega s \cos \omega t)$$
$$= \sigma^2 \cos \omega (t-s).$$

Therefore, the covariance function depends only on the difference $\tau = t-s$:

$$C(\tau) = \sigma^2 \cos \omega \tau.$$

Thus, the process $\{X(t), t \in (-\infty, +\infty)\}$ is weakly stationary. □

Example 2.8 (***randomly delayed pulse code modulation***) Based on the stochastic process $\{X(t), t \in (-\infty, +\infty)\}$ defined in example 2.4, the stochastic process

$$\{Y(t), t \in (-\infty, +\infty)\} \text{ with } Y(t) = X(t-Z)$$

is introduced, where Z is uniformly distributed over $[0,1]$. Thus, when shifting the sample paths of the process $\{X(t), t \in (-\infty, +\infty)\}$ exactly Z time units to the right, one obtains the corresponding sample paths of the process $\{Y(t), t \in (-\infty, +\infty)\}$. For instance, shifting the section of the sample path shown in Figure 2.3 exactly $Z = z$ time units to the right yields the corresponding section of the sample path of the process $\{Y(t), t \in (-\infty, +\infty)\}$ shown in Figure 2.4.

The trend function of the process $\{Y(t), t \in (-\infty, +\infty)\}$ is

$$m(t) \equiv a(1-p).$$

To determine the covariance function, let $B = B(s,t)$ denote the random event that $X(s)$ and $X(t)$ are separated by a switching point $n+Z$; $n = 0, \pm 1, \pm 2, \ldots$ Then

$$P(B) = |t-s|, \quad P(\overline{B}) = 1 - |t-s|.$$

The random variables $X(s)$ and $X(t)$ are independent if $|t-s| > 1$ and/or B occurs. Therefore,

$$C(s,t) = 0 \text{ if } |t-s| > 1 \text{ and/or } B \text{ occurs.}$$

If $|t-s| \le 1$, $X(s)$ and $X(t)$ are only then independent if B occurs. Hence, the covariance function of $\{Y(t), t \in (-\infty, +\infty)\}$ given $|t-s| \le 1$ can be obtained as follows:

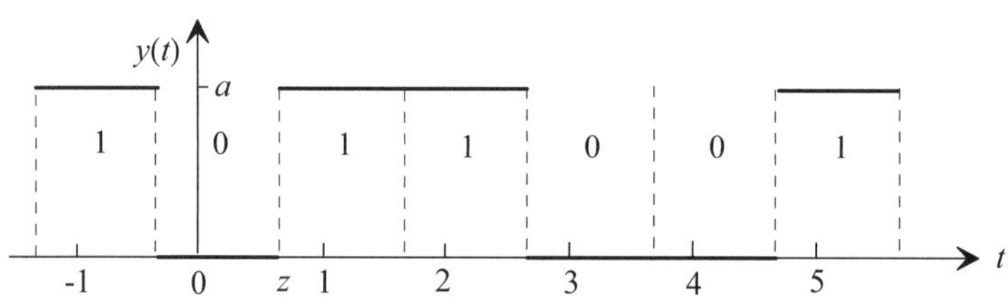

Figure 2.4 Randomly delayed pulse code modulation

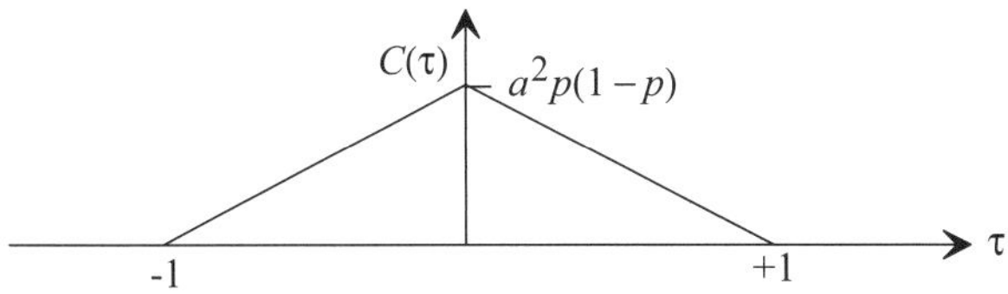

Figure 2.5 Covariance function of the randomly delayed pulse code modulation

$$\begin{aligned} C(s,t) &= E(X(s)X(t)|B)P(B) + E(X(s)X(t)|\overline{B})\,P(\overline{B}) - m(s)\,m(t) \\ &= E(X(s))\,E(X(t))\,P(B) + E([X(s)]^2)\,P(\overline{B}) - m(s)\,m(t) \\ &= [a(1-p)]^2|t-s| + a^2(1-p)(1-|t-s|) - [a(1-p)]^2 . \end{aligned}$$

Finally, with $\tau = t - s$, the covariance function becomes

$$C(\tau) = \begin{cases} a^2 p(1-p)(1-|\tau|) & \text{for } |\tau| \le 1 \\ 0 & \text{elsewhere} \end{cases} .$$

Thus, the process $\{Y(t), t \in (-\infty, +\infty)\}$ is weakly stationary. Analogously to the transition from example 2.3 to example 2.6, stationarity is achieved by introducing a uniformly distributed phase shift in the pulse code modulation of example 2.4. □

2.4 EXERCISES

2.1) A stochastic process $\{X(t), t > 0\}$ has the one-dimensional distribution

$$F_t(x) = P(X(t) \le x) = 1 - e^{-(x/t)^2}, \; x \ge 0.$$

Is this process weakly stationary?

2.2) The one-dimensional distribution of the stochastic process $\{X(t), t > 0\}$ is

$$F_t(x) = P(X(t) \le x) = \frac{1}{\sqrt{2\pi t}\,\sigma} \int_{-\infty}^{x} e^{-\frac{(u-\mu t)^2}{2\sigma^2 t}}\,du$$

with $\mu > 0,\ \sigma > 0;\ x \in (-\infty + \infty)$.

Determine its trend function $m(t)$ and, for $\mu = 2$ and $\sigma = 0.5$, sketch the functions

$$y_1(t) = m(t) + \sqrt{Var(X(t))} \quad \text{and} \quad y_2(t) = m(t) - \sqrt{Var(X(t))}\,, \quad 0 \le t \le 10.$$

2.3) Let $X(t) = A\,\sin(\omega t + \Phi)$, where A and Φ are independent, nonnegative random variables with Φ being uniformly distributed over $[0, 2\pi]$ and $E(A^2) < \infty$.

(1) Determine trend-, covariance- and correlation function of $\{X(t), t \in (-\infty, +\infty)\}$.

(2) Is the stochastic process $\{X(t), t \in (-\infty, +\infty)\}$ weakly and/or strongly stationary?

2.4) Let $X(t) = A(t)\sin(\omega t + \Phi)$, where $A(t)$ and Φ are independent, nonnegative random variables for all t, and let Φ be uniformly distributed over $[0, 2\pi]$.

Verify: If $\{A(t), t \in (-\infty, +\infty)\}$ is a weakly stationary process, then the stochastic process $\{X(t), t \in (-\infty, +\infty)\}$ is also weakly stationary.

2.5) Let $\{a_1, a_2, ..., a_n\}$ be a sequence of real numbers and $\{\Phi_1, \Phi_2, ..., \Phi_n\}$ a sequence of independent random variables which are uniformly distributed over the interval $[0, 2\pi]$. Determine covariance- and correlation function of the stochastic process $\{X(t), t \in (-\infty, +\infty)\}$ given by

$$X(t) = \Sigma_{i=1}^{n}\, a_i \sin(\omega t + \Phi_i).$$

2.6)* A modulated signal (pulse code modulation) $\{X(t), t \in (-\infty, +\infty)\}$ is given by

$$X(t) = \Sigma_{-\infty}^{+\infty} A_n\, h(t - n),$$

where the A_n are independent and identically distributed random variables which can only take on values -1 and $+1$ and have mean value 0. Further, let

$$h(t) = \begin{cases} 1 & \text{for } 0 \le t < 1/2 \\ 0 & \text{elsewhere} \end{cases}.$$

1) Sketch a section of a possible sample path of the process $\{X(t), t \in (-\infty, +\infty)\}$.

2) Determine the covariance function of this process.

3) Let $Y(t) = X(t - Z)$, where the random variable Z has a uniform distribution over $[0, 1]$. Is the stochastic process $\{Y(t), t \in (-\infty, +\infty)\}$ weakly stationary?

2.7) Let $\{X(t), t \in (-\infty, +\infty)\}$ and $\{Y(t), t \in (-\infty, +\infty)\}$ be two independent, weakly stationary stochastic processes, whose trend functions are identically 0 and which have the same covariance function $C(\tau)$.

Prove: The stochastic process $\{Z(t), t \in (-\infty, +\infty)\}$ with

$$Z(t) = X(t)\cos\omega t - Y(t)\sin\omega t$$

is weakly stationary.

2.8) Let $X(t) = \sin \Phi t$, where Φ is uniformly distributed over the interval $[0, 2\pi]$.

Verify: (1) The discrete-time stochastic process $\{X(t);\ t = 1, 2, ...\}$ is weakly, but not strongly stationary. (2) The continuous-time stochastic process $\{X(t), t \ge 0\}$ is neither weakly nor strongly stationary.

2.9) Let $\{X(t), t \in (-\infty, +\infty)\}$ and $\{Y(t), t \in (-\infty, +\infty)\}$ be two independent stochastic processes with trend- and covariance functions $m_X(t)$, $m_Y(t)$ and $C_X(s, t)$, $C_Y(s, t)$, respectively. Further, let $U(t) = X(t) + Y(t)$ and $V(t) = X(t) - Y(t)$, $t \in (-\infty, +\infty)$.

Determine the covariance functions of the stochastic processes $\{U(t), t \in (-\infty, +\infty)\}$ and $\{V(t), t \in (-\infty, +\infty)\}$.

CHAPTER 3

Random Point Processes

3.1 BASIC CONCEPTS

A *point process* is a sequence of real numbers $\{t_1, t_2, ...\}$ with properties

$$t_1 < t_2 < \cdots \quad \text{and} \quad \lim_{i \to \infty} t_i = +\infty. \tag{3.1}$$

That means, a point process is a strictly increasing sequence of real numbers, which does not have a finite limit point. In practice, point processes occur in numerous situations: arrival time points of customers at service stations (workshops, filling stations, supermarkets, ...), failure time points of machines, time points of traffic accidents, occurrence of nature catastrophies, occurrence of supernovas,... Generally, at time point t_i a certain *event* happens. Hence, the t_i are called *event times*. With regard to the arrival of customers at service stations, the t_i are also called *arrival times*. If not stated otherwise, the assumption $t_1 \geq 0$ is made.

Although the majority of applications of point processes refer to sequences of time points, there are other interpretations as well. For instance, sequences $\{t_1, t_2, ...\}$ can be generated by the location of potholes in a road. Then t_i denotes the distance of the ith pothole from the beginning of the road. Or, the location is measured, at which an imaginary straight line, which runs through a forest stand, hits trees. (This is the base of the well-known *Bitterlich method* for estimating the total number of trees in a forest stand.) Strictly speaking, since both road and straight line through a forest stand have finite lengths, to meet assumption (3.1), they have to be considered finite samples from a point process.

A point process $\{t_1, t_2, ...\}$ can equivalently be represented by the sequence of its *interevent* (*interarrival*) times

$$\{y_1, y_2, ...\} \text{ with } y_i = t_i - t_{i-1};\ i = 1, 2, ...;\ t_0 = 0.$$

Counting Process Frequently, the event times are of less interest than the number of events, which occur in an interval $(0, t]$, $t > 0$. This number is denoted as $n(t)$:

$$n(t) = \max\{n,\ t_n \leq t\}.$$

For obvious reasons, $\{n(t), t \geq 0\}$ is said to be the *counting process* belonging to the point process $\{t_1, t_2, ...\}$. Here and in what follows, it is assumed that more than one event cannot occur at a time. Point processes with this property are called *simple.* The number of events, which occur in an interval $(s, t]$, $s < t$, is

$$n(s, t) = n(t) - n(s).$$

To be able to count the number $n(A)$ of events which occur in an arbitrary subset A of $[0,\infty)$ the indicator function of the event 't_i belongs to A' is introduced:

$$I_i(A)=\begin{cases}1 & \text{if } t_i\in A\\ 0 & \text{otherwise}\end{cases}. \tag{3.2}$$

Then,

$$n(A)=\Sigma_{i=0}^{\infty} I_i(A).$$

Example 3.1 Let be given a finite sample from a point process:

$$\{2,4,10,18,24,31,35,38,40,44,45,51,57,59\}$$

The figures indicate the times (in seconds) at which within a time span of a minute a car passes a control point. Then, within the first 16 seconds, $n(16)=3$ cars passed the control point, and in the interval $(31,49]$ exactly $n(31,49)=n(49)-n(30)=5$ cars passed the control point. In terms of the indicator function (3.2), given the time span $A=(10,20]\cup[51,60]$

$$I_{18}(A)=I_{24}(A)=I_{51}(A)=I_{57}(A)=I_{59}(A)=1,$$

$$I_i(A)=0 \text{ for } i\neq 18,\ 24,\ 51,\ 57,\ 59.$$

Hence,

$$n(A)=\Sigma_{i=0}^{\infty} I_i(A)=\Sigma_{i=0}^{60} I_i(A)=5.$$

□

Recurrence Times The *forward recurrence time* of a point process $\{t_1,t_2,...\}$ with respect to time point t is defined as

$$a(t)=t_{n+1}-t \quad \text{for} \quad t_n\le t<t_{n+1};\ n=0,1,...,\ t_0=0. \tag{3.3}$$

Hence, $a(t)$ is the time span from t (usually interpreted as the 'presence') to the occurrence of the next event. A simpler way of characterizing $a(t)$ is

$$a(t)=t_{n(t)+1}-t. \tag{3.4}$$

$t_{n(t)}$ is the largest event time before t and $t_{n(t)+1}$ is the smallest event time after t.

The *backward recurrence time* $b(t)$ with respect to time point t is

$$b(t)=t-t_{n(t)}. \tag{3.5}$$

Thus, $b(t)$ is the time which has elapsed from the last event time before t to time t.

Marked Point Processes Frequently, in addition to their arrival times, events come with another piece of information. For instance: If t_i is the time point the ith customer arrives at a supermarket, then the customer will spend there a certain amount of money m_i. If t_i is the failure time point of a machine, then the time (or cost) m_i necessary for removing the failure may be assigned to t_i. If t_i denotes the time of the ith bank robbery in a town, then the amount m_i the robbers got away with is of interest. If t_i is the arrival time of the ith claim at an insurance company, then the size

m_i of this claim is of particular importance to the company. If t_i is the time of the ith supernova in a century, then its light intensity m_i is of interest to astronomers, and so on. This leads to the concept of a marked point process: Given a point process $\{t_1, t_2, ...\}$, a sequence of two-dimensional vectors

$$\{(t_1, m_1), (t_2, m_2), ...\} \tag{3.6}$$

with m_i being an element of a *mark space* **M** is called a *marked point process*. In most applications, as in the four examples above, the mark space **M** is a subset of the real axis $(-\infty, +\infty)$ with the respective unites of measurements attached.

Random Point Processes Usually the event times are random variables. A sequence of random variables $\{T_1, T_2, ...\}$ with

$$T_1 < T_2 < \cdots \quad \text{and} \quad P(\lim_{i\to\infty} T_i = +\infty) = 1 \tag{3.7}$$

is a *random point process*. By introducing the *random interevent* (*interarrival*) *times*

$$Y_i = T_i - T_{i-1}; \; i = 1, 2, ...; \; T_0 = 0,$$

a random point process can equivalently be defined as a sequence of positive random variables $\{Y_1, Y_2, ...\}$ with property

$$P(\lim_{n\to\infty} \Sigma_{i=0}^{n} Y_i = \infty) = 1.$$

In either case, with the terminology introduced in section 2.1, a random point process is a discrete-time stochastic process with state space $\mathbf{Z} = [0, +\infty)$. Thus, a point process (3.1) is a *sample path*, a *realization* or a *trajectory* of a random point process. A point process is called *simple* if at any time point t not more than one event can occur.

Recurrent Point Processes A random point process $\{T_1, T_2, ...\}$ is said to be *recurrent* if its corresponding sequence of interarrival times $\{Y_1, Y_2, ...\}$ is a sequence of independent, identically distributed random variables. The most important recurrent point processes are homogenous Poisson processess and renewal processes (sections 3.2.1 and 3.3).

Random Counting Processes Let

$$N(t) = \max\{n, \; T_n \le t\}$$

be the random number of events occurring in the interval $(0, t]$. Then the continuous-time stochastic process $\{N(t), t \ge 0\}$ with state space $\mathbf{Z} = \{0, 1, ...\}$ is called the *random counting process* belonging to the random point process $\{T_1, T_2, ...\}$. Any counting process $\{N(t), t \ge 0\}$ has properties

1) $N(0) = 0$,

2) $N(s) \le N(t)$ for $s \le t$,

3) For any s, t with $0 \le s < t$, the increment $N(s, t) = N(t) - N(s)$ is equal to the number of events which occur in $(s, t]$.

Conversely, every stochastic process $\{N(t), t \ge 0\}$ in continuous time having these three properties is the counting process of a certain point process $\{T_1, T_2, ...\}$. Thus, from the statistical point of view, the stochastic processes

$$\{T_1, T_2, ...\},\ \{Y_1, Y_2, ...\},\ \text{and}\ \{N(t), t \ge 0\}$$

are equivalent. For that reason, a random point process is frequently defined as a continuous-time stochastic process $\{N(t), t \ge 0\}$ with properties 1 to 3. Note that

$$N(t) = N(0, t).$$

The most important characteristic of a counting process $\{N(t), t \ge 0\}$ is the probability distribution of its increments $N(s, t) = N(t) - N(s)$, which determines for all intervals $[s, t),\ s < t,$ the probabilities

$$p_k(s, t) = P(N(s, t) = k); \quad k = 0, 1, ...$$

The mean numbers of events in $(s, t]$ is

$$m(s, t) = m(t) - m(s) = E(N(s, t)) = \Sigma_{k=0}^{\infty}\, k p_k(s, t). \tag{3.8}$$

With

$$p_k(t) = p_k(0, t),$$

the trend function of the counting process $\{N(t), t \ge 0\}$ is

$$m(t) = E(N(t)) = \Sigma_{k=0}^{\infty}\, k p_k(t), \quad t \ge 0. \tag{3.9}$$

A random counting process is called *simple* if the underlying point process is simple. Figure 3.1 shows a possible sample path of a simple random counting process.

Note In what follows the attribute 'random' is usually omitted if it is obvious from the notation or the context that random point processes or random counting processes are being dealt with.

Definition 3.1 **(*stationarity*)** A point process $\{T_1, T_2, ...\}$ is called *stationary* if its sequence of interarrival times $\{Y_1, Y_2, ...\}$ is strongly stationary (section 2.3), that is if for any sequence of integers $i_1, i_2, ..., i_k$ with $1 \le i_1 < i_2 < \cdots < i_k,\ k = 1, 2, ...$ and for any $\tau = 0, 1, 2, ...,$ the joint distribution functions of the following two random vectors coincide:

$$\{Y_{i_1}, Y_{i_2}, ..., Y_{i_k}\} \ \text{and}\ \{Y_{i_1+\tau}, Y_{i_2+\tau}, ..., Y_{i_k+\tau}\}.$$ ●

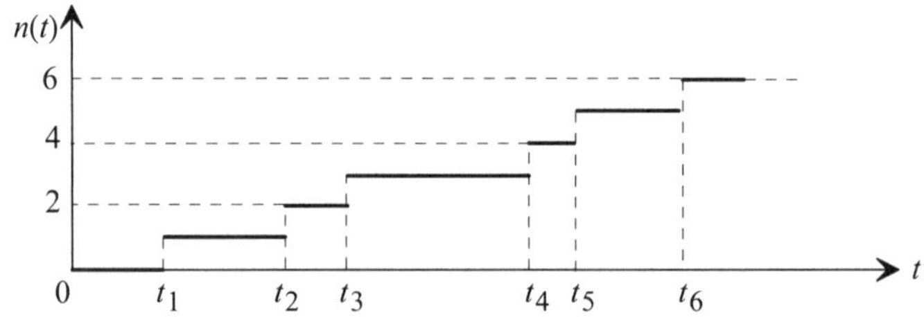

Figure 3.1 Sample path of a simple counting process

It is an easy exercise to show that if the sequence $\{Y_1, Y_2, ...\}$ is strongly stationary, the corresponding counting process $\{N(t), t \geq 0\}$ has homogeneous increments and vice versa. This implies the following corollary from definition 3.1:

Corollary A point process $\{T_1, T_2, ...\}$ is *stationary* if and only if its corresponding counting process $\{N(t), t \geq 0\}$ has homogeneous increments.

Hence, for a stationary point process, the probability distribution of any increment $N(s,t)$ depends only on the difference $\tau = t - s$:

$$p_k(\tau) = P(N(s, s+\tau) = k); \quad k = 0, 1, ...; \ s \geq 0, \ \tau > 0. \tag{3.10}$$

Thus, for a stationary point process,

$$m(\tau) = m(s, s+\tau) = m(s+\tau) - m(s) \quad \text{for all } s \geq 0, \ \tau \geq 0. \tag{3.11}$$

For having nondecreasing sample paths, neither the point process $\{T_1, T_2, ...\}$ nor its corresponding counting process $\{N(t), t \geq 0\}$ can be stationary as defined in section 2.3. In particular, since only simple point processes are considered, the sample paths of $\{N(t), t \geq 0\}$ are step functions with jump heights being equal to 1.

Remark Sometimes it is more convenient or even necessary to define random point processes as doubly infinite sequences

$$\{..., T_{-2}, T_1, T_0, T_1, T_2, ...\},$$

which tend to infinity to the left and to the right with probability 1. Then their sample paths are also doubly infinite sequences: $\{..., t_{-2}, t_1, t_0, t_1, t_2, ...\}$ and only the increments of the corresponding counting process over finite intervals are finite.

Intensity of Random Point Processes For stationary point processes, the mean number of events occurring in [0, 1] is called the *intensity* of the process and will be denoted as λ. By making use of notation (3.9),

$$\lambda = m(1) = \Sigma_{k=0}^{\infty} k p_k(1). \tag{3.12}$$

In view of the stationarity, λ is equal to the mean number of events occurring in any interval of length 1:

$$\lambda = m(s, s+1), \ s \geq 0.$$

Hence, the mean number of events occurring in any interval $(s, t]$ of length $\tau = t - s$ is

$$m(s,t) = \lambda(t-s) = \lambda\tau.$$

Given a sample path $\{t_1, t_2, ...\}$ of a stationary random point process, λ is estimated by the number of events occurring in $[0, t]$ divided by the length of this interval:

$$\hat{\lambda} = n(t)/t,$$

In example 3.1, an estimate of the intensity of the underlying point process (assumed to be stationary) is $\hat{\lambda} = 14/60 \approx 0.233$.

In case of a nonstationary point process, the role of the constant intensity λ is taken over by an *intensity function* $\lambda(t)$. This function allows to determine the mean number of events $m(s,t)$ occurring in an interval $(s,t]$: For any s,t with $0 \le s < t$,

$$m(s,t)) = \int_s^t \lambda(x)\,dx.$$

Specifically, the mean number of events in $[0,t]$ is the trend function of the corresponding counting process:

$$m(t) = m(0,t) = \int_0^t \lambda(x)\,dx, \quad t \ge 0. \tag{3.13}$$

Hence, for $\Delta t \to 0$,

$$\Delta m(t) = \lambda(t)\,\Delta t + o(\Delta t), \tag{3.14}$$

so that for small Δt the product $\lambda(t)\,\Delta t$ is approximately the mean number of events in $(t, t+\Delta t]$. Another interpretation of (3.14) is: If Δt is sufficiently small, then $\lambda(t)\,\Delta t$ is approximately the probability of the occurrence of an event in the interval $[t, t+\Delta t]$. Hence, the intensity function $\lambda(t)$ is the *arrival rate* of events at time t. (For *Landau's order symbol* $o(x)$, see (1.41).)

Random Marked Point Processes Let $\{T_1, T_2, ...\}$ be a random point process with random marks M_i assigned to the event times T_i. Then the sequence

$$\{(T_1, M_1), (T_2, M_2), ...\} \tag{3.15}$$

is called a *random marked point process*. Its (2-dimensional) sample paths are given by (3.6). The pulse process $\{(T_n, A_n);\ n = 1, 2, ...\}$ considered in example 2.5 is a special marked point processes.

Random marked point processes are dealt with in full generality in Matthes, Kerstan, and Mecke [60]. For other mathematically prestigious treatments, see, for instance, König and Schmidt [51] or Stigman [78].

Compound Stochastic Processes Let $\{(T_1, M_1), (T_2, M_2), ...\}$ be a random marked point process and $\{N(t), t \ge 0\}$ be the counting process belonging to the point process $\{T_1, T_2, ...\}$. The stochastic process $\{C(t), t \ge 0\}$ defined by

$$C(t) = \begin{cases} 0 & \text{for } 0 \le t < T_1 \\ \sum_{i=1}^{N(t)} M_i & \text{for } \quad t \ge T_1 \end{cases}$$

is called a *compound* (*cumulative, aggregate*) *stochastic process*. According to the underlying point process, there are, for instance, compound Poisson processes and compound renewal processes. If $\{T_1, T_2, ...\}$ is a claim arrival process and M_i the size of the ith claim, then $C(t)$ is the total claim amount in $[0, t)$. If T_i is the time of the ith breakdown of a machine and M_i the corresponding repair cost, then $C(t)$ is the total repair cost in $[0, t)$.

3.2 POISSON PROCESSES

3.2.1 Homogeneous Poisson Processes

3.2.1.1 Definition and Properties

In the theory of stochastic processes, and maybe even more in its applications, the homogeneous Poisson process is just as popular as the exponential distribution in probability theory. Moreover, there is a close relationship between the homogeneous Poisson process and the exponential distribution (theorem 3.2).

Definition 3.2 (*homogeneous Poisson process*) A counting process $\{N(t), t \geq 0\}$ is a *homogeneous Poisson process with intensity* λ, $\lambda > 0$, if it has the following properties:

1) $N(0) = 0$,

2) $\{N(t), t \geq 0\}$ is a stochastic process with independent increments.

3) Its increments $N(s,t) = N(t) - N(s)$, $0 \leq s < t$, have a Poisson distribution with parameter $\lambda(t-s)$:

$$P(N(s,t) = i) = \frac{(\lambda(t-s))^i}{i!} e^{-\lambda(t-s)}; \quad i = 0, 1, \ldots, \tag{3.16}$$

or, equivalently, introducing the length $\tau = t - s$ of the interval $[s,t]$, for all $\tau > 0$,

$$P(N(s, s+\tau) = i) = \frac{(\lambda\tau)^i}{i!} e^{-\lambda\tau}; \quad i = 0, 1, \ldots \tag{3.17}$$

●

(3.16) implies that the homogeneous Poisson process has homogeneous increments. Thus, the corresponding *Poisson point process* $\{T_1, T_2, \ldots\}$ is stationary in the sense of definition 3.1.

Theorem 3.1 A counting process $\{N(t), t \geq 0\}$ with $N(0) = 0$ is a homogeneous Poisson process with intensity λ if and only if it has the following properties:

a) $\{N(t), t \geq 0\}$ has homogeneous and independent increments.

b) The process is *simple*, i.e. $P(N(t, t+h) \geq 2) = o(h)$.

c) $P(N(t, t+h) = 1) = \lambda h + o(h)$.

Proof To prove that definition 3.2 implies properties a), b) and c), it is only necessary to show that a homogeneous Poisson process satisfies properties b) and c).

The simplicity of the Poisson process easily results from (3.17):

$$P(N(t, t+h) \geq 2) = e^{-\lambda h} \sum_{i=2}^{\infty} \frac{(\lambda h)^i}{i!}$$

$$= \lambda^2 h^2 e^{-\lambda h} \sum_{i=0}^{\infty} \frac{(\lambda h)^i}{(i+2)!} \leq \lambda^2 h^2 = o(h).$$

Another application of (3.17) and the simplicity of the Poisson process proves c):

$$P(N(t,t+h)=1)=1-P(N(t,t+h)=0)-P(N(t,t+h)\geq 2)$$
$$=1-e^{-\lambda h}+o(h)=1-(1-\lambda h)+o(h)$$
$$=\lambda h+o(h).$$

Conversely, it needs to be shown that a stochastic process with properties a), b) and c) is a homogeneous Poisson process. In view of the assumed homogeneity of the increments, it is sufficient to prove the validity of (3.17) for $s=0$. Thus, letting

$$p_i(t)=P(N(0,t)=i)=P(N(t)=i)\,;\;\; i=0,1,...$$

it is to show that

$$p_i(t)=\frac{(\lambda t)^i}{i!}e^{-\lambda t};\;\; i=0,1,.... \tag{3.18}$$

From a),

$$p_0(t+h)=P(N(t+h)=0)=P(N(t)=0,\,N(t,t+h)=0)$$
$$=P(N(t)=0)\,P(N(t,t+h)=0)=p_0(t)p_0(h).$$

In view of b) and c), this result implies

$$p_0(t+h)=p_0(t)(1-\lambda h)+o(h)$$

or, equivalently,

$$\frac{p_0(t+h)-p_0(t)}{h}=-\lambda p_0(t)+o(h).$$

Taking the limit as $h\to 0$ yields

$$p_0'(t)=-\lambda p_0(t).$$

Since $p_0(0)=1$, the solution of this differential equation is

$$p_0(t)=e^{-\lambda t},\;\; t\geq 0,$$

so that (3.18) holds for $i=0$.

Analogously, for $i\geq 1$,

$$p_i(t+h)=P(N(t+h)=i)$$
$$=P(N(t)=i,\,N(t+h)-N(t)=0)+P(N(t)=i-1,\,N(t+h)-N(t)=1)$$
$$+\Sigma_{k=2}^{i}\;P(N(t)=k,\,N(t+h)-N(t)=i-k).$$

Because of c), the sum in the last row is $o(h)$. Using properties a) and b),

$$p_i(t+h)=p_i(t)p_0(h)+p_{i-1}(t)p_1(h)+o(h)$$
$$=p_i(t)\,(1-\lambda h)+p_{i-1}(t)\,\lambda h+o(h),$$

or, equivalently,

$$\frac{p_i(t+h)-p_i(t)}{h} = -\lambda\,[p_i(t)-p_{i-1}(t)]+o(h).$$

Taking the limit as $h \to 0$ yields a system of linear differential equations in the $p_i(t)$

$$p_i'(t) = -\lambda[p_i(t)-p_{i-1}(t)];\ \ i=1,2,... \tag{3.19}$$

Starting with $p_0(t)=e^{-\lambda t}$, the solution (3.18) is obtained by induction. ■

The practical importance of theorem 3.1 is that the properties a), b) and c) can be verified without any quantitative investigations, only by qualitative reasoning based on the physical or other nature of the process. In particular, the simplicity of the homogeneous Poisson process implies that the occurrence of more than one event at the same time has probability 0.

Note Throughout this chapter, those events, which are counted by a Poisson process $\{N(t),\, t \geq 0\}$, will be called *Poisson events.*

Let $\{T_1, T_2, ...\}$ be the point process, which belongs to the homogeneous Poisson process $\{N(t),\, t \geq 0\}$, i.e. T_n is the random time point at which the nth Poisson event occurs. The obvious relationship

$$T_n \leq t \text{ if and only if } N(t) \geq n$$

implies

$$P(T_n \leq t) = P(N(t) \geq n)\,. \tag{3.20}$$

Therefore, T_n has distribution function

$$F_{T_n}(t) = P(N(t) \geq n) = \sum_{i=n}^{\infty} \frac{(\lambda t)^i}{i!}\, e^{-\lambda t};\quad n=1,2,... \tag{3.21}$$

Differentiation of $F_{T_n}(t)$ with respect to t yields the density of T_n:

$$f_{T_n}(t) = \lambda\, e^{-\lambda t} \sum_{i=n}^{\infty} \frac{(\lambda t)^{i-1}}{(i-1)!} - \lambda e^{-\lambda t} \sum_{i=n}^{\infty} \frac{(\lambda t)^i}{i!}\,.$$

On the right-hand side of this equation, all terms but one cancel:

$$f_{T_n}(t) = \lambda \frac{(\lambda t)^{n-1}}{(n-1)!} e^{-\lambda t};\ \ t \geq 0,\ \ n=1,2,... \tag{3.22}$$

Thus, T_n has an Erlang distribution with parameters n and λ. In particular, T_1 has an exponential distribution with parameter λ and the interevent times

$$Y_i = T_i - T_{i-1};\ \ i=1,2,...;\ k=1,2,...;\ T_0 = 0.$$

are independent and identically distributed as T_1 (see example 1.23). Moreover,

$$T_n = \Sigma_{i=1}^{n}\, Y_i.$$

These results yield the most simple and, at the same time, the most important characterization of the homogeneous Poisson process:

Theorem 3.2 Let $\{N(t), t \geq 0\}$ be a counting process and $\{Y_1, Y_2, ...\}$ be the corresponding sequence of interarrival times. Then $\{N(t), t \geq 0\}$ is a homogeneous Poisson process with intensity λ if and only if the $Y_1, Y_2, ...$ are independent, exponentially with parameter λ distributed random variables. ■

The counting process $\{N(t), t \geq 0\}$ is statistically equivalent to both its corresponding point process $\{T_1, T_2, ...\}$ of event times and the sequence of interarrival times $\{Y_1, Y_2, ...\}$. Hence, $\{T_1, T_2, ...\}$ and $\{Y_1, Y_2, ...\}$ are sometimes also called Poisson processes.

Example 3.2 From previous observations it is known that the number of traffic accidents $N(t)$ in an area over the time interval $[0, t)$ can be described by a homogeneous Poisson process $\{N(t), t \geq 0\}$. On an average, there is one accident within 4 hours, i.e. the intensity of the process is

$$\lambda = 0.25\,[h^{-1}].$$

(1) What is the probability p of the event (time unit: hour)

"at most one accident in [0, 10), at least two accidents in [10, 16), and no accident in [16, 24)"?

This probability is

$$p = P(N(10) - N(0) \leq 1,\ N(16) - N(10) \geq 2,\ N(24) - N(16) = 0).$$

In view of the independence and the homogeneity of the increments of $\{N(t), t \geq 0\}$, p can be determined as follows:

$$\begin{aligned} p &= P(N(10) - N(0) \leq 1)\,P(N(16) - N(10) \geq 2)\,P(N(24) - N(16) = 0) \\ &= P(N(10) \leq 1)\,P(N(6) \geq 2)\,P(N(8) = 0)\,. \end{aligned}$$

Now,

$$\begin{aligned} P(N(10) \leq 1) &= P(N(10) = 0) + P(N(10) = 1) \\ &= e^{-0.25\cdot 10} + 0.25 \cdot 10 \cdot e^{-0.25\cdot 10} = 0.2873, \\ P(N(6) \geq 2) &= 1 - e^{-0.25\cdot 6} - 0.25 \cdot 6 \cdot e^{0.25\cdot 6} = 0.4422, \\ P(N(8) = 0) &= e^{-0.25\cdot 8} = 0.1353. \end{aligned}$$

Hence, the desired probability is $p = 0.0172$.

(2) What is the probability that the 2 nd accident occurs not before 5 hours?

Since T_2, the random time of the occurrence of the second accident, has an Erlang distribution with parameters $n = 2$ and $\lambda = 0.25$,

$$P(T_2 > 5) = 1 - F_{T_2}(5) = e^{-0.25\cdot 5}(1 + 0.25 \cdot 5).$$

Thus, $P(T_2 > 5) = 0.6446$. □

The following examples make use of the hyperbolic sine and cosine functions:

$$\sinh x = \frac{e^x - e^{-x}}{2}, \quad \cosh x = \frac{e^x + e^{-x}}{2}, \quad x \in (-\infty, +\infty).$$

Example 3.3 (*random telegraph signal*) A random signal $X(t)$ have structure

$$X(t) = Y(-1)^{N(t)}, \; t \geq 0, \tag{3.23}$$

where $\{N(t), \; t \geq 0\}$ is a homogeneous Poisson process with intensity λ and Y is a binary random variable with

$$P(Y = 1) = P(Y = -1) = 1/2,$$

which is independent of $N(t)$ for all t. Signals of this structure are called *random telegraph signals.* Random telegraph signals are basic modules for generating signals with a more complicated structure. Obviously, $X(t) = 1$ or $X(t) = -1$ and Y determines the sign of $X(0)$. Figure 3.2 shows a sample path $x = x(t)$ of the stochastic process $\{X(t), \; t \geq 0\}$ on condition $Y = 1$ and $T_n = t_n; \; n = 1, 2, ...$

$\{X(t), \; t \geq 0\}$ is wide-sense stationary. To see this, firstly note that

$$|X(t)|^2 = 1 < \infty \quad \text{for all } t \geq 0.$$

Hence, $\{X(t), t \geq 0\}$ is a second-order process. With

$$I(t) = (-1)^{N(t)},$$

its trend function is $m(t) = E(X(t)) = E(Y)\,E(I(t))$. Hence, since $E(Y) = 0$,

$$m(t) \equiv 0.$$

It remains to show that the covariance function $C(s, t)$ of this process depends only on $|t - s|$. This requires knowledge of the probability distribution of $I(t)$: A transition from $I(t) = -1$ to $I(t) = +1$ or, conversely, from $I(t) = +1$ to $I(t) = -1$ occurs at those time points, at which Poisson events occur, i.e. when $N(t)$ jumps:

$$P(I(t) = 1) = P(\text{even number of jumps in } [0, t])$$

$$= e^{-\lambda t} \sum_{i=0}^{\infty} \frac{(\lambda t)^{2i}}{(2i)!} = e^{-\lambda t} \cosh \lambda t,$$

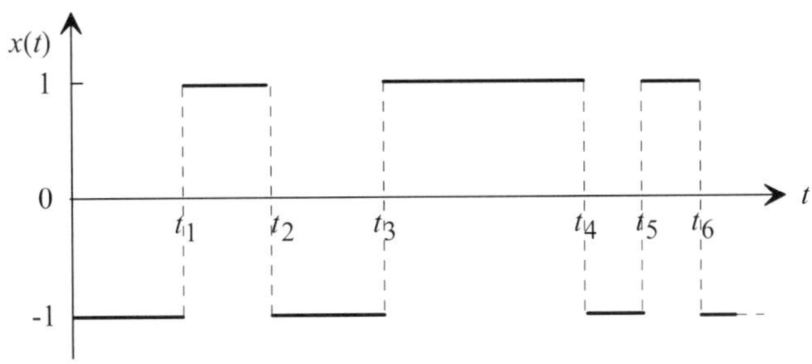

Figure 3.2 Sample path of the random telegraph signal

Analogously,

$$P(I(t)=-1)=P(\text{odd number of jumps in }[0,t])$$

$$=e^{-\lambda t}\sum_{i=0}^{\infty}\frac{(\lambda t)^{2i+1}}{(2i+1)!}=e^{-\lambda t}\sinh\lambda t\,.$$

Hence the mean value of $I(t)$ is

$$E[I(t)]=1\cdot P(I(t)=1)+(-1)\cdot P(I(t)=-1)$$

$$=e^{-\lambda t}[\cosh\lambda t-\sinh\lambda t]=e^{-2\lambda t}.$$

Since

$$C(s,t)=Cov\,[X(s),X(t)]$$

$$=E[(X(s)\,X(t))]=E[Y\,I(s)\,Y\,I(t)]$$

$$=E[Y^2\,I(s)\,I(t)]=E(Y^2)\,E[I(s)\,I(t)]$$

and $E(Y^2)=1$, the covariance function of $\{X(t),\ t\geq 0\}$ has structure

$$C(s,t)=E[I(s)\,I(t)]\,.$$

Thus, in order to evaluate $C(s,t)$, the joint distribution of $(I(s),I(t))$ has to be determined: From (1.6) and the homogeneity of the increments of $\{N(t),\,t\geq 0\}$, assuming $s<t$,

$$p_{1,1}=P(I(s)=1,\,I(t)=1)=P(I(s)=1)P(\,I(t)=1\,|\,I(s)=1)$$

$$=e^{-\lambda s}\cosh\lambda s\,P(\text{even number of jumps in }(s,t])$$

$$=e^{-\lambda s}\cosh\lambda s\,e^{-\lambda(t-s)}\cosh\lambda(t-s)$$

$$=e^{-\lambda t}\cosh\lambda s\ \cosh\lambda(t-s)\,.$$

Analogously,

$$p_{1,-1}=P(I(s)=1,I(t)=-1)\quad=e^{-\lambda t}\cosh\lambda s\ \sinh\lambda(t-s)\,,$$

$$p_{-1,1}=P(I(s)=-1,I(t)=1)\quad=e^{-\lambda t}\sinh\lambda s\ \sinh\lambda(t-s)\,,$$

$$p_{-1,-1}=P(I(s)=-1,I(t)=-1)=e^{-\lambda t}\sinh\lambda s\ \cosh\lambda(t-s)\,.$$

Now

$$E[I(s)I(t)]=p_{1,1}+p_{-1,-1}-p_{1,-1}-p_{-1,1}\,,$$

so that

$$C(s,t)=e^{-2\lambda(t-s)},\quad s<t.$$

Since the order of s and t can be changed,

$$C(s,t)=e^{-2\lambda|t-s|}\,.$$

Hence, the random telegraph signal $\{X(t),\,t\geq 0\}$ is a weakly stationary process. □

Theorem 3.3 Let $\{N(t), t \geq 0\}$ be a homogeneous Poisson process with intensity λ. Then the random number of Poisson events which occur in the interval $[0, s]$ on condition that exactly n events occur in $[0, t]$, $s < t$; $i = 0, 1, ..., n$; has a binomial distribution with parameters $p = s/t$ and n.

Proof In view of the homogeneity and independence of the increments of the Poisson process $\{N(t), t \geq 0\}$,

$$P(N(s) = i | N(t) = n) = \frac{P(N(s) = i, N(t) = n)}{P(N(t) = n)}$$

$$= \frac{P(N(s) = i, N(s, t) = n - i)}{P(N(t) = n)}$$

$$= \frac{P(N(s) = i)\, P(N(s, t) = n - i)}{P(N(t) = n)} = \frac{\frac{(\lambda s)^i}{i!} e^{-\lambda s} \frac{[\lambda(t-s)]^{n-i}}{(n-i)!} e^{-\lambda(t-s)}}{\frac{(\lambda s)^n}{n!} e^{-\lambda t}}$$

$$= \binom{n}{i} \left(\frac{s}{t}\right)^i \left(1 - \frac{s}{t}\right)^{n-i}; \quad i = 0, 1, ..., n. \tag{3.24}$$

This proves the theorem. ■

3.2.1.2 Homogeneous Poisson Process and Uniform Distribution

Theorem 3.3 implies that on condition '$N(t) = 1$' the random time T_1 to the first and only event occuring in $[0, t]$ is uniformly distributed over this interval, since, from (3.24), for $s < t$,

$$P(T_1 \leq s \,|\, T_1 \leq t) = P(N(s) = 1 | N(t) = 1) = \frac{s}{t}.$$

This relationship between the homogeneous Poisson process and the uniform distribution is a special case of a more general result. To prove it, the joint probability density of the random vector $(T_1, T_2, ..., T_n)$ is needed.

Theorem 3.4 The joint probability density of the random vector $(T_1, T_2, ..., T_n)$ is

$$f(t_1, t_2, ..., t_n) = \begin{cases} \lambda^n e^{-\lambda t_n} & \text{for } 0 \leq t_1 < t_2 < \cdots < t_n \\ 0 & \text{elsewhere} \end{cases}. \tag{3.25}$$

Proof For $0 \leq t_1 < t_2$, the joint distribution function of (T_1, T_2) is given by

$$P(T_1 \leq t_1, T_2 \leq t_2) = \int_0^{t_1} P(T_2 \leq t_2 | T_1 = t)\, f_{T_1}(t)\, dt.$$

According to theorem 3.2, the interarrival times

$$Y_i = T_i - T_{i-1}; \; i = 1, 2, ...,$$

are independent, identically distributed random variables which have an exponential distribution with parameter λ.

Hence, since $T_1 = Y_1$,

$$P(T_1 \le t_1, T_2 \le t_2) = \int_0^{t_1} P(T_2 \le t_2 | T_1 = t)\, \lambda e^{-\lambda t} dt.$$

Given '$T_1 = t$', the random event

$$'T_2 \le t_2' \text{ and } 'Y_2 \le t_2 - t'$$

are equivalent. Thus, the desired two-dimensional distribution function is

$$F(t_1, t_2) = P(T_1 \le t_1, T_2 \le t_2) = \int_0^{t_1} (1 - e^{-\lambda(t_2 - t)})\, \lambda e^{-\lambda t} dt$$

$$= \lambda \int_0^{t_1} (e^{-\lambda t} - e^{-\lambda t_2})\, dt.$$

Hence,

$$F(t_1, t_2) = 1 - e^{-\lambda t_1} - \lambda t_1 e^{-\lambda t_2}, \quad t_1 < t_2.$$

Partial differentiation yields the corresponding two-dimensional probability density

$$f(t_1, t_2) = \begin{cases} \lambda^2 e^{-\lambda t_2} & \text{for } 0 \le t_1 < t_2 \\ 0 & \text{elsewhere} \end{cases}.$$

The proof of the theorem is now easily completed by induction. ■

The formulation of the following theorem requires a result from the theory of ordered samples: Let $\{X_1, X_2, ..., X_n\}$ be a random sample taken from X, i.e. the X_i are independent, identically as X distributed random variables. The corresponding ordered sample is denoted as

$$(X_1^*, X_2^*, \cdots, X_n^*), \quad 0 \le X_1^* \le X_2^* \le \cdots \le X_n^*.$$

Given that X has a uniform distribution over $[0, x]$, the joint probability density of the random vector $\{X_1^*, X_2^*, ..., X_n^*\}$ is

$$f^*(x_1^*, x_2^*, ..., x_n^*) = \begin{cases} n!/x^n, & 0 \le x_1^* < x_2^* < \cdots < x_n^* \le x, \\ 0\,, & \text{elsewhere} \end{cases}. \tag{3.26}$$

For the sake of comparison: The joint probability density of the original (unordered) sample $\{X_1, X_2, ..., X_n\}$ is

$$f(x_1, x_2, ..., x_n) = \begin{cases} 1/x^n, & 0 \le x_i \le x \\ 0\quad, & \text{elsewhere} \end{cases}. \tag{3.27}$$

Theorem 3.5 Let $\{N(t), t \ge 0\}$ be a homogeneous Poisson process with intensity λ, and let T_i be ith event time; $i = 1, 2, ...$; $T_0 = 0$. Given $N(t) = n$; $n = 1, 2, ...,$ the random vector $\{T_1, T_2, ..., T_n\}$ has the same joint probability density as an ordered random sample taken from a uniform distribution over $[0, t]$.

Proof By definition, for disjoint, but otherwise arbitrary subintervals $[t_i, t_i + h_i]$ of $[0, t]$, the joint probability density of $\{T_1, T_2, ..., T_n\}$ on condition $N(t) = n$ is

$$f(t_1, t_2, ..., t_n | N(t) = n)$$

$$= \lim_{\max(h_1, h_2, ..., h_n) \to 0} \frac{P(t_i \le T_i < t_i + h_i;\ i = 1, 2, ..., n | N(t) = n)}{h_1 h_2 \cdots h_n}.$$

Since the event '$N(t) = n$' is equivalent to $T_n \le t < T_{n+1}$,

$$P(t_i \le T_i < t_i + h_i;\ i = 1, 2, ..., n | N(t) = n)$$

$$= \frac{P(t_i \le T_i < t_i + h_i,\ i = 1, 2, ..., n;\ t < T_{n+1})}{P(N(t) = n)}$$

$$= \frac{\int_t^\infty \int_{t_n}^{t_n + h_n} \int_{t_{n-1}}^{t_{n-1} + h_{n-1}} \cdots \int_{t_1}^{t_1 + h_1} \lambda^{n+1} e^{-\lambda x_{n+1}} dx_1 \cdots dx_n \, dx_{n+1}}{\frac{(\lambda t)^n}{n!} e^{-\lambda t}}$$

$$= \frac{h_1 h_2 \cdots h_n \lambda^n e^{-\lambda t}}{\frac{(\lambda t)^n}{n!} e^{-\lambda t}} = \frac{h_1 h_2 \cdots h_n}{t^n} n!.$$

Hence, the desired conditional joint probability density is

$$f(t_1, t_2, ..., t_n | N(t) = n) = \begin{cases} n!/t^n, & 0 \le t_1 < t_2 < \cdots < t_n \le t \\ 0, & \text{elsewhere} \end{cases}. \tag{3.28}$$

Apart from the notation of the variables, this is the joint density (3.26). ■

The relationship between homogeneous Poisson processes and the uniform distribution proved in this theorem motivates the common phrase that a homogeneous Poisson process is a *purely random process*, since, given $N(t) = n$, the event times $T_1, T_2, ..., T_n$ are 'purely randomly' distributed over $[0, t]$.

Example 3.4 (*shot noise*) Shot noise processes have been formally introduced in example 2.5. Now an application is discussed in detail: In the circuit, depicted in Figure 3.3, a light source is switched on at time $t = 0$. A current pulse is initiated in the circuit as soon as the cathode emits a photoelectron due to the light falling on it. Such a current pulse can be quantified by a function $h(t)$ with properties

$$h(t) \ge 0, \quad h(t) = 0 \text{ for } t < 0 \text{ and } \int_0^\infty h(t)\, dt < \infty. \tag{3.29}$$

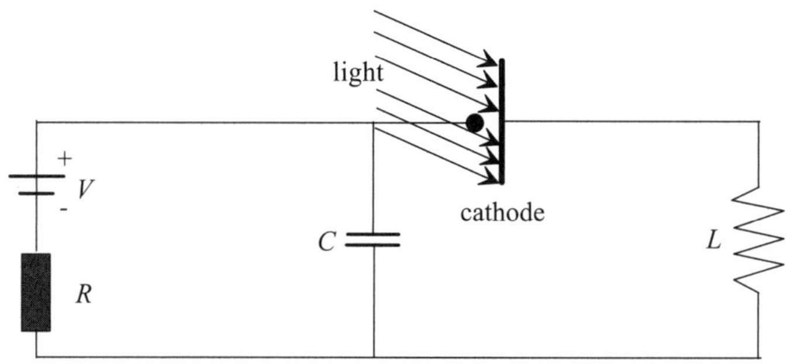

Figure 3.3 Photodetection circuit (Example 3.4)

Let $T_1, T_2, \ldots$ be the sequence of random time points, at which the cathode emits photoelectrons and $\{N(t), t \geq 0\}$ be the corresponding counting process. Then the total current flowing in the circuit at time t is

$$X(t) = \sum_{i=1}^{\infty} h(t - T_i). \tag{3.30}$$

In view of (3.29), $X(t)$ can also be written in the form

$$X(t) = \sum_{i=1}^{N(t)} h(t - T_i).$$

In what follows, $\{N(t), t \geq 0\}$ is assumed to be a homogeneous Poisson process with parameter λ. For determining the trend function of the shot noise $\{X(t), t \geq 0\}$, note that according to theorem 3.5, on condition '$N(t) = n$', the $T_1, T_2, \ldots, T_n$ are uniformly distributed over $[0, t]$. Hence,

$$E(h(t - T_i)|N(t) = n) = \frac{1}{t}\int_0^t h(t - x)\,dx = \frac{1}{t}\int_0^t h(x)\,dx.$$

Therefore,

$$\begin{aligned} E(X(t)|N(t) = n) &= E\left(\sum_{i=1}^{n} h(t - T_i)\,\middle|\,N(t) = n\right) \\ &= \sum_{i=1}^{n} E(h(t - T_i)|N(t) = n) \\ &= \left(\frac{1}{t}\int_0^t h(x)\,dx\right) n. \end{aligned}$$

The total probability rule (1.7) yields

$$\begin{aligned} E(X(t)) &= \sum_{n=0}^{\infty} E(X(t)|N(t) = n)\, P(N(t) = n) \\ &= \frac{1}{t}\int_0^t h(x)\,dx \sum_{n=0}^{\infty} n\,\frac{(\lambda t)^n}{n!}\,e^{-\lambda t} \\ &= \left(\frac{1}{t}\int_0^t h(x)\,dx\right) E(N(t)) = \left(\frac{1}{t}\int_0^t h(x)\,dx\right)(\lambda t). \end{aligned}$$

Therefore, the trend function of the shot noise process is

$$m(t) = \lambda \int_0^t h(x)\,dx. \tag{3.31}$$

In order to obtain the covariance variance function, the mean value of the product $X(s)\,X(t)$ has to be determined:

$$\begin{aligned} E(X(s)X(t)) &= \Sigma_{i,j=1}^{\infty} E[h(s-T_i)\,h(t-T_j)] \\ &= \Sigma_{i=1}^{\infty} E(h(s-T_i)\,h(t-T_i)) \\ &+ \sum_{i,j=1,\,i\neq j}^{\infty} E\Big[h(s-T_i)\,h(t-T_j)\Big]. \end{aligned}$$

Since, on condition '$N(t)=n$', the $T_1, T_2, ..., T_n$ are uniformly distributed over $[0, t]$,

$$E(h(s-T_i)\,h(t-T_i)|N(t)=n) = \frac{1}{t}\int_0^t h(s-y)h(t-y)\,dy.$$

Thus, for $s < t$, substituting $x = s - y$,

$$E(h(s-T_i)\,h(t-T_i)|N(t)=n) = \frac{1}{t}\int_0^s h(x)\,h(t-s+x)\,dx.$$

Moreover, by theorem 3.5, on condition '$N(t)=n$' the $T_1, T_2, ..., T_n$ are independent. Hence,

$$\begin{aligned} E(h(s-T_i)\,h(t-T_j)|N(t)=n) &= E(h(s-T_i)|N(t)=n)\,E(h(t-T_j)|N(t)=n) \\ &= \left(\frac{1}{t}\int_0^s h(s-x)\,dx\right)\left(\frac{1}{t}\int_0^t h(t-x)\,dx\right) \\ &= \left(\frac{1}{t}\int_0^s h(x)\,dx\right)\left(\frac{1}{t}\int_0^t h(x)\,dx\right). \end{aligned}$$

Thus, for $s < t$,

$$\begin{aligned} E(X(s)\,X(t)|N(t)=n) &= \left(\frac{1}{t}\int_0^s h(x)h(t-s+x)\,dx\right)n \\ &+ \left(\frac{1}{t}\int_0^s h(x)\,dx\right)\left(\frac{1}{t}\int_0^t h(x)\,dx\right)(n-1)\,n. \end{aligned}$$

Applying once more the total probability rule,

$$\begin{aligned} E(X(s)\,X(t)) &= \left(\frac{1}{t}\int_0^s h(x)h(t-s+x)\,dx\right)E(N(t)) \\ &+ \left(\frac{1}{t}\int_0^s h(x)\,dx\right)\left(\frac{1}{t}\int_0^t h(x)\,dx\right)\Big[E(N^2(t)) - E(N(t))\Big]. \end{aligned}$$

In view of

$$E(N(t)) = \lambda t \quad \text{and} \quad E(N^2(t)) = \lambda t\,(\lambda t + 1),$$

making use of (3.31) and (2.4) yields the covariance function:

$$C(s,t) = \lambda \int_0^s h(x)h(t-s+x)\,dx, \quad s < t.$$

More generally, for any s and t, $C(s,t)$ can be written in the form

$$C(s,t) = \lambda \int_0^{\min(s,t)} h(x)\, h(|t-s|+x)\, dx.$$

Letting $s = t$ yields the variance of $X(t)$:

$$Var(X(t)) = \lambda \int_0^t h^2(x)\, dx.$$

By letting $s \to \infty$, keeping $|\tau| = t - s$ constant, trend- and covariance function become

$$m = \lambda \int_0^\infty h(x)\, dx, \tag{3.32}$$

$$C(\tau) = \lambda \int_0^\infty h(x)\, h(|\tau|+x)\, dx. \tag{3.33}$$

These two formulas are known as *Campell's theorem.* They imply that, for large t, the shot noise process $\{X(t),\, t \ge 0\}$ is approximately weakly stationary. (For another proof of Campell's theorem see exercise 3.7, and for more general formulations of this theorem see, for instance, Brandt, Franken, and Lisek [13], Stigman [78].)

If the current impulses induced by photoelectrons have random intensities A_i, then the total current flowing in the circuit at time t is

$$X(t) = \sum_{i=1}^{N(t)} A_i\, h(t - T_i).$$

Provided the A_i are identically distributed as A, independent of each other, and independent of all T_k, then determining trend- and covariance function of the generalized shot noise $\{X(t),\, t \ge 0\}$ does not give rise to principally new problems. Provided the first two moments of A exist, one obtains

$$m(t) = \lambda E(A) \int_0^t h(x)\, dx, \tag{3.34}$$

$$C(s,t) = \lambda E(A^2) \int_0^{\min(s,t)} h(x)\, h(|t-s|+x)\, dx. \tag{3.35}$$

If the process of inducing current impulses by photoelectrons has already been operating for an unboundedly long time (the circuit was switched on a sufficiently long time ago), then the underlying shot noise process $\{X(t),\, t \in (-\infty, +\infty)\}$ is given by

$$X(t) = \sum_{-\infty}^{+\infty} A_i\, h(t - T_i).$$

In this case the process is a priori stationary. □

Example 3.5 Customers arrive at a service station (service system, queueing system) according to a homogeneous Poisson process $\{N(t),\, t \ge 0\}$ with intensity λ. Hence, the arrival of a customer is a Poisson event. The number of servers in the system is assumed to be so large that an incoming customer will always find an available server. To cope with this situation, the service system must be modeled as having an infinite number of servers. The service times of all customers are assumed to be independent random variables, which are identically distributed as Z.

Let $G(t) = P(Z \le t)$ be the distribution function of Z, and $X(t)$ be the random number of customers in the system at time t, $X(0) = 0$. The aim is to determine the *state probabilities* $p_i(t)$ of the system:

$$p_i(t) = P(X(t) = i); \quad i = 0, 1, ...; \ t \ge 0.$$

A customer arriving at time x is still in the system at time t, $t > x$, with probability $1 - G(t - x)$, i.e. its service has not yet been finished by t. Given $N(t) = n$, the arrival times $T_1, T_2, ..., T_n$ of the n customers in the system are, by theorem 3.4, independent and uniformly distributed over $[0, t]$. For calculating the state probabilities, the order of the T_i is not relevant. Thus, the probability that any of the n customers who arrived in $[0, t]$ is still in the system at time t, is

$$p(t) = \int_0^t (1 - G(t-x)) \frac{1}{t} dx = \frac{1}{t} \int_0^t (1 - G(x)) \, dx .$$

Since, by assumption, the service times are independent of each other,

$$P(X(t) = i \,|\, N(t) = n) = \binom{n}{i} [p(t)]^i [1 - p(t)]^{n-i}; \ i = 0, 1, ..., n .$$

By the total probability rule (1.7),

$$p_i(t) = \sum_{n=i}^{\infty} P(X(t) = i | N(t) = n) \cdot P(N(t) = n)$$

$$= \sum_{n=i}^{\infty} \binom{n}{i} [p(t)]^i [1 - p(t)]^{n-i} \cdot \frac{(\lambda t)^n}{n!} e^{-\lambda t}.$$

This is a mixture of binomial distributions with regard to a Poisson structure distribution. Thus, from example 1.8 (there the parameter λ has to be replaced with λt), the state probabilities of the system are

$$p_i(t) = \frac{[\lambda t p(t)]^i}{i!} \cdot e^{-\lambda t p(t)}; \quad i = 0, 1, ...$$

Hence, $X(t)$ has a Poisson distribution with parameter

$$E(X(t)) = \lambda t p(t).$$

Consequently, the trend function of the stochastic process $\{X(t), t \ge 0\}$ is

$$m(t) = \lambda \int_0^t (1 - G(x)) \, dx , \ t \ge 0.$$

For $t \to \infty$ the trend function tends to

$$\lim_{t \to \infty} m(t) = \frac{E(Z)}{E(Y)}, \tag{3.36}$$

where $E(Y) = 1/\lambda$ is the mean interarrival time and $E(Z)$ the mean service time of a customer:

$$E(Z) = \int_0^\infty (1 - G(x)) \, dx .$$

By letting

$$\rho = E(Z)/E(Y),$$

the *stationary state probabilities* of the system become

$$p_i = \lim_{t\to\infty} p_i(t) = \frac{\rho^i}{i!} e^{-\rho}; \quad i = 0, 1, ... \tag{3.37}$$

If Z has an exponential distribution with parameter μ, then

$$m(t) = \lambda \int_0^t e^{-\mu x}\, dx = \frac{\lambda}{\mu}\left(1 - e^{-\mu t}\right).$$

In this case, $\rho = \lambda/\mu$. □

3.2.2 Nonhomogeneous Poisson Processses

In this section a stochastic process is investigated, which, except for the homogeneity of its increments, has all the other properties listed in theorem 3.1. Abandoning the assumption of homogeneous increments implies that a time-dependent intensity function $\lambda = \lambda(t)$ takes over the role of λ. This leads to the concept of a nonhomogeneous Poisson process. As in section 3.1, the following notation will be used:

$$N(s,t) = N(t) - N(s), \quad 0 \le s < t,$$

Definition 3.3 A counting process $\{N(t), t \ge 0\}$ satisfying $N(0) = 0$ is called a *nonhomogeneous Poisson process* with *intensity function* $\lambda(t)$ if it has properties

(1) $\{N(t), t \ge 0\}$ has independent increments,

(2) $P(N(t, t+h) \ge 2) = o(h)$,

(3) $P(N(t, t+h) = 1) = \lambda(t)\, h + o(h)$. ●

Three problems will be considered:

1) Computation of the probability distribution of increments $N(s,t)$:

$$p_i(s,t) = P(N(s,t) = i); \quad 0 \le s < t, \quad i = 0, 1, ...$$

2) Computation of the probability density of the random event time T_i (time point at which the ith Poisson event occurs).

3) Computation of the joint probability density of $(T_1, T_2, ..., T_n)$; $n = 1, 2, ...$

1) In view of the assumed independence of the increments, for $h > 0$,

$$\begin{aligned} p_0(s, t+h) &= P(N(s, t+h) = 0) \\ &= P(N(s,t) = 0,\ N(t, t+h) = 0) \\ &= P(N(s,t) = 0) \cdot P(N(t, t+h) = 0) \\ &= p_0(s,t)\,[1 - \lambda(t)\, h + o(h)]. \end{aligned}$$

Thus,

$$\frac{p_0(s,t+h)-p_0(s,t)}{h}=-\lambda(t)\,p_0(s,t)+\frac{o(h)}{h}.$$

Letting $h \to 0$ yields a partial differential equation of the first order:

$$\frac{\partial}{\partial t}p_0(s,t)=-\lambda(t)\,p_0(s,t).$$

Since $N(0)=0$ or, equivalently, $p_0(0,0)=1$, the solution is

$$p_0(s,t)=e^{-[\Lambda(t)-\Lambda(s)]}, \tag{3.38}$$

where

$$\Lambda(x)=\int_0^x \lambda(u)\,du\,;\;\; x\geq 0. \tag{3.39}$$

Starting with $p_0(s,t)$, the probabilities $p_i(s,t)$ for $i\geq 1$ can be determined by induction:

$$p_i(s,t)=\frac{[\Lambda(t)-\Lambda(s)]^i}{i!}\,e^{-[\Lambda(t)-\Lambda(s)]};\quad i=0,1,2,\ldots \tag{3.40}$$

In particular, the absolute state probabilities

$$p_i(t)=p_i(0,t)=P(N(t)=i)$$

of the nonhomogeneous Poisson process at time t are

$$p_i(t)=\frac{[\Lambda(t)]^i}{i!}\,e^{-\Lambda(t)};\quad i=0,1,2,\ldots \tag{3.41}$$

Hence, the mean number of Poisson events $m(s,t)=E(N(s,t))$ occurring in the interval $[s,t],\ s<t,$ is

$$m(s,t)=\Lambda(t)-\Lambda(s)=\int_s^t \lambda(x)\,dx\,. \tag{3.42}$$

In particular, the trend function of the nonhomogeneous Poisson process is

$$m(t)=\Lambda(t)=\int_0^t \lambda(x)\,dx,\;\; t\geq 0.$$

2) Let $F_{T_1}(t)=P(T_1\leq t)$ be the distribution function and $f_{T_1}(t)$ the probability density of the random time T_1 to the occurrence of the first Poisson event. Then

$$p_0(t)=p_0(0,t)=P(T_1>t)=1-F_{T_1}(t)\,.$$

From (3.38),

$$p_0(t)=e^{-\Lambda(t)}.$$

Hence,

$$F_{T_1}(t)=1-e^{-\int_0^t \lambda(x)\,dx},\quad f_{T_1}(t)=\lambda(t)e^{-\int_0^t \lambda(x)\,dx},\quad t\geq 0. \tag{3.43}$$

A comparison of (3.43) with (1.40) shows that the intensity function $\lambda(t)$ of the nonhomogeneous Poisson process $\{N(t), t \geq 0\}$ is identical to the failure rate belonging to T_1. Since

$$F_{T_n}(t) = P(T_n \leq t) = P(N(t) \geq n), \tag{3.44}$$

the distribution function of the nth event time T_n is

$$F_{T_n}(t) = \sum_{i=n}^{\infty} \frac{[\Lambda(t)]^i}{i!} e^{-\Lambda(t)}, \quad n = 1, 2, ... \tag{3.45}$$

Differentiation with respect to t yields the probability density of T_n:

$$f_{T_n}(t) = \frac{[\Lambda(t)]^{n-1}}{(n-1)!} \lambda(t)\, e^{-\Lambda(t)}; \quad t \geq 0, \quad n = 1, 2, ... \tag{3.46}$$

Equivalently,

$$f_{T_n}(t) = \frac{[\Lambda(t)]^{n-1}}{(n-1)!} f_{T_1}(t); \quad t \geq 0, \quad n = 1, 2, ...$$

By (1.17), the mean value of T_n is

$$E(T_n) = \int_0^{\infty} e^{-\Lambda(t)} \left(\sum_{i=0}^{n-1} \frac{[\Lambda(t)]^i}{i!} \right) dt. \tag{3.47}$$

Hence, the mean time

$$E(Y_n) = E(T_n) - E(T_{n-1})$$

between the $(n-1)$th and the nth event is

$$E(Y_n) = \frac{1}{(n-1)!} \int_0^{\infty} [\Lambda(t)]^{n-1} e^{-\Lambda(t)}\, dt; \quad n = 1, 2, ... \tag{3.48}$$

Letting $\lambda(t) \equiv \lambda$ and $\Lambda(t) \equiv \lambda t$ yields the corresponding characteristics for the homogeneous Poisson process, in particular $E(Y_n) = 1/\lambda$.

3) The conditional probability $P(T_2 \leq t_2 \,|\, T_1 = t_1$, is equal to the probability that at least one Poisson event occurs in $(t_1, t_2]$, $t_1 < t_2$. Thus, from (3.40),

$$F_{T_2}(t_2 | t_1) = 1 - p_0(t_1, t_2) = 1 - e^{-[\Lambda(t_2) - \Lambda(t_1)]}.$$

Differentiation with respect to t_2 yields the corresponding probability density:

$$f_{T_2}(t_2 | t_1) = \lambda(t_2)\, e^{-[\Lambda(t_2) - \Lambda(t_1)]}, \quad 0 \leq t_1 < t_2.$$

By (1.59), the joint probability density of (T_1, T_2) is

$$f(t_1, t_2) = \begin{cases} \lambda(t_1)\, f_{T_1}(t_2) & \text{for } t_1 < t_2 \\ 0, & \text{elsewhere} \end{cases}.$$

Starting with $f(t_1, t_2)$, one inductively obtains the joint density of $(T_1, T_2, ..., T_n)$:

$$f(t_1, t_2, ..., t_n) = \begin{cases} \lambda(t_1)\lambda(t_2)\cdots\lambda(t_{n-1})f_{T_1}(t_n) & \text{for } 0 \le t_1 < t_2 < \cdots < t_n \\ 0, & \text{elsewhere} \end{cases} \tag{3.49}$$

This result includes as a special case formula (3.25).

As with the homogeneous Poisson process, the nonhomogeneous Poisson counting process $\{N(t),\, t \ge 0\}$, the corresponding point process of Poisson event times $\{T_1, T_2, ...\}$ and the sequence of interevent times $\{Y_1, Y_2, ...\}$ are statistically equivalent stochastic processes.

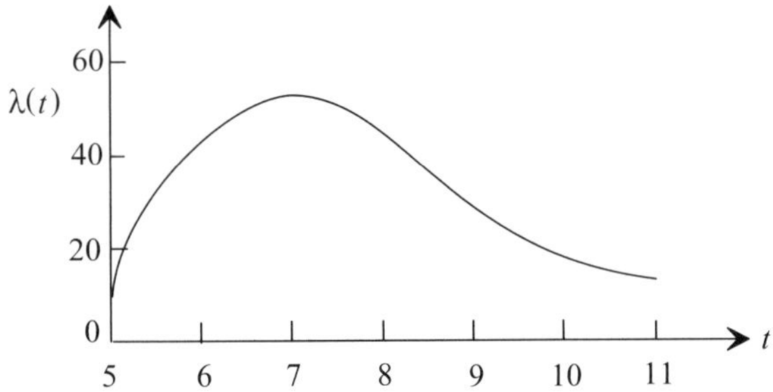

Figure 3.4 Intensity of the arrival of cars at a filling station

Example 3.6 From historical observations it is known that the number of cars arriving for petrol at a particular filling station weekdays between 5:00 and 11:00 a.m. can be modeled by an nonhomogeneous Poisson process $\{N(t), t \ge 0\}$ with intensity function (Figure 3.4)

$$\lambda(t) = 10 + 35.4\,(t-5)\,e^{-(t-5)^2/8}, \quad 5 \le t \le 11.$$

1) What is the mean number of cars arriving for petrol weekdays between 5:00 and 11:00? According to (3.42), this mean number is

$$E(N(5,11)) = \int_5^{11} \lambda(t)\,dt = \int_0^6 \left(10 + 35.4\,t\,e^{-t^2/8}\right) dt$$

$$= \left[10\,t - 141.6\,e^{-t^2/8}\right]_0^6 = 200.$$

2) What is the probability that at least 90 cars arrive for petrol weekdays between 6:00 and 8:00? The mean number of cars arriving between 6:00 and 8:00 is

$$\int_6^8 \lambda(t)\,dt = \int_1^3 (10 + 35.4\,t\,e^{-t^2/8})\,dt$$

$$= \left[10\,t - 141.6\,e^{-t^2/8}\right]_1^3 = 99.$$

Hence, the random number of cars $N(6,8) = N(8) - N(6)$ arriving between 6:00 and 8:00 has a Poisson distribution with parameter 99. Thus, desired probability is

$$P(N(6,8) \geq 90) = \sum_{n=90}^{\infty} \frac{99^n}{n!} e^{-0.99}.$$

By using the normal approximation to the Poisson distribution (section 1.9.3):

$$\sum_{n=90}^{\infty} \frac{99^n}{n!} e^{-0.99} \approx 1 - \Phi\left(\frac{90-99}{\sqrt{99}}\right) \approx 1 - 0.1827.$$

Hence,

$$P(N(6,8) \geq 90) = 0.8173.$$

□

3.2.3 Mixed Poisson Processes

Mixed Poisson processes were already introduced by Dubourdieu [24] for modeling claim number processes in accident and sickness insurance. In view of their flexibility, they are now a favourite point process model for many other applications. A recent monograph on mixed Poisson processes is Grandell [35].

Let $\{N(t), t \geq 0\}$ be a homogeneous Poisson process with parameter λ. To explicitely express the dependence of this process on λ, in this section the notation $\{N_\lambda(t), t \geq 0\}$ for the process $\{N(t), t \geq 0\}$ is adopted. The basic idea of Dubourdieu was to consider λ a realization of a positive random variable L, which is called the (random) *structure* or *mixing parameter*. Correspondingly, the probability distribution of L is called the *structure* or *mixing distribution* (see section 1.2.4).

Definition 3.4 Let L be a positive random variable with range $\mathbf{R}_L$. Then the counting process $\{N_L(t), t \geq 0\}$ is said to be a *mixed Poisson process* with structure parameter L if it has the following properties:

(1) $\{N_{L|L=\lambda}(t), t \geq 0\}$ has independent, homogeneous increments for all $\lambda \in \mathbf{R}_L$.

(2) $P\left(N_{L|L=\lambda}(t) = i\right) = \frac{(\lambda t)^i}{i!} e^{-\lambda t}$ for all $\lambda \in \mathbf{R}_L$, $i = 0, 1, ...$ •

Thus, on condition $L = \lambda$, the mixed Poisson process is a homogeneous Poisson process with parameter λ:

$$\{N_{L|L=\lambda}(t), t \geq 0\} = \{N_\lambda(t), t \geq 0\}.$$

The absolute state probabilities $p_i(t) = P(N_L(t) = i)$ of the mixed Poisson process at time t are

$$P\left(N_L(t) = i\right) = E\left(\frac{(L\,t)^i}{i!} e^{-L\,t}\right); \quad i = 0, 1, .. \tag{3.50}$$

If L is a discrete random variable with $P(L=\lambda_k)=\pi_k$; $k=0,1,...$; then

$$P\Big(N_L(t)=i\Big) = \sum_{k=0}^{\infty}\frac{(\lambda_k t)^i}{i!}e^{-\lambda_k t}\pi_k . \tag{3.51}$$

In applications, a binary structure parameter L is particularly important. In this case,

$$P\Big(N_L(t)=i\Big) = \frac{(\lambda_1 t)^i}{i!}e^{-\lambda_1 t}\pi + \frac{(\lambda_2 t)^i}{i!}e^{-\lambda_2 t}(1-\pi) \tag{3.52}$$

for $0 \le \pi \le 1,\ \lambda_1 \ne \lambda_2$.

The basic results, obtained in what follows, do not depend on the probability distribution of L. Hence, for convenience, throughout this section the assumption is made that L is a continuous random variable with density $f_L(\lambda)$. Then,

$$p_i(t) = \int_0^{\infty}\frac{(\lambda t)^i}{i!}e^{-\lambda t}f_L(\lambda)\,d\lambda;\quad i=0,1,...$$

Obviously, the probability $p_0(t)=P(N_L(t)=0)$ is the Laplace transform of $f_L(\lambda)$ with parameter $s=t$ (section 1.3.2):

$$p_0(t)=\hat{f}_L(t)=E(e^{-L\,t})=\int_0^{\infty}e^{-\lambda t}f_L(\lambda)\,d\lambda .$$

The ith derivative of $p_0(t)$ is

$$\frac{d^i p_0(t)}{d^i t}=p_0^{(i)}(t)=\int_0^{\infty}(-\lambda)^i e^{-\lambda t}f_L(\lambda)\,d\lambda .$$

Therefore, all state probabilities of a mixed Poisson process can be written in terms of $p_0(t)$:

$$p_i(t)=P(N_L(t)=i)=(-1)^i\frac{t^i}{i!}p_0^{(i)}(t);\quad i=1,2,... \tag{3.53}$$

Mean value and variance of $N_L(t)$ are (compare with the parameters of the mixed Poisson distribution given in section 1.2.4)):

$$E(N_L(t))=t\,E(L),\qquad Var(N_L(t))=t\,E(L)+t^2\,Var(L). \tag{3.54}$$

The following theorem lists two important properties of mixed Poisson processes.

Theorem 3.6 (1) A mixed Poisson process $\{N_L(t),\ t\ge 0\}$ has homogeneous increments.

(2) If L is not a constant (i.e. the structure distribution is not *degenerate*), then the increments of the mixed Poisson process $\{N_L(t),\ t\ge 0\}$ are not independent.

Proof (1) Let $0=t_0<t_1<\cdots<t_n$; $n=1,2,...$ Then, for any nonnegative integers $i_1,i_2,...,i_n$,

$$P(N_L(t_{k-1}+\tau,\ t_k+\tau)=i_k;\ k=1,2,...,n)$$

$$=\int_0^\infty P(N_\lambda(t_{k-1}+\tau,\ t_k+\tau)=i_k;\ k=1,2,...,n)f_L(\lambda)\,d\lambda$$

$$=\int_0^\infty P(N_\lambda(t_{k-1},\ t_k)=i_k;\ k=1,2,...,n)f_L(\lambda)d\lambda$$

$$=P(N_L(t_{k-1},\ t_k)=i_k;\ k=1,2,...,n).$$

(2) Let $0\le t_1<t_2<t_3$. Then,

$$P(N_L(t_1,t_2)=i_1,\ N_L(t_2,t_3)=i_2)$$

$$=\int_0^\infty P(N_\lambda(t_1,t_2)=i_1,\ N_\lambda(t_2,t_3)=i_2)f_L(\lambda)d\lambda$$

$$=\int_0^\infty P(N_\lambda(t_1,t_2)=i_1)\ P(N_\lambda(t_2,t_3)=i_2)f_L(\lambda)d\lambda$$

$$\ne\int_0^\infty P(N_\lambda(t_1,t_2)=i_1)f_L(\lambda)d\lambda\ \int_0^\infty P(N_\lambda(t_2,t_3)=i_2)f_L(\lambda)d\lambda$$

$$=P(N_L(t_1,t_2)=i_1)P(N_L(t_2,t_3)=i_2).$$

This proves the theorem if the mixing parameter L is a continuous random variable. If L is discrete, the same pattern applies. ■

Multinomial Criterion Let $0=t_0<t_1<\cdots<t_n;\ \ n=1,2,...$ Then, for any nonnegative integers $i_1,i_2,...,i_n$ with $i=i_1+i_2+\cdots+i_n$,

$$P(N_L(t_{k-1},\ t_k)=i_k;\ k=1,2,...,n\,|\,N_L(t_n)=i)$$

$$=\frac{i!}{i_1!\,i_2!\cdots i_n!}\left(\frac{t_1}{t_n}\right)^{i_1}\left(\frac{t_2-t_1}{t_n}\right)^{i_2}\cdots\left(\frac{t_n-t_{n-1}}{t_n}\right)^{i_n}. \tag{3.55}$$

Interestingly, this conditional probability does not depend on the structure distribution (compare to theorem 3.4). Although the derivation of the multinomial criterion is elementary, it is not done here (exercise 3.15).

As an application of the multinomial criterion (3.55), the joint distibution of the increments $N_L(0,t)=N_L(t)$ and $N_L(t,t+\tau)$ will be derived:

$$P(N_L(t)=i,\ N_L(t,t+\tau)=k)$$

$$=P(N_L(t)=i\,|\,N_L(t+\tau)=i+k)\,P(N_L(t+\tau)=i+k)$$

$$=\frac{(i+k)!}{i!\,k!}\left(\frac{t}{t+\tau}\right)^i\left(\frac{\tau}{t+\tau}\right)^k\int_0^\infty\frac{[\lambda(t+\tau)]^{i+k}}{(i+k)!}e^{-\lambda(t+\tau)}f_L(\lambda)\,d\lambda.$$

Hence, the joint distribution is

$$P(N_L(0,t)=i,\ N_L(t,t+\tau)=k)=\frac{t^i\tau^k}{i!\,k!}\int_0^\infty \lambda^{i+k}e^{-\lambda(t+\tau)}f_L(\lambda)\,d\lambda \tag{3.56}$$

for $i,k=0,1,...$

Since a mixed Poisson process has dependent increments, it is important to get information on the nature and strength of the statistical dependence between two neighbouring increments. As a first step into this direction, the mean value of the product of the increments $N_L(t)=N_L(0,t)$ and $N_L(t,t+\tau)$ has to be determined. From (3.56),

$$\begin{aligned}E([N_L(t)]\,[N_L(t,t+\tau)]) &= \sum_{i=1}^{\infty}\sum_{k=1}^{\infty} ik\frac{t^i\tau^k}{i!\,k!}\int_0^\infty \lambda^{i+k}e^{-\lambda(t+\tau)}f_L(\lambda)\,d\lambda\\ &= t\tau\int_0^\infty \lambda^2\sum_{i=0}^{\infty}\frac{(\lambda t)^i}{i!}\sum_{k=0}^{\infty}\frac{(\lambda\tau)^k}{k!}e^{-\lambda(t+\tau)}f_L(\lambda)\,d\lambda\\ &= t\tau\int_0^\infty \Sigma_{i=0}^{\infty}\lambda^2 e^{\lambda t}e^{\lambda\tau}e^{-\lambda(t+\tau)}f_L(\lambda)\,d\lambda\\ &= t\tau\int_0^\infty \lambda^2 f_L(\lambda)\,d\lambda .\end{aligned}$$

Thus,

$$E([N_L(t)]\,[N_L(t,t+\tau)])=t\tau\,E(L^2). \tag{3.57}$$

Hence, in view of (2.4) and (3.57),

$$Cov(N_L(\tau),N_L(\tau,\tau+t))=t\tau\,Var(L).$$

Thus, two neighbouring increments of a mixed Poisson process are positively correlated. Consequently, a large number of events in an interval will on average induce a large number of events in the following interval ('large' relative to the respective lengths of these intervals). This property of a stochastic process is also called *positive contagion.*

Polya Process A mixed Poisson process with a gamma distributed structure parameter L is called a *Polya process* (or *Polya-Lundberg process*).

Let the gamma density of L be

$$f_L(\lambda)=\frac{\beta^\alpha}{\Gamma(\alpha)}\lambda^{\alpha-1}e^{-\beta\lambda},\quad \lambda>0,\ \alpha>0,\ \beta>0.$$

Then, proceeding as in example 1.9 (section 1.2.4) yields

$$\begin{aligned}P(N_L(t)=i) &= \int_0^\infty \frac{(\lambda t)^i}{i!}e^{-\lambda t}\frac{\beta^\alpha}{\Gamma(\alpha)}\lambda^{\alpha-1}e^{-\beta\lambda}\,d\lambda\\ &= \frac{\Gamma(i+\alpha)}{i!\,\Gamma(\alpha)}\,\frac{t^i\beta^\alpha}{(\beta+t)^{i+\alpha}} .\end{aligned}$$

Hence,

$$P\big(N_L(t) = i\big) = \binom{i-1+\alpha}{i}\left(\frac{t}{\beta+t}\right)^i\left(\frac{\beta}{\beta+t}\right)^\alpha; \quad i = 0, 1, ... \tag{3.58}$$

Thus, the one-dimensional distribution of the Polya process $\{N_L(t), t \geq 0\}$ is a negative binomial distribution with parameters $r = \alpha$ and $p = t/(\beta + t)$. In particular, for an exponential structure distribution $(\alpha = 1)$, $N_L(t)$ has a geometric distribution with parameter $p = t/(t+\beta)$.

To determine the n-dimensional distribution of the Polya process, (3.58) and the multinomial criterion (3.55) are used:

For $0 = t_0 < t_1 < \cdots < t_n;\ n = 1, 2, ...$ and $i_0 = 0$,

$$P(N_L(t_k) = i_k;\ k = 1, 2, ..., n)$$

$$= P(N_L(t_k) = i_k;\ k = 1, 2, ..., n \mid N_L(t_n) = i_n)\, P(N_L(t_n) = i_n)$$

$$= P(N_L(t_{k-1}, t_k) = i_k - i_{k-1};\ k = 1, 2, ..., n \mid N_L(t_n) = i_n)\ P(N_L(t_n) = i_n)$$

$$= \frac{i_n!}{\Pi_{k=1}^n (i_k - i_{k-1})!} \prod_{k=1}^n \left(\frac{t_k - t_{k-1}}{t_n}\right)^{i_k - i_{k-1}} \binom{i_n - 1 + \alpha}{i_n}\left(\frac{t_n}{\beta + t_n}\right)^{i_n}\left(\frac{\beta}{\beta + t_n}\right)^\alpha.$$

After some algebra, the n-dimensional distribution of the Polya process becomes

$$P(N_L(t_k) = i_k;\ k = 1, 2, ..., n)$$

$$= \frac{i_n!}{\Pi_{k=1}^n (i_k - i_{k-1})!} \binom{i_n - 1 + \alpha}{i_n}\left(\frac{\beta}{\beta + t_n}\right)^\alpha \prod_{k=1}^n \left(\frac{t_k - t_{k-1}}{\beta + t_n}\right)^{i_k - i_{k-1}}. \tag{3.59}$$

For the following three reasons it is not surprising that the Polya process is increasingly used for modeling real-life point processes, in particular customer flows:

1) The finite dimensional distributions of this process are explicitely available.

2) Dependent increments occur more frequently than independent ones.

3) The two free parameters α and β of this process allow its adaptation to a wide variety of data sets.

Example 3.7 An insurance company analyzed the incoming flow of claims and found that the arrival intensity λ is subjected to random fluctuations, which can be modeled by the probability density $f_L(\lambda)$ of a gamma distributed random variable L with mean value $E(L) = 0.24$ and variance $Var(L) = 0.16$ (unit: working hour). The parameters α and β of this gamma distribution can be obtained from

$$E(L) = 0.24 = \alpha/\beta, \quad Var(L) = 0.16 = \alpha/\beta^2.$$

Hence, $\alpha = 0.36$ and $\beta = 1.5$. Thus, L has density

$$f_L(\lambda) = \frac{(1.5)^{0.36}}{\Gamma(0.36)} \lambda^{-0.64} e^{-(1.5)\lambda}, \quad \lambda > 0.$$

In time intervals, in which the arrival rate was nearly constant, the flow of claims behaved like a homogeneous Poisson process. Hence, the insurance company modeled the incoming flow of claims by a Polya process $\{N_L(t), t \ge 0\}$ with the one-dimensional probability distribution

$$P\left(N_L(t) = i\right) = \binom{i-0.64}{i}\left(\frac{t}{1.5+t}\right)^i \left(\frac{1.5}{1.5+t}\right)^{0.36}; \quad i = 0, 1, ...$$

According to (3.54), mean value and variance of $N_L(t)$ are

$$E(N_L(t)) = 0.24\,t, \quad Var(N_L(t)) = 0.24\,t + 0.16\,t^2.$$

As illustrated by this example, the Polya process (as any other mixed Poisson process) is a more appropriate model than a homogeneous Poisson process with intensity $\lambda = E(L)$ for fitting claim number developments, which exhibit a greater variability with increasing t. □

Doubly Stochastic Poisson Process The mixed Poisson process generalizes the homogeneous Poisson process by replacing its parameter λ with a random variable L. The corresponding generalization of the nonhomogeneous Poisson process leads to the concept of a doubly stochastic Poisson process. A doubly stochastic Poisson process $\{N_{L(\cdot)}(t), t \ge 0\}$ can be thought of as a nonhomogeneous Poisson process the intensity function $\lambda(t)$ of which has been replaced with a stochastic process $\{L(t), t \ge 0\}$ called *intensity process*. Thus, a sample path of a doubly stochastic Poisson process $\{N_{L(\cdot)}(t), t \ge 0\}$ can be generated as follows:

1) A sample path $\{\lambda(t), t \ge 0\}$ of a given intensity process $\{L(t), t \ge 0\}$ is simulated according to the probability distribution of $\{L(t), t \ge 0\}$.

2) Given $\{\lambda(t), t \ge 0\}$, the process $\{N_{L(\cdot)}(t), t \ge 0\}$ evolves like a nonhomogeneous Poisson process with intensity function $\lambda(t)$.

Thus, a doubly stochastic Poisson process $\{N_{L(\cdot)}(t), t \ge 0\}$ is generated by two independent 'stochastic mechanisms'.

The absolute state probabilities of the doubly stochastic Poisson process at time t are

$$P(N_{L(\cdot)}(t) = i) = \frac{1}{i!} E\left(\left[\int_0^t L(x)\,dx\right]^i e^{-\int_0^t L(x)\,dx}\right); \quad i = 0, 1, ... \tag{3.60}$$

In this formula, the mean value operation 'E' eliminates the randomness generated by the intensity process in $[0, t]$.

The trend function of $\{N_{L(\cdot)}(t), t \geq 0\}$ is

$$m(t) = E\left(\int_0^t L(x)\,dx\right) = \int_0^t E(L(x))\,dx, \quad t \geq 0.$$

A nonhomogeneous Poisson process with intensity function $\lambda(t) = E(L(t))$ can be used as an approximation to the doubly stochastic Poisson process $\{N_{L(\cdot)}(t), t \geq 0\}$.

The doubly stochastic Poisson process becomes

1. the homogeneous Poisson process if $L(t)$ is a constant λ,
2. the nonhomogeneous process if $L(t)$ is a nonrandom function $\lambda(t),\ t \geq 0$,
3. the mixed Poisson process if $L(t)$ is a random variable L, which does not depend on t.

The two 'degrees of freedom' a doubly stochastic Poisson process has make this process a universal point process model. The term 'doubly stochastic Poisson process' was introduced by Cox [21], who was the first to investigate this class of point processes. Hence, these processes are also called *Cox processes*. For detailed treatments and applications in engineering and insurance, respectively, see, for instance, Snyder [76] and Grandell [34].

3.2.4 Superposition and Thinning of Poisson Processes

3.2.4.1 Superposition

Assume that a service station recruits its customers from n different sources. For instance, a branch bank serves customers from n different towns, or a car workshop repairs and maintains n different makes of cars. Each town or each make of cars, respectively, generates its own arrival process (flow of demands). Let

$$\{N_i(t), t \geq 0\}; \quad i = 1, 2, ..., n,$$

be the corresponding counting processes. Then, the total number of customers arriving at the service station in $[0, t]$ is

$$N(t) = N_1(t) + N_2(t) + \cdots + N_n(t).$$

Note that $\{N(t), t \geq 0\}$ can be thought of as the counting process of a marked point process, where the marks indicate from which source the 'customers' come.

On condition that $\{N_i(t), t \geq 0\}$ is a homogeneous Poisson process with parameter $\lambda_i;\ i = 1, 2, ..., n,$ what type of counting process is $\{N(t), t \geq 0\}$?

From example 1.22 (section 1.7.1) it is known that the z-transform of $N(t)$ is

$$M_{N(t)}(z) = e^{-(\lambda_1+\lambda_2+\cdots+\lambda_n)\,t\,(z-1)}.$$

Therefore, $N(t)$ has a Poisson distribution with parameter

$$(\lambda_1 + \lambda_2 + \cdots + \lambda_n)\,t.$$

Since the counting processes $\{N_i(t), t \ge 0\}$ have homogeneous and independent increments, their additive *superposition* $\{N(t), t \ge 0\}$ also has homogeneous and independent increments. This proves the following theorem:

Theorem 3.7 The additive superposition $\{N(t), t \ge 0\}$ of independent homogeneous Poisson processes $\{N_i(t), t \ge 0\}$ with intensities λ_i; $i = 1, 2, ..., n$; is a homogeneous Poisson process with intensity $\lambda = \lambda_1 + \lambda_2 + \cdots + \lambda_n$. ■

Quite analogously, if the $\{N_i(t), t \ge 0\}$ are independent nonhomogeneous Poisson processes with intensity functions $\lambda_i(t)$; $i = 1, 2, ..., n$; then their additive superposition $\{N(t), t \ge 0\}$ is a nonhomogeneous Poisson process with intensity function

$$\lambda(t) = \lambda_1(t) + \lambda_2(t) + \cdots + \lambda_n(t).$$

3.2.4.2 Thinning

There are many situations in which not superposition, but the opposite operation, namely *thinning* or *splitting*, of a Poisson process occurs. For instance, a cosmic particle counter registers only α-particles and ignores other types of particles. Or, a reinsurance company is only interested in claims, the size of which exceeds, say, one million dollars. Formally, a marked point process $\{(T_1, M_1), (T_2, M_2), ...\}$ arrives and only events with special marks will be taken into account. It is assumed that the marks M_i are independent of each other and independent of $\{T_1, T_2, ...\}$, and that they are identically distributed as

$$M = \begin{cases} m_1 & \text{with probability } 1-p \\ m_2 & \text{with probability } \quad p \end{cases},$$

i.e. the mark space only consists of two elements: $\mathbf{M} = \{m_1, m_2\}$. In this case, there are two different types of events, type 1-events (attached with mark m_1) and type 2-events (attached with mark m_2). If only type 1-events are counted, of what kind is the arising point process?

Let Y be the first event time with mark m_2. Note that if $t < T_1$, then there is surely no type 2-event in $[0, t]$, and if $T_n \le t < T_{n+1}$, then there are exactly n events in $[0, t]$ and $(1-p)^n$ is the probability that none of them is a type 2-event. Hence,

$$P(Y > t) = P(0 < t < T_1) + \sum_{n=1}^{\infty} P(T_n \le t < T_{n+1})(1-p)^n.$$

Since $P(T_n \le t < T_{n+1}) = P(N(t) = n)$,

$$P(Y > t) = e^{-\lambda t} + \sum_{n=1}^{\infty} \left(\frac{(\lambda t)^n}{n!} e^{-\lambda t} \right) (1-p)^n$$

$$= e^{-\lambda t} + e^{-\lambda t} \sum_{n=1}^{\infty} \frac{[\lambda(1-p)t]^n}{n!} = e^{-\lambda t} + e^{-\lambda t}\left[e^{\lambda(1-p)t} - 1\right].$$

Hence,

$$P(Y > t) = e^{-\lambda p t}, \quad t \ge 0.$$

Hence, the interevent times between type 2-events have an exponential distribution with parameter $p\lambda$. Moreover, in view of our assumptions, these interevent times are independent. By changing the roles of type 1 and type 2-events, theorem 3.2 implies theorem 3.8:

Theorem 3.8 Given a homogeneous Poisson process $\{N(t), t \ge 0\}$ with intensity λ and two types of Poisson events 1 and 2, which occur independently with respective probabilities $1-p$ and p. Then $N(t)$ can be represented in the form

$$N(t) = N_1(t) + N_2(t), \tag{3.61}$$

where $\{N_1(t),\, t \ge 0\}$ and $\{N_2(t),\, t \ge 0\}$ are two independent homogeneous Poisson processes with respective intensities $(1-p)\lambda$ and $p\lambda$, which count only type 1- and type 2-events, respectively. ■

Nonhomogeneous Poisson Process Now the situation is somewhat generalized by assuming that the underlying counting process $\{N(t), t \ge 0\}$ is a nonhomogeneous Poisson process with intensity function $\lambda(t)$ and that an event, occurring at time t, is of type 1 with probability $1-p(t)$ and of type 2 with probability $p(t)$. Let Y be the time to the first occurrence of a type 2-event,

$$G(t) = P(Y \le t)$$

its distribution function, and $\overline{G}(t) = 1 - G(t)$. Then the relationship

$$P(t < Y \le t + \Delta t \,|\, Y > t) = p(t)\,\lambda(t)\,\Delta t + o(\Delta t)\,.$$

implies

$$\frac{1}{\overline{G}(t)} \cdot \frac{G(t+\Delta t) - G(t)}{\Delta t} = p(t)\,\lambda(t) \, + \frac{o(\Delta t)}{\Delta t}\,.$$

Letting Δt tend to 0 yields,

$$\frac{G'(t)}{\overline{G}(t)} = p(t)\,\lambda(t)\,.$$

By integration,

$$\overline{G}(t) = e^{-\int_0^t p(x)\,\lambda(x)\,dx}, \quad t \ge 0. \tag{3.62}$$

If $p(t) \equiv 1$, then $\overline{G}$ is the survival function of the system.

Theorem 3.9 Given a nonhomogeneous Poisson process $\{N(t), t \ge 0\}$ with intensity function $\lambda(t)$ and two types of events 1 and 2, which occur independently with respective probabilities $1-p(t)$ and $p(t)$ if t is an event time. Then $N(t)$ can be represented in the form

$$N(t) = N_1(t) + N_2(t),$$

where $\{N_1(t), t \geq 0\}$ and $\{N_2(t), t \geq 0\}$ are independent nonhomogeneous Poisson processes with respective intensity functions

$$(1-p(t))\lambda(t) \text{ and } p(t)\lambda(t),$$

which count only type 1 and type 2-events, respectively. ■

In section 3.2.6.3, some more sophisticated results will be needed: Let Z be the random number of type 1-events to the occurrence of the first type 2-event. Then

$$P(Z=0) = \int_0^\infty p(t) f(t)\, dt,$$

where $f(t) = f_{T_1}(t)$ is the density of the first event time T_1 as given by (3.43):

$$f(t) = \lambda(t)\, e^{-\int_0^t \lambda(x)\, dx}, \quad t \geq 0.$$

From (3.49), for $k \geq 1$,

$$P(Z=k) = \int_0^\infty \int_0^{x_{k+1}} \cdots \int_0^{x_3} \int_0^{x_2} \prod_{i=1}^k \bar{p}(x_i)\, \lambda(x_i)\, dx_i\, p(x_{k+1}) f(x_{k+1})\, dx_{k+1}.$$

By making use of the well-known formula

$$\int_0^t \int_0^{x_n} \cdots \int_0^{x_3} \int_0^{x_2} \prod_{i=1}^n g(x_i)\, dx_1 dx_2 \cdots dx_n = \frac{1}{n!}\left[\int_0^t g(x)\, dx\right]^n, \quad n \geq 2, \tag{3.63}$$

the desired probability is seen to be

$$P(Z=k) = \frac{1}{k!} \int_0^\infty \left(\int_0^t \bar{p}(x)\, \lambda(x)\, dx\right)^k p(t) f(t)\, dt, \quad k = 0, 1, \ldots \tag{3.64}$$

After some algebra,

$$E(Z) = \sum_{k=1}^\infty k\, P(Z=k) = \int_0^\infty \Lambda(t)\, dG(t) - 1. \tag{3.65}$$

If $p(t) \equiv p > 0$, then Z has a geometric distribution with parameter p so that

$$E(Z) = \frac{1-p}{p} \tag{3.66}$$

and $\overline{G}(t)$ has structure

$$\overline{G}(t) = [\overline{F}(t)]^p; \quad t \geq 0. \tag{3.67}$$

Now, let Z_t be the random number of type 1-events in $(0, \min(Y, t))$ and

$$r_t(k) = P(Z_t = k \mid Y = t); \quad k = 0, 1, \ldots$$

Then, by (1.6),

$$r_t(k) = \lim_{\Delta t \to 0} \frac{P(Z_t = k \cap t \leq Y \leq t + \Delta t)}{P(t \leq Y \leq t + \Delta t)}$$

$$= \lim_{\Delta t \to 0} \frac{P(t \leq Y = X_{k+1} \leq t + \Delta t)}{G(t+\Delta t) - G(t)}. \tag{3.68}$$

From (3.49) and (3.63), the numerator in (3.68) becomes

$$P(t \le Y = X_{k+1} \le t + \Delta t)$$

$$= \int_t^{t+\Delta t} \int_0^{x_{k+1}} \cdots \int_0^{x_3} \int_0^{x_2} \prod_{i=1}^k \bar{p}(x_i)\,\lambda(x_i)\,dx_i\, p(x_{k+1}) f(x_{k+1})\,dx_{k+1}$$

$$= \frac{1}{k!} \int_t^{t+\Delta t} \left(\int_0^y \bar{p}(x)\,\lambda(x)\,dx \right)^k p(y) f(y)\,dy.$$

Dividing numerator and denominator in (3.68) by Δt and taking the limit as $\Delta t \to 0$ yields

$$r_t(k) = \frac{1}{k!} \left(\int_0^t \bar{p}(x)\,\lambda(x)\,dx \right)^k e^{-\int_0^t \bar{p}(x)\,\lambda(x)\,dx}; \quad k = 0, 1, \ldots$$

Hence, given $Y = t$, the random variable Z_t has a Poisson distribution with mean

$$E(Z_t | Y = t) = \int_0^t \bar{p}(x)\,\lambda(x)\,dx\,, \tag{3.69}$$

so that

$$E(Z_t | Y < t) = \int_0^t E(Z_x | Y = x)\,dG(x)/G(t)$$

$$= \int_0^t \int_0^x \bar{p}(y)\,\lambda(y)\,dy\,dG(x)/G(t) \tag{3.70}$$

and

$$E(Z_t | Y \ge t) = E(Z_t | Y = t) = \int_0^t \bar{p}(x)\,\lambda(x)\,dx. \tag{3.71}$$

Now the (unconditional) mean value of Z_t can be otained from

$$E(Z_t) = E(Z_t | Y < t)\,G(t) + E(Z_t | Y \ge t)\,\overline{G}(t).$$

The result is

$$E(Z_t) = \int_0^t \overline{G}(x)\,\lambda(x)\,dx - G(t). \tag{3.72}$$

For these and related results see Beichelt [5].

3.2.5 Compound Poisson Processes

Let $\{(T_i, M_i);\ i = 1, 2, \ldots\}$ be a marked point process, where $\{T_i;\ i = 1, 2, \ldots\}$ is a Poisson point process with corresponding counting process $\{N(t),\ t \ge 0\}$. Then the stochastic process $\{C(t),\ t \ge 0\}$ defined by

$$C(t) = \sum_{i=0}^{N(t)} M_i$$

with $M_0 = 0$ is called a *compound* (*cumulative*, *aggregate*) *Poisson process.*

Compound Poisson processes occur in many situations: 1) If T_i is the time point at which the ith customer arrives at an insurance company and M_i its claim size, then $C(t)$ is the total claim amount the company is confronted with in time interval $[0, t]$. 2) If T_i is the time of the ith breakdown of a machine and M_i the corresponding repair cost, then $C(t)$ is the total repair cost in $[0, t]$. 3) If T_i is the time point the ith shock occurs and M_i the amount of (mechanical) wear this shock contributes to the degree of wear of an item, then $C(t)$ is the degree of wear of the item at time t. (For the brake discs of a car, every application of the brakes is a shock, which increases their degree of mechanical wear. For the tires of the undercarriage of an aircraft, every takeoff and touchdown is a shock, which diminishes their tread depth.)

In what follows, $\{N(t), t \geq 0\}$ is assumed to be a homogeneous Poisson process with intensity λ. If the M_i are independent and identically distributed as M and independent of $\{T_1, T_2, ...\}$, then $\{C(t), t \geq 0\}$ has the following properties:

1) $\{C(t), t \geq 0\}$ has independent, homogeneous increments.

2) The Laplace transform of $C(t)$ is

$$\hat{C}_t(s) = e^{\lambda t [\hat{M}(s) - 1]}, \tag{3.73}$$

where

$$\hat{M}(s) = E\left(e^{-s M}\right)$$

is the Laplace transform of M. The proof of (3.73) is straightforward: By (1.27),

$$\hat{C}_t(s) = E\left(e^{-s\, C(t)}\right) = E\left(e^{-s\,(M_0 + M_1 + M_2 + \cdots + M_{N(t)})}\right)$$

$$= \sum_{n=0}^{\infty} E\left(e^{-s\,(M_0 + M_1 + M_2 + \cdots + M_n)}\right) P(N(t) = n)$$

$$= \sum_{n=0}^{\infty} E\left(e^{-s\,M}\right)^n \frac{(\lambda t)^n}{n!} e^{-\lambda t}$$

$$= e^{-\lambda t} \sum_{n=0}^{\infty} \frac{[\lambda t\, \hat{M}(s)]^n}{n!} = e^{\lambda t [\hat{M}(s) - 1]}.$$

From $\hat{C}_t(s)$, all the moments of $C(t)$ can be obtained by making use of (1.28). In particular, mean value and variance of $C(t)$ are

$$E(C(t)) = \lambda t\, E(M), \quad Var(C(t)) = \lambda t\, E(M^2). \tag{3.74}$$

These formulas also follow from (1.125) and (1.126).

Now the compound Poisson process is considered on condition that M has a Bernoulli distribution:

$$M = \begin{cases} 1 & \text{with probability } \quad p \\ 0 & \text{with probability } 1 - p \end{cases}.$$

Then $M_1 + M_2 + \cdots + M_n$ as a sum of independent and Bernoulli distributed random variables is binomially distributed with parameters n and p (section 1.2.2.2). Hence,

$$P(C(t) = k) = \Sigma_{0=1}^{n} P(M_0 + M_1 + \cdots + M_n = k | N(t) = n)\, P(N(t) = n)$$

$$= \sum_{n=0}^{\infty} \binom{n}{k} p^k (1-p)^{n-k} \frac{(\lambda t)^n}{n!} e^{-\lambda t}.$$

This is a mixture of binomial distributions with regard to a Poisson structure distribution. Hence, by example 1.8 (section 1.2.4), $C(t)$ has a Poisson distribution with parameter $\lambda p t$:

$$P(C(t) = k) = \frac{(\lambda p t)^n}{n!} e^{-\lambda p t}; \quad k = 0, 1, ...$$

Corollary If the marks of a compound Poisson process $\{C(t), t \geq 0\}$ have a Bernoulli distribution with parameter p, then $\{C(t), t \geq 0\}$ is a thinned homogeneous Poisson process with parameter λp.

If the underlying counting process $\{N(t), t \geq 0\}$ is a nonhomogeneous Poisson process with intensity function

$$\lambda(t) \text{ and } \Lambda(t) = \int_0^t \lambda(x)\, dx,$$

then (3.73) and (3.74) become

$$\hat{C}_t(s) = e^{\Lambda(t)\, [\, \hat{M}(s) - 1]}$$

and

$$E(C(t)) = \Lambda(t)\, E(M), \quad Var(C(t)) = \Lambda(t)\, E(M^2). \tag{3.75}$$

Again, formulas (3.75) are an immediate consequence of (1.125) and (1.126). For compound renewal processes, see section 3.3.7.

3.2.6 Applications to Maintenance

3.2.6.1 Nonhomogeneous Poisson Process and Minimal Repair

The nonhomogeneous Poisson process is an important mathematical tool for optimizing the maintenance of technical systems with respect to cost and reliability criteria by applying proper *maintenance policies* (*strategies*). *Maintenance policies* prescribe when to carry out (preventive) repairs, replacements or other maintenance measures. *Repairs* after system failures usually only tackle the causes which triggered off the failures. A *minimal repair* performed after a failure enables the system to continue its work but does not affect the failure rate of the system. In other words, after a minimal repair the failure rate of the system has the same value as immediately before the failure. For example, if a failure of a complicated electronic system is caused by a defective plug and socket connection, then removing this cause of failure can be considered a minimal repair. *Preventive replacements* (*renewals*) and *preventive repairs* are not initiated by system failures, but they are carried out to prevent or at least to postpone future ones. Of course, preventive minimal repairs make no sense.

In what follows it is assumed that all renewals and repairs take only negligibly small times and that, after completing a renewal or a repair, the system immediately resumes its work. The random lifetime T of the system has probability density $f(t)$, distribution function $F(t)$, survival probability $\overline{F}(t) = 1 - F(t)$, and failure rate $\lambda(t)$. For a recent survey see Kapur, Garg, and Kumar [44]. The following maintenance policy is directly related to a nonhomogeneous Poisson process.

Basic Policy Every system failure is (and can be) removed by a minimal repair.

Let T_n be the random time point, at which the nth system failure (minimal repair) occurs. Then

$$Y_n = T_n - T_{n-1}$$

is the length of the time span between the $(n-1)$th and the nth system failure, $n = 1, 2, ...;\ T_0 = 0$. The first failure of the system after starting to work at time $t = 0$ occurs at time $T = T_1$. Given $T_1 = t,$ the failure rate of the system after completion of the repair is $\lambda(t)$. Hence, the future failure behaviour of the system is the same as that of a system which has worked up to time point t without failing. Therefore, from (1.34), the time between the first and the second system failure $Y_2 = T_2 - T_1$ given $T_1 = t,$ has distribution function

$$F_t(y) = P(Y_2 \le y) = \frac{F(t+y) - F(t)}{\overline{F}(t)}.$$

According to (1.40) and (3.38), equivalent representations of $F_t(y)$ are

$$F_t(y) = 1 - e^{-[\Lambda(t+y) - \Lambda(t)]} \tag{3.76}$$

and

$$F_t(y) = 1 - p_0(t, t+y).$$

Obviously, these equations are also valid if t is not the time point of the first failure, but the time point of any failure, for instance the nth failure. Then $F_t(y)$ is the distribution function of the $(n+1)$th interarrival time $Y_n = T_{n+1} - T_n$ given that $T_n = t$. The occurrence of system failures (minimal repairs) is, therefore, governed by the same probability distribution as the occurrence of Poisson events generated by a nonhomogeneous Poisson process with intensity function $\lambda(t)$. Specifically, the random vector $(T_1, T_2, ..., T_n)$ has the joint probability density (3.49) for all $n = 1, 2, ...$ Therefore, if $N(t)$ denotes the number of system failures (minimal repairs) in $[0, t]$, then $\{N(t), t \ge 0\}$ is a nonhomogeneous Poisson process with intensity function $\lambda(t)$.

In particular, $N(t)$ has a Poisson distribution with parameter $\Lambda(t)$:

$$E(N(t)) = \Lambda(t) = \int_0^t \lambda(x)\, dx. \tag{3.77}$$

The nonhomogeneous Poisson point process $\{T_1, T_2, ...\}$ is an ingredient to a marked point process $\{(T_1, M_1), (T_2, M_2), ...\}$, where M_i denotes the cost of the ith minimal repair. The corresponding compound process $\{M(t), t \geq 0\}$ is given by

$$M(t) = \Sigma_{i=0}^{N(t)} M_i, \quad M_0 = 0,$$

where $M(t)$ is the total repair cost in $[0, t]$. The $M_1, M_2, ...$ are assumed to be independent of each other, independent of $N(t)$, and identically distributed as M with $c_m = E(M) < \infty$. Then the trend function of $\{M(t), t \geq 0\}$ is

$$m(t) = E(M(t)) = c_m \, \Lambda(t) . \tag{3.78}$$

3.2.6.2 Standard Replacement Policies with Minimal Repair

The basic policy discussed in the previous section provides the theoretical fundament for analyzing a number of more sophisticated maintenance policies. In what follows, four policies of this kind will be considered. To justify preventive replacements, the assumption has to be made that the underlying system is aging (section 1.4, definition 1.1), i.e. its failure rate $\lambda(t)$ is increasing. In addition, all replacement and repair times are assumed to be negligibly small. The latter assumption is merely a matter of convenience.

The criterion for evaluating the efficiency of maintenance policies will be the average maintenance cost per unit time over an infinite time span. To establish this criterion, the time axis is partitioned into *replacement cycles*, i.e. into the times between two neighbouring replacements. Let L_i be the random length of the i th replacement cycle and C_i the total random maintenance cost (replacement + repair cost) in the ith replacement cycle. It is assumed that the $L_1, L_2, ...$ are independent and identically distributed as L. This assumption implies that a replaced system is statistically as good as the previous one ('as good as new') from the point of view of its lifetime. The $C_1, C_2, ...$ are assumed to be independent, identically distributed as C, and independent on the L_i. Then the *maintenance cost per unit time over an infinite time span* is

$$K = \lim_{n \to \infty} \frac{\Sigma_{i=1}^n C_i}{\Sigma_{i=1}^n L_i} .$$

The strong law of the large numbers implies that

$$K = \frac{E(C)}{E(L)} . \tag{3.79}$$

For the sake of brevity, K is referred to as the (*long-run*) *maintenance cost rate.* Thus, the maintenance cost rate is equal to the mean maintenance cost per cycle divided by the mean cycle length. In what follows, c_p denotes the cost of a preventive replacement, and c_m is the cost of a minimal repair; c_p, c_m constant.

Policy 1 A system is preventively replaced at fixed times $\tau, 2\tau, ...$ Failures between replacements are removed by minimal repairs.

This policy reflects the common approach of preventively overhauling complicated systems after fixed time periods whilst in between only the absolutely necessary repairs are done. With this policy, all cycle lengths are equal to τ, and, in view of (3.77), the mean cost per cycle is equal to

$$c_p + c_m \Lambda(\tau).$$

Hence, the corresponding maintenance cost rate is

$$K_1(\tau) = \frac{c_p + c_m \Lambda(\tau)}{\tau}.$$

A replacement interval $\tau = \tau*$, which minimizes $K_1(\tau)$, satisfies condition

$$\tau \lambda(\tau) - \Lambda(\tau) = c_p / c_m .$$

If $\lambda(t)$ tends to ∞ as $t \to \infty$, there exists a unique solution $\tau = \tau*$ of this equation. The corresponding minimal maintenance cost rate is

$$K_1(\tau*) = c_m \lambda(\tau*) .$$

Policy 2 A system is replaced at the first failure which occurs <u>after</u> a fixed time τ. Failures which occur between replacements are removed by minimal repairs.

This policy fully makes use of the system lifetime so that, from this point of view, it is preferable to policy 1. However, the partial uncertainty about the times of replacements leads to larger replacement costs than with policy 1. Thus, in practice the maintenance cost rate of policy 2 may actually exceed the one of policy 1.

The residual lifetime T_τ of the system after time point τ, when having survived interval $[0, \tau]$, has, according to (1.36), the mean value

$$\mu(\tau) = E(T_\tau) = e^{\Lambda(\tau)} \int_\tau^\infty e^{-\Lambda(t)} dt . \tag{3.80}$$

The mean maintenance cost per cycle is, from the notational point of view, equal to that of policy 1. Thus, the maintenance cost rate is

$$K_2(\tau) = \frac{c_p + c_m \Lambda(\tau)}{\tau + \mu(\tau)},$$

since $\tau + \mu(\tau)$ is the mean cycle length. An optimal renewal interval $\tau = \tau*$ satisfies the necessary condition $dK_2(\tau)/d\tau = 0$, i.e.

$$(\Lambda(\tau) + \frac{c_p}{c_m} - 1)\, \mu(\tau) = \tau .$$

If $\tau*$ exists, then the minimal maintenance cost rate is

$$K_2(\tau*) = \frac{c_p + c_m[\Lambda(\tau*) - 1]}{\tau*}.$$

Example 3.8 Let the system lifetime T have a Rayleigh distribution with failure rate $\lambda(t) = 2t/\theta^2$. The corresponding mean residual lifetime of the system after having survived $[0, \tau]$ is

$$\mu(\tau) = \theta\sqrt{\pi}\, e^{(\tau/\theta)^2}\left[1 - \Phi\left(\frac{\sqrt{2}}{\theta}\tau\right)\right].$$

In particular, if $\theta = 100\,[h^{-1}]$, $c_m = 1$, and $c_p = 5$, the optimal parameters are

$$\tau^* = 180\,[h], \quad K_2(\tau^*) = 0.0402.$$

□

Policy 3 Each failure is removed by a minimal repair. On the first failure after a given time τ_1, an unscheduled replacement is carried out. However, if there is no replacement in $[\tau_1, \tau_2]$, $\tau_1 < \tau_2$, then at time point τ_2 a preventive replacement is done.

Under this policy, the random cycle length is

$$L = \tau_1 + \min(T_{\tau_1}, \tau_2 - \tau_1),$$

so that the mean cycle length is

$$E(L) = \tau_1 + \mu(\tau_1, \tau_2) \quad \text{with} \quad \mu(\tau_1, \tau_2) = \int_0^{\tau_2 - \tau_1} \overline{F}_{\tau_1}(t)\, dt.$$

Hence, if c_r is the cost of an unscheduled replacement, the maintenance cost rate is

$$K_3(\tau_1, \tau_2) = \frac{c_m \Lambda(\tau_1) + c_r F_{\tau_1}(\tau_2 - \tau_1) + c_p \overline{F}_{\tau_1}(\tau_2 - \tau_1)}{\tau_1 + \mu(\tau_1, \tau_2)}.$$

An optimal pair $(\tau_1, \tau_2) = (\tau_1^*, \tau_2^*)$ is solution of the equation system

$$\lambda(\tau_2)\,\mu(\tau_1, \tau_2) + \overline{F}_{\tau_1}(\tau_2 - \tau_1) - c_m/(c_r - c_p) = 0,$$

$$\lambda(\tau_2) - \frac{c_m \Lambda(\tau_1) + c_r - c_m}{(c_r - c_p)\tau_1} = 0.$$

A sufficient condition for the existence of a unique solution (τ_1^*, τ_2^*) which satisfies the condition $0 \le \tau_1^* < \tau_2^*$ is

$$\lambda(t) \to \infty \quad \text{and} \quad 0 < c_r - c_p < c_m < c_r < \infty.$$

In this case, the minimal maintenance cost rate is

$$K_3(\tau_1^*, \tau_2^*) = (c_r - c_p)\,\lambda(\tau_2^*).$$

Policy 4 The first $n-1$ failures are removed by minimal repairs. At the time point of the nth failure, an (unscheduled) replacement is carried out.

The random cycle length is $L = T_n$. Hence, the maintenance cost rate is

$$K_4(n) = \frac{c_r + (n-1)c_m}{E(T_n)},$$

where the mean cycle length $E(T_n)$ is given by (3.47).

By analyzing the behaviour of the difference $K_4(n) - K_4(n-1)$, an optimal $n = n*$ is seen to be the smallest integer k satisfying

$$E(T_n) - [n - 1 + c_r/c_m]\, E(Y_{n+1}) \geq 0; \quad n = 1, 2, ..., \tag{3.81}$$

where the mean time $E(Y_n)$ between the $(n-1)$th and the nth minimal repair is given by formula (3.48).

Example 3.9 Let the system lifetime T have a Weibull distribution:

$$\lambda(t) = \frac{\beta}{\theta}\left(\frac{t}{\theta}\right)^{\beta-1}, \quad \Lambda(t) = \left(\frac{t}{\theta}\right)^{\beta}, \quad \beta > 1. \tag{3.82}$$

Under this assumption, condition (3.82) becomes

$$\beta n - [n - 1 + c_r/c_m] \geq 0\,.$$

Hence, if $c_r > c_m$,

$$n* = \left\| \frac{1}{\beta-1}\left(\frac{c_r}{c_m} - 1\right) \right\| + 1,$$

where $\|x\|$ is the largest integer being less than or equal to x. (If $x < 0$, then $\|x\| = 0$.) □

3.2.6.3 Replacement Policies for Systems with two Failure Types

So far, it has been assumed that every system failure can be removed by a minimal repair. This is not always possible. For example, the restoration of the roadworthiness of a car after a serious traffic accident can surely not be achieved by a minimal repair. To be able to model such situations, two failure types are introduced:

Type 1: Failures of this type are (and can be) removed by minimal repairs.

Type 2: Failures of this type are removed by replacements.

Type 1 failures are minor ones, which can be removed without much effort, whereas type 2 failures may be complete system breakdowns. A failure occuring at system age t is a type 2 failure with probability $p(t)$ and a type 1 failure with probability $1 - p(t)$. The types of failures are assumed to occur independently of each other. Obviously, this is the same situation as discussed in section 3.2.4.2: The type 1 (type 2) Poisson events introduced there are here interpreted as type 1 (type 2) failures.

Policy 5 The system is maintained according to the failure type.

Under policy 5, a replacement cycle is the time between two neighbouring type 2 failures. Hence, according to (3.62), the distribution function of the cycle length L is

$$G(t) = 1 - e^{-\int_0^t p(x)\lambda(x)dx}, \quad t \geq 0. \tag{3.83}$$

The random number Z of minimal repairs between neighbouring replacements has mean value (3.65). Thus, the maintenance cost rate is

$$K_5 = \frac{\left[\int_0^\infty \Lambda(t)dG(t) - 1\right]c_m + c_r}{\int_0^\infty e^{-\int_0^t p(x)\lambda(x)dx}dt}. \tag{3.84}$$

In the special case $p(t) \equiv p > 0$, by (3.66) and (3.67),

$$K_5 = \frac{[(1-p)/p]\,c_m + c_r}{\int_0^\infty \left[\bar{F}(t)\right]^p dt}. \tag{3.85}$$

Policy 6 The system is maintained according to the failure type. In addition, preventive replacements are carried out τ time units after the previous replacement.

Let c_m, c_r, and c_p with $0 < c_m < c_p < c_r$ denote the cost of a minimal repair, a replacement after a type 2 failure (unscheduled replacement) and a preventive replacement, respectively. Then

$$L_\tau = \min(Y, \tau)$$

is the random length of a replacement cycle (time between successive replacements of any type) and, if Z_τ denotes the random number of minimal repairs in a replacement cycle, the maintenance cost rate has structure

$$K_6(\tau) = \frac{c_m\, E(Z_\tau)) + c_r\, G(\tau) + c_p\, \bar{G}(\tau)}{E(L_\tau)}.$$

In view of (3.72) and

$$E(L_\tau) = \int_0^\tau \bar{G}(t)\, dt,$$

the maintenance cost rate becomes

$$K_6(\tau) = \frac{c_m\left[\int_0^\tau \bar{G}(t)\lambda(t)dt - G(\tau)\right] + c_r\, G(\tau) + c_p\, \bar{G}(\tau)}{\int_0^\tau \bar{G}(t)\, dt}. \tag{3.86}$$

In particular, for $p(t) \equiv p$,

$$K_6(\tau) = \frac{\{c_r + [(1-p)/p]\,c_m\}G(\tau) + c_p\, \bar{G}(\tau)}{\int_0^\tau \bar{G}(t)\, dt}. \tag{3.87}$$

If there exists an optimal preventive replacement interval $\tau = \tau^*$ with regard to the maintenance cost rate $K_6(\tau)$, then it is solution of the equation

$$p\,\lambda(\tau)\int_0^\tau \bar{G}(t)\, dt - G(\tau) = \frac{p\, c_p}{(c_r - c_p - c_m)p + c_m}.$$

As proved in [5], a unique solution $\tau*$ exists if $\lambda(t)$ is strictly increasing to infinity and $c_r - c_p > c_m(1+p)/p$. If there is no preventive maintenance, i.e. $\tau = \infty$, then (3.86) and (3.87) reduce to (3.84) and (3.85), respectively.

Minimal repairs and replacements are extreme maintenance actions in the sense that they have no influence at all at the system reliability, or they restore the initial reliability level, respectively. Beginning with the papers of Uematsu and Nishida [83] and Kijma, Morimura and Suzuki [49], approaches to modeling general degrees of repairs have been suggested which take into account the intermediate stages. For a recent, comprehensive survey see Guo, Ascher and Love [37].

3.2.6.4 Repair Cost Limit Replacement Policies with Minimal Repair

Replacement policies based on repair cost limits are widely acknowledged as particularly userfriendly and efficient strategies for organizing the maintenance of complex systems. Different from the maintenance policies considered so far, repair cost limit replacement policies explicitely take into account that repair costs are random variables. The theoretical basis for the analysis of the repair cost limit replacement policy considered in this section is the two failure type model introduced in the previous section.

Policy 7 (*Repair cost limit replacement policy*) After a system failure, the necessary repair cost is estimated. The system is replaced by an equivalent new one if the repair cost exceeds a given level $c(t)$, where t is the age of the system at the time of failure. Otherwise, a minimal repair is carried out.

Let C_t be the random repair cost of the system if it fails at age t. Then the two failure type model applies to policy 7 if the failure types are generated by C_t in the following way: A system failure at time t is of type 1 (type 2) if

$$C_t \le c(t) \;\; (C_t > c(t)) .$$

Thus, if

$$R_t(x) = P(C_t \le x)$$

denotes the distribution function of C_t and if $\overline{R}_t(x) = 1 - R_t(x)$, then the respective probabilities of type 1 and type 2 failures are

$$1 - p(t) = R_t(c(t)), \quad p(t) = \overline{R}_t(c(t)). \tag{3.88}$$

As before, let c_r be the cost of a replacement after a type 2 failure. It is reasonable to assume that, for all $t \ge 0$,

$$0 < c(t) < c_r \quad \text{and} \quad R_t(x) = \begin{cases} 1 & \text{if } x \ge c_r \\ 0 & \text{if } x \le 0 \end{cases} .$$

With the failure type probabilities given by (3.88), the length L of a replacement cycle has, according to (3.83), distribution function

$$G(t) = 1 - e^{-\int_0^t \overline{R}_x(c(x))\lambda(x)\,dx}, \quad t \ge 0. \tag{3.89}$$

By (3.86), the corresponding maintenance cost rate is

$$K_7 = \frac{\left[\int_0^\infty \Lambda(t)\lambda(t)\overline{R}_t(c(t)\, e^{-\int_0^t \overline{R}_x(c(x))\lambda(x)\,dx} dt - 1.\right] c_m + c_r}{\int_0^\infty e^{-\int_0^t \overline{R}_x(c(x))\lambda(x)\,dx} dt.}. \tag{3.90}$$

The problem consists in finding a repair cost limit function $c = c(t)$ which minimizes (3.66). Generally, an explicit analytical solution cannot be given. Hence, some special cases will be discussed. In particular, the system lifetime X is assumed to be Weibull distributed:

$$F(t) = P(X \le t) = 1 - e^{-(t/\theta)^\beta}, \quad t \ge 0,\ \beta > 1,\ \theta > 0. \tag{3.91}$$

The respective failure rate and integrated failure rate are given by (3.82).

Constant Repair Cost Limit For the sake of comparison, next the case is considered that the repair cost limit is constant and that the cost of a repair C does not depend on t, i.e.

$$c(t) \equiv c \quad \text{and} \quad R_t(x) = R(x) \ \text{ for all } x \text{ and } t.$$

In this case, the probability $p = \overline{R}(c)$ does not depend on time so that the length of a replacement cycle has distribution function

$$G(t) = 1 - e^{-\overline{R}(c)\,(t/\theta)^\beta}, \quad t \ge 0.$$

Hence, the mean cycle length is

$$E(L) = \theta\,\Gamma(1 + 1/\beta)\left[\,\overline{R}(c)\right]^{-1/\beta}.$$

The corresponding maintenance cost rate can immediately be obtained from (3.87):

$$K_7(c) = \frac{\frac{R(c)}{\overline{R}(c)} c_m + c_r}{\theta\,\Gamma(1 + 1/\beta)\left[\,\overline{R}(c)\right]^{-1/\beta}}.$$

This maintenance cost rate depends on c only via $R(c)$. The value of $y = \overline{R}(c)$ minimizing $K_7(c)$ is easily seen to be

$$y^* = \overline{R}(c^*) = \frac{\beta - 1}{k - 1} \quad \text{with } k = c_r/c_m.$$

By assumption, $k > 1$ and $\beta > 1$. Hence, since $0 < y^* < 1$, an additional assumption has to be made:

$$1 < \beta < k. \tag{3.92}$$

Given (3.92), for any $\overline{R}$ with inverse function $\overline{R}^{-1}$, the optimal limit $c = c^*$ is

$$c^* = \overline{R}^{-1}\left(\frac{\beta - 1}{k - 1}\right). \tag{3.93}$$

Its application yields the smallest possible maintenance cost rate, which can be achieved with a constant repair cost limit:

$$K_7(c^*) = \frac{\beta c_m}{\theta\Gamma(1+1/\beta)}\left(\frac{k-1}{\beta-1}\right)^{1-1/\beta}.$$

In particular, for $\beta = 2$ (Rayleigh distribution),

$$\Gamma 1 + 1/\beta) = \Gamma(3/2) = \sqrt{\pi/4}$$

so that

$$K_7(c^*) = \frac{4\,c_m}{\theta}\sqrt{\frac{k-1}{\pi}} \approx 2.2568\,\frac{c_m}{\theta}\sqrt{k-1}\,. \tag{3.94}$$

Hyperbolic Repair Cost Limit Function System aging implies an increase in the mean failure frequency and in the mean repair cost with increasing system age t. Thus, a decreasing repair cost limit $c(t)$ is supposed to lead to a lower maintenance cost rate than a constant repair cost limit or an increasing repair cost limit function. To demonstrate this, the efficiency of the following nonincreasing repair cost limit function will be investigated in conjunction with a repair cost C being uniformly distributed over the interval $[0, c_r]$:

$$c(t) = \begin{cases} c_r, & 0 \le t < d/(c_r - c) \\ c + d/t, & d/(c_r - c) \le t < \infty \end{cases}, \quad 0 \le c < c_r, \tag{3.95}$$

$$R(x) = P(C \le x) = \begin{cases} x/c_r, & 0 \le x \le c_r \\ 1, & c_r < x < \infty \end{cases}. \tag{3.96}$$

Combining (3.95) and (3.96) gives the probability that a system failure, which occurs at age t, implies a replacement:

$$\overline{R}(c(t)) = \begin{cases} 0, & 0 \le t < d/(c_r - c) \\ \frac{c_r - c}{c_r} - \frac{d}{c_r t}, & \frac{d}{c_r - c} \le t < \infty \end{cases}, \quad 0 \le c < c_r.$$

Letting

$$r = d/c,\ s = (c_r - c)/c_r,\ z = r/s \tag{3.97}$$

yields

$$\overline{R}(c(t)) = \begin{cases} 0, & 0 \le t < z \\ s\,(1 - z/t), & z \le t < \infty \end{cases}. \tag{3.98}$$

Scheduling replacements based on (3.98) is well motivated: Replacements of systems in the first period of their useful life will not be scheduled. After this period, a system failure makes a replacement more and more likely with increasing system age.

To obtain tractable formulas, the system lifetime is assumed to have a Rayleigh distribution (distribution function (3.91) with $\beta = 2$):

$$\lambda(t) = 2\,t/\theta^2, \quad \Lambda(t) = (t/\theta)^2\,. \tag{3.99}$$

Under these assumptions, the maintenance cost rate (3.90) can explicitely be evaluated by making use of the following three basic integrals:

$$\int_0^\infty x^3 e^{-\lambda s x^2}\,dx = \frac{1}{2(\lambda s)^2},$$

$$\int_0^\infty x^2 e^{-\lambda s x^2}\,dx = \frac{1}{4\lambda s}\sqrt{\frac{\pi}{\lambda s}}, \tag{3.100}$$

$$\int_0^\infty x e^{-\lambda s x^2}\,dx = \frac{1}{2\lambda s}.$$

The result is

$$K_7(r,s) = \frac{2}{2r+\theta\sqrt{\pi s}}\left(1 - s + \frac{1}{\theta}\sqrt{\frac{\pi r}{s}} + \frac{1}{s}\left(\frac{r}{\theta}\right)^2 + k\right) c_m .$$

In order to minimize $K_7(r,s)$ with respect to r and s, in a first step $K_7(r,s)$ is minimized with respect to r with s fixed. The corresponding optimal value of r, denoted as $r^* = r^*(s)$, is solution of the quadratic equation $\partial K_7(r,s)/\partial r = 0$:

$$\left(r + \frac{\theta}{2}\sqrt{\pi s}\right)^2 = \frac{\theta^2 s}{4}\,[4s(k-1) + 4 - \pi] .$$

Since, by assumption, $k = c_r/c_m > 1$, the right-hand side of this equation is positive. Hence, a solution exists:

$$r^*(s) = \frac{\theta}{2}\sqrt{s}\left[\sqrt{4s(k-1)+4-\pi} - \sqrt{\pi}\right].$$

To make sure that $r^*(s) > 0$, an additional assumption has to be made:

$$k > \frac{\pi - 2}{2s} + 1. \tag{3.101}$$

The corresponding maintenance cost rate is

$$K_7(r^*(s), s) = \frac{c_m}{\theta}\sqrt{4(k-1) + \frac{4-\pi}{s}} . \tag{3.102}$$

Since $s \le 1$, the function $K_7(r^*(s), s)$ assumes its minimum at $s = 1$. Hence, $c = 0$. With $s = 1$, condition (3.101) holds if and only if

$$k > \pi/2 \approx 1.57.$$

Since replacement costs are usually much higher than repair costs, this condition hardly imposes a restriction on the application of the repair cost limit function (3.95).

Summarizing: If $k > \pi/2$, the optimal repair cost limit function of structure (3.95) is

$$c = 0 \quad \text{and} \quad d^* = \frac{\theta}{2}\left[\sqrt{4k-\pi} - \sqrt{\pi}\right] c_r ,$$

and the corresponding minimal maintenance cost rate is

$$K_7(d^*) = \frac{c_m}{\theta}\sqrt{4k - \pi} . \tag{3.103}$$

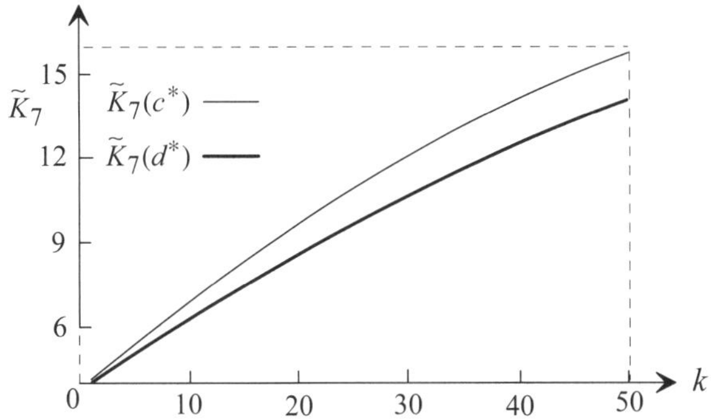

Figure 3.5 Cost comparison constant-decreasing repair cost limit

Under what condition is $K_7(d^*)$ smaller than $K_7(c^*)$? The inequality

$$K_7(d^*) = \frac{c_m}{\theta}\sqrt{4k-\pi} < K_7(c*) = \frac{4\,c_m}{\theta}\sqrt{\frac{k-1}{\pi}}$$

holds if and only if

$$k > k^* = \frac{16-\pi^2}{16-4\pi}.$$

Since $1.785 < k^* < 1.786$, this restriction is slightly stronger than $k > \pi/2$, but for the same reasons as given above, will have no negative impact on practical applications. Figure 3.5 compares the relative cost criteria

$$\widetilde{K}_7(d^*) = \frac{\theta}{c_m} K_7(d^*) \quad \text{and} \quad \widetilde{K}_7(c^*) = \frac{\theta}{c_m} K_7(c^*)$$

in dependence on k, $k \geq k^*$.

Age Dependent Repair Cost Till now it has been assumed that the repair cost C does not depend on the failure time. However, it is more realistic to assume that on average repair costs increase with increasing system age. Hence, let the cost of a repair, occurring at system age t, have a uniform distribution over $[a, a+bt]$ with $a \geq 0$ and $b > 0$. Then,

$$R_t(x) = P(C_t \leq x) = \begin{cases} 1, & 0 \leq t < \frac{x-a}{b} \\ \frac{x-a}{bt}, & \frac{x-a}{b} < t \end{cases}. \tag{3.104}$$

Constant repair cost limit For the sake of comparison, next a constant repair cost c limit is applied. Then, a failure at time t implies a replacement with probability

$$\overline{R}_t(c) = P(C_t > c) = \begin{cases} 0, & 0 \leq t < r \\ 1 - r/t, & r \leq t \end{cases},$$

where $r = (c-a)/b$. With the lifetime characteristics (3.99) and again making use of the integrals (3.100), the maintenance cost rate (3.90) reduces to

$$K_7(r) = \frac{\left[(r/\theta)^2 + \frac{r}{\theta}\sqrt{\pi}\,\right] c_m + c_r}{r + \theta\sqrt{\pi/4}}.$$

The value $r = r^*$ minimizing $K_7(r)$ is

$$r^* = \frac{\theta}{2}\left[\sqrt{4k - \pi} - \sqrt{\pi}\,\right],$$

where, as before, $k = c_r/c_m$. To make sure that $r^* > 0$, the inequality $k > \pi/2$ must be satisfied. Then corresponding optimal repair cost limit is $c^* = a + b\,r^*$. Its application yields

$$K_7(c^*) = \frac{c_m}{\theta}\sqrt{4k - \pi}\,. \tag{3.105}$$

Decreasing repair cost limit Let

$$c(t) = \begin{cases} c - dt, & 0 \le t \le c/d \\ 0, & c/d < t \end{cases}, \quad a < c, \ d > 0,$$

be a linearly decreasing repair cost limit and

$$r = (c - a)/b, \quad s = (b + d)/b, \quad y = (c - a)/d, \quad z = r/s\,.$$

Then, from (3.104), a system failure at age t implies a replacement with probability

$$\overline{R}_t(c(t)) = \begin{cases} 0, & 0 \le t < z \\ s\,(1 - z/t), & z \le t \le y \\ 1, & y < t \end{cases}.$$

If d is assumed to be sufficiently small, then letting $y = \infty$ will only have a negligibly small effect on the maintenance cost rate (3.90). Moreover, the replacement probability $\overline{R}_t(c(t))$ has the same functional structure as (3.98). Thus, for small d the minimal maintenance cost rate is again given by (3.102):

$$K_7(r^*(s), s) = \frac{c_m}{\theta}\sqrt{4(k-1) + \frac{4 - \pi}{s}}\,. \tag{3.106}$$

Note that, different to the definition of s by (3.97), now $s > 1$. Hence, with $K_7(c^*)$ given by (3.105), one easily verifies that

$$K_7(c^*) > K_7(r^*(s), s).$$

Thus, a linearly decreasing repair cost limit function must exist, which is more efficient than a constant repair cost limit. However, an optimal parameter $s = s^*$ cannot be constructed by minimizing (3.106), since $K_7(r^*(s), s)$ decreases with increasing s, but for (3.106) to be approximately valid, the assumption 's is sufficiently near to 1' had to be made.

The results obtained in this section indicate that the application of decreasing repair cost limits leads to lower maintenance cost rates than the application of constant repair cost limits if the total repair cost is progressively increasing in time.

3.3 RENEWAL PROCESSES

3.3.1 Definitions and Examples

The motivation for this chapter is a simple maintenance policy: A system is replaced on every failure by a statistically equivalent new one in negligible time and, after that, the new system (or the 'renewed system') immediately starts operating. In this context, the replacements of failed systems are also called *renewals*. The sequence of the system lifetimes after renewals generates a renewal process.

Definition 3.5 An *ordinary renewal process* is a sequence of nonnegative, independent, and identically distributed random variables $\{Y_i;\ i = 1, 2, ...\}$. ●

Thus, Y_i is the time between the $(i-1)$th and the ith renewal; $i = 1, 2, ...,\ Y_0 = 0$. Renewal processes do not only play an important role in engineering, but also in the natural, economical and social sciences. They are a basic stochastic tool for modeling particle counting, population development, and arrivals of customers at a service station. In the latter context, Y_i is the time between the arrival of the $(i-1)$th and the ith customer. Renewal processes are particularly important in actuarial risk analysis, namely for modeling the arrival of claims at an insurance company (section 3.5). In this chapter a terminology is adopted which refers to the 'simple maintenance policy'.

If the observation of a renewal process starts at time $t = 0$ and the process has already been operating for a while, then the lifetime of the system operating at time $t = 0$ is a 'residual lifetime' and will, therefore, usually not have the same probability distribution as the lifetime of a system after a renewal. Hence it makes sense to define a generalized renewal process by assuming that only the $Y_2, Y_3, ...$ are identically distributed. This leads to the following definition:

Definition 3.6 Let $\{Y_1, Y_2, ...\}$ be a sequence of nonnegative, independent random variables with property that Y_1 has distribution function

$$F_1(t) = P(Y_1 \le t),$$

whereas the random variables $Y_2, Y_3, ...$ are identically distributed as Y with distribution function

$$F(t) = P(Y \le t),\ F_1(t) \not\equiv F(t).$$

Then $\{Y_1, Y_2, ...\}$ is called a *delayed renewal process.* ●

The random time point at which the nth renewal takes place is

$$T_n = \Sigma_{i=1}^n Y_i; \quad n = 1, 2, ...$$

The random point process $\{T_1, T_2, ...\}$ is called the process of the *time points of renewals*. The time intervals between two neighbouring renewals are *renewal cycles*.

The corresponding counting process $\{N(t),\ t \ge 0\}$, defined by

$$N(t) = \begin{cases} \max(n;\ T_n \le t) \\ 0 \quad \text{for} \quad t < T_1 \end{cases},$$

is called *renewal counting process*. Note that $N(t)$ is the random number of renewals in $(0, t]$. The relationship

$$N(t) \ge n \ \text{ if and only if } \ T_n \le t, \tag{3.107}$$

implies

$$F_{T_n}(t) = P(T_n \le t) = P(N(t) \ge n). \tag{3.108}$$

Because of the independence of the Y_i, the distribution function $F_{T_n}(t)$ is the convolution of $F_1(t)$ with the $(n-1)$ th convolution power of F (see section 1.7.2):

$$F_{T_n}(t) = F_1 * F^{*(n-1)}(t), \ \ F^{*(0)}(t) \equiv 1, \ \ t \ge 0;\ \ n = 1, 2, \ldots \tag{3.109}$$

If the densities

$$f_1(t) = F_1'(t) \ \text{ and } \ f(t) = F'(t)$$

exist, then the density of T_n is

$$f_{T_n}(t) = f_1 * f^{*(n-1)}(t), \ \ f^{*(0)}(t) \equiv 1, \ \ t \ge 0;\ \ n = 1, 2, \ldots \tag{3.110}$$

Using (3.108) and

$$P(N(t) \ge n) = P(N(t) = n) + P(N(t) \ge n+1),$$

the probability distribution of $N(t)$ is seen to be

$$P(N(t) = n) = F_{T_n}(t) - F_{T_{n+1}}(t), \ \ F_{T_0}(t) \equiv 1;\ \ n = 0, 1, \ldots. \tag{3.111}$$

Example 3.10 Let $\{Y_1, Y_2, \ldots\}$ be an ordinary renewal process with property that the renewal cycle lengths Y_i have an exponential distribution with parameter λ :

$$F(t) = P(Y \le t) = 1 - e^{-\lambda t}, \ \ t \ge 0.$$

Then, by theorem 3.2, the corresponding counting process $\{ N(t),\ t \ge 0\}$ is the homogeneous Poisson process with intensity λ. In particular, by (3.21), T_n has an Erlang distribution with parameters n and λ:

$$F_{T_n}(t) = P(T_n \le t) = e^{-\lambda t} \sum_{i=n}^{\infty} \frac{(\lambda t)^i}{i!}. \qquad \square$$

Apart from the homogeneous Poisson process, there are two other important ordinary renewal processes for which the convolution powers of the renewal cycle length distributions explicitely exist so that the distribution functions of the renewal times T_n can be given:

1) Erlang Distribution The renewal cycle length Y has an Erlang distribution with parameters m and λ. Then, T_n is a sum of mn independent, identically distributed exponential random variables with parameter λ (example 1.23, section 1.7.2). Therefore, T_n has an Erlang distribution with parameters mn and λ :

$$F^{*(n)}(t) = P(T_n \le t) = e^{-\lambda t} \sum_{i=mn}^{\infty} \frac{(\lambda t)^i}{i!}, \quad t \ge 0. \tag{3.112}$$

This result is of general importance, since the probability distribution of any nonnegative random variable can be arbitrarily accurately approximated by an Erlang distribution by proper choice of the parameters of this distribution.

2) Normal Distribution Let the renewal cycle length Y have a normal distribution with parameters μ and σ, $\mu > 3\sigma$. (The assumption $\mu > 3\sigma$ is necesssary for making sure that the cycle lengths are practically nonnegative. However, renewal theory has been extended to negative 'cycle lengths'.) Since the sum of independent, normally distributed random variables is again normally distributed, where the parameters of the sum are obtained by summing up the parameters of the summands (example 1.24, section 1.7.2), T_n has distribution function

$$F^{*(n)}(t) = P(T_n \le t) = \Phi\left(\frac{t - n\mu}{\sigma\sqrt{n}}\right), \quad t \ge 0. \tag{3.113}$$

This result also has a more general meaning: Since T_n is the sum of n independent, identically distributed random variables, then, by the central limit theorem 1.9, T_n has approximately the distribution function (3.113) if n is sufficiently large, i.e.

$$T_n \approx N(n\mu, \sigma^2 n) \text{ if } n \ge 20.$$

Example 3.11 The distribution function of T_n can be used to solve the so-called *spare part problem*: How many spare parts (spare systems) are absolutely necessary for making sure that the renewal process can be maintained over the intervall $[0, t]$ with probability $1 - \alpha$?

This requires the computation of the smallest integer n satisfying

$$1 - F_{T_n}(t) = P(N(t) \le n) \ge 1 - \alpha\,.$$

For instance, let be

$$\mu = E(Y) = 8 \text{ and } \sigma^2 = Var(Y) = 25.$$

If $t = 200$ and $1 - \alpha = 0.99$, then

$$1 - F_{T_n}(200) = 1 - \Phi\left(\frac{200-8\,n}{5\sqrt{n}}\right) \ge 1 - \alpha = 0.99$$

is equivalent to

$$z_{0.01} = 2.32 \le \frac{8\,n-200}{5\sqrt{n}}\,.$$

Thus, at least $n_{min} = 34$ spare parts have to be in stock to ensure that with probability 0.99 every failed part can be replaced by a new one over the interval $(0, 200]$. In view of $n_{min} \geq 20$, the application of the normal approximation to the distribution of T_n is justified. □

3.3.2 Renewal Function

3.3.2.1 Renewal Equations

The mean number of renewals which occur in a given time interval is of great practical and theoretical importance.

Definition 3.7 The mean value of the random number $N(t)$ of renewals occurring in $(0, t]$ as a function of t is called *renewal function.* ●

Thus, with the terminology and the notation introduced in section 2.2, the renewal function is the trend function of the renewal counting process $\{N(t), t \geq 0\}$:

$$m(t) = E(N(t)).$$

However, to be in line with the majority of publications on renewal theory, in what follows, the renewal functions belonging to an ordinary and a delayed renewal process are denoted as $H(t)$ and $H_1(t)$, respectively.

If not stated otherwise, it is assumed throughout section 3 that the densities of Y and Y_1 exist. Hence,

$$dF(t) = f(t)\,dt \quad \text{and} \quad dF_1(t) = f_1(t)\,dt.$$

In this case, the first derivatives of $H_1(t)$ and $H(t)$ also exist:

$$h_1(t) = \frac{dH_1(t)}{dt}, \quad h(t) = \frac{dH(t)}{dt}.$$

The functions $h_1(t)$ and $h(t)$ are the *renewal densities* of a delayed and of an ordinary renewal process, respectively. From (1.15), a sum representation of the renewal function is

$$H_1(t) = E(N(t)) = \Sigma_{n=1}^{\infty} P(N(t) \geq n). \tag{3.114}$$

In view of (3.108) and (3.109),

$$H_1(t) = \Sigma_{n=1}^{\infty} F_1 * F^{*(n-1)}(t). \tag{3.115}$$

In particular, the renewal function of an ordinary renewal process is

$$H(t) = \Sigma_{n=1}^{\infty} F^{*(n)}(t). \tag{3.116}$$

By differentiation of (3.114) and (3.115) with respect to t, one obtains sum representations of the respective renewal densities:

$$h_1(t) = \Sigma_{n=1}^{\infty} f_1 * f^{*(n-1)}(t), \quad h(t) = \Sigma_{n=1}^{\infty} f^{*(n)}(t).$$

Remark These sum representations allow a useful probabilistic interpretation of the renewal density: For Δt sufficiently small,

$$h_1(t)\,\Delta t$$

is approximately equal to the probability of the occurrence of a renewal in the interval $[t, t+\Delta t]$.

In view of (3.115) and the definition of the convolution power of distribution functions,

$$\begin{aligned} H_1(t) &= \Sigma_{n=0}^{\infty} F_1 * F^{*(n)}(t) \\ &= F_1(t) + \Sigma_{n=1}^{\infty} \int_0^t F_1 * F^{*(n-1)}(t-x)\, dF(x) \\ &= F_1(t) + \int_0^t \Sigma_{n=1}^{\infty} \left(F_1 * F^{*(n-1)}(t-x)\right) dF(x). \end{aligned}$$

Again by (3.115), the integrand is equal to $H_1(t-x)$. Hence, $H_1(t)$ satisfies

$$H_1(t) = F_1(t) + \int_0^t H_1(t-x)\, dF(x). \tag{3.117}$$

According to (1.32), the integral in (3.117) is the convolution $H_1 * f$ of the renewal function H_1 with f. In particular, the renewal function $H(t)$ of an ordinary renewal process satisfies

$$H(t) = F(t) + \int_0^t H(t-x)\, dF(x). \tag{3.118}$$

Another derivation of formula (3.118) can be done by conditioning with regard to the time point of the first renewal: Given that the first renewal occurs at time x, the mean number of renewals in $[0, t]$ is

$$[1 + H(t-x)], \quad 0 < x \le t.$$

Since the first renewal occurs at time x with 'probability' $dF(x) = f(x)\, dx$, taking into account all possible values of x, yields (3.118). The same argument yields an integral equation for the renewal function of a delayed renewal process:

$$H_1(t) = F_1(t) + \int_0^t H(t-x)\, dF_1(x). \tag{3.119}$$

This is because after the first renewal at time x the process develops in $(x, t]$ as an ordinary renewal process. By partial integration of the convolutions, the renewal equations can be rewritten. For instance, integral equation (3.117) is equivalent to

$$H_1(t) = F_1(t) + \int_0^t F(t-x)\, dH_1(x). \tag{3.120}$$

By differentiating the renewal equations (3.117) to (3.119) with respect to t, one obtains the following integral equations of renewal type for $h_1(t)$ and $h(t)$:

$$h_1(t) = f_1(t) + \int_0^t h_1(t-x)f(x)\,dx, \tag{3.121}$$

$$h(t) = f(t) + \int_0^t h(t-x)f(x)\,dx, \tag{3.122}$$

$$h_1(t) = f_1(t) + \int_0^t h(t-x)f_1(x)\,dx. \tag{3.123}$$

Generally, solutions of integral equations of renewal type can only be obtained by numerical methods. However, since all these integral equations involve convolutions, it is easily possible to find their solutions in the image space of the Laplace transformation: Let $\hat{h}_1(s)$, $\hat{h}(s)$, $\hat{f}_1(s)$ and $\hat{f}(s)$ be the respective Laplace transforms of $h_1(t)$, $h(t)$, $f_1(t)$ and $f(t)$. Then, by (1.33), applying the Laplace transform to the integral equations (3.121) and (3.122) yields algebraic equations for $\hat{h}_1(s)$ and $h(s)$:

$$\hat{h}_1(s) = \hat{f}_1(s) + \hat{h}_1(s)\cdot\hat{f}(s), \qquad \hat{h}(s) = \hat{f}(s) + \hat{h}(s)\cdot\hat{f}(s).$$

The solutions are

$$\hat{h}_1(s) = \frac{\hat{f}_1(s)}{1-\hat{f}(s)}, \qquad \hat{h}(s) = \frac{\hat{f}(s)}{1-\hat{f}(s)}. \tag{3.124}$$

Thus, for ordinary renewal processes there is a one-to-one correspondence between the renewal function and the probability distribution of the cycle length. By (1.29), the Laplace transforms of the corresponding renewal functions are

$$\hat{H}_1(s) = \frac{\hat{f}_1(s)}{s\,(1-\hat{f}(s))}, \qquad \hat{H}(s) = \frac{\hat{f}(s)}{s\,(1-\hat{f}(s))}. \tag{3.125}$$

Integral Equations of Renewal Type The integral equations (3.117) to (3.119) and the equivalent ones derived from these are called *renewal equations.* They belong to the broader class of integral equations of renewal type. A function $Z(t)$ is said to satisfy an *integral equation of renewal type* if for any function $g(t)$, which is bounded on intervals of finite length, and for any distribution function $F(t)$ with probability density $f(t)$,

$$Z(t) = g(t) + \int_0^t Z(t-x)f(x)dx. \tag{3.126}$$

The unique solution of this integral equation is

$$Z(t) = g(t) + \int_0^t g(t-x)h(x)\,dx, \tag{3.127}$$

where $h(t)$ is the renewal density of the ordinary renewal process generated by $f(t)$. For a proof, see Feller [28]. A function $Z(t)$ given by (3.127) need not be the trend function of a renewal counting process.

Example 3.12 Let

$$f_1(t) = f(t) = \lambda e^{-\lambda t}, \ t \ge 0.$$

The Laplace transform of $f(t)$ is

$$\hat{f}(s) = \frac{\lambda}{s+\lambda}.$$

By (3.125),

$$\hat{H}(s) = \frac{\lambda}{s+\lambda} \Big/ \left(s - \frac{\lambda s}{s+\lambda}\right) = \frac{\lambda}{s^2}.$$

The corresponding pre-image is

$$H(t) = \lambda t.$$

Thus, an ordinary renewal process has exponentially with parameter λ distributed cycle lengths if and only if its renewal function is given by $H(t) = \lambda t$. □

Example 3.13 Let the cycle length of an ordinary renewal process have distribution function

$$F(t) = (1 - e^{-t})^2, \ t \ge 0.$$

Thus, $\overline{F}(t) = 1 - F(t)$ can be thought of the survival function of a parallel system consisting of two subsystems, whose lifetimes are independent, identically distributed exponential random variables with parameter $\lambda = 1$. The corresponding probability density and its Laplace transform are

$$f(t) = 2\,(e^{-t} - e^{-2t}) \quad \text{and} \quad \hat{f}(s) = \frac{2}{(s+1)(s+2)}.$$

From (3.124), the Laplace transform of the corresponding renewal density is

$$\hat{h}(s) = \frac{2}{s\,(s+3)}.$$

By splitting the fraction into partial fractions, the pre-image of $\hat{h}(s)$ is seen to be

$$h(t) = \frac{2}{3}\,(1 - e^{-3t}).$$

Integration yields the renewal function:

$$H(t) = \frac{2}{3}\left[t + \frac{1}{3}\left(e^{-3t} - 1\right)\right].$$ □

Explicit formulas for the renewal function of ordinary renewal processes exist for the following two classes of cycle length distributions:

1) Erlang Distribution Let the cycle lengths be Erlang distributed with parameters m and λ. Then, by (3.108), (3.112), and (3.116),

$$H(t) = e^{-\lambda t} \sum_{n=1}^{\infty} \sum_{i=mn}^{\infty} \frac{(\lambda t)^i}{i!}.$$

Special cases are:

$m = 1: \quad H(t) = \lambda t$ (*homogeneous Poisson process*)

$$m = 2: \quad H(t) = \frac{1}{2}\left[\lambda t - \frac{1}{2} + \frac{1}{2}e^{-2\lambda t}\right]$$

$$m = 3: \quad H(t) = \frac{1}{3}\left[\lambda t - 1 + \frac{2}{\sqrt{3}}e^{-1,5\lambda t}\sin\left(\frac{\sqrt{3}}{2}\lambda t + \frac{\pi}{3}\right)\right]$$

$$m = 4: \quad H(t) = \frac{1}{4}\left[\lambda t - \frac{3}{2} + \frac{1}{2}e^{-2\lambda t} + \sqrt{2}\,e^{-\lambda t}\sin\left(\lambda t + \frac{\pi}{4}\right)\right].$$

2) Normal distribution Let the cycle lengths be normally distributed with mean value μ and variance σ^2, $\mu > 3\sigma^2$. From (3.108), (3.113) and (3.116),

$$H(t) = \sum_{n=1}^{\infty} \Phi\left(\frac{t - n\mu}{\sigma\sqrt{n}}\right). \tag{3.128}$$

This sum representation is very convenient for numerical computations, since only the sum of the first few terms approximates the renewal function with sufficient accuracy.

As shown in example 3.12, an ordinary renewal process has renewal function

$$H(t) = \lambda t = t/\mu \quad \text{if and only if } f(t) = \lambda e^{-\lambda t},\ t \geq 0,$$

where $\mu = E(Y)$. Hence an interesting question is, whether, for given $F(t)$, a delayed renewal process exists which also has renewal function $H_1(t) = t/\mu$.

Theorem 3.10 Let $\{Y_1, Y_2, ...\}$ be a delayed renewal process with cycle lengths $Y_2, Y_3, ...$ being identically distributed as Y. If Y has finite mean value μ and distribution function $F(t) = P(Y \leq t)$, then $\{Y_1, Y_2, ...\}$ has renewal function

$$H_1(t) = t/\mu \tag{3.129}$$

if and only if the length of the first renewal cycle Y_1 has density $f_1(t) \equiv f_S(t)$, where

$$f_S(t) = \frac{1}{\mu}(1 - F(t)), \quad t \geq 0. \tag{3.130}$$

Equivalently, $\{Y_1, Y_2, ...\}$ has renewal function (3.129) if and only if Y_1 has distribution function $F_1(t) \equiv F_S(t)$ with

$$F_S(t) = \frac{1}{\mu}\int_0^t (1 - F(x))\,dx, \quad t \geq 0. \tag{3.131}$$

Proof Let $\hat{f}(s)$ and $\hat{f}_S(s)$ be the respective Laplace transforms of $f(t)$ and $f_S(t)$. Then, by applying the Laplace transformation to both sides of (3.130) and taking into account (1.29),

$$\hat{f}_S(s) = \frac{1}{\mu s}(1 - \hat{f}(s)).$$

Replacing in the first equation of (3.125) $\hat{f}_1(s)$ with $\hat{f}_S(s)$ yields the Laplace transform of the corresponding renewal function $H_1(t) = H_S(t)$:

$$\hat{H}_S(s) = 1/(\mu s^2).$$

Retransformation of $\hat{H}_S(s)$ gives the desired result: $H_S(t) = t/\mu$. ■

The random variable S with density (3.130) (distribution function (3.131)) plays an important role in characterizing stationary renewal processes (section 3.3.5). Moreover, this distribution type already occurred in section 1.4 in connection with the *NBUE*-distribution (formula (1.45). The first two moments of S are (exercise 3.24)

$$E(S) = \frac{\mu^2 + \sigma^2}{2\mu} \quad \text{and} \quad E(S^2) = \frac{\mu_3}{3\mu}, \tag{3.132}$$

where

$$\sigma^2 = Var(Y) \text{ and } \mu_3 = E(Y^3).$$

Higher Moments of *N(t)* Apart from the renewal function, which is the first moment of $N(t)$, higher moments of $N(t)$ also have some importance, in particular when investigating the behaviour of the renewal function as $t \to \infty$.

Let $\{Y_1, Y_2, ...\}$ an ordinary renewal process and $\{N(t), t \geq 0\}$ its corresponding renewal counting process. Then, moments of higher order can be derived from binomial moments of $N(t)$. The *binomial moment of the order n* of $N(t)$ is defined as

$$E\binom{N(t)}{n} = \frac{1}{n!} E\{[N(t)][N(t)-1]\cdots[N(t)-(n-1)]\}. \tag{3.133}$$

The binomial moment of order n of $N(t)$ is equal to the nth convolution power of the renewal function:

$$E\binom{N(t)}{n} = H^{*(n)}(t).$$

Specifically, for $n = 2$,

$$E\binom{N(t)}{2} = \frac{1}{2} E\{[N(t)][N(t)-1]\} = \frac{1}{2}\left\{E[N(t)]^2 - H(t)\right\} = H^{*(2)}(t)$$

so that the variance of $N(t)$ is equal to

$$Var(N(t)) = 2\int_0^t H(t-x)\, dH(x) + H(t) - [H(t)]^2.$$

Since

$$H(t-x) \leq H(t) \quad \text{for } 0 \leq x \leq t,$$

this equation implies an upper bound for the variance of $N(t)$:

$$Var(N(t)) \leq [H(t)]^2 + H(t).$$

3.3.2.2 Bounds on the Renewal Function

Generally, integral equations of renewal type have to be solved by numerical methods. Hence, bounds on $H(t)$, which only require information on one or more numerical parameters of the cycle length distribution, are of special interest. This section presents bounds on the renewal function of ordinary renewal processes.

1) Elementary Bounds By definition of T_n,

$$\max_{1 \le i \le n} Y_i \le \Sigma_{i=1}^{n} Y_i = T_n .$$

Hence, for any t with $F(t) < 1$,

$$F^{*(n)}(t) = P(T_n \le t) \le P(\max_{1 \le i \le n} Y_i \le t) = [F(t)]^n .$$

Summing from $n = 1$ to ∞ on both sides of this inequality, the sum representation of the renewal function (3.116) and the geometric series yield

$$F(t) \le H(t) \le \frac{F(t)}{1 - F(t)} .$$

Note that the left-hand side of this inequality is the first term of the sum (3.116). These 'elementary bounds' are only useful for small t.

2) Linear Bounds Let $\mathbf{F} = \{t;\ t \ge 0,\ F(t) < 1\}$, $\mu = E(Y)$, $\overline{F}(t) = 1 - F(t)$, and

$$a_0 = \inf_{t \in \mathbf{F}} \frac{F(t) - F_S(t)}{\overline{F}(t)}, \qquad a_1 = \sup_{t \in \mathbf{F}} \frac{F(t) - F_S(t)}{\overline{F}(t)},$$

where $F_S(t)$ is given by (3.131). Then (Marshall [59])

$$\frac{t}{\mu} + a_0 \le H(t) \le \frac{t}{\mu} + a_1 . \tag{3.134}$$

The derivation of these bounds is straightforward and very instructive: According to the definition of a_0 and a_1,

$$a_0 \overline{F}(t) \le F(t) - F_S(t) \le a_1 \overline{F}(t) .$$

Convolution of both sides with $F^{*(n)}(t)$ leads to

$$a_0\Big[F^{*(n)}(t) - F^{*(n+1)}(t)\Big] \le F^{*(n+1)}(t) - F_S * F^{*(n)}(t) \le a_1\Big[F^{*(n)}(t) - F^{*(n+1)}(t)\Big].$$

In view of (3.116) and theorem 3.10, summing up from $n = 0$ to ∞ on both sides of this inequality proves (3.134). Since

$$\frac{F(t) - F_S(t)}{\overline{F}(t)} \ge -F_S(t) \ge -1 \quad \text{for all } t \ge 0,$$

formula (3.134) implies a simpler lower bound on $H(t)$:

$$H(t) \ge \frac{t}{\mu} - F_S(t) \ge \frac{t}{\mu} - 1 .$$

Let

$$\lambda_S(t) = f_S(t)/\overline{F}_S(t)$$

be the failure rate belonging to $F_S(t)$:

$$\lambda_S(t) = \frac{\overline{F}(t)}{\int_t^\infty \overline{F}(x)\,dx}.$$

Then a_0 and a_1 can be rewritten as follows:

$$a_0 = \frac{1}{\mu} \inf_{t \in \mathbf{F}} \frac{1}{\lambda_S(t)} - 1 \quad \text{and} \quad a_1 = \frac{1}{\mu} \sup_{t \in \mathbf{F}} \frac{1}{\lambda_S(t)} - 1.$$

Thus, (3.134) becomes

$$\frac{t}{\mu} + \frac{1}{\mu} \inf_{t \in \mathbf{F}} \frac{1}{\lambda_S(t)} - 1 \le H(t) \le \frac{t}{\mu} + \frac{1}{\mu} \sup_{t \in \mathbf{F}} \frac{1}{\lambda_S(t)} - 1. \tag{3.135}$$

Since

$$\inf_{t \in \mathbf{F}} \lambda(t) \le \inf_{t \in \mathbf{F}} \lambda_S(t) \text{ and } \sup_{t \in \mathbf{F}} \lambda(t) \ge \sup_{t \in \mathbf{F}} \lambda_S(t),$$

the bounds (3.135) can be simplified:

$$\frac{t}{\mu} + \frac{1}{\mu} \inf_{t \in \mathbf{F}} \frac{1}{\lambda(t)} - 1 \le H(t) \le \frac{t}{\mu} + \frac{1}{\mu} \sup_{t \in \mathbf{F}} \frac{1}{\lambda(t)} - 1. \tag{3.136}$$

3) Upper Bound If $\mu = E(Y)$ and $\mu_2 = E(Y^2)$, then (Lorden [55])

$$H(t) \le \frac{t}{\mu} + \frac{\mu_2}{\mu^2} - 1. \tag{3.137}$$

4) Upper Bound for IFR If $F(t)$ is *IFR*, then (3.137) can be improved (Brown [14]):

$$H(t) \le \frac{t}{\mu} + \frac{\mu_2}{2\mu^2} - 1.$$

5) Two-Sided Bounds for IFR If $F(t)$ is *IFR*, then (Barlow and Proschan [4])

$$\frac{t}{\int_0^t \overline{F}(x)\,dx} - 1 \le H(t) \le \frac{t\,F(t)}{\int_0^t \overline{F}(x)\,dx}, \quad t > 0. \tag{3.138}$$

Example 3.14 As in example 3.13, let be

$$F(t) = (1 - e^{-t})^2, \quad t \ge 0,$$

the distribution function of the cycle length Y of an ordinary renewal process. In this case, $\mu = E(Y) = 3/2$ and

$$\overline{F}_S(t) = \frac{1}{\mu} \int_t^\infty \overline{F}(x)\,dx = \frac{2}{3}\left(2 - \frac{1}{2}e^{-t}\right)e^{-t}, \quad t \ge 0.$$

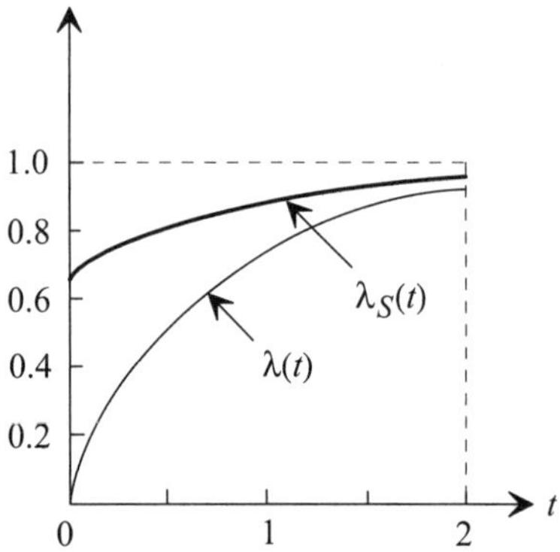

Figure 3.6 Failure rates

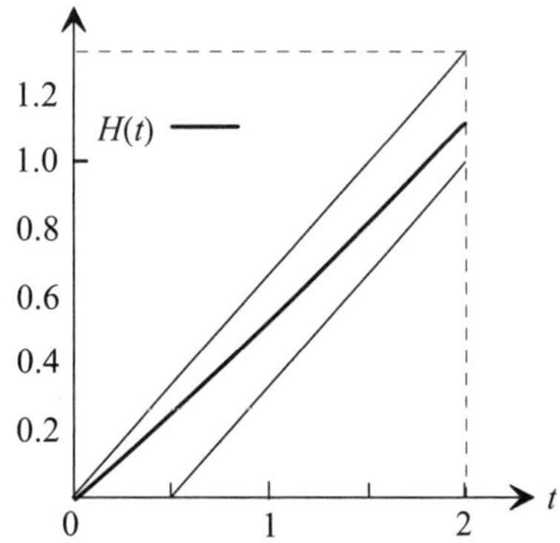

Figure 3.7 Bounds for the renewal function

Therefore, the failure rates belonging to $F(t)$ and $F_S(t)$ are (Figure 3.6)

$$\lambda(t) = \frac{2(1-e^{-t})}{2-e^{-t}}, \quad \lambda_S(t) = 2\,\frac{2-e^{-t}}{4-e^{-t}}, \quad t \geq 0.$$

Both failure rates are strictly increasing in t and have, moreover, the properties

$$\lambda(0) = 0, \quad \lambda(\infty) = 1,$$
$$\lambda_S(0) = 2/3, \quad \lambda_S(\infty) = 1.$$

Hence, the respective bounds (3.135) and (3.136) are

$$\frac{2}{3}t - \frac{1}{3} \leq H(t) \leq \frac{2}{3}t \ \text{ and } \ \frac{2}{3}t - \frac{1}{3} \leq H(t) \leq \infty.$$

In this case, the upper bound in (3.136) contains no information on the renewal function. Figure 3.7 compares the bounds (3.135) with the exact graph of the renewal function given in example 3.13. The deviation of the lower bound from $H(t)$ is negligibly small for $t \geq 3$. □

3.3.3 Asymptotic Behaviour

This section investigates the behaviour of the renewal counting process $\{N(t), t \geq 0\}$ and its trend function as $t \to \infty$. The results allow the construction of estimates of the renewal function and of the probability distribution of $N(t)$ if t is sufficiently large. Throughout this section, it is assumed that both $E(Y_1)$ and $E(Y) = \mu$ are finite. Some of the key results require that the cycle length Y or, equivalently, its distribution function, is *nonarithmetic*, i.e. that there is no positive constant a with property that the possible values of Y are multiples of a. Correspondingly, Y is called *arithmetic* if there is a constant a so that Y has range $\mathbf{R} = \{0, a, 2a, ...\}$. (The set $\mathbf{R}$ consists of all possible values, which Y can assume.) A continuous random variable is always nonarithmetic.

A simple consequence of the strong law of the large numbers is

$$P\left(\lim_{t\to\infty}\frac{N(t)}{t}=\frac{1}{\mu}\right)=1. \tag{3.139}$$

To avoid technicalities, the proof is given for an ordinary renewal process: The inequality

$$T_{N(t)}\le t<T_{N(t)+1}$$

implies that

$$\frac{T_{N(t)}}{N(t)}\le\frac{t}{N(t)}<\frac{T_{N(t)+1}}{N(t)}=\frac{T_{N(t)+1}}{N(t)+1}\,\frac{N(t)+1}{N(t)}$$

or, equivalently, that

$$\frac{1}{N(t)}\sum_{i=1}^{N(t)}Y_i\le\frac{t}{N(t)}<\left[\frac{1}{N(t)+1}\sum_{i=1}^{N(t)+1}Y_i\right]\frac{N(t)+1}{N(t)}.$$

Since by assumption $\mu=E(Y)<\infty$, $N(t)$ tends to infinity as $t\to\infty$. Hence, theorem 1.8 yields the desired result (3.139). For μ being the mean distance between two renewals, this result is quite intuitive. The following theorem considers the corresponding limit behaviour of the mean value of $N(t)$. As with theorems 3.12 and 3.13, no proof is given.

Theorem 3.11 (*elementary renewal theorem*) The renewal function satisfies

$$\lim_{t\to\infty}\frac{H_1(t)}{t}=\frac{1}{\mu}. \qquad \blacksquare$$

Thus, for large t, $H_1(t)\approx t/\mu$. The theorem shows that for $t\to\infty$ the influence of the first renewal interval with possibly $E(Y_1)\neq\mu$ fades away. (For this property to be valid, the assumption $E(Y_1)<\infty$ had to be made.) In terms of the renewal density, the analogue to theorem 3.11 is

$$\lim_{t\to\infty}h_1(t)=\frac{1}{\mu}.$$

Note that (1.139) does not imply theorem 3.11. The following theorem was called the *fundamental renewal theorem* by its discoverer *W. L. Smith.*

Theorem 3.12 (*fundamental renewal theorem*) If $F(t)$ is nonarithmetic and $g(t)$ an integrable function on $[0,\infty)$, then

$$\lim_{t\to\infty}\int_0^t g(t-x)\,dH_1(x)=\frac{1}{\mu}\int_0^\infty g(x)\,dx. \qquad \blacksquare$$

The fundamental renewal theorem (or *key renewal theorem*, *theorem of Smith*) has proved a useful tool for solving many problems in applied probability theory and stochastic modeling. Theorem 3.13 gives another variant of the fundamental renewal theorem. It refers to the integral equation of renewal type (3.126).

Theorem 3.13 Let $g(t)$ be an integrable function on $[0, \infty)$ and $f(t)$ a probability density. If $Z(t)$ satisfies the equation of renewal type (3.126), namely

$$Z(t) = g(t) + \int_0^t Z(t-x) f(x)\, dx,$$

then

$$\lim_{t \to \infty} Z(t) = \frac{1}{\mu} \int_0^\infty g(x)\, dx. \tag{3.140}$$

■

Proofs of the now 'classic' theorems 3.11 to 3.13 can be found in [28]. The equivalence of the theorems 3.12 and 3.12 results from the structure (3.127) of $Z(t)$.

Blackwell's renewal theorem Let

$$g(x) = \begin{cases} 1 & \text{for } 0 \le x \le h \\ 0 & \text{elsewhere} \end{cases}.$$

Then the fundamental renewal theorem implies *Blackwell's renewal theorem*: If $F(t)$ is nonarithmetic, then, for any $h > 0$,

$$\lim_{t \to \infty} [H_1(t+h) - H_1(t)] = \frac{h}{\mu}. \tag{3.141}$$

Whereas the elementary renewal theorem refers to 'a global transition' into the stationary regime, Blackwell's renewal theorem refers to the corresponding 'local behaviour' in a time interval of length h.

Theorem 3.14 If $F(t)$ is nonarithmetic and $\sigma^2 = Var(Y) < \infty$, then

$$\lim_{t \to \infty} \left(H_1(t) - \frac{t}{\mu} \right) = \frac{\sigma^2}{2\mu^2} - \frac{E(Y_1)}{\mu} + \frac{1}{2}. \tag{3.142}$$

Proof The renewal equation (3.120) is equivalent to

$$H_1(t) = F_1(t) + \int_0^t F_1(t-x)\, dH(x). \tag{3.143}$$

If $F_1(t) \equiv F_S(t)$, then, by theorem 3.10, this integral equation becomes

$$\frac{t}{\mu} = F_S(t) + \int_0^t F_S(t-x)\, dH(x). \tag{3.144}$$

By subtracting integral equation (3.144) from integral equation (3.143),

$$H_1(t) - \frac{t}{\mu} = \overline{F}_S(t) - \overline{F}_1(t) + \int_0^t \overline{F}_S(t-x)\, dH(x) - \int_0^t \overline{F}_1(t-x)\, dH(x).$$

Applying the fundamental renewal theorem yields

$$\lim_{t \to \infty} \left(H_1(t) - \frac{t}{\mu} \right) = \frac{1}{\mu} \int_0^\infty \overline{F}_S(x)\, d(x) - \frac{1}{\mu} \int_0^\infty \overline{F}_1(x)\, d(x).$$

Now the desired results follows from (1.17) and (3.132). ■

For ordinary renewal processes, (3.142) simplifies to

$$\lim_{t\to\infty}\left(H(t)-\frac{t}{\mu}\right)=\frac{1}{2}\left(\frac{\sigma^2}{\mu^2}-1\right). \tag{3.145}$$

Corollary Under the assumptions of theorem 3.14, the fundamental renewal theorem implies the elementary renewal theorem.

Theorem 3.15 For an ordinary renewal process, the integrated renewal function has property

$$\lim_{t\to\infty}\left\{\int_0^t H(x)\,dx-\left[\frac{t^2}{2\mu}+\left(\frac{\mu_2}{2\mu^2}-1\right)t\right]\right\}=\frac{\mu_2^2}{4\mu^3}-\frac{\mu_3}{6\mu^2}$$

with $\mu_2=E(Y^2)$ and $\mu_3=E(Y^3)$. ■

For a proof see, for instance, Tijms [81]. The following theorem is basically a consequence of the central limit theorem (for details see Karlin and Taylor [45]).

Theorem 3.16 The random number $N(t)$ of renewals in $[0,t]$ satisfies

$$\lim_{t\to\infty}P\left(\frac{N(t)-t/\mu}{\sigma\sqrt{t\mu^{-3}}}\le x\right)=\Phi(x).$$ ■

Thus, for t sufficiently large, $N(t)$ is approximately normally distributed with mean value t/μ and variance $\sigma^2 t/\mu^3$:

$$N(t)\approx N(t/\mu,\ \sigma^2 t/\mu^3). \tag{3.146}$$

Hence, theorem 3.16 can be used to construct approximate intervals, which contain $N(t)$ with a given probability: If t is sufficiently large, then

$$P\left(\frac{t}{\mu}-z_{\alpha/2}\,\sigma\sqrt{t\mu^{-3}}\le N(t)\le\frac{t}{\mu}+z_{\alpha/2}\,\sigma\sqrt{t\mu^{-3}}\right)=1-\alpha. \tag{3.147}$$

As usual, $z_{\alpha/2}$ is the $(1-\alpha/2)$- percentile of the standard normal distribution.

Example 3.15 Let $t=1000,\ \mu=10,\ \sigma=2,$ and $\alpha=0.05$. Since $z_{0.025}\approx 2$,

$$P(96\le N(t)\le 104)=0.95.$$ □

Knowledge of the asymptotic distribution of $N(t)$ makes it possible, without knowing the exact distribution of Y, to approximately answer a question which already arose in section 3.3.1: How many spare systems (spare parts) are necessary for guaranteeing that the (ordinary) renewal process can be maintained over an interval $[0,t]$ with a given probability of $1-\alpha$?

Since with probability $1-\alpha$,

$$\frac{N(t)-t/\mu}{\sigma\sqrt{t\mu^{-3}}} \le z_\alpha ,$$

for large t the required number $n_{\min}$ is approximately equal to

$$n_{\min} \approx \frac{t}{\mu} + z_\alpha \sigma \sqrt{t\mu^{-3}} \,. \tag{3.148}$$

Example 3.16 The same numerical parameters as in example 3.11 are considered:

$$t = 200, \ \mu = 8, \ \sigma^2 = 25, \ \text{and} \ \alpha = 0.01.$$

Since $z_{0.01} = 2.32$,

$$n_{\min} \ge \frac{200}{8} + 2.32 \cdot 5\sqrt{200 \cdot 8^{-3}} \ = 32.25.$$

Hence, about 33 spare parts are needed to make sure that with probability 0.99 the renewal process can be maintained over a period of 200 time units. (Formula (3.113) applied in example 3.11 yielded $n_{\min} = 34$.) □

3.3.4 Recurrence Times

For any point processes, recurrence times have been defined by (3.3) and (3.5). In particular, if $\{Y_1, Y_2, ...\}$ is a renewal process and $\{T_1, T_2, ...\}$ is the corresponding process of renewal time points, then its (random) *forward recurrence time* $A(t)$ is

$$A(t) = T_{N(t)+1} - t$$

and its (random) *backward recurrence time* $B(t)$ is

$$B(t) = t - T_{N(t)}.$$

$A(t)$ is the residual lifetime and $B(t)$ the age of the system operating at time t.

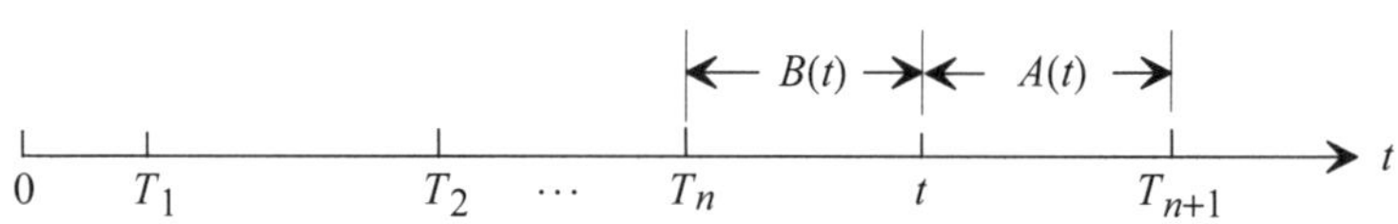

Figure 3.8 Illustration of the recurrence times

The stochastic processes

$$\{Y_1, Y_2, ...\}, \{T_1, T_2, ...\}, \{N(t), t \ge 0\}, \ \{A(t), t \ge 0\}, \ \text{and} \ \{B(t), t \ge 0\}$$

are statistically equivalent, since there is a one to one correspondence between their sample paths, i.e. each of these five processes can be used to define a renewal process (Figures 3.8 and 3.9).

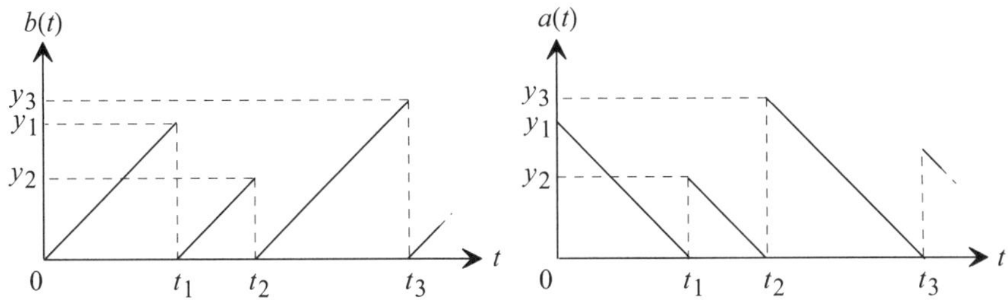

Figure 3.9 Sample paths of the backward and forward recurrence times processes

Let

$$F_{A(t)}(x) = P(A(t) \le x) \quad \text{and} \quad F_{B(t)}(x) = P(B(t) \le x)$$

be the distribution functions of the forward and the backward recurrence times, respectively. Then, for $0 < x < t$, by making use of (3.115),

$$\begin{aligned}
F_{A(t)}(x) &= P(T_{N(t)+1} - t \le x) \\
&= \textstyle\sum_{n=0}^{\infty} P(T_{N(t)+1} \le t+x,\, N(t)=n) \\
&= F_1(t+x) - F_1(t) + \textstyle\sum_{n=1}^{\infty} P(T_n \le t < T_{n+1} \le t+x) \\
&= F_1(t+x) - F_1(t) + \textstyle\sum_{n=1}^{\infty} \int_0^t [F(x+t-y) - F(t-y)]\, dF_{T_n}(y) \\
&= F_1(t+x) - F_1(t) + \int_0^t [F(x+t-y) - F(t-y)] \textstyle\sum_{n=1}^{\infty} dF_{T_n}(y) \\
&= F_1(t+x) - F_1(t) + \int_0^t [F(x+t-y) - F(t-y)] \textstyle\sum_{n=1}^{\infty} d(F_1 * F^{*(n-1)}(y)) \\
&= F_1(t+x) - F_1(t) + \int_0^t [F(x+t-y) - F(t-y)]\, d\left(\textstyle\sum_{n=1}^{\infty} F_1 * F^{*(n-1)}(y)\right) \\
&= F_1(t+x) - F_1(t) + \int_0^t [F(x+t-y) - F(t-y)]\, dH_1(y).
\end{aligned}$$

This representation of $F_{A(t)}$ can be simplified by combining it with (3.120). The result is

$$F_{A(t)}(x) = F_1(t+x) - \int_0^t \overline{F}(x+t-y)\, dH_1(y); \quad x, t \ge 0. \tag{3.149}$$

Differentiation yields the probability density of $A(t)$:

$$f_{A(t)}(x) = f_1(t+x) + \int_0^t f(x+t-y)\, h_1(y)\, dy; \quad x, t \ge 0. \tag{3.150}$$

$\overline{F}_{A(t)}(x) = 1 - F_{A(t)}(x)$ is the probability that the system, which is working at time t, does not fail in $(t, t+x]$. Therefore, $\overline{F}_{A(t)}(x)$ is sometimes called *interval reliability*.

For determining the mean value of the forward recurrence time of an ordinary renewal process, $A(t)$ is written in the form

$$A(t) = \Sigma_{i=1}^{N(t)+1} Y_i - t,$$

where the $Y_1, Y_2, ...$ are independent and identically distributed as Y with $\mu = E(Y)$. Wald's identity (1.125) cannot be applied to obtain $E(A(t))$, since $N(t)+1$ is surely not independent of the sequence $Y_1, Y_2, ...$ However, $N(t)+1$ is a stopping time for the sequence $Y_1, Y_2, ...$:

$$"N(t)+1 = n" = "N(t) = n-1" = "Y_1 + Y_2 + \cdots + Y_{n-1} \le t < Y_1 + Y_2 + \cdots + Y_n".$$

Thus, the event $"N(t)+1 = n"$ is independent of all $Y_{n+1}, Y_{n+2}, ...$ so that, by definition 1.2, $N(t)+1$ is a stopping time for the sequence $Y_1, Y_2, ...$ Hence, the mean value of $A(t)$ can be obtained from (1.127) with $N = N(t)+1$:

$$E(A(t)) = \mu\,[H_1(t)+1] - t.$$

Thus, the mean forward recurrence time of an ordinary renewal process is

$$E(A(t)) = \mu\,[H(t)+1] - t.$$

The second moment of the forward recurrence time of an ordinary renewal process is given without proof:

$$E((A(t))^2) = E(Y^2)[1 + H(t)] - 2\,E(Y)[t + \textstyle\int_0^t H(x)\,dx] + t^2, \quad t \ge 0.$$

The probability distribution of the backward recurrence time is obtained as follows:

$$\begin{aligned} F_{B(t)}(x) &= P(t - x \le T_{N(t)}) \\ &= \Sigma_{n=1}^{\infty} P(t-x \le T_n,\, N(t) = n) \\ &= \Sigma_{n=1}^{\infty} P(t-x \le T_n \le t < T_{n+1}) \\ &= \Sigma_{n=1}^{\infty} \int_{t-x}^{t} \overline{F}(t-u)\, dF_{T_n}(u) \\ &= \int_{t-x}^{t} \overline{F}(t-u)\, d\Big(\Sigma_{n=1}^{\infty} F_1 * F^{*(n)}\Big) \\ &= \int_{t-x}^{t} \overline{F}(t-u)\, dH_1(u). \end{aligned}$$

Hence, the distribution function of $B(t)$ is

$$F_{B(t)}(x) = \begin{cases} \int_{t-x}^{t} \overline{F}(t-u)\, dH_1(u) & \text{for } \; 0 \le x \le t \\ 1 & \text{for } \; t > x \end{cases}. \tag{3.151}$$

Differentiation yields the probability density of $B(t)$:

$$f_{B(t)}(x) = \begin{cases} \overline{F}(x)\, h_1(t-x) & \text{for } \; 0 \le x \le t \\ 0 & \text{for } \; t < x \end{cases}. \tag{3.152}$$

One easily verifies that the forward and backward recurrence times of an ordinary renewal process, whose cycle lengths are exponentially distributed with parameter λ, are also exponentially distributed with parameter λ :

$$f_{A(t)}(x) = f_{B(t)}(x) = \lambda e^{-\lambda x} \quad \text{for all } t \geq 0.$$

In view of the *memoryless property* of the exponential distribution (example 1.14, section 1.4), this result is not surprising. A direct consequence of the fundamental renewal theorem is that $F_S(t)$, as defined by (3.131), is the limiting distribution function of both backward and forward recurrence time as t tends to infinity:

$$\lim_{t\to\infty} F_{A(t)}(x) = \lim_{t\to\infty} F_{B(t)}(x) = F_S(x), \quad x \geq 0. \tag{3.153}$$

Paradox of Renewal Theory In view of the definition of the forward recurrence time, one supposes that the following equation is true:

$$\lim_{t\to\infty} E(A(t)) = \mu/2 .$$

However, according to (3.153) and (3.132),

$$\lim_{t\to\infty} E(A(t)) = \int_0^\infty \overline{F}_S(t)\, dt = E(S) = \frac{\mu^2 + \sigma^2}{2\mu} > \frac{\mu}{2}.$$

This 'contradiction' is known as the *paradox of renewal theory.* The intuitive explanation of this phenomenon is that on average the 'reference time point' t is to be found more frequently in longer renewal cycles than in shorter ones.

3.3.5 Stationary Renewal Processes

By definition 3.1, a renewal process $\{Y_1, Y_2, ...\}$ is stationary if for all $k = 1, 2, ...$ and any sequence of integers $i_1, i_2, ..., i_k$ with $1 \leq i_1 < i_2 < \cdots < i_k$ and any $\tau = 0, 1, ...$ the joint distribution functions of the vectors

$$(Y_{i_1}, Y_{i_2}, ..., Y_{i_k}) \quad \text{and} \quad (Y_{i_1+\tau}, Y_{i_2+\tau}, ..., Y_{i_k+\tau})$$

coincide, $k = 1, 2, ...$ According to the corollary after definition 3.1, $\{Y_1, Y_2, ...\}$ is stationary if and only if the corresponding renewal counting process $\{N(t), t \geq 0\}$ has homogeneous increments. A third way of defining the stationarity of a renewal process $\{Y_1, Y_2, ...\}$ makes use of the statistical equivalence between $\{Y_1, Y_2, ...\}$ and the corresponding process $\{A(t), t \geq 0\}$ of its forward recurrence times:

> A renewal process is stationary if and only if the process of its forward recurrence times $\{A(t), t \geq 0\}$ is strongly stationary.

Of course, the process of backward recurrence times $\{B(t), t \geq 0\}$ would do as well:

A renewal process is stationary if and only if the process of its backward recurrence times $\{B(t),\ t \geq 0\}$ is strongly stationary.

The stochastic process in continuous time $\{B(t),\ t \geq 0\}$ is a Markov process. This is quite intuitive, but a strict proof will not be given here. By theorem 2.1, a Markov process $\{X(t), t \in \mathbf{T}\}$ is strongly stationary if and only if its one-dimensional distribution functions

$$F_t(x) = P(X(t) \leq x)$$

do not depend on t. Hence, a renewal process is stationary if and only if there is a distribution function $F(x)$ so that

$$F_{A(t)}(x) = P(A(t) \leq x) = F(x) \quad \text{for all } x \geq 0 \text{ and } t \geq 0.$$

The following theorem yields a simple criterion for the stationarity of renewal processes.

Theorem 3.17 Let $F(x) = P(Y \leq x)$ be nonarithmetic and $\mu = E(Y) < \infty$. Then a delayed renewal process given by $F_1(x)$ and $F(x)$ is stationary if and only if

$$H_1(t) = t/\mu . \tag{3.154}$$

Equivalently, as a consequence of theorem 3.10, a delayed renewal process is stationary if and only if

$$F_1(x) = F_S(x) = \frac{1}{\mu} \int_0^x \overline{F}(y)\, dy \quad \text{for all } x \geq 0. \tag{3.155}$$

Proof If (3.154) holds, then (3.155) as well, so that, from (3.149),

$$\begin{aligned} F_{A(t)}(x) &= \frac{1}{\mu} \int_0^{t+x} \overline{F}(y)\, dy - \frac{1}{\mu} \int_0^t \overline{F}(x+t-y)\, dy \\ &= \frac{1}{\mu} \int_0^{t+x} \overline{F}(y)\, dy - \frac{1}{\mu} \int_x^{t+x} \overline{F}(y)\, dy \\ &= \frac{1}{\mu} \int_0^x \overline{F}(y)\, dy . \end{aligned}$$

Hence, $F_{A(t)}(x)$ does not depend on t.

Conversely, if $F_{A(t)}(x)$ does not depend on t, then (3.153) implies

$$F_{A(t)}(x) \equiv F_S(x) \ \text{ for all } t .$$

This completes the proof of the theorem. ■

As a consequence from theorem 3.17 and the elementary renewal theorem: After a sufficiently large time span (*transient response time*) every renewal process with nonarithmetic distribution function $F(t)$ and finite mean cycle length $\mu = E(Y)$ behaves as a stationary renewal process.

3.3.6 Alternating Renewal Processes

So far it has been assumed that renewals take only negligibly small amounts of time. In order to be able to model practical situations, in which this assumption is not fulfilled, the concept of a renewal process has to be generalized in the following way: The renewal time of the system after its ith failure is assumed to be a positive random variable Z_i; $i = 1, 2, ...$ Immediately after a renewal the system starts operating. In this way, a sequence of two-dimensional random vectors $\{(Y_i, Z_i); i = 1, 2, ...\}$ is generated, where Y_i denotes the lifetime of the system after the ith renewal.

Definition 3.8 (*alternating renewal process*) If $\{Y_1, Y_2, ...\}$ and $\{Z_1, Z_2, ...\}$ are two independent sequences of independent, nonnegative random variables, then the sequence of two-dimensional random vectors $\{(Y_1, Z_1), (Y_2, Z_2), ...\}$ is said to be an *alternating renewal process*. ●

The random variables

$$S_1 = Y_1; \quad S_n = \Sigma_{i=1}^{n-1}(Y_i + Z_i) + Y_n; \ n = 2, 3, ...,$$

are the time points at which failures occur, and the random variables

$$T_n = \Sigma_{i=1}^{n-1}(Y_i + Z_i); \ n = 1, 2, ...$$

are the time points at which a renewed system starts operating. If an operating system is assigned a '1' and a failed system a '0', then a binary indicator variable of the system state is

$$X(t) = \begin{cases} 0, & \text{if } t \in [S_n, T_n), \ n = 1, 2, ... \\ 1, & \text{elsewhere} \end{cases} \tag{3.156}$$

Obviously, an alternating renewal process can equivalently be defined by the stochastic process in continuous time $\{X(t), t \geq 0\}$ with $X(t)$ given by (3.156) (Figure 3.10). In what follows, all Y_i and Z_i are assumed to be distributed as Y and Z with distribution functions $F_Y(y) = P(Y \leq y)$ and $F_Z(z) = P(Z \leq z)$, , respectively. By agreement,

$$P(X(+0) = 1) = 1.$$

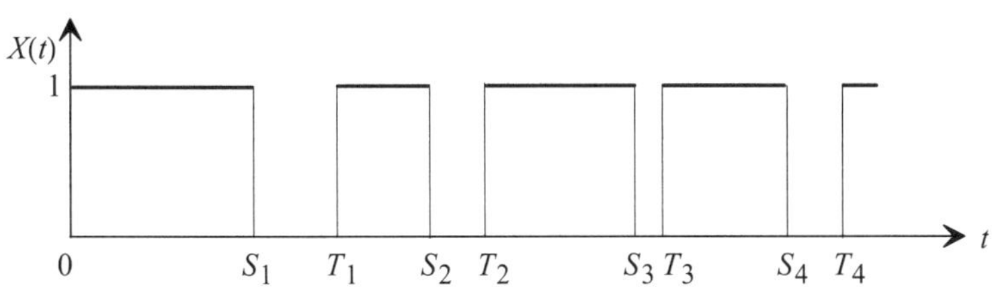

Figure 3.10 Sample path of an alternating renewal process

Analogously to the concept of a delayed renewal process, the alternating renewal process can be generalized by assigning the random lifetime Y_1 a probability distribution different from that of Y. However, this way of generalization and some other possibilities will not be discussed here, although no principal difficulties would arise.
Let $N_f(t)$ and $N_r(t)$ be the respective numbers of failures and renewals in $(0, t]$. Since S_n and T_n are sums of independent random variables (compare to (3.109)),

$$F_{S_n}(t) = P(S_n \le t) = P(N_f(t) \ge n) = F_Y * (F_Y * F_Z)^{*(n-1)}(t), \tag{3.157}$$

$$F_{T_n}(t) = P(T_n \le t) = P(N_r(t) \ge n) = (F_Y * F_Z)^{*(n)}(t). \tag{3.158}$$

Hence, analogously to the formulas (3.115) and (3.116), sum representations of

$$H_f(t) = E(N_f(t)) \text{ and } H_r(t) = E(N_r(t))$$

are

$$H_f(t) = \sum_{n=1}^{\infty} F_Y * (F_Y * F_Z)^{*(n-1)}(t),$$

$$H_r(t) = \sum_{n=1}^{\infty} (F_Y * F_Z)^{*(n)}(t).$$

$H_f(t)$ and $H_r(t)$ are referred to as the *renewal functions* of the alternating renewal process. Since $H_f(t)$ can be interpreted as the renewal function of a delayed renewal process, whose first system lifetime is distributed as Y, whereas the following 'system lifetimes' are identically distributed as $Y+Z$, it satisfies renewal equation (3.117) with

$$F_1(t) \equiv F_Y(t) \text{ and } F(t) = F_Y * F_Z(t).$$

Analogously, $H_r(t)$ can be interpreted as the renewal function of an ordinary renewal process whose cycle lengths are identically distributed as $Y+Z$. Therefore, $H_r(t)$ satisfies renewal equation (3.118) with $F(t)$ replaced by $F_Y * F_Z(t)$.

Let R_t be the residual lifetime of the system if it is operating at time t. Then

$$P(X(t) = 1, R_t > x)$$

is the probability that the system is working at time t and does not fail in the interval $(t, t+x]$. This probability is called *interval availability* (or *interval reliability*) and is denoted as $A_x(t)$. It can be obtained as follows:

$$\begin{aligned} A_x(t) &= P(X(t) = 1, R_t > x) \\ &= \sum_{n=0}^{\infty} P(T_n \le t,\ T_n + Y_{n+1} > t+x) \\ &= \overline{F}_Y(t+x) + \sum_{n=1}^{\infty} \int_0^t P(t+x < T_n + Y_{n+1} \mid T_n = u)\, dF_{T_n}(u) \\ &= \overline{F}_Y(t+x) + \int_0^t P(t+x-u < Y)\, d\sum_{n=1}^{\infty} (F_Y * F_Z)^{*(n)}(u). \end{aligned}$$

Hence,

$$A_X(t) = \overline{F}_Y(t+x) + \int_0^t \overline{F}_Y(t+x-u)\, dH_r(u)\,. \tag{3.159}$$

Note In this section 'A' no longer refers to 'forward recurrence time'.

Let $A(t)$ be the probability that the system is operating at time t, or, more generally, that it is available at time t:

$$A(t) = P(X(t) = 1). \tag{3.160}$$

This important characteristic of an alternating renewal process is obtained from (3.159) by letting there $x = 0$:

$$A(t) = \overline{F}_Y(t) + \int_0^t \overline{F}_Y(t-u)\, dH_r(u)\,. \tag{3.161}$$

$A(t)$ is called *availability* of the system (*system availability*) or, more exactly, *point availability* of the system, since it refers to a specific time point t. It is equal to the mean value of the indicator variable of the system state:

$$E(X(t)) = 1 \cdot P(X(t) = 1) + 0 \cdot P(X(t) = 0) = P(X(t) = 1) = A(t).$$

The *average availability* of the system in the interval $[0, t]$ is

$$\overline{A}(t) = \frac{1}{t} \int_0^t A(x)\, dx\,.$$

The random *total operating time* $U(t)$ of the system in the interval $[0, t]$ is

$$U(t) = \int_0^t X(x)\, dx\,. \tag{3.162}$$

By changing the order of integration,

$$E(U(t)) = E\left(\int_0^t X(x)\, dx\right) = \int_0^t E(X(x))\, dx\,.$$

Thus,

$$E(U(t)) = \int_0^t A(x)\, dx = t\, \overline{A}(t)\,.$$

The following theorem provides information on the limiting behaviour of the interval reliability and the point availability as t tends to infinity. A proof of the assertions need not be given since they are an immediate consequence of the fundamental renewal theorem 3.12.

Theorem 3.18 If $E(Y) + E(Z) < \infty$ and the distribution function $(F_Y * F_Z)(t)$ of the sum $Y + Z$ is nonarithmetic, then

$$A_X = \lim_{t\to\infty} A_X(t) = \frac{1}{E(Y) + E(Z)} \int_x^\infty \overline{F}_Y(u)\, du\,,$$

$$A = \lim_{t\to\infty} A(t) = \lim_{t\to\infty} \overline{A}(t) = \frac{E(Y)}{E(Y) + E(Z)}\,. \tag{3.163}$$

■

A_x is said to be the *long-run* or *stationary interval availability* (*reliability*) with regard to an interval of length x, and A is called the *long-run* or *stationary availability*. Clearly, it is $A = A_0$. If, analogously to renewal processes, the time between two neighbouring time points at which a new system starts operating is called a *renewal cycle*, then the long-run availability is equal to the mean share of the operating time of a system in the mean renewal cycle length.

It should be mentioned that equation (3.163) is also valid if within renewal cycles Y_i and Z_i depend on each other. As illustrated by the following example, in general,

$$E\left(\frac{Y}{Y+Z}\right) \neq \frac{E(Y)}{E(Y)+E(Z)}. \tag{3.164}$$

Example 3.17 Let life and renewal times have exponential distributions:

$$f_Y(y) = \lambda e^{-\lambda y}, \ y \geq 0; \quad f_Z(z) = \nu e^{-\nu z}, \ z \geq 0.$$

Application of the Laplace transform to (3.161) yields

$$\hat{A}(s) = \hat{F}_Y(s) + \hat{F}_Y(s) \cdot \hat{h}_r(s) = \frac{1}{s+\lambda}\left[1 + \hat{h}_r(s)\right]. \tag{3.165}$$

The Laplace transform of the convolution of f_Y and f_Z is

$$L\{f_Y * f_Z\} = \frac{\lambda\nu}{(s+\lambda)(s+\nu)}.$$

Hence, from the second equation of (3.126),

$$\hat{h}_r(s) = \frac{\lambda\nu}{s(s+\lambda+\nu)}.$$

By inserting $\hat{h}_r(s)$ into (3.165) and expanding $\hat{A}(s)$ into partial fractions,

$$\hat{A}(s) = \frac{1}{s+\lambda} + \frac{\lambda}{s(s+\lambda)} - \frac{\lambda}{s(s+\lambda+\nu)}.$$

Retransformation yields the point availability:

$$A(t) = \frac{\nu}{\lambda+\nu} + \frac{\lambda}{\lambda+\nu} e^{-(\lambda+\nu)t}, \ t \geq 0. \tag{3.166}$$

Since

$$E(Y) = 1/\lambda \text{ and } E(Z) = 1/\nu,$$

taking in (3.166) the limit as $t \to \infty$ verifies relationship (3.163). On the other hand, if $\lambda \neq \nu$, then, from example 1.20,

$$E\left(\frac{Y}{Y+Z}\right) = \frac{\nu}{\nu-\lambda}\left(1 + \frac{\lambda}{\nu-\lambda} \ln\frac{\lambda}{\nu}\right).$$

For instance, if $E(Z) = 0.25\,E(Y)$, then

$$A = \frac{E(Y)}{E(Y)+E(Z)} = 0.800 \quad \text{and} \quad E\left(\frac{Y}{Y+Z}\right) = 0.717. \qquad \square$$

Generally, numerical methods have to be applied to determine interval and point availability when applying formulas (3.159) and (3.161). This is again due to the fact that there are either no explicit or rather complicated representations of the renewal function for most of the common lifetime distributions. However, formulas (3.159) and (3.161) can be applied for obtaining approximate values for interval and point availability if they are used in conjunction with the bounds and approximations for the renewal function given in sections 3.3.2.2 and 3.3.3.

3.3.7 Compound Renewal Processes

3.3.7.1 Definition and Properties

Compound stochastic processes arise by additive superposition of random variables at random time points. (For motivation, see section 3.2.5.)

Definition 3.9 Let $\{(T_1, M_1), (T_2, M_2), ...\}$ be a random marked point process with property that $\{T_1, T_2, ...\}$ is the sequence of renewal times of a renewal process $\{Y_1, Y_2, ...\}$, and let $\{N(t), t \geq 0\}$ be the corresponding renewal counting process. Then the stochastic process $\{C(t), t \geq 0\}$ defined by

$$C(t) = \begin{cases} \sum_{i=1}^{N(t)} M_i & \text{if } N(t) \geq 1 \\ 0 & \text{if } N(t) = 0 \end{cases} . \tag{3.167}$$

is called a *compound* (*aggregate*, *cumulative*) *renewal process*, and $C(t)$ is called a *compound random variable.* ●

The compound Poisson process defined in section 3.2.5 is a compound renewal process with property that the renewal cycle lengths $Y_i = T_i - T_{i-1},\ i = 1, 2, ...,$ are independent and identically exponentially distributed (theorem 3.2).

A compound renewal process is also called a *renewal reward process*, in particular, if M_i is a 'profit' of any kind made at the renewal time points. In most applications,

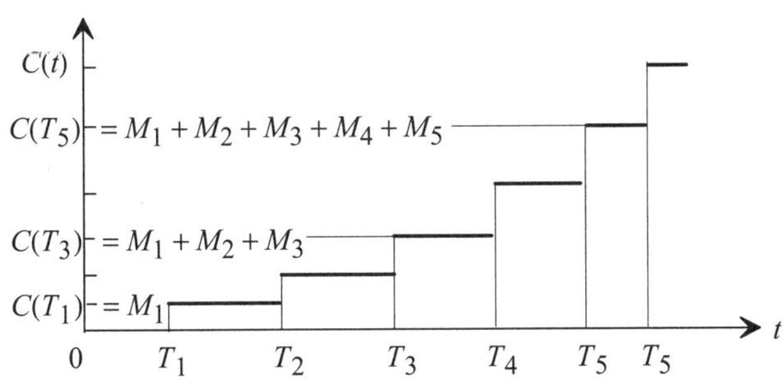

Figure 3.11 Sample path of a compound process with positive increments

however, M_i is a 'loss', for instance, replacement cost or claim size. M_i also can represent a 'loss' or 'gain' which accumulates over the ith renewal cycle (maintenance cost, profit by operating the system). In any case, $C(t)$ is the total loss (gain), which has accumulated over the interval $(0, t]$. The sample paths of a compound renewal process are step functions. Jumps occur at times T_i and the respective jump heights are M_i (Figure 3.11).

Compound renewal processes are considered in this section under the following assumptions:

1) $\{N(t), t \geq 0\}$ is a renewal counting process, which belongs to an <u>ordinary</u> renewal process $\{Y_1, Y_2, ...\}$.

2) The sequences are $\{M_1, M_2, ...\}$ and $\{Y_1, Y_2, ...\}$ are independent of each other and consist each of independent, nonnegative random variables, which are identically distributed as M and Y, respectively. However, M_i and Y_j may depend on each other if $i = j$, i.e. if they refer to the same renewal cycle.

3) The mean values of Y and M are finite and positive.

Under these assumptions, Wald's equation (1.125) yields the trend function $m(t) = E(C(t))$ of a compound renewal process:

$$m(t) = E(M)\,H(t), \tag{3.168}$$

where $H(t) = E(N(t))$ is the renewal function of the underlying renewal process $\{Y_1, Y_2, ...\}$. Formula (3.168) and the elementary renewal theorem (theorem 3.11) imply an important asymptotic property of the trend function of compound renewal processes:

$$\lim_{t\to\infty} \frac{E(C(t))}{t} = \frac{E(M)}{E(Y)}. \tag{3.169}$$

Equation (3.169) means that the average long-run (stationary) loss or profit per unit time is equal to the average loss or profit per unit time within a renewal cycle. The 'stochastic analogue' to (3.169) is: With probability 1,

$$\lim_{t\to\infty} \frac{C(t)}{t} = \frac{E(M)}{E(Y)}. \tag{3.170}$$

To verify (3.170), consider the obvious relationship

$$\Sigma_{i=1}^{N(t)} M_i \leq C(t) \leq \Sigma_{i=1}^{N(t)+1} M_i .$$

From this,

$$\left(\frac{1}{N(t)} \Sigma_{i=1}^{N(t)} M_i\right) \frac{N(t)}{t} \leq \frac{C(t)}{t} \leq \left(\frac{1}{N(t)+1} \Sigma_{i=1}^{N(t)} M_i\right) \frac{N(t)+1}{t}.$$

Now the strong law of the large numbers (theorem 1.8) and (3.139) imply (3.170). The relationships (3.169) and (3.170) are called *renewal reward theorems.*

Distribution of $C(t)$ If M has distribution function $G(t)$, then, given $N(t)=n$, the compound random variable $C(t)$ has distribution function

$$P(C(t)\le x|N(t)=n)=G^{*(n)}(x),$$

where $G^{*(n)}(x)$ is the nth convolution power of $G(t)$. Hence, by the total probability rule,

$$F_{C(t)}(x)=P(C(t)\le x)=\Sigma_{n=1}^{\infty}G^{*(n)}(x)\,P(N(t)=n), \tag{3.171}$$

where the probabilities $P(N(t)=n)$ are given by (3.111). (In the light of section 1.2.4, $F_{C(t)}$ is a mixture of the probability distribution functions $G^{*(1)}, G^{*(2)},...$) If Y has an exponential distribution with parameter λ, then $C(t)$ has distribution function

$$F_{C(t)}(x)=e^{-\lambda t}\sum_{n=0}^{\infty}G^{*(n)}(x)\frac{(\lambda t)^n}{n!};\quad G^{*(0)}(x)\equiv 1,\ x>0,\ t>0. \tag{3.172}$$

If, in addition, M has a normal distribution with $E(M)\ge 3\sqrt{Var(M)}$, then

$$F_{C(t)}(x)=e^{-\lambda t}\left[1+\sum_{n=1}^{\infty}\Phi\left(\frac{x-n\,E(M)}{\sqrt{n\,Var(M)}}\right)\frac{(\lambda t)^n}{n!}\right];\quad x>0,\ t>0. \tag{3.173}$$

The distribution function $F_{C(t)}$, for being composed of convolution powers of G and F, is usually not tractable and useful for numerical applications. Hence, much effort has been put into constructing bounds on $F_{C(t)}$ and into establishing asymptotic expansions. For surveys, see, e.g. [67, 89]. The following result of Gut [38] is particularly useful.

Theorem 3.19 If

$$\gamma^2=Var(E(Y)M-E(M)Y)>0, \tag{3.174}$$

then

$$\lim_{t\to\infty}P\left(\frac{C(t)-\frac{E(M)}{E(Y)}t}{[E(Y)]^{-3/2}\gamma\sqrt{t}}\le x\right)=\Phi(x),$$

where $\Phi(x)$ is the distribution function of the standardized normal distribution. ■

This theorem implies that for large t the compound variable $C(t)$ has approximately a normal distribution with mean value $\frac{E(M)}{E(Y)}t$ and variance $[E(Y)]^{-3}\gamma^2 t$, i.e.

$$C(t)\approx N\left(\frac{E(M)}{E(Y)}t,\ [E(Y)]^{-3}\gamma^2 t\right). \tag{3.175}$$

If M and Y are independent, then the parameter γ^2 can be written in the following form:

$$\gamma^2 = [E(Y)]^2\, Var(M) + [E(M)]^2\, Var(Y). \tag{3.176}$$

In this case, in view of assumption 3, condition (3.174) is always fulfilled. Condition (3.174) actually only excludes the case $\gamma^2 = 0$, i.e. linear dependence between Y and M. The following example presents an application of theorem 3.19. Its application to actuarial risk analysis is illustrated in section 3.4.4.

Example 3.18 For an alternating renewal process $\{(Y_i, Z_i);\ i = 1, 2, ...\}$, the total renewal time in $(0, t]$ is given by (a possible renewal time running at time t is neglected)

$$C(t) = \Sigma_{i=1}^{N(t)} Z_i,$$

where

$$N(t) = \max_n \{n,\ T_n < t\}.$$

(Notation and assumptions as in section 3.3.6.) Hence, the development of the total renewal time is governed by a compound stochastic process. In order to investigate the asymptotic behaviour of $C(t)$ as $t \to \infty$ by means of theorem 3.19, M has to be replaced with Z and Y with $Y + Z$. Consequently, if t is sufficiently large, then $C(t)$ has approximately a normal distribution with parameters

$$E(X(t)) = \frac{E(Z)}{E(Y)+E(Z)}\, t \quad \text{and} \quad Var(X(t)) = \frac{\gamma^2}{[E(Y)+E(Z)]^3}\, t.$$

Because of the independence of Y and Z,

$$\begin{aligned}\gamma^2 &= Var[Z\,E(Y+Z) - (Y+Z)\,E(Z)] \\ &= Var[Z\,E(Y) - Y\,E(Z)] \\ &= [E(Y)]^2\, Var(Z) + [E(Z)]^2\, Var(Y) > 0\end{aligned}$$

so that assumption (3.174) is satisfied. In particular, let (all parameters in *hours*)

$$E(Y) = 120,\ \sqrt{Var(Y)} = 40 \quad \text{and} \quad E(Z) = 4,\ \sqrt{Var(Z)} = 2.$$

Then,

$$\gamma^2 = 120^2 \cdot 4 + 16 \cdot 1600 = 83200 \quad \text{and} \quad \gamma = 288.4.$$

Consider, for example, the total renewal time in the interval $[0,\ 10^4$ *hours*]. The probability that $C(10^4)$ does not exceed a nominal value of 350 *hours* is

$$P(C(10^4) \le 350) = \Phi\left(\frac{350 - \frac{4}{124}\, 10^4}{124^{-3/2} \cdot 288.4 \cdot \sqrt{10^4}}\right) = \Phi(1.313).$$

Hence,

$$P(C(10^4) \le 350) = 0.905.$$

□

3.3.7.2 First Passage Time

The previous example motivates an investigation of the random time $L(x)$, at which the compound renewal process $\{C(t), t \geq 0\}$ exceeds a given nominal value x for the first time:

$$L(x) = \inf_t \{t, C(t) > x\}. \tag{3.177}$$

If, for instance, x is the critical wear limit of an item, then crossing level x is commonly referred to as the occurrence of a *drift failure*. Hence, it is justified to denote L as the lifetime of the system. Since, by assumption 2, the M_i are nonnegative random variables, the compound renewal process $\{C(t), t \geq 0\}$ has nondecreasing sample paths. In such a case, the following relationship between the distribution function of the *first passage time* $L(x)$ and the distribution function of the compound random variable $C(t)$ is obvious (Figure 3.12):

$$P(L(x) \leq t) = P(C(t) > x). \tag{3.178}$$

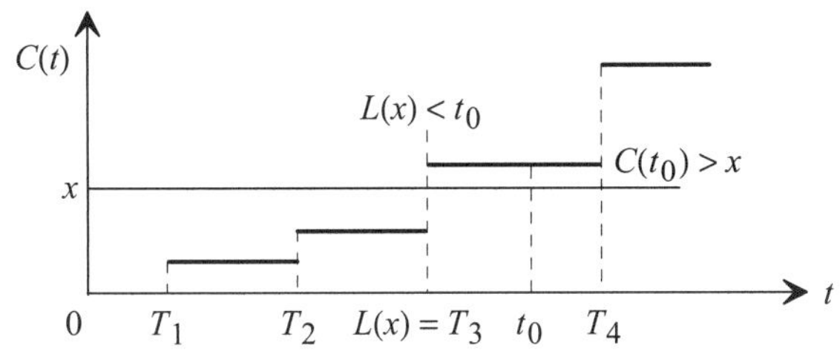

Figure 3.12 Level crossing of a compound stochastic process

Specifically, if $\{N(t), t \geq 0\}$ is the homogeneous Poisson process, then, from formulas (3.172) and (3.177),

$$P(L(x) > t) = e^{-\lambda t} \sum_{n=0}^{\infty} G^{*(n)}(x) \frac{(\lambda t)^n}{n!}; \quad t \geq 0$$

with x, $x > 0$, fixed. The probability distribution of $L(x)$ is generally not explicitly available. Hence the following theorem (Gut [38]) is important for practical applications, since it provides information on the asymptotic behaviour of the distribution of $L(x)$ as $x \to \infty$. The analogy of this theorem to theorem 3.19 is obvious.

Theorem 3.20 If $\gamma^2 = Var[E(Y)\, M - E(M)\, Y] > 0$, then

$$\lim_{x \to \infty} P\left(\frac{L(x) - \dfrac{E(Y)}{E(M)}\, x}{[E(M)]^{-3/2}\, \gamma \sqrt{x}} \leq t \right) = \Phi(t),$$

where $\Phi(t)$ is the distribution function of the standardized normal distribution. ■

Actually, in view of our assumption that the compound process $\{C(t), t \geq 0\}$ has nondecreasing sample paths, condition (3.178) implies that theorems 3.19 and 3.20 are equivalent.

A consequence of theorem 3.20 is that, for large x, the first passage time $L = L(x)$ has approximately a normal distribution with parameters

$$E(L(x)) = \frac{E(Y)}{E(M)}x \quad \text{and} \quad Var(L(x)) = [E(M)]^{-3}\gamma^2 x,$$

i.e.

$$L(x) \approx N\left(\frac{E(Y)}{E(M)}x,\, [E(M)]^{-3}\gamma^2 x\right), \quad x > 0. \tag{3.179}$$

The probability distribution given by (3.179) is called *Birnbaum-Saunders distribution.*

Example 3.19 Mechanical wear of an item is caused by shocks. (For instance, for the brake discs of a car, every application of the brakes is a shock.) After the ith shock the degree of wear of the item increases by M_i units. The $M_1, M_2, \ldots$ are supposed to be independent random variables, which are identically normally distributed as M with parameters

$$E(M) = 9.2 \text{ and } \sqrt{Var(M)} = 2.8 \text{ [in } 10^{-4}\text{mm]}.$$

The initial degree of wear of the item is zero. The item is replaced by an equivalent new one if the total degree of wear exceeds a critical level of 0.1 mm.

(1) What is the probability p_{100} that the item has to be replaced before or at the occurrence of the 100th shock? The degree of wear after 100 shocks is

$$C_{100} = \Sigma_{i=1}^{100} M_i$$

and has approximately the distribution function (unit of x: 10^{-4} mm)

$$P(C_{100} \leq x) = \Phi\left(\frac{x - 9.2 \cdot 100}{\sqrt{2.8^2 \cdot 100}}\right) = \Phi\left(\frac{x - 920}{28}\right).$$

Thus, the item survives the first 100 shocks with probability

$$p_{100} = P(C_{100} \leq 1000) = \Phi(2.86).$$

Hence, $p_{100} = 0.979$.

(2) In addition to the parameters of M, the random cycle Y is assumed to have mean value and variance

$$E(Y) = 6 \text{ and } \sqrt{Var(Y)} = 2 \text{ [}hours\text{]}.$$

What is the probability that the nominal value of 0.1 mm is not exceeded within the time interval [0, 600] (*hours*)?

To answer this question, theorem 3.20 can be applied since 0.1 mm is sufficiently large in comparison to the shock parameter $E(M)$. Provided M and Y are independent, $\gamma = 0.0024916$. Hence,

$$P(L(0.1) > 600) = 1 - \Phi\left(\frac{600 - \frac{6}{9.2}\,10^3}{(9.2)^{-3/2} \cdot 2491.6 \cdot \sqrt{0.1}}\right) = 1 - \Phi(-1.848)\,.$$

Thus, the desired probability is $P(L(0.1) > 600) = 0.967$. □

3.3.8 Regenerative Stochastic Processes

At the beginning of this chapter on renewal theory it has been pointed out that, apart from its own significance, renewal theory provides mathematical foundations for analyzing the behaviour of complicated systems which have renewal points imbedded in their running times. This is always the case if the running time of a system is partitioned by so-called *regeneration points* into *regeneration cycles* with the following characteristic properties:

1) After every regeneration point the future operation of the system is independent of its past operation.

2) Within every regeneration cycle the operation of the system is governed by the same stochastic rules.

Thus, regeneration points are nothing but renewal points of a system and, hence, generate a renewal process. However, now it is not only the distance between regeneration points that is interesting, but also the behaviour of the system within a regeneration cycle.

For a mathematical definition of a regenerative stochastic process, an ordinary renewal process $\{L_1, L_2, ...\}$ is introduced, where L_i is the random length of the ith regeneration cycle. Thus, the L_i are independent and identically distributed as L. The time points

$$T_n = \Sigma_{i=1}^{n} L_i\,; \quad n = 1, 2, ...$$

are now called *regeneration points* of the system. The ith *regeneration cycle* is given by

$$\{(L_i,\, W_i(x)),\, 0 \le x < L_i\},$$

where $W_i(x)$ denotes the state of the system at time x (with respect to the preceeding regeneration point). The verbally given properties of regeneration points and regeneration cycles become mathematically precise by assuming that the regeneration cycles are independent of each other and are identically distributed as the *typical regeneration cycle* $\{(L, W(x)),\ 0 \le x < L\}$. The probability distribution of the typical regeneration cycle is called the *cycle distribution.*

Definition 3.10 Let $\{N(t), t \geq 0\}$ be the renewal counting process, which belongs to the ordinary renewal process $\{L_1, L_2, ...\}$. Then the stochastic process $\{X(t), t \geq 0\}$ defined by

$$X(t) = W_{N(t)}(t - T_{N(t)}) \tag{3.180}$$

is said to be a *regenerative stochastic process*. The time points T_n; $n = 1, 2, ...$; are its *regeneration points*. ●

Intuitively speaking, definition 3.10 means that $T_{N(t)}$, the regeneration point before t, is declared to be the new origin. After $T_{N(t)}$ the process

$$\{W_{N(t)}(x), x \geq 0\} \text{ with } x = t - T_{N(t)}$$

evolves from $x = 0$ to the following regeneration point $T_{N(t)+1}$, which is reached at 'cycle time'

$$x = L_{N(t)+1} = T_{N(t)+1} - T_{N(t)}.$$

Thus, a regenerative process restarts at every regeneration point.

Example 3.20 The alternating renewal process $\{(Y_i, Z_i);\ i = 1, 2, ...\}$ is a simple example of a regenerative process. In this special case the cycle length L_i is given by the sum of life- plus renewal time $L_i = Y_i + Z_i$, where the random vectors (Y_i, Z_i) are independent of each other and identically distributed as (Y, Z). The stochastic process $\{W(x), x \geq 0\}$ indicates the working and renewal phases within a cycle:

$$W(x) = \begin{cases} 1 & \text{for} \quad 0 \leq x < Y \\ 0 & \text{for} \quad Y \leq x < Y + Z \end{cases}.$$

Therefore, the typical regeneration cycle is

$$\{(L, W(x)), 0 \leq x < L\}$$

with $L = Y + Z$. Thus, not only the lengths L_i of the regeneration cycles are of interest, but also the working and renewal phases within a cycle. □

Let B be a subset of the state space of $\{W(x), x \geq 0\}$ and $H(t)$ be the renewal function belonging to the ordinary renewal process $\{L_1, L_2, ...\}$. Analogously to the derivation of (3.159) it can be shown that the one-dimensional probability distribution of the regenerative stochastic process $\{X(t), t \geq 0\}$ is given by

$$P(X(t) \in B) = Q(t, B) + \int_0^t Q(t - x, B)\, dH(x), \tag{3.181}$$

where

$$Q(x, B) = P(W(x) \in B, L > x).$$

The following theorem considers the behaviour of the probability (3.181) as $t \to \infty$.

Theorem 3.21 (Theorem of *Smith*) If L is nonarithmetic and $E(L) > 0$, then

$$\lim_{t\to\infty} P(X(t) \in B) = \frac{1}{E(L)} \int_0^\infty Q(x, B)\, dx \tag{3.182}$$

and

$$\lim_{t\to\infty} \frac{1}{t} \int_0^t P(X(x) \in B)\, dx = \frac{1}{E(L)} \int_0^\infty Q(x, B)\, dx\,. \tag{3.183}$$

■

This theorem is an immediate consequence of the fundamental renewal theorem 3.12. The practical application of the *stationary state probabilities* (3.182) and (3.183) of a regenerative stochastic process is illustrated by analyzing a standard maintenance policy. This policy is a special case of policy 6 in section 3.2.6.3.

Example 3.21 (***age replacement policy***) The system is replaced upon failure or at age τ by a preventive renewal, whichever occurs first.

After a replacement the system has the same lifetime distribution as the original one, i.e. it is 'as good as new'. Unscheduled and preventive replacements require the constant times d_r and d_p, respectively. Furthermore, let $F(t) = P(T \le t)$ be the distribution function of the system lifetime T, $\overline{F}(t) = 1 - F(t)$ the survival probability and $\lambda(t)$ the failure rate of the system.

To specify an underlying regenerative stochastic process, the time points at which a system starts resuming its work are declared to be the regeneration points. Therefore, the random length L of the typical renewal (replacement) cycle has structure

$$L = \min\,(T, \tau) + Z,$$

where the random replacement time Z is

$$Z = \begin{cases} d_r & \text{for } \; T < \tau \\ d_p & \text{for } \; T \ge \tau \end{cases} \quad \text{or} \quad Z = \begin{cases} d_r & \text{with probability } F(\tau) \\ d_p & \text{with probability } \overline{F}(\tau) \end{cases}.$$

Since

$$E\{\min(T, \tau)\} = \int_0^\tau \overline{F}(t)\, dt,$$

the mean length of a regeneration cycle is

$$E(L) = \int_0^\tau \overline{F}(t)\, dt + d_r F(\tau) + d_p \overline{F}(\tau)\,.$$

Let

$$W(x) = \begin{cases} 1 & \text{if the system is working} \\ 0 & \text{otherwise} \end{cases}.$$

Then, for $B = \{1\}$,

$$Q(x, B) = P(W(x) = 1,\, L > x) = \begin{cases} 0 & \text{for } \; \tau < x \le L \\ \overline{F}(x) & \text{for } \; 0 \le x \le \tau \end{cases}.$$

Thus,

$$\int_0^\infty Q(x, B)\, dx = \int_0^\tau \overline{F}(x)\, dx .$$

Now (3.182) yields the stationary availability of the system:

$$A(\tau) = \lim_{t\to\infty} P(X(t)=1) = \frac{\int_0^\tau \overline{F}(x)\, dx}{\int_0^\tau \overline{F}(x)\, dx + d_e F(\tau) + d_p \overline{F}(\tau)} .$$

The age replacement policy can also be described by an alternating renewal process. Applying formula (3.163) would yield the same result.

Let τ^* denote a renewal interval τ which maximizes $A(\tau)$. Then τ^* satisfies the necessary condition

$$\lambda(\tau)\int_0^\tau \overline{F}(x)\, dx - F(\tau) = \frac{d}{1-d}$$

with $d = d_p/d_e$. A unique solution τ^* exists if $\lambda(t)$ is strictly increasing to infinity and $d<1$. The corresponding maximum availability is

$$A(\tau *) = \frac{1}{1+(d_e - d_p)\lambda(\tau *)} .$$

□

3.4 APPLICATIONS TO ACTUARIAL RISK ANALYSIS

3.4.1 Basic Concepts

Random point processes are key tools for quantifying risk in the insurance industry. (Principally, the following results are applicable to analyzing financial risk in many other branches as well.) A risky situation for an insurance company arises if it has to pay out a total claim amount, which tends to exceed the total premium income plus its initial capital. To be able to establish the corresponding mathematical risk model, next the concept of a risk process has to be introduced: An insurance company starts its business at time $t=0$. Claims arrive at random time points $T_1, T_2, ...$ and come with the respective random claim sizes $M_1, M_2, ...$ Thus, the insurance company is subjected to a random marked point process $\{(T_1, M_1), (T_2, M_2), \cdots\}$ called *risk process*. The two components of the risk process are the *claim arrival process* $\{T_1, T_2, ...\}$ and the *claim size process* $\{M_1, M_2, ...\}$. Let $\{N(t), t \ge 0\}$ be the random counting process which belongs to the claim arrival process. Then the total claim size $C(t)$, the company is faced with in the interval $[0, t]$, is a compound random variable:

$$C(t) = \begin{cases} \sum_{i=1}^{N(t)} M_i & \text{if } N(t) \ge 1 \\ 0 & \text{if } N(t) = 0 \end{cases} . \tag{3.184}$$

The compound stochastic process

$$\{C(t), t \geq 0\}$$

is the main ingredient of the risk model. With the terminology introduced in sections 3.2.5 and 3.3.7, $\{C(t), t \geq 0\}$ is a compound Poisson process if $\{N(t), t \geq 0\}$ is a Poisson process and a compound renewal process if $\{N(t), t \geq 0\}$ is a renewal process.

To equalize the loss caused by claims and to eventually make profit, an insurance company imposes a premium on its clients. Let $\kappa(t)$ be the total premium income of the insurance company in $[0, t]$. In case $C(t) < \kappa(t)$, the company has made a profit of

$$\kappa(t) - C(t)$$

in the interval $[0, t]$. With an *initial capital* or an *initial reserve* of x, $x \geq 0$, which the company has at its disposal at the start, the *risk reserve* at time t is defined as

$$R(t) = x + \kappa(t) - C(t). \tag{3.185}$$

The corresponding *risk reserve process* is

$$\{R(t), t \geq 0\}.$$

If the sample path is negative at a time t_0, the financial expenses of the company in $[0, t_0]$ exceed its available capital at time t_0. This leads to the definition of the *ruin probability* $p(x)$ of the company (Figure 3.13):

$$p(x) = P(\text{there is a positive, finite } t \text{ so that } R(t) < 0). \tag{3.186}$$

Consequently, the *non-ruin probability* or *survival probability* of the company is

$$q(x) = 1 - p(x).$$

The probabilities $p(x)$ and $q(x)$ refer to an infinite time horizon.

The *ruin probability* of the company with regard to a finite time horizon τ is

$$p(x, \tau) = P(\text{there is a finite } t \text{ with } 0 < t \leq \tau \text{ so that } R(t) < 0).$$

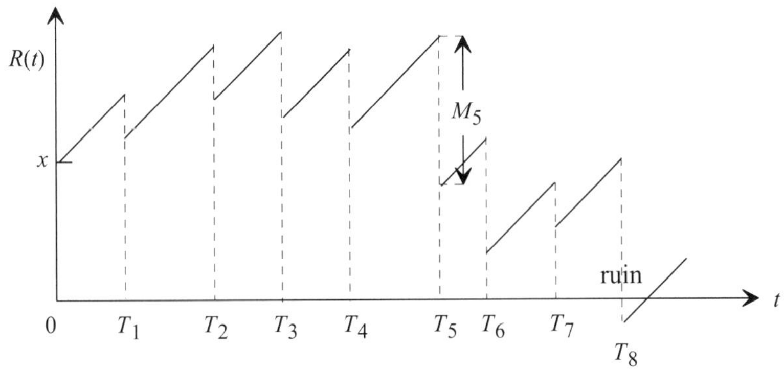

Figure 3.13 Sample path of a risk process leading to ruin

Of course, the ruin probabilities $p(x)$ and $p(x,\tau)$ decrease with increasing initial capital x. Since ruin can only occur at the arrival times of claims (Figure 3.13), $p(x)$ and $p(x,\tau)$ can also be defined in the following way:

$$p(x) = P(\text{there is a finite, positive integer } n \text{ so that } R(T_n) < 0), \qquad (3.187)$$

and

$$p(x,\tau) = P(\text{there is a finite, positive integer } n \text{ with } T_n \leq \tau \text{ so that } R(T_n) < 0),$$

where $R(T_n)$ is understood to be $R(+T_n)$, i.e. the value of the risk reserve process including the effect of the nth claim. (In the actuarial literature, claim sizes are frequently denoted as U_i, the initial capital as u, and the ruin probability as $\psi(u)$.)

3.4.2 Poisson Claim Arrival Process

In this section, the problem of determining the ruin probability is considered under the following 'classical assumptions':

1) $\{N(t), t \geq 0\}$ is a homogeneous Poisson process with parameter $\lambda = 1/\mu$.

2) The claim sizes $M_1, M_2, ...$ are independent, identically as M distributed random variables. The M_i are independent of $\{N(t), t \geq 0\}$.

3) The premium income is a linear function in t:

$$\kappa(t) = \kappa t, \ \ \kappa > 0, \ \ t \geq 0.$$

The parameter κ is the *premium rate.*

4) The time horizon is infinite ($\tau = \infty$).

Under asumptions 1 and 2, risk analysis is subjected to a homogeneous portfolio, i.e. claim sizes are independent, differences in the claim sizes are purely random and the arrival rate of claims is constant. For instance, consider a portfolio which comprises policies covering burgleries in houses. If the houses are in a demarcated area, have about the same security standard and comparable valuables inside, then this portfolio may be considered a homogeneous one. Generally, an insurance company tries to establish its portfolios in such a way that they are approximately homogeneous. Regardless of the terminology adopted, the subsequent risk analysis will not apply to an insurance company as a whole, but to its basic operating blocks, the homogeneous portfolios.

By assumption 1, the interarrival time Y of claims has an exponential distribution with parameter $\lambda = 1/\mu$. The mean claim size is denoted as ν. Hence,

$$\mu = E(Y) \ \text{ and } \ \nu = E(M). \qquad (3.188)$$

By (3.74) or (3.168), under the assumptions 1 and 2, the trend function of the total claim size process $\{C(t), t \geq 0\}$ is a linear function in time:

$$E(C(t)) = \frac{\nu}{\mu} t, \ \ t \geq 0. \qquad (3.189)$$

This justifies assumption 3, namely a linear premium income in time. In the long-run, an insurance company, however large its initial capital may be, cannot be successful if the average total claim cost in any interval $[0, t]$ exceeds the premium income in $[0, t]$. Hence, in what follows, let

$$\kappa - \frac{\nu}{\mu} = \frac{\kappa\mu - \nu}{\mu} > 0.$$

The positive difference $\kappa\mu - \nu$ is called *safety loading* and will be denoted as σ:

$$\sigma = \kappa\mu - \nu. \tag{3.190}$$

Let distribution function and density of the claim size M be

$$B(y) = P(M \le y) \quad \text{and} \quad b(y) = dB(y)/dy.$$

To derive an integro-differential equation for $q(x)$, consider what may happen in the time interval $[0, \Delta t]$:

1) No claim arrives in $[0, \Delta t]$. Under this condition, the survival probability is

$$q(x + \kappa\Delta t).$$

2) One claim arrives in $[0, \Delta t]$, the risk reserve remains positive. Under this condition, the survival probability is

$$\int_0^{x+\kappa\Delta t} q(x + \kappa\Delta t - y)\, b(y)\, dy.$$

3) One claim arrives in $[0, \Delta t]$, the risk reserve becomes negative (ruin occurs). Under this condition, the survival probability is 0.

4) At least 2 claims arrive in $[0, \Delta t]$. Since the Poisson process is ordinary, the probability of this event is $o(\Delta t)$.

Therefore, given the initial capital x,

$$q(x) = [1 - \lambda\Delta t + o(\Delta t)]\, q(x + \kappa\Delta t)$$

$$+[\lambda\Delta t + o(\Delta t)] \int_0^{x+\kappa\Delta t} q(x + \kappa\Delta t - y)\, b(y)\, dy + o(\Delta t).$$

From this, letting $h = \kappa\Delta t$,

$$\frac{q(x+h) - q(x)}{h} = \frac{\lambda}{\kappa} q(x+h) - \frac{\lambda}{\kappa} \int_0^{x+h} q(x+h-y)\, b(y)\, dy + \frac{o(h)}{h}.$$

Assuming that $q(x)$ is differentiable, letting $h \to 0$ yields

$$q'(x) = \frac{\lambda}{\kappa}\left[q(x) - \int_0^x q(x-y)\, b(y)\, dy \right]. \tag{3.191}$$

A solution can be obtained in terms of Laplace transforms: Let $\hat{q}(s)$ and $\hat{b}(s)$ be the Laplace transforms of $q(x)$ and $b(s)$. Then, applying the Laplace transformation to (3.191), using its properties (1.30) and (1.33), and replacing λ with $1/\mu$ yields

$$\hat{q}(s) = \frac{1}{s - \frac{1}{\kappa\mu}[1 - \hat{b}(s)]} q(0). \tag{3.192}$$

This representation of $\hat{q}(s)$ involves the survival probability $q(0)$ on condition that there is no initial capital.

Example 3.22 Let the claim size M have an exponential distribution with parameter $1/\nu$. Then,

$$\hat{b}(s) = \int_0^\infty e^{-sy} \frac{1}{\nu} e^{-(1/\nu)y} dy = \frac{1}{\nu s + 1}.$$

Hence,

$$\hat{q}(s) = \frac{\nu s + 1}{\mu\kappa s(\nu s + 1) - \nu s} q(0)\mu\kappa.$$

By introducing the coefficient

$$\alpha = \frac{\mu\kappa - \nu}{\mu\kappa} = \frac{\sigma}{\mu\kappa}, \quad 0 < \alpha < 1, \tag{3.193}$$

$\hat{q}(s)$ simplifies to

$$\hat{q}(s) = \left[\frac{1}{s + \alpha/\nu} + \frac{1}{\nu s} \cdot \frac{1}{s + \alpha/\nu}\right] q(0).$$

Retransformation yields (mind formula (1.29))

$$q(x) = \left[e^{-\frac{\alpha}{\nu}x} + \frac{1}{\alpha} - \frac{1}{\alpha} e^{-\frac{\alpha}{\nu}x}\right] q(0).$$

Condition $q(+\infty) = 1$ yields the survival- and ruin probabilities in case $x = 0$:

$$q(0) = \alpha, \quad p(0) = 1 - \alpha \tag{3.194}$$

so that the parameter α is the company's probability to survive without any initial capital. Thus, survival and ruin probability are for $x \geq 0$

$$q(x) = 1 - (1 - \alpha)e^{-\frac{\alpha}{\nu}x}, \quad p(x) = (1 - \alpha)e^{-\frac{\alpha}{\nu}x}. \tag{3.195}$$

Other explicit results can be obtained for mixed exponential claim size distributions, for instance

$$b(y) = \varepsilon\lambda_1 e^{-\lambda_1 y} + (1 - \varepsilon)\lambda_2 e^{-\lambda_2 y}; \quad y \geq 0, \ 0 < \varepsilon < 1.$$ □

Renewal Equation for $q(x)$ To be able to construct an approximation for $q(x)$ for large x, the integro-differential equation (3.191) needs to be transformed into an integral equation of renewal type, i.e. into an integral equation of type (3.126):

$$q(x) = a(x) + \int_0^x q(x - y)g(y)\,dy, \tag{3.196}$$

where $g(y)$ is a probability density and $a(x)$ an integrable function on $[0, \infty)$.

1) Firstly, an integral equation for $q(x)$ will be constructed. Integrating (3.191) from $x=0$ to $x=t$ yields

$$q(t)-q(0)=\frac{1}{\mu\kappa}\left[\int_0^t q(x)\,dx-\int_0^t\int_0^x q(x-y)\,b(y)\,dy\,dx\right]. \tag{3.197}$$

By partial integration and application of Dirichlet's formula (1.33), the double integral in (3.197) becomes

$$\begin{aligned}
&\int_0^t\int_0^x q(x-y)\,b(y)\,dy\,dx\\
&=\int_0^t q(x)dx-q(0)\int_0^t \overline{B}(x)dx-\int_0^t\int_0^x q'(x-y)\,\overline{B}(y)\,dy\,dx\\
&=\int_0^t q(x)dx-q(0)\int_0^t \overline{B}(x)dx-\int_0^t \overline{B}(y)\,q(t-y)\,dy+q(0)\int_0^t \overline{B}(x)dx\\
&=\int_0^t q(x)dx-\int_0^t \overline{B}(y)\,q(t-y)\,dy\,.
\end{aligned}$$

By combining this result with (3.197) and replacing t with x,

$$q(x)=q(0)+\frac{1}{\mu\kappa}\left[\int_0^x q(x-y)\,\overline{B}(y)\,dy\right]. \tag{3.198}$$

Letting $x\to\infty$ in (3.198) yields

$$q(\infty)=q(0)+\frac{1}{\mu\kappa}\,\nu\, q(\infty)\,.$$

Since $q(\infty)=1$,

$$q(0)=1-\frac{\nu}{\mu\kappa}=\alpha\,. \tag{3.199}$$

Interestingly, this probability depends on the probability distributions of the random variables involved only via their mean values. Hence, its is not surprising that formulas (3.194) and (3.199) coincide.

2) To establish an integro-differential equation for the ruin probability $p(x)$, in formula (3.198) the survival probability $q(x)$ is replaced with $1-p(x)$:

$$\begin{aligned}
1-p(x)&=\alpha+\frac{1}{\mu\kappa}\left[\int_0^x [1-p(x-y)]\,\overline{B}(y)\,dy\right]\\
&=\alpha+\frac{1}{\mu\kappa}\int_0^x \overline{B}(y)\,dy-\frac{1}{\mu\kappa}\int_0^x p(x-y)\,\overline{B}(y)\,dy\,.
\end{aligned}$$

Hence,

$$p(x)=1-\alpha-\frac{1}{\mu\kappa}\int_0^x \overline{B}(y)\,dy+\int_0^x p(x-y)\,\frac{1}{\mu\kappa}\overline{B}(y)\,dy. \tag{3.200}$$

Formally, this integral equation in the ruin probability $p(x)$ looks like the integral equation of renewal type (3.196) with functions $a(x)$ and $g(y)$ given by

$$a(x)=1-\alpha-\frac{1}{\mu\kappa}\int_0^x \overline{B}(y)\,dy,\quad g(y)=\frac{1}{\mu\kappa}\overline{B}(y),\qquad x\ge 0,\ y\ge 0. \tag{3.201}$$

The function $g(y)$ is nonnegative, but it is not a probability density since

$$\int_0^\infty g(y)dy = \frac{1}{\mu\kappa}\int_0^\infty \overline{B}(y)\,dy = \frac{\nu}{\mu\kappa} = 1-\alpha < 1.$$

However, $g(y)$ can be thought of as characterizing a *defective probability distribution* with a defect of α. Hence, integral equation (3.200) is called a *defective integral equation of renewal type.*

3) Now a proper integral equation of renewal type for $p(x)$ will be constructed: The procedure is simple: The integral equation (3.200) will be multiplied by a factor e^{ry} so that the product $e^{ry}g(y)$ is a probability density. Hence, the parameter r has to be chosen such that

$$\frac{1}{\mu\kappa}\int_0^\infty e^{ry}\,\overline{B}(y)\,dy = 1. \tag{3.202}$$

The unique constant r satisfying (3.202) is called a *Lundberg exponent.* It exists for claim size probability densities with a sufficiently short tail, which implies that large claim sizes occur very seldom. With $a(x)$ and $g(y)$ given by (3.201), let

$$a_r(x) = e^{rx}a(x), \quad g_r(y) = e^{ry}g(y), \quad p_r(x) = e^{rx}p(x).$$

Then, multiplying (3.200) by $e^{rx} = e^{r(x-y)}\cdot e^{ry}$, where r satisfies (3.202), gives an integral equation of renewal type for the function $p_r(x)$:

$$p_r(x) = a_r(x) + \int_0^x p_r(x-y)\,g_r(y)\,dy. \tag{3.203}$$

This integral equation can easily be solved in the image space of the Laplace transformation (just as (3.192)). When doing this, note that, for instance, the Laplace transform of a_r is given by

$$L(a_r) = \hat{a}(s-r),$$

where $\hat{a}$ is the Laplace transform of a.

Approximation of the Ruin Probability For being able to apply theorem 3.13 to the integral equation of renewal type (3.203), the following integrals have to be determined:

$$\int_0^\infty a_r(x)\,dx \quad \text{and} \quad \int_0^\infty y\,g_r(y)\,dy.$$

Since $1-\alpha = \frac{\nu}{\mu\kappa}$,

$$\begin{aligned}\int_0^\infty a_r(x)\,dx &= \int_0^\infty e^{rx}\left[1-\alpha-\frac{1}{\mu\kappa}\int_0^x \overline{B}(y)\,dy\right]dx\\ &= \int_0^\infty e^{rx}\left[1-\alpha-\frac{1}{\mu\kappa}\left(\nu-\int_x^\infty \overline{B}(y)\,dy\right)\right]dx\\ &= \frac{1}{\mu\kappa}\int_0^\infty e^{rx}\left(\int_x^\infty \overline{B}(y)\,dy\right)dx.\end{aligned}$$

Now, changing the order of integration according to Dirichlet's formula (1.33) and making use of (3.202) yields

$$\int_0^\infty a_r(x)\,dx = \frac{1}{\mu\kappa}\int_0^\infty \overline{B}(y)\left[\int_0^y e^{rx}\,dx\right]dy$$

$$= \frac{1}{r\mu\kappa}\int_0^\infty \overline{B}(y)\,[e^{ry}-1]\,dy$$

$$= \frac{1}{r\mu\kappa}\left[\int_0^\infty e^{ry}\,\overline{B}(y)dy - \nu\right].$$

Hence,

$$\int_0^\infty a_r(x)\,dx = \frac{\alpha}{r}.$$

The mean value, which belongs to the density $g_r(y)$, is

$$m = \int_0^\infty y\,g_r(y)\,dy = \frac{1}{\mu\kappa}\int_0^\infty y\,e^{ry}\,\overline{B}(y)\,dy. \tag{3.204}$$

Now, from theorem 3.13 (the constant μ which occurs in theorem 3.13 is here denoted as m),

$$\lim_{x\to\infty} p_r(x) = \lim_{x\to\infty} e^{rx}\,p(x) = \frac{\alpha}{m\,r}.$$

Hence, for large values of the initial capital x,

$$p(x) \approx \frac{\alpha}{m\,r}\,e^{-rx}, \tag{3.205}$$

where the parameters r and m are given by (3.202) and (3.204), respectively. This approximation frequently yields excellent results even for small values of x. Formula (3.205) is called the *Cramér-Lundberg approximation* to the ruin probability. Under the assumptions stated, the ruin probability is bounded by

$$p(x) \le e^{-rx}. \tag{3.206}$$

This is the famous *Lundberg inequality.* A proof will be given in section 6.2 by using martingale based methods. Both *H. Cramér* and *F. Lundberg* did their pioneering research in *collective risk analysis* in the first third of the 20th century.

Continuation of example 3.22 It is interesting to evaluate the Cramér-Lundberg approximation to the ruin probability if the claim size M has an exponential distribution, since in this case the exact value of the ruin probability is known. Thus, let M have distribution function

$$F(y) = 1 - e^{-(1/\nu)y}, \quad y \ge 0.$$

According to (3.202), the corresponding Lundberg exponent r is given by

$$\int_0^\infty e^{ry}\,e^{-(1/\nu)y}\,dy = \mu\kappa.$$

Hence,

$$r = \alpha/\nu.$$

By (3.204), the parameter m is obtained as follows:

$$m = \frac{1}{\mu\kappa}\int_0^\infty y e^{ry} e^{-(1/\nu)y}\, dy = \frac{1}{\mu\kappa}\cdot\frac{\nu}{1-\nu r}\int_0^\infty y\left(\tfrac{1}{\nu}-r\right) e^{-\left(\frac{1}{\nu}-r\right)y} dy$$

$$= \frac{1}{\mu\kappa}\cdot\left(\frac{\nu}{1-\nu r}\right)^2 .$$

Hence,

$$\frac{\alpha}{m r} = 1-\alpha .$$

By comparing these results with (3.150): In case of exponentially distributed claim sizes, the Cramér-Lundberg approximation gives the exact formula for the ruin probability. □

3.4.3 Renewal Claim Arrival Process

Much effort has been put into determining the ruin probability under more general assumptions than the 'classical' assumptions 1 to 4 stated in section 3.4.2. In what follows, some results are listed on condition that, whilst retaining assumptions 2 to 3, assumption 1 is replaced by assuming that claims arrive according to a renewal process. Thus, the interarrival times need no longer be exponentially distributed. For proofs and surveys on the state of art of actuarial risk theory, including first-passage time behaviour of random walks, see Feller [28], Grandell [34, 35], Asmussen [1], and Rolski et al. [67].

Ordinary Renewal Process Let the sequence $\{Y_1, Y_2, ...\}$ of the claim interarrival times Y_i be an ordinary renewal process. In the ith cycle, the company makes the random 'profit' (notation as introduced before)

$$Z_i = \kappa Y_i - M_i ;\ \ i = 1, 2, ...$$

The $Z_1, Z_2, ...$ are independent, identically as $Z = \kappa Y - M$ distributed random variables. Hence, the discrete-time stochastic process $\{S_1, S_2, ...\}$ with

$$S_n = \sum_{i=1}^n Z_i = \kappa T_n - C(T_n) \tag{3.207}$$

is a *random walk* with independent, identically distributed increments Z_i. Let $L(a)$ be the first passage time of this random walk with regard to a negative level a:

$$L(a) = \min_{n=1,2,...} \{n,\ S_n < a\}.$$

Ruin will occur at time $L(-x)$ if x is the initial capital of the company. Thus, determining the ruin probability is closely related to the first passage time behaviour of random walks. In particular, the ruin probability is given by

$$p(x) = P(L(-x) < \infty).$$

As in section 3.4.2, to make sure that $p(x) < 1$, a positive *safety loading* $\sigma = \kappa\mu - \nu$ is required. In this case, by (3.168), the stochastic process $\{S_1, S_2, ...\}$ has a nonnegative, linearly increasing trend function:

$$m(t) = E(S_t) = (\kappa\mu - \nu)\,t; \quad t = 1, 2, ...$$

Let $\hat{z}(s)$ be the Laplace transform of $Z = \kappa Y - M$:

$$\hat{z}(s) = Ee^{-sZ})$$

Since M and Y are independent,

$$\hat{z}(s) = E(e^{-s\kappa Y})\,E(e^{sM}).$$

In terms of the Laplace transforms of the densities of Y and M,

$$\hat{z}(s) = \hat{f}_Y(s\kappa)\,\hat{b}(-s).$$

The *Lundberg exponent* r is now the positive solution of

$$\hat{z}(r) = 1. \tag{3.208}$$

As under the assumption of a homogeneous Poisson claim arrival process, an explicit formula for the ruin probability exists if M has an exponential distribution:

$$p(x) = (1 - r\nu)\,e^{-rx}, \quad x \geq 0. \tag{3.209}$$

Given r as solution of (3.208), the *Lundberg inequality* has the same structure as (3.206):

$$p(x) \leq e^{-rx}.$$

For large x, there is also a Cramér-Lundberg approximation for the ruin probability:

$$p(x) \approx c\,e^{-rx}.$$

However, the value of the constant c cannot be given here (see the references given above).

Stationary Renewal Process Let the sequence $\{Y_1, Y_2, ...\}$ of the claim interarrival times Y_i be a stationary renewal process. Then, by theorem 3.17, if the $Y_2, Y_3, ...$ are identically distributed with distribution function $F(t)$, Y_1 has distribution function (3.155). Again by theorem 3.17 (formula (3.154)), the trend function of the total claim size process $\{C(t), t \geq 0\}$ is a linear function in time:

$$E(C(t)) = E\left(\sum_{i=1}^{N(t)} M_i\right) = E(N(t))\,E(M) = \frac{t}{\mu}\,\nu = \frac{\nu}{\mu}\,t.$$

In what follows, the ruin probability referring to a stationary renewal claim arrival process is denoted as $p_S(x)$, whereas $p(x)$ refers to the ordinary renewal claim arrival process. With a positive safety loading $\sigma = \kappa\mu - \nu$, there is the following relationship between $p_S(x)$ and $p(x)$:

$$p_S(x) = \frac{1}{\kappa\mu}\left[\int_x^{\infty} \overline{B}(y)\,dy + \int_0^x p(x-y)\,\overline{B}(y)\,dy\right]. \tag{3.210}$$

In particular, survival and ruin probabilities on condition $x = 0$ (no initial capital) are

$$q_S(0) = 1 - \frac{\nu}{\kappa\mu}, \quad p_S(0) = \frac{\nu}{\kappa\mu}.$$

These probabilities do not depend on the type of the distributions of Y and M, but only on their mean values (insensitivity). Since in case of exponentially distributed claim interarrival times $F(t) \equiv F_S(t)$, on condition $x = 0$ the probabilities $q_S(0)$ and $p_S(0)$ coincide with the 'classical' survival and ruin probabilities (3.194).
For exponentially distributed claim sizes, inserting (3.209) in (3.210) yields

$$p_S(x) = \frac{\nu}{\kappa\mu} e^{-rx}, \quad x \geq 0.$$

3.4.4 Normal Approximations for Risk Processes

Let the process of the claim interarrival times $\{Y_1, Y_2, ...\}$ be an ordinary renewal process. Otherwise, assumptions 2 to 4 of section 3.4.2 will be retained. Then, by theorem 3.19, if t is sufficiently large compared to μ, the total claim size in $[0, t]$ has approximately a normal distribution with mean value $\frac{\nu}{\mu} t$ and variance $\mu^{-3}\gamma^2 t$:

$$C(t) \approx N\left(\frac{\nu}{\mu} t, \mu^{-3}\gamma^2 t\right), \tag{3.211}$$

where

$$\gamma^2 = \mu^2 Var(M) + \nu^2 Var(Y).$$

The random profit the insurance company has made in $[0, t]$ is given by

$$G(t) = R(t) - x = \kappa t - C(t).$$

By (3.211), $G(t)$ has approximately a normal distribution with parameters

$$E(G(t)) = (\kappa - \tfrac{\nu}{\mu})t \quad \text{and} \quad Var(G(t)) = \mu^{-3}\gamma^2 t.$$

The application of this result is illustrated by two examples. Note that examples 3.23 and 3.24 refer to the situation that, when being 'in red numbers' (ruin has happened), the company continues operating until it reaches a profitable time period and so on. In case of a positive safety loading, it will leave 'loss periods' with probability 1.

Example 3.23 Given a risk process $\{(Y_1, M_1), (Y_2, M_2), ...\}$ with

$$\mu = E(Y) = 2\ [h], \quad Var(Y) = 3\ [h^2],$$
$$\nu = E(M) = 900\ [\$], \quad Var(M) = 360\,000\ [\$^2].$$

(1) What minimal premium per hour κ_α has the insurance company to take in so that it will achieve a profit of at least 10^6 [\$] within 10^3 *hours* with probability $\alpha = 0.95$?

Since $\gamma = 1967.2$,

$$P(G(10^4) \geq 10^6) = P(C(t) < 10^4(\kappa_{0.95} - 100))$$

$$= \Phi\left(\frac{(\kappa_{0.95} - 100) - 450}{2^{-1.5} \cdot 19.672}\right).$$

Since the 0.95-percentile of the standardized normal distribution is $z_{0.95} = 1.64$, the desired premium per hour $\kappa_{0.95}$ satisfies equation

$$\frac{\kappa_{0.95} - 550}{6.955} = 1.64\,.$$

Hence,

$$\kappa_{0.95} = 561 \ [\$/h].$$

Of course, this result does not take into account the fact that the premium size has an influence on the claim flow.

(2) Let the premium income of the company be $\kappa = 460\,[\$/h]$. Thus, the company has a positive safety loading of $\sigma = 10\,[\$]$. Given an initial capital of $x = 10^4\,[\$]$, what is the probability of the company to be in the state of ruin at time $t = 1000\,[h]$?
This probability is given by

$$P(G(10^3) < -10^{-4}) = \Phi\left(\frac{-10^4 - (460 - 450)\,10^3}{2^{-1.5} \cdot 1967.2 \cdot \sqrt{1000}}\right)$$

$$= \Phi(-0.910) = 0.181. \qquad \square$$

The following example uses the approximate distribution of the first passage time $L(a)$ of the compound claim size process $\{C(t),\, t \geq 0\}$ with respect to level a as given by theorem (3.20):

$$L(a) \approx N(\tfrac{\mu}{\nu}\, a,\, \nu^{-3}\gamma^2 a)\,.$$

Example 3.24 Let the parameters of a risk process $\{(Y_1, M_1), (Y_2, M_2), ...\}$ be

$$\mu = E(Y) = 5 \ [h], \quad Var(Y) = 25 \ [h^2],$$
$$\nu = E(M) = 1000 \ [\$], \quad Var(M) = 640\,000 \ [\$^2].$$

What is the probability that the total claim reaches level $a = 10^6$ [\$] before the time point $t = 5500\,[h]$?

Since $\gamma = 6403$,

$$P(L(10^6) < 6000) = \Phi\left(\frac{5500 - 5000}{1000^{-1.5} \cdot 6403 \cdot 10^3}\right)$$

$$= \Phi(2.45) = 0.993. \qquad \square$$

3.5 EXERCISES

Sections 3.1 and 3.2

3.1) The number of catastrophic accidents at Sosal & Sons can be described by a homogeneous Poisson process with intensity $\lambda = 3$ a year.

(1) What is the probability $p_{\geq 2}$ that at least two catastrophic accidents will occur in the second half of the current year?

(2) Determine the same probability given that two catastrophic accidents have occurred in the first half of the current year.

3.2) By making use of the independence and homogeneity of the increments of a homogeneous Poisson process $\{N(t), t \geq 0\}$ with intensity λ show that its covariance function is given by

$$C(s,t) = \lambda \min(s,t).$$

3.3) The number of cars which pass a certain intersection daily between 12:00 and 14:00, follows a homogeneous Poisson process with intensity $\lambda = 40$ per hour. Among these there are 0.8% which disregard the STOP-sign.

What is the probability $p_{\geq 1}$ that at least one car disregards the STOP-sign between 12:00 and 13:00?

3.4) A Geiger counter is struck by radioactive particles according to a homogeneous Poisson process with intensity $\lambda = 1$ per 12 seconds. On average, the Geiger counter only records 4 out of 5 particles.

(1) What is the probability $p_{\geq 2}$ that the Geiger counter records at least 2 particles a minute?

(2) What are mean value [min] and variance [min^2] of the random time Y between the occurrence of two successively recorded particles?

3.5) An electronic system is subject to two types of shocks which arrive independently of each other according to homogeneous Poisson processes with intensities

$$\lambda_1 = 0.002 \text{ and } \lambda_2 = 0.01 \text{ per hour,}$$

respectively. A shock of type 1 always causes a system failure, whereas a shock of type 2 causes a system failure with probability 0.4.

What is the probability of the event A that the system fails within 24 *hours* due to a shock?

3.6) Consider two independent homogeneous Poisson processes 1 and 2 with respective intensities λ_1 and λ_2. Determine the mean value of the random number of events of process 2 (type 2-events) which occur between any two successive events of process 1 (type 1-events).

3.7) Let $\{N(t), t \ge 0\}$ be a homogeneous Poisson process with intensity λ. Prove that for an arbitrary, but fixed positive h the stochastic process $\{X(t), t \ge 0\}$ defined by

$$X(t) = N(t+h) - N(t)$$

is weakly stationary.

3.8) Let $\{N(t), t \ge 0\}$ be a homogeneous Poisson process with intensity λ and let $\{T_1, T_2, ...\}$ be the associated point process. For $t \to \infty$, determine and sketch the covariance function $C(\tau)$ of the (stochastic) shot noise process $\{X(t), t \ge 0\}$ given by

$$X(t) = \sum_{i=1}^{N(t)} h(t - T_i) \quad \text{with} \quad h(t) = \begin{cases} \sin t & \text{for } 0 \le t \le \pi \\ 0, & \text{elsewhere} \end{cases}.$$

3.9)* Let $\{N(t), t \ge 0\}$ be a homogeneous Poisson process with intensity λ and let $\{T_1, T_2, ...\}$ be the associated random point process. Derive trend function $m(t)$ and covariance function $C(s, t)$ of the shot noise process $\{X(t), t \ge 0\}$ defined by

$$X(t) = \sum_{i=1}^{N(t)} h(t - T_i) \quad \text{with} \quad h(t) = 0 \text{ for } t < 0, \quad \int_0^\infty h(x)\, dx < \infty,$$

by partitioning the positive half axis $[0, \infty)$ into intervals of length Δx and making use of the homogeneity and independence of the increments of a homogeneous Poisson process.

Note that $\{X(t), t \ge 0\}$ is the same process as the one analyzed in example 3.4 with another technique.

3.10) At a used car dealer, cars of a specific type arrive according to a homogeneous Poisson process $\{N(t), t \ge 0\}$ with intensity λ. Let $\{T_1, T_2, ...\}$ be the corresponding arrival time process. The car arriving at time T_i can immediately be resaled by the dealer at price C_i, where the $C_1, C_2, ...$ are assumed to be independent and identically distributed as C. However, if a buyer acquires the car, which arrived at T_i, at time $T_i + \tau$, then he only has to pay an amount of

$$e^{-\alpha\tau} C_i \text{ with } \alpha > 0.$$

At time t, the dealer is in a position to sell all cars of this type to a customer. What will be the mean total price $E(K)$ the car dealer achieves?

3.11) Statistical evaluation of a large sample justifies to model the number of cars which arrive daily for petrol between 12:00 a.m. and 4:00 a.m. at a particular filling station by a nonhomogeneous Poisson process $\{N(t), t \ge 0\}$ with intensity function

$$\lambda(t) = 8 - 4t + 3t^2 \quad [h^{-1}], \quad 0 \le t \le 4.$$

(1) How many cars arrive on average between 12:00 a.m. and 4:00 a.m.?

(2) What is the probability that at least 40 cars arrive between 2:00 and 4.00 a.m.?

3.12)* Let $\{N(t), t \ge 0\}$ be a nonhomogeneous Poisson process with intensity function $\lambda(t)$, trend function $\Lambda(t) = \int_0^t \lambda(x)\, dx$ and arrival time point T_i of the ith Poisson event. Show that, given $N(t) = n$, the random vector $(T_1, T_2, ..., T_n)$ has the same probability distribution as n ordered, independent, and identically distributed random variables with distribution function

$$F(x) = \begin{cases} \dfrac{\Lambda(x)}{\Lambda(t)} & \text{for } 0 \le x < t \\ 1, & t \le x \end{cases} .$$

Hint Compare to theorem 3.5.

3.13) Determine the optimal renewal interval τ^* and the corresponding maintenance cost rate $K(\tau)$ for policy 1 (section 3.2.6.2) given that the system lifetime has a Weibull distribution with form parameter β and scale parameter θ; $\beta > 1, \theta > 0$.

3.14) Clients arrive at an insurance company according to a mixed Poisson process the structure parameter L of which has a uniform distribution over the interval $[0, 1]$.

(1) Determine the state probabilities of this process at time t.

(2) Determine trend and variance function of this process.

(3) For what values of α and β are trend and variance function of a Polya arrival process identical to the ones obtained under (2)?

3.15)* Prove the multinomial criterion (formula 3.55). Assume that L has density f_L.

3.16)* A system is maintained according to policy 7 (section 3.2.6.4). The repair cost of a system failing at time t has a uniform distribution over the interval $[a, a+bt]$ with $a \ge 0$ and $b > 0$.

Under the same assumptions as in section 3.2.6.4 (in particular assumptions (3.96) and (3.99)), show that every linearly increasing repair cost limit

$$c(t) = c + dt \quad \text{with } a < c \text{ and } d < b$$

leads to a higher maintenance cost rate than $K_7(c^*)$ given by (3.105).

3.17) A system is maintained according to policy 7 with a constant repair cost limit c. System lifetime L and repair cost C have the respective distribution functions $F(t)$ and $R(x)$. The cost of a minimal repair is assumed (quite naturally) to depend on c as follows: $c_m = c_m(c) = E(C | C \le c)$.

(1) Determine the corresponding maintenance cost rate via formula (3.85) for any distribution function $F(t)$ and for any distribution function $R(x) = P(C \le x)$ with density $r(x)$ and property $R(c_r) = 1$.

(2) Determine the optimal repair cost limit with $F(t)$ given by (3.91) and $R(x)$ given by (3.96)

Sections 3.3 and 3.4

Note Exercises 3.18 to 3.29 refer to ordinary renewal processes. The functions $f(t)$ and $F(t)$ denote density and distribution function; the parameters μ and μ_2 are mean value and second moment of the cycle length Y. $N(t)$ is the (random) renewal counting function and $H(t)$ denotes the corresponding renewal function.

3.18) A system starts working at time $t = 0$. Its lifetime has approximately a normal distribution with mean value $\mu = 120$ and standard deviation $\sigma = 24$ [hours]. After a failure, the system is replaced by an equivalent new one in negligible time and immediately resumes its work. How many spare systems must be available in order to be able maintain the replacement process over an interval of length 10,000 *hours*

(1) with probability 0.90,

(2) with probability 0.99 ?

3.19) (1) Use the Laplace transformation to find the renewal function $H(t)$ of an ordinary renewal process whose cycle lengths have an Erlang distribution with parameters $n = 2$ and λ.

(2) For $\lambda = 1$, sketch the exact graph of the renewal function and the bounds (3.138) in the interval $0 \le t \le 6$. (Make sure that the bounds (3.138) are applicable.)

3.20) The probability density function of the cycle lengths of an ordinary renewal process is the mixture of two exponential distributions:

$$f(t) = p\lambda_1 e^{-\lambda_1 t} + (1-p)\lambda_2 e^{-\lambda_2 t}, \quad 0 \le p \le 1, \; t \ge 0.$$

By means of the Laplace transformation, determine the associate renewal function.

3.21)* (1) Verify that the probability

$$p(t) = P(N(t) \text{ is odd})$$

satisfies the integral equation

$$p(t) = F(t) - \int_0^t p(t-x) f(x)\, dx, \quad f(x) = F'(x).$$

(2) Determine $p(t)$ if the cycle lengths are exponential with parameter λ.

3.22) An ordinary renewal process has the renewal function $H(t) = t/10$. Determine the probability $P(N(10) \ge 2)$.

3.23)* Verify that $H_2(t) = E(N^2(t))$ satisfies the integral equation

$$H_2(t) = 2H(t) - F(t) + \int_0^t H_2(t-x) f(x)\, dx.$$

3.24) Given the existence of the first 3 moments of the cycle length Y, prove equations (3.132).

3.25) The cycle length Y of an ordinary renewal process is a discrete random variable with probability distribution $p_k = P(Y = k);\; k = 0, 1, 2, ...$

(1) Show that the corresponding renewal function $H(n);\; n = 0, 1, ...$ satisfies

$$H(n) = q_n + H(0)\,p_n + H(1)\,p_{n-1} + \cdots + H(n)\,p_0$$

with $q_n = P(Y \le n) = p_0 + p_1 + \cdots + p_n;\; n = 0, 1, ...$

(2) Consider the special cycle length distribution

$$P(Y = 0) = p,\; P(Y = 1) = 1 - p$$

and determine the corresponding renewal function. (This special renewal process is sometimes referred to as the *negative binomial process*.)

3.26) Consider an ordinary renewal process the cycle length Y of which has the distribution function

$$F(t) = 1 - e^{-t^2},\quad t \ge 0.$$

(1) What is the statement of theorem 3.12 if $g(x) = (x+1)^{-2},\; x \ge 0$?

(2) What is the statement of theorem 3.14 (formula (3.145))?

3.27) The time intervals between the arrivals of successive particles at a counter generate an ordinary renewal process. After having recorded 10 particles, the counter is blocked for τ time units. Particles arriving during a blocked period are not registered. What is the distribution function of the time from the end of a blocked period to the arrival of the first particle after this period if $\tau \to \infty$?

3.28) Let $A(t)$ be the forward and $B(t)$ the backward recurrence times of an ordinary renewal process at time t. For $x > y/2$, determine functional relationships between $F(t)$ and the conditional probabilities

(1) $P(A(t) > y - t \,|\, B(t) = t - x),\quad 0 \le x < t < y,$

(2) $P(A(t) \le y \,|\, B(t) = x)$.

3.29)* Prove formula (3.145) by means of theorem 3.13.

Hint Let $Z(t) = H(t) - t/\mu$.

3.30) Let (Y, Z) be the typical cycle of an alternating renewal process, where Y and Z have an Erlang distribution with joint parameter λ and parameters $n = 2$ and $n = 1$, respectively.

For $t \to \infty$, determine the probability that the system is in state 1 at time t and that it stays in this state over the entire interval $[t, t + x],\; x > 0$.

Hint Process states as introduced in section 3.3.6.

3.31) The time intervals between successive repairs of a system generate an ordinary renewal process $\{Y_1, Y_2, ...\}$ with typical cycle length Y. The costs of repairs are mutually independent, independent of $\{Y_1, Y_2, ...\}$, and identically distributed as M. The random variables Y and M have parameters

$$\mu = E(Y) = 180\,[days], \quad \sigma = \sqrt{Var(Y)} = 30,$$
$$\nu = E(M) = \$\,200, \quad \sqrt{Var(M)} = 40.$$

Determine approximately the probabilities that

(1) the total repair cost arising in [0, 3600 *days*] does not exceed \$ 4500,

(2) a total repair cost of \$ 3000 is not exceeded before 2200 *days*.

3.32) A system is subjected to an age renewal policy with renewal interval τ as described in example 3.21. Determine the stationary availability of the system by modeling its operation by an alternating renewal process.

3.33) A system is subjected to an age renewal policy with renewal interval τ. Contrary to example 3.21, it is assumed that renewals occur in negligible time and that preventive and emergency renewals give rise to the respective constant costs c_p and c_e with $0 < c_p < c_e$. Further, let $F(t)$ be the distribution function of the system lifetime T and $\lambda(t)$ be the corresponding failure rate.

(1) Determine the maintenance cost rate (total maintenance cost per unit time) $K(\tau)$ for an unbounded running time of the system. (*Note* Total maintenance cost' includes replacement and repair costs.)

(2) Give a necessary and sufficient condition for the existence of an optimal renewal interval τ^*.

(3) Determine τ^* if T has a uniform distribution over the interval $[0, z]$.

3.34) A system is preventively renewed at fixed time points $\tau, 2\tau, ...$ Failures between these time points are removed by emergency renewals. (This replacement policy is called *block replacement.*)

(1) With the notation and assumptions of the previous problem, determine the maintenance cost rate $K(\tau)$.

(2) On condition that the system lifetime has distribution function

$$F(t) = (1 - e^{-\lambda t})^2, \; t \ge 0,$$

give a necessary condition for a renewal interval $\tau = \tau^*$ which is optimal with respect to $K(\tau)$. (*Hint* Make use of the renewal function obtained in example 3.13.)

3.35) Under the model assumptions of example 3.22,

(1) determine the ruin probability $p(x)$ of an insurance company with an initial capital of $x = \$\,20,000$ and operating parameters

$$1/\mu = 2\,[h^{-1}],\ \ \nu = \$\,800,\ \text{and}\ \kappa = 1700\,[\$/h],$$

(2) with the numerical parameters given under (1), determine the upper bound e^{-rx} of the Lundberg inequality (3.206),

(3) under otherwise the same conditions, draw the respective graphs of the ruin probability $p(x)$ for $x = 20,000$ and $x = 0$ (no initial capital) in dependence on κ over the interval $1600 \le \kappa \le 1800$,

3.36) Under otherwise the same assumptions and numerical parameters as made in exercise 3.35 (1),

(1) determine the ruin probability if claims arrive according to an ordinary renewal process the typical cycle length of which has an Erlang distribution with parameters $n = 2$ and $\lambda = 4$,

(2) determine the ruin probability if claims arrive according to the corresponding stationary renewal process.

3.37) Under otherwise the same assumptions as made in example 3.22, determine the ruin probability if the claim size M has density

$$b(y) = a^2 y\, e^{-ay},\quad a > 0,\ y > 0.$$

3.38) Claims arrive at an insurance company according to an ordinary renewal process $\{Y_1, Y_2, ...\}$. The corresponding claim sizes $M_1, M_2, ...$ are independent and identically distributed as M and independent of $\{Y_1, Y_2, ...\}$. Let the Y_i be distributed as Y; i.e. Y is the typical interarrival interval. Then (Y, M) is the typical interarrival cycle. From historical observations it is known that

$$\mu = E(Y) = 2\,[h],\ \ Var(Y) = 3,\quad \nu = E(M) = \$\,900,\ \ Var(M) = 360,000.$$

Find approximate answers to the following problems:

(1) What minimum premium per unit time $\kappa_{\text{min},\alpha}$ has the insurance company to take in so that it will make a profit of at least $\$\,10^6$ within $10,000$ hours with probability $\alpha = 0.95$?

(2) What is the probability that the total claim amount hits level $\$\,4 \cdot 10^6$ in the interval [0, 7,000 hours]?

(Before possibly reaching its goals the insurance company may have experienced one or more ruins with subsequent 'red number periods'.)

CHAPTER 4

Discrete-Time Markov Chains

4.1 FOUNDATIONS AND EXAMPLES

This chapter is subjected to discrete-time stochastic processes $\{X_0, X_1, ...\}$ with discrete state space **Z** which have the Markov property. That is, on condition $X_n = x_n$ the random variable X_{n+1} is independent of all $X_0, X_1, ..., X_{n-1}$. However, without this condition, X_{n+1} may very well depend on all the other X_i, $i \le n$.

Definition 4.1 Let $\{X_0, X_1, ...\}$ be a stochastic process in discrete-time with discrete state space **Z**. Then $\{X_0, X_1, ...\}$ is a *discrete-time Markov chain* if for all vectors $x_0, x_1, ..., x_{n+1}$ with $x_k \in \mathbf{Z}$ and for all $n = 1, 2, ...,$

$$P(X_{n+1} = x_{n+1} \mid X_n = x_n, ..., X_1 = x_1, X_0 = x_0) = P(X_{n+1} = x_{n+1} \mid X_n = x_n) \quad (4.1)$$

●

Condtion (4.1) is called the the *Markov property*. It can be interpreted as follows: If time point $t = n$ is the present, then $t = n + 1$ is a future time point and the time points $t = n - 1, ..., 1, 0$ are in the past. Thus,

> The future development of a discrete-time Markov chain depends only on its present state, but not on its evolution in the past.

Note that for the special class of stochastic processes considered in this chapter definition 4.1 is equivalent to the definition of the Markov property via (2.19) in chapter 2. It usually requires much effort to check by statistical methods, whether a particular stochastic process has the Markov property (4.1). Hence one should first try to confirm or to reject this hypothesis by considering properties of the underlying technical, physical, economical or other practical situation. For instance, the final profit of a gambler usually depends on his present profit, but not on the way he has obtained it. If it is known that at the end of the nth month a manufacturer has sold a total of $X_n = x_n$ personal computers, then for predicting thc total numbcr of computers X_{n+1}, sold a month later, knowledge about the number of computers sold within the first $n - 1$ months will make no difference. A car driver checks the tread depth of his tires after every 5000 *km*. For predicting the tread depth after a further 5000 *km*, the driver will only need the present tread depth, not how the tread depth has evolved to its present value. On the other hand, for predicting the future concentration of noxious substances in the air, it has been proved necessary to take into account not only the present value of the concentration, but also the past development leading to this value. In this chapter it will be assumed that the state space of the Markov chain is given by $\mathbf{Z} = \{0, \pm 1, \pm 2, ...\}$. Hence, states will be denoted as $i, j, k, ...$

Transition Probabilities The conditional probabilities

$$p_{ij}(n) = P(X_{n+1} = j | X_n = i); \quad n = 0, 1, ...$$

are the *one-step transition probabilities* of the Markov chain. A Markov chain is said to be *homogeneous* if it has homogeneous increments. Thus, a Markov chain is homogeneous if and only if its one-step transition probabilities do not depend on n:

$$p_{ij}(n) = p_{ij} \quad \text{for all} \quad n = 0, 1, ...$$

Note This chapter only deals with homogeneous Markov chains. For the sake of brevity, the attribute *homogeneous* is generally omitted.

The one-step transition probabilities are combined in the *matrix of the one-step transition probabilities* (shortly: *transition matrix*) **P**:

$$\mathbf{P} = \begin{pmatrix} p_{00} & p_{01} & p_{02} & \cdots \\ p_{10} & p_{11} & p_{12} & \cdots \\ \vdots & \vdots & \vdots & \cdots \\ p_{i0} & p_{i1} & p_{i2} & \cdots \\ \vdots & \vdots & \vdots & \cdots \end{pmatrix}.$$

p_{ij} is the probability of a transition from state i to state j in one step (or, equivalently, *in one time unit, in one jump*). With probability p_{ii} the Markov chain remains in state i for another time unit. The one-step transition probabilities have some obvious properties:

$$p_{ij} \geq 0, \quad \sum_{j \in \mathbf{Z}} p_{ij} = 1; \quad i, j \in \mathbf{Z}. \tag{4.2}$$

The *m-step transition probabilities* of a Markov chain are defined as

$$p_{ij}^{(m)} = P(X_{n+m} = j | X_n = i); \quad m = 1, 2, ... \tag{4.3}$$

Thus, $p_{ij}^{(m)}$ is the probability that the Markov chain, starting from state i, will be after m steps in state j. However, in between the Markov chain may already have arrived at state j. Note that $p_{ij} = p_{ij}^{(1)}$. It is convenient to introduce the notation

$$p_{ij}^{(0)} = \delta_{ij} = \begin{cases} 1 & \text{if } i = j \\ 0 & \text{if } i \neq j \end{cases}. \tag{4.4}$$

δ_{ij} defined in this way is called the *Kronecker symbol.*

The following relationship between the multi-step transition probabilities of a discrete-time Markov chain is called ***Chapman-Kolmogorov equations***:

$$p_{ij}^{(m)} = \sum_{k \in \mathbf{Z}} p_{ik}^{(r)} p_{kj}^{(m-r)}; \quad r = 0, 1, ..., m. \tag{4.5}$$

The proof is easy: Conditioning with regard to the state, which the Markov chain assumes after r time units, $0 \le r \le m$, and making use of the Markov property yields

$$p_{ij}^{(m)} = P(X_m = j | X_0 = i) = \sum_{k \in \mathbf{Z}} P(X_m = j, X_r = k | X_0 = i)$$
$$= \sum_{k \in \mathbf{Z}} P(X_m = j | X_r = k, X_0 = i) P(X_r = k | X_0 = i)$$
$$= \sum_{k \in \mathbf{Z}} P(X_m = j | X_r = k) P(X_r = k | X_0 = i)$$
$$= \sum_{k \in \mathbf{Z}} p_{ik}^{(r)} p_{kj}^{(m-r)} .$$

This proves formula (4.5).

It simplifies notation, when making use of the *matrix of the m-step transition probabilities* of the Markov chain:

$$\mathbf{P}^{(m)} = \left(\left(p_{ij}^{(m)}\right)\right); \quad m = 0, 1, ...$$

Then Chapman-Kolmogorov's equations can be written in the elegant form

$$\mathbf{P}^{(m)} = \mathbf{P}^{(r)} \mathbf{P}^{(m-r)}; \quad r = 0, 1, ..., m.$$

This relationship implies that

$$\mathbf{P}^{(m)} = \mathbf{P}^{m}.$$

Thus, the matrix of the m-step transition probabilities is equal to the m-fold product of the matrix of the one-step transition probabilities.

A probability distribution $\mathbf{p}^{(0)}$ of X_0 is said to be an *initial distribution* of the Markov chain:

$$\mathbf{p}^{(0)} = \left\{ p_i^{(0)} = P(X_0 = i), \ i \in \mathbf{Z}, \ \sum_{i \in \mathbf{Z}} p_i^{(0)} = 1 \right\}. \tag{4.6}$$

A Markov chain is completely characterized by its transition matrix $\mathbf{P}$ and an initial distribution $\mathbf{p}^{(0)}$. To prove this, one has to show that, given $\mathbf{P}$ and $\mathbf{p}^{(0)}$, all finite-dimensional probabilities can be determined: By the Markov property, for any finite set of states $i_0, i_1, ..., i_n$,

$$P(X_0 = i_0, X_1 = i_1, ..., X_n = i_n)$$
$$= P(X_n = i_n | X_0 = i_0, X_1 = i_1, ..., X_{n-1} = i_{n-1}) \cdot P(X_0 = i_0, X_1 = i_1, ..., X_{n-1} = i_{n-1})$$
$$= P(X_n = i_n | X_{n-1} = i_{n-1}) \cdot P(X_0 = i_0, X_1 = i_1, ..., X_{n-1} = i_{n-1})$$
$$= p_{i_{n-1} i_n} \cdot P(X_0 = i_0, X_1 = i_1, ..., X_{n-1} = i_{n-1}).$$

The second factor in the last line is now treated in the same way. Continuing in this way yields

$$P(X_0 = i_0, X_1 = i_1, ..., X_n = i_n) = p_{i_0}^{(0)} \cdot p_{i_0 i_1} \cdot p_{i_1 i_2} \cdot \cdots \cdot p_{i_{n-1} i_n}. \tag{4.7}$$

This proves the assertion.

The *absolute* or *one-dimensional state probabilities* of the Markov chain after m steps are

$$p_j^{(m)} = P(X_m = j), \quad j \in \mathbf{Z}.$$

Given an initial distribution $\mathbf{p}^{(0)} = \left\{p_i^{(0)}, \ i \in \mathbf{Z}\right\}$, by the total probability rule,

$$p_j^{(m)} = \sum_{i \in \mathbf{Z}} p_i^{(0)} \, p_{ij}^{(m)}, \quad m = 1, 2, ... \tag{4.8}$$

Definition 4.2 An initial distribution $\{\pi_i = P(X_0 = i);\ i \in \mathbf{Z}\}$ is called *stationary* if it satisfies the system of linear equations

$$\pi_j = \sum_{i \in \mathbf{Z}} \pi_i \, p_{ij}; \quad j \in \mathbf{Z}. \tag{4.9}$$

●

Furthermore, it can be shown by induction that in this case even the absolute state probabilities after any number of steps are the same as in the beginning:

$$p_j^{(m)} = \sum_{i \in \mathbf{Z}} \pi_i \, p_{ij}^{(m)} = \pi_j, \quad m = 1, 2, ... \tag{4.10}$$

Thus, state probabilities π_i satisfying (4.9) are time-independent absolute state probabilities, which, together with the transition matrix $\mathbf{P}$ fully characterize a stationary probability distribution of the Markov chain. They are also called *equilibrium state probabilities* of the Markov chain. Moreover, in this particular case, the structure (4.7) of the n-dimensional state probabilities verifies theorem 2.1: A Markov chain is strictly stationary if and only if its (one-dimensional) absolute state probabilities do not depend on time.

Markov chains in discrete time virtually occur in all fields of science, engineering, operations research, economics, risk analysis and finance. In what follows, this will be illustrated by some examples. More examples will be given in the text.

Example 4.1 (*random walk*) A particle moves along the real axis in one step from an integer-valued coordinate i either to $i+1$ or to $i-1$ with equal probabilities. The steps occur independently of each other. If X_0 is the starting position of the particle and X_n the position of the particle after n steps, then $\{X_0, X_1, ...\}$ is a discrete-time Markov chain with state space $\mathbf{Z} = \{0, \pm 1, \pm 2, \cdots\}$ and one-step transition probabilities

$$p_{ij} = \begin{cases} 1/2 & \text{for } j = i+1 \text{ or } j = i-1 \\ 0 & \text{otherwise} \end{cases}.$$

□

Example 4.2 (***random walk with absorbing barriers***) Example 4.1 is modified in the following way: The starting position of the particle is restricted to $0 < X_0 < 5$. There are absorbing barriers at $x = 0$ and $x = 6$, i.e. if the particle arrives at state 0 or at state 6, it cannot leave these states anymore. The state space of the corresponding Markov chain $\{X_0, X_1, ...\}$ is $\mathbf{Z} = \{0, 1, ..., 6\}$ and the transition probabilities are

$$p_{ij} = \begin{cases} 1/2 & \text{for } j = i+1 \text{ or } j = i-1 \text{ and } 1 \le i \le 5 \\ 1 & \text{for } i = j = 0 \text{ or } i = j = 6 \\ 0 & \text{otherwise} \end{cases}.$$

The matrices of the one and two-step transition probabilities are

$$\mathbf{P} = \begin{pmatrix} 1 & 0 & 0 & 0 & 0 & 0 & 0 \\ 1/2 & 0 & 1/2 & 0 & 0 & 0 & 0 \\ 0 & 1/2 & 0 & 1/2 & 0 & 0 & 0 \\ 0 & 0 & 1/2 & 0 & 1/2 & 0 & 0 \\ 0 & 0 & 0 & 1/2 & 0 & 1/2 & 0 \\ 0 & 0 & 0 & 0 & 1/2 & 0 & 1/2 \\ 0 & 0 & 0 & 0 & 0 & 0 & 1 \end{pmatrix},$$

$$\mathbf{P}^{(2)} = \begin{pmatrix} 1 & 0 & 0 & 0 & 0 & 0 & 0 \\ 1/2 & 1/4 & 0 & 1/4 & 0 & 0 & 0 \\ 1/4 & 0 & 1/2 & 0 & 1/4 & 0 & 0 \\ 0 & 1/4 & 0 & 1/2 & 0 & 1/4 & 0 \\ 0 & 0 & 1/4 & 0 & 1/2 & 0 & 1/4 \\ 0 & 0 & 0 & 1/4 & 0 & 1/4 & 1/2 \\ 0 & 0 & 0 & 0 & 0 & 0 & 1 \end{pmatrix}.$$

If the starting position of the particle X_0 is uniformly distributed over $\{1, 2, ..., 5\}$,

$$p_i^{(0)} = P(X_0 = i) = 1/5; \; i = 1, 2, ..., 5;$$

then, by (4.8), the absolute distribution of the position of the particle after 2 steps is

$$\mathbf{p}^{(2)} = \left\{ \frac{3}{20}, \frac{2}{20}, \frac{3}{20}, \frac{3}{20}, \frac{3}{20}, \frac{3}{20}, \frac{3}{20} \right\}.$$ □

Example 4.3 (***random walk with reflecting barriers***) For a given positive integer z, the state space of a Markov chain is $\mathbf{Z} = \{0, 1, \cdots, 2z\}$. A particle moves from position i to position j in one step with probability

$$p_{ij} = \begin{cases} \frac{2z-i}{2z} & \text{for } j = i+1 \\ \frac{i}{2z} & \text{for } j = i-1 \\ 0 & \text{otherwise} \end{cases}. \tag{4.11}$$

Thus, the greater the distance of the particle from the central point z of $\mathbf{Z}$, the greater the probability that the particle moves in the next step into the direction of the central point. Once the particle has arrived at one of the end points $x = 0$ or $x = 2z$, it will return in the next step with probability 1 to position $x = 1$ or $x = 2z - 1$, respectively. (Hence the terminology *reflecting barriers.*) If the particle is at $x = z$, then the probabilities of moving to the left or to the right in the next step are equal, namely 1/2. In this sense, the particle is at $x = z$ in an *equilibrium state.* This situation may be thought of as caused by a force, which is situated at the central point. Its attraction to a particle increases with the particle's distance from this point. □

Example 4.4 (***Ehrenfest's diffusion model***) *P.* and *T. Ehrenfest* came across a random walk with reflecting barriers as early as 1907 whilst investigating the following diffusion model: In a closed container there are exactly $2z$ molecules of a particular type. The container is separated into two equal parts by a membrane, which is permeable to these molecules. Let X_n be the random number of the molecules in one part of the container after n transitions of any molecule from one part of the container to the other. If X_0 denotes the initial number of molecules in the specified part of the container, then they observed that the random sequence $\{X_0, X_1, ...\}$ behaves approximately as a Markov chain with transition probabilities (4.11). Thus, the more molecules are in one part of the container, the more they want to move into the other part. In other words, the system tends to the equilibrium state, i.e. to equal numbers of particles in each part of the container. The system of linear equations (4.9) for the stationary state probabilities is

$$\pi_0 = \pi_1 p_{10},$$

$$\pi_j = \pi_{j-1} p_{j-1,j} + \pi_{j+1} p_{j+1,j}; \quad j = 1, 2, ..., 2z - 1.$$

$$\pi_{2z} = \pi_{2z-1} p_{2z-1,2z}$$

The solution is

$$\pi_j = \binom{2z}{j} 2^{-2z}; \quad j = 0, 1, ..., 2z.$$

As expected, state z has the greatest stationary probability. □

Example 4.5 (***electron orbits***) Depending on its energy, an electron circles around the atomic nucleus in one of the countably infinite set of trajectories $\{1, 2, ...\}$. The one-step transition from trajectory i to trajectory j occurs with probability

$$p_{ij} = a_i e^{-b|i-j|}, \quad b > 0.$$

Hence, the two-step transition probabilities are

$$p_{ij}^{(2)} = a_i \sum_{k=1}^{\infty} a_k e^{-b(|i-k|+|k-j|)}.$$

The a_i cannot be chosen arbitrarily. In view of (4.2), they must satisfy condition

$$a_i\left(e^{-b(i-1)}+e^{-b(i-2)}+\cdots+e^{-b}\right)+a_i\Sigma_{k=0}^{\infty}e^{-bk}=1,$$

or, equivalently,

$$a_i\left(e^{-b}\,\frac{1-e^{-b(i-1)}}{1-e^{-b}}+\frac{1}{1-e^{-b}}\right)=1\,.$$

Therefore,

$$a_i=\frac{e^b-1}{1+e^b-e^{-b(i-1)}}\,;\quad i=1,2,...$$

The structure of the p_{ij} implies that $a_i=p_{ii}$ for all $i=1,2,...$ □

Example 4.6 (*occurrence of traffic accidents*) Let X_n denote the number of traffic accidents over a period of n weeks in a particular area, and let Y_i be the corresponding number in the ith week. Then,

$$X_n=\Sigma_{i=1}^{n}Y_i\,.$$

The Y_i are assumed to be independent and identically distributed as a random variable Y with probability distribution $\{q_k=P(Y=k);\ k=0,1,...\}$. Then $\{X_1,X_2,...\}$ is a Markov chain with state space $\mathbf{Z}=\{0,1,...\}$ and transition probabilities

$$p_{ij}=\begin{cases} q_k & \text{if } j=i+k;\ \ k=0,1,... \\ 0 & \text{otherwise} \end{cases}\,.$$

Example 4.7 (*sequence of moving averages*) Let $\{Y_i;\ i=0,1,...\}$ be a sequence of independent, identically distributed binary random variables with

$$P(Y_i=1)=P(Y_i=-1)=1/2\,.$$

Moving averages X_n are defined as follows:

$$X_n=\tfrac{1}{2}(Y_n+Y_{n-1})\,;\quad n=1,2,...$$

X_n has range $\{-1,\,0,\,+1\}$ and probability distribution

$$\left\{P(X_n=-1)=\tfrac{1}{4},\ P(X_n=0)=\tfrac{1}{2},\ P(X_n=+1)=\tfrac{1}{4}\right\}.$$

Since X_n and X_{n+m} are independent for $m>1$, the corresponding matrix of the m-step transition probabilities $p_{ij}^{(m)}=P(X_{n+m}=j\,|\,X_n=i)$ is

$$\mathbf{P}^{(m)}=\begin{array}{c} \\ -1 \\ 0 \\ +1 \end{array}\begin{array}{c} \begin{array}{ccc} -1 & 0 & +1 \end{array} \\ \begin{pmatrix} 1/4 & 1/2 & 1/4 \\ 1/4 & 1/2 & 1/4 \\ 1/4 & 1/2 & 1/4 \end{pmatrix} \end{array}.$$

The matrix of the one-step transition probabilities $p_{ij}=P(X_{n+1}=j\,|\,X_n=i)$ is

$$\mathbf{P}^{(1)} = \mathbf{P} = \begin{pmatrix} 1/2 & 1/2 & 0 \\ 1/4 & 1/2 & 1/4 \\ 0 & 1/2 & 1/2 \end{pmatrix}.$$

Since $\mathbf{P}^{(1)} \cdot \mathbf{P}^{(1)} \neq \mathbf{P}^{(2)}$, the Chapman-Kolmogorov equations do not hold. Therefore, the sequence of moving averages $\{X_1, X_2, ...\}$ cannot be a Markov chain. □

4.2 CLASSIFICATION OF STATES

4.2.1 Closed Sets of States

A subset $\mathbf{C}$ of the state space $\mathbf{Z}$ of a Markov chain is said to be *closed* if

$$\sum_{j \in \mathbf{C}} p_{ij} = 1 \quad \text{for all } i \in \mathbf{C} \tag{4.12}$$

If a Markov chain is in a closed set of states, then it cannot leave this set since (4.12) is equivalent to $p_{ij} = 0$ for all $i \in \mathbf{C}$, $j \notin \mathbf{C}$. Furthermore, (4.12) implies that

$$p_{ij}^{(m)} = 0 \quad \text{for all} \quad i \in \mathbf{C}, j \notin \mathbf{C} \text{ and } m \geq 1. \tag{4.13}$$

For $m = 2$ formula (4.12) can be proved as follows: From (4.5),

$$p_{ij}^{(2)} = \sum_{k \in \mathbf{C}} p_{ik} p_{kj} + \sum_{k \notin \mathbf{C}} p_{ik} p_{kj} = 0,$$

since $j \notin \mathbf{C}$ implies $p_{kj} = 0$ in the first sum and $p_{ik} = 0$ in the second sum. Now formula (4.13) follows inductively from the Chapman-Kolmogorov equations.

A closed set of states is called *minimal* if it does not contain a proper closed subset. In particular, a Markov chain is said to be *irreducible* if its state space $\mathbf{Z}$ is minimal. Otherwise the Markov chain is *reducible*.

A state i is said to be absorbing if $p_{ii} = 1$. Thus, if a Markov chain has arrived in an absorbing state, it cannot leave this state anymore. Hence, an absorbing state is a minimal closed set of states. Absorbing barriers of a random walk (example 4.2) are absorbing states.

Example 4.8 Let $\mathbf{Z} = \{1, 2, 3, 4, 5\}$ be the state space of a Markov chain with transition matrix

$$\mathbf{P} = \begin{pmatrix} 0.2 & 0 & 0.5 & 0.3 & 0 \\ 0.1 & 0 & 0.9 & 0 & 0 \\ 0 & 1 & 0 & 0 & 0 \\ 0.4 & 0.1 & 0.2 & 0 & 0.3 \\ 0 & 0 & 0 & 0 & 1 \end{pmatrix}.$$

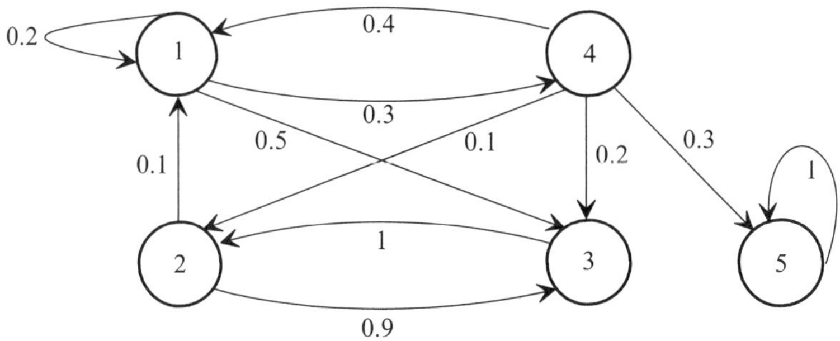

Figure 4.1 Transition graph in example 4.8

It is helpful to illustrate the possible transitions between the states of a Markov chain by *transition graphs*. The nodes of these graphs represent the states of the Markov chain. A directed edge from node i to node j exists if and only if $p_{ij} > 0$, that is, if a one-step transition from state i to state j is possible. The corresponding one-step transition probabilities are attached to the edges. Figure 4.1 shows that $\{1,2,3,4\}$ is not a closed set of states since condition (4.12) is not fulfilled for $i = 4$. State 5 is absorbing so that $\{4\}$ is a minimal closed set of states. This Markov chain is, therefore, reducible. □

4.2.2 Equivalence Classes

State j is said to be *accessible* from state i (symbolically: $i \Rightarrow j$) if there exists an $m \ge 1$ such that $p_{ij}^{(m)} > 0$. The relation '$\Rightarrow$' is transitive: If $i \Rightarrow k$ and $k \Rightarrow j$, there exist $m > 0$ and $n > 0$ with $p_{ik}^{(m)} > 0$ and $p_{kj}^{(n)} > 0$. Therefore,

$$p_{ij}^{(m+n)} = \sum_{r \in \mathbf{Z}} p_{ir}^{(m)} p_{rj}^{(n)} \ge p_{ik}^{(m)} p_{kj}^{(n)} > 0.$$

Consequently, $i \Rightarrow k$ and $k \Rightarrow j$ imply $i \Rightarrow j$, that is, the transitivity of '$\Rightarrow$'.

The set $\mathbf{M}(i) = \{k,\ i \Rightarrow k\}$ consisting of all those states which are accessible from i is closed. In order to prove this assertion it is to show that $k \in \mathbf{M}(i)$ and $j \notin \mathbf{M}(i)$ imply $k \not\Rightarrow j$. The proof is carried out indirectly: If under the assumptions stated $k \Rightarrow j$, then $i \Rightarrow k$ and the transitivity would imply $i \Rightarrow j$. But this contradicts the definition of $\mathbf{M}(i)$.

If both $i \Rightarrow j$ and $j \Rightarrow i$ hold, then i and j are said to *communicate* (symbolically: $i \Leftrightarrow j$). Communication '$\Leftrightarrow$' is an *equivalence relation* since it satisfies the three characteristic properties:

(1) $i \Leftrightarrow i$. *reflexivity*

(2) If $i \Leftrightarrow j$, then $j \Leftrightarrow i$. *commutativity*

(3) If $i \Leftrightarrow j$ and $j \Leftrightarrow k$, then $i \Leftrightarrow k$. *associativity*

Properties (1) and (2) are an immediate consequence of the definition of '$\Leftrightarrow$'. To verify property (3), note that $i \Leftrightarrow j$ and $j \Leftrightarrow k$ imply the existence of m and n so that $p_{ij}^{(m)} > 0$ and $p_{jk}^{(n)} > 0$, respectively. Hence, by (4.5),

$$p_{ik}^{(m+n)} = \sum_{r \in \mathbf{Z}} p_{ir}^{(m)} p_{rk}^{(n)} \geq p_{ij}^{(m)} p_{jk}^{(n)} > 0 .$$

Likewise, there exist M and N with

$$p_{ki}^{(M+N)} \geq p_{kj}^{(M)} p_{ji}^{(N)} > 0$$

so that the associativity is proved.

The equivalence relation '$\Leftrightarrow$' partitions state space $\mathbf{Z}$ into disjoint, but not necessarily closed classes in the following way: *Two states i and j belong to the same class if and only if they communicate.* In what follows, the class containing state i is denoted as $\mathbf{C}(i)$. Clearly, any state in a class can be used to characterize this class. All properties of states introduced in what follows will be *class properties*, i.e. if state i has one of these properties, all states in $\mathbf{C}(i)$ have this property as well.

A state i is called *essential* if any state j which is accessible from i has the property that i is also accessible from j. In this case, $\mathbf{C}(i)$ is called an *essential class*.

A state i is called *inessential* if it is not essential. In this case, $\mathbf{C}(i)$ is called an *inessential class*. If i is inessential, then there exists a state j for which $i \Rightarrow j$ and $j \not\Rightarrow i$.

It is easily verified that *essential* and *inessential* are indeed class properties. In example 4.8, the states 1, 2, 3 and 4 are inessential since state 5 is accessible from each of these states but none of the states 1, 2, 3 or 4 is accessible from state 5.

Theorem 4.1 (1) Essential classes are minimal closed classes. (2) Inessential classes are not closed.

Proof (1) The assertion is a direct consequence of the definition of essential classes.
(2) If i is inessential, then there is a state j with $i \Rightarrow j$ and $j \not\Rightarrow i$. Hence, $j \notin \mathbf{C}(i)$. Assuming $\mathbf{C}(i)$ is closed implies that $p_{kj}^{(m)} = 0$ for all $m \geq 1$, $k \in \mathbf{C}(i)$ and $j \notin \mathbf{C}(i)$. Therefore, $\mathbf{C}(i)$ cannot be closed. (According to the definition of the relation $i \Rightarrow j$, there exists a positive integer m with $p_{ij}^{(m)} > 0$.) ■

Let $p_i^{(m)}(\mathbf{C})$ be the probability that the Markov chain, starting from state i, is in state set $\mathbf{C}$ after m time units:

$$p_i^{(m)}(\mathbf{C}) = \sum_{j \in \mathbf{C}} p_{ij}^{(m)} .$$

Furthermore, let $\mathbf{C}_w$ and $\mathbf{C}_u$ be the sets of all essential and inessential states of a Markov chain. The following theorem asserts that a Markov chain with finite state space, which starts from an inessential state, will leave the set of inessential states with probability 1 and never return (for a proof see Chung [19]. This theorem justifies the notation *essential* and *inessential states.* However, depending on the transition probabilites, the Markov chain may in the initial phase return more or less frequently to the set of inessential states if it has started there.

Theorem 4.2 Let the state space set $\mathbf{Z}$ be finite. Then,

$$\lim_{m\to\infty} p_i^{(m)}(\mathbf{C}_u) = 0 .$$ ■

Example 4.9 If the number of states in a Markov chain is small, the essential and inessential states can immediately be identified from the transition matrix. However, it may be useful to create a more suitable form of this matrix by rearranging its rows and colums, or, equivalently, by changing the notation of the states, For instance, consider a Markov chain with state space $\mathbf{Z} = \{0, 1, 2, 3\}$ and transition matrix

$$\mathbf{P} = \begin{pmatrix} 3/5 & 0 & 2/5 & 0 \\ 0 & 3/4 & 0 & 1/4 \\ 1/3 & 0 & 2/3 & 0 \\ 0 & 1/2 & 0 & 1/2 \end{pmatrix}.$$

By changing the order of rows and columns, an equivalent representation of $\mathbf{P}$ is

$$\mathbf{P} = \begin{pmatrix} 3/5 & 2/5 & 0 & 0 \\ 1/3 & 2/3 & 0 & 0 \\ 0 & 0 & 3/4 & 1/4 \\ 0 & 0 & 1/2 & 1/2 \end{pmatrix} = \begin{pmatrix} \mathbf{Q}_{11} & \mathbf{0} \\ \mathbf{0} & \mathbf{Q}_{22} \end{pmatrix},$$

where $\mathbf{Q}_{11}$ and $\mathbf{Q}_{22}$ are square matrices of order 2 and $\mathbf{0}$ is a square matrix with all elements equal to zero. Hence this Markov chain is reducible. Its state space (in new notation) consists of two essential classes $\mathbf{C}(0) = \{0, 1\}$ and $\mathbf{C}(2) = \{2, 3\}$ with transition matrices $\mathbf{Q}_{11}$ and $\mathbf{Q}_{22}$, respectively. □

Example 4.10 Let $\mathbf{Z} = \{0, 1, ..., 5\}$ be the state space of a Markov chain with transition matrix

$$\mathbf{P} = \begin{pmatrix} 1/3 & 2/3 & 0 & 0 & 0 & 0 \\ 1/2 & 1/2 & 0 & 0 & 0 & 0 \\ 0 & 0 & 1/3 & 2/3 & 0 & 0 \\ 0 & 0 & 2/3 & 1/3 & 0 & 0 \\ 0.4 & 0 & 0.2 & 0.1 & 0.1 & 0.2 \\ 0.1 & 0.2 & 0.1 & 0.2 & 0.3 & 0.1 \end{pmatrix} = \begin{pmatrix} \mathbf{Q}_{11} & \mathbf{0} & \mathbf{0} \\ \mathbf{0} & \mathbf{Q}_{22} & \mathbf{0} \\ \mathbf{Q}_{31} & \mathbf{Q}_{32} & \mathbf{Q}_{33} \end{pmatrix},$$

where the symbolic representation of the transition matrix, introduced in the previous example, is used. This Markov chain has the two essential classes

$$\mathbf{C}(0) = \{0, 1\} \text{ and } \mathbf{C}(2) = \{2, 3\}$$

and the inessential class

$$\mathbf{C}(4) = \{4, 5\}.$$

It is evident that, from the class of inessential states, transitions both to essential and inessential states are possible. However, according to theorem 4.2, the Markov chain will sooner or later leave the inessential class for one of the essential classes and never return. □

4.2.3 Periodicity

Let d_i be the greatest common divisor of those indices $m \geq 1$ for which $p_{ii}^{(m)} > 0$. Then d_i is said to be the *period* of state *i*. If $p_{ii}^{(m)} = 0$ for all $m > 0$, then the period of *i* is defined to be infinite. A state *i* is said to be *aperiodic* if $d_i = 1$.

If *i* has period d_i, then $p_{ii}^{(m)} > 0$ holds if and only if *m* can be represented in the form

$$m = n \cdot d_i; \quad n = 1, 2, \ldots$$

Hence, returning to state *i* is only possible after such a number of steps which is a multiple of d_i. The following theorem shows that the period is a class property.

Theorem 4.3 All states of a class have the same period.

Proof Let $i \Leftrightarrow j$. Then there exist integers *m* and *n* with $p_{ij}^{(m)} > 0$ and $p_{ji}^{(n)} > 0$. If the inequality $p_{ii}^{(r)} > 0$ holds for a positive integer *r*, then, from (4.5),

$$p_{jj}^{(n+r+m)} \geq p_{ji}^{(n)} p_{ii}^{(r)} p_{ij}^{(m)} > 0.$$

Since

$$p_{ii}^{(2r)} \geq p_{ii}^{(r)} \cdot p_{ii}^{(r)} > 0,$$

this inequality also holds if *r* is replaced with $2r$:

$$p_{jj}^{(n+2r+m)} > 0.$$

Thus, d_j divides the difference

$$(n + 2r + m) - (n + r + m) = r.$$

Since this holds for all *r* for which $p_{ii}^{(r)} > 0$, d_j must divide d_i. Changing the roles of *i* and *j* shows that d_i also divides d_j. Thus, $d_i = d_j$. ■

Example 4.11 Let a Markov chain have state space $\mathbf{Z} = \{0, 1, ..., 6\}$ and transition matrix

$$\mathbf{P} = \begin{pmatrix} 1/3 & 2/3 & 0 & 0 & 0 & 0 & 0 \\ 1/3 & 1/3 & 1/3 & 0 & 0 & 0 & 0 \\ 1 & 0 & 0 & 0 & 0 & 0 & 0 \\ 0 & 1/3 & 0 & 1/3 & 1/3 & 0 & 0 \\ 0 & 0 & 0 & 0 & 1 & 0 & 0 \\ 0 & 0 & 0 & 0 & 0 & 1/2 & 1/2 \\ 0 & 0 & 0 & 0 & 1/2 & 0 & 1/2 \end{pmatrix}.$$

Clearly, {0, 1, 2} is a closed set of essential states. State 4 is absorbing, so {4} is another closed set. Having once arrived in a closed set of states the Markov chain cannot leave it again. {3, 5, 6} is a set of inessential states. When starting in one of its sets of inessential states, the Markov chain will at some stage leave this set and never return. All states in {0, 1, 2} have period 1. □

Theorem 4.4 (Chung [19]) The state space $\mathbf{Z}$ of an irreducible Markov chain with period d, $d > 1$, can be partitioned into disjoint subsets $\mathbf{Z}_1, \mathbf{Z}_1, ..., \mathbf{Z}_d$ with

$$\mathbf{Z} = \bigcup_{k=1}^{d} \mathbf{Z}_k$$

such that from any state $i \in \mathbf{Z}_k$ a transition can only be made to a state $j \in \mathbf{Z}_{k+1}$. (By agreement, $j \in \mathbf{Z}_1$ if $i \in \mathbf{Z}_d$). ■

This theorem implies a characteristic structure of the transition matrix of a periodic Markov chain. For instance, if $d = 3$, then the transition matrix $\mathbf{P}$ looks like

$$\mathbf{P} = \begin{matrix} \\ \mathbf{Z}_1 \\ \mathbf{Z}_2 \\ \mathbf{Z}_3 \end{matrix} \begin{matrix} \begin{matrix} \mathbf{Z}_1 & \mathbf{Z}_2 & \mathbf{Z}_3 \end{matrix} \\ \begin{pmatrix} \mathbf{0} & \mathbf{Q}_1 & \mathbf{0} \\ \mathbf{0} & \mathbf{0} & \mathbf{Q}_2 \\ \mathbf{Q}_3 & \mathbf{0} & \mathbf{0} \end{pmatrix} \end{matrix},$$

where $\mathbf{P}$ may be rotated by 90^0. ($\mathbf{Q_i}$ and $\mathbf{0}$ refer to the notation introduced in example 4.10.) According to the definition of a period, if a Markov chain with period d starts in $\mathbf{Z_i}$, it will again be in $\mathbf{Z_i}$ after d transitions. Hence the corresponding d-step transition matrix is

$$\mathbf{P}^{(d)} = \begin{matrix} \\ \mathbf{Z}_1 \\ \mathbf{Z}_2 \\ \mathbf{Z}_3 \end{matrix} \begin{matrix} \begin{matrix} \mathbf{Z}_1 & \mathbf{Z}_2 & \mathbf{Z}_3 \end{matrix} \\ \begin{pmatrix} \mathbf{R}_1 & \mathbf{0} & \mathbf{0} \\ \mathbf{0} & \mathbf{R}_2 & \mathbf{0} \\ \mathbf{0} & \mathbf{0} & \mathbf{R}_3 \end{pmatrix} \end{matrix}.$$

This structure of the transition matrix allows the following interpretation: A Markov chain $\{X_0, X_1, ...\}$ with period d becomes a Markov chain with period 1 and closed equivalence classes $\mathbf{Z}_1, \mathbf{Z}_2, ..., \mathbf{Z}_d$ if, with respect to transitions within the Markov chain $\{X_0, X_1, ...\}$, only the states after every d steps are registered.

Example 4.12 Let a Markov chain have state space $\mathbf{Z} = \{0, 1, ..., 5\}$ and transition matrix

$$\mathbf{P} = \begin{pmatrix} 0 & 0 & 2/5 & 3/5 & 0 & 0 \\ 0 & 0 & 1 & 0 & 0 & 0 \\ 0 & 0 & 0 & 0 & 1/2 & 1/2 \\ 0 & 0 & 0 & 0 & 2/3 & 1/3 \\ 1/2 & 1/2 & 0 & 0 & 0 & 0 \\ 1/4 & 3/4 & 0 & 0 & 0 & 0 \end{pmatrix}.$$

This Markov chain has period $d = 3$. One-step transitions are possible in the order

$$\mathbf{Z}_1 = \{0, 1\} \to \mathbf{Z}_2 = \{2, 3\} \to \mathbf{Z}_1 = \{4, 5\} \to \mathbf{Z}_1.$$

The 3-step transition matrix is

$$\mathbf{P}^{(3)} = \begin{pmatrix} 2/5 & 3/5 & 0 & 0 & 0 & 0 \\ 3/8 & 5/8 & 0 & 0 & 0 & 0 \\ 0 & 0 & 31/40 & 9/40 & 0 & 0 \\ 0 & 0 & 3/4 & 1/4 & 0 & 0 \\ 0 & 0 & 0 & 0 & 11/20 & 9/20 \\ 0 & 0 & 0 & 0 & 21/40 & 19/40 \end{pmatrix}. \qquad \square$$

4.2.4 Recurrence and Transience

This section deals with the return of a Markov chain to an initial state. Such returns are controlled by the *first-passage time probabilities*

$$f_{ij}^{(m)} = P(X_m = j;\, X_k \neq j;\, k = 1, 2, ..., m-1 \,\big|\, X_0 = i\big)\ ; \quad i, j \in \mathbf{Z}$$

Thus, $f_{ij}^{(m)}$ is the probability that the Markov chain, starting from state i, makes its first transition into state j after m steps. Recall that $p_{ij}^{(m)}$ is the probability that the Markov chain, starting from state i, is in state j after m steps, but it may have been in state j in between. For $m = 1$,

$$f_{ij}^{(1)} = p_{ij}^{(1)} = p_{ij}.$$

The total probability rule yields a relationship between the m-step transition probabilities and the first-passage time probabilities

$$p_{ij}^{(m)} = \sum_{k=1}^{m} f_{ij}^{(k)} p_{jj}^{(m-k)}, \tag{4.14}$$

where, by convention, $p_{jj}^{(0)} = 1$ for all $j \in \mathbf{Z}$. Thus, the first-passage time probabilities can be determined recursivley from the following formula:

$$f_{ij}^{(m)} = p_{ij}^{(m)} - \sum_{k=1}^{m-1} f_{ij}^{(k)} p_{jj}^{(m-k)}; \quad m = 2, 3, ... \tag{4.15}$$

The random variable L_{ij} with probability distribution

$$\left\{f_{ij}^{(m)};\ m = 1, 2, ...\right\},$$

is a *first-passage time*. Its mean value is

$$\mu_{ij} = E(L_{ij}) = \sum_{m=1}^{\infty} m\, f_{ij}^{(m)}.$$

The probability of ever making a transition into state j if the process starts in state i is

$$f_{ij} = \sum_{m=1}^{\infty} f_{ij}^{(m)}. \tag{4.16}$$

In particular, f_{ii} is the probability of ever returning to state i. This motivates the introduction of the following concepts:

A state i is said to be *recurrent* if $f_{ii} = 1$ and *transient* if $f_{ii} < 1$.

Clearly, if state i is transient, then $\mu_{ii} = \infty$. But, if i is recurrent, then $\mu_{ii} = \infty$ is also possible. Therefore, recurrent states are classified as follows:

A recurrent state i is said to be *positive recurrent* if $\mu_{ii} < \infty$ and *null-recurrent* if $\mu_{ii} = \infty$. An aperiodic and positive recurrent state is called *ergodic*.

The random time points $T_{i,n};\ n = 1, 2, ...;$ at which the nth return into starting state i occurs, are *regeneration points* of the Markov chain (see definition 3.10, section 3.3.8). By convention, $T_{i,0} = 0$. The time spans between neighbouring regeneration points $T_{i,n} - T_{i,n-1};\ n = 1, 2, ...;$ are called *recurrence times*. They are independent and identically distributed as L_{ii}. Hence the sequence of recurrence times constitutes an ordinary renewal process. Let

$$N_i(t) = \max(n;\ T_{i,n} \le t), \quad H_i(t) = E(N_i(t)),$$

$$N_i(\infty) = \lim_{t \to \infty} N_i(t), \quad H_i(\infty) = \lim_{t \to \infty} H_i(t).$$

Theorem 4.5 State i is recurrent if and only if

(1) $H_i(\infty) = \infty$, or

(2) $\sum_{m=1}^{\infty} p_{ii}^{(m)} = \infty$.

Proof (1) If i is recurrent, then $P(T_{i,n} = \infty) = 0$ for $n = 1, 2, ...$ The limit $N_i(\infty)$ is finite if and only if there is an n with $T_{i,n} = \infty$. Therefore,

$$P(N_i(\infty) < \infty) \le \sum_{i=1}^{\infty} P(T_{i,n} = \infty) = 0 .$$

Thus, assumption $f_{ii} = 1$ implies $N_i(\infty) = \infty$ and, therefore, $H_i(\infty) = \infty$ is true with probability 1.

On the other hand, if $f_{ii} < 1$, then the Markov chain will not return to state i with positive probability $1 - f_{ii}$. In this case, $N_i(\infty)$ has a geometric distribution with mean value (section 1.2.2.2)

$$E(N_i(\infty)) = H_i(\infty) = \frac{f_{ii}}{1 - f_{ii}} < \infty.$$

Both results together prove part (1) of the theorem.

(2) Let the indicator variable for the random event that the Markov chain is in state i at time $t = m$ be

$$I_{m,i} = \begin{cases} 1 & \text{for } X_m = i \\ 0 & \text{for } X_m \ne i \end{cases} ; \quad m = 1, 2, ...$$

Then,

$$N_i(\infty) = \sum_{m=1}^{\infty} I_{m,i} .$$

Hence,

$$H_i(\infty) = E\left(\sum_{m=1}^{\infty} I_{m,i} \right) = \sum_{m=1}^{\infty} E(I_{m,i})$$

$$= \sum_{m=1}^{\infty} P(I_{m,i} = 1) = \sum_{m=1}^{\infty} p_{ii}^{(m)} .$$

Now assertion (2) follows from (1). ■

By adding up both sides of (4.15) from $m = 1$ to ∞ and changing the order of summation according to formula (1.25), theorem 4.5 implies the following corollary.

Corollary If state j is transient, then, for any $i \in \mathbf{Z}$,

$$\sum_{m=1}^{\infty} p_{ij}^{(m)} < \infty$$

and, therefore,

$$\lim_{m \to \infty} p_{ij}^{(m)} = 0 . \tag{4.17}$$

Theorem 4.6 Let i be a recurrent state and $i \Leftrightarrow j$. Then state j is also recurrent.

Proof By definition of the equivalence relation '$\Leftrightarrow$', there are integers m and n with

$$p_{ij}^{(m)} > 0 \text{ and } p_{ji}^{(n)} > 0.$$

By (4.5),

$$p_{jj}^{n+r+m} \geq p_{ji}^{(n)} \, p_{ii}^{(r)} \, p_{ij}^{(m)},$$

so that

$$\sum_{r=1}^{\infty} p_{jj}^{n+r+m} \geq p_{ij}^{(m)} \, p_{ji}^{(n)} \sum_{r=1}^{\infty} p_{ii}^{(r)} = \infty .$$

The assertion is now a consequence of theorem 4.5. ■

Corollary Recurrence and transience are class properties. Hence, an irreducible Markov chain is either *recurrent* or *transient*.

The following statement is elementary, but important.

An irreducible Markov chain with finite state space is recurrent.

It is easy to see that an inessential state is transient. Therefore, each recurrent state is essential. But not each essential state is recurrent. This assertion is proved by the following example.

Example 4.13 (*unbounded random walk*) Starting from $x = 0$, a particle jumps a unit distance along the x-axis to the right with probability p or to the left with probability $1-p$. The transitions occur independently of each other. Let X_n denote the location of the particle after the nth jump. Then the Markov chain $\{X_0, X_1, ...\}$ with $X_0 = 0$ has period $d = 2$. Thus,

$$p_{00}^{(2m+1)} = 0; \quad m = 0, 1, ...$$

In order to be back in state $x = 0$ after $2m$ steps, the particle must jump m times to the left and m times to the right. There are $\binom{2m}{m}$ sample paths which satisfy this condition. Hence,

$$p_{00}^{(2m)} = \binom{2m}{m} p^m (1-p)^m; \quad m = 1, 2, ...$$

Letting $x = p(1-p)$ and making use of the well-known series

$$\sum_{m=0}^{\infty} \binom{2m}{m} x^m = \frac{1}{\sqrt{1-4x}}, \qquad -1/4 < x < 1/4,$$

yields

$$\sum_{m=0}^{\infty} p_{00}^{(m)} = \frac{1}{\sqrt{(1-2p)^2}} = \frac{1}{|1-2p|}, \quad p \neq 1/2 .$$

Thus, the sum

$$\Sigma_{m=0}^{\infty} p_{00}^{(m)}$$

is finite for all $p \neq 1/2$. Hence, by theorem 4.5, state 0 is transient. Consequently, by the corollary from theorem 4.6, the Markov chain is transient, since it is irreducible.

If $p = 1/2$ (*symmetric random walk*), then

$$\sum_{m=0}^{\infty} p_{00}^{(m)} = \lim_{p \to 1/2} \frac{1}{|1-2p|} = \infty. \tag{4.18}$$

Therefore, in this case all states are recurrent. However, for any p with $0 < p < 1$, all states are essential since there is always a positive probability of making a transition to any state irrespective of the starting position. □

The symmetric random walk along a straight line can easily be generalized to n-dimensional Euclidian spaces: In the plane, the particle jumps one unit to the West, South, East, or North, respectively, each with probability 1/4. In the 3-dimensional Euclidian space, the particle jumps one unit to the West, South, East, North, upward, or downward, respectively, each with probability 1/6. When analyzing these random walks analogously to the one-dimensional case, an interesting phenomenon becomes visible: the symmetric two-dimensional random walk (more exactly, the underlying Markov chain) is recurrent like the one-dimensional symmetric random walk, but all n-dimensional symmetric random walks with $n > 2$ are transient. Thus, there is a positive probability that somebody who randomly chooses one of the six possibilities in a 3-dimensional labyrinth, each with probability 1/6, will never return to its starting position.

Example 4.14 A particle jumps from $x = i$ to $x = 0$ with probability p_i or to $i+1$ with probability $1-p_i$; $0 < p_i < 1$, $i = 0, 1, ...$ The jumps are independent of each other. Let X_n denote the position of the particle after the nth jump. Then the transition matrix of the Markov chain $\{X_0, X_1, ...\}$ is

$$\mathbf{P} = \begin{pmatrix} p_0 & 1-p_0 & 0 & 0 & 0 & \cdots & 0 & 0 & \cdots \\ p_1 & 0 & 1-p_1 & 0 & 0 & \cdots & 0 & 0 & \cdots \\ p_2 & 0 & 0 & 1-p_2 & 0 & \cdots & 0 & 0 & \cdots \\ \vdots & \vdots & \vdots & \vdots & \vdots & \cdots & 0 & 0 & \cdots \\ p_i & 0 & \cdots & \cdots & 0 & \cdots & 1-p_i & 0 & \cdots \\ \vdots & \vdots & \vdots & \vdots & \vdots & \cdots & \vdots & \vdots & \vdots \end{pmatrix}.$$

The Markov chain $\{X_0, X_1, ...\}$ is irreducible and aperiodic. Hence, for finding the conditions under which this Markov chain is recurrent or transient it is sufficient to consider state 0, say. It is not difficult to determine $f_{00}^{(m)}$:

$$f_{00}^{(1)} = p_0 ,$$

$$f_{00}^{(m)} = \left(\prod_{i=0}^{m-2} (1 - p_i) \right) p_{m-1} ; \quad m = 2, 3, \ldots$$

If p_{m-1} is replaced with $(1-(1-p_{m-1}))$, then $f_{00}^{(m)}$ becomes

$$f_{00}^{(m)} = \left(\prod_{i=0}^{m-2} (1 - p_i) \right) - \left(\prod_{i=0}^{m-1} (1 - p_i) \right) ; \quad m = 2, 3, \ldots$$

Hence,

$$\sum_{n=1}^{m+1} f_{00}^{(n)} = 1 - \left(\prod_{i=0}^{m} (1 - p_i) \right) , \quad m = 1, 2, \ldots$$

Thus, state 0 is recurrent if and only if

$$\lim_{m \to \infty} \prod_{i=0}^{m} (1 - p_i) = 0 . \tag{4.19}$$

Proposition Condition (4.19) is true if and only if

$$\Sigma_{i=0}^{\infty} p_i = \infty . \tag{4.20}$$

To prove this proposition, note that

$$1 - p_i \le e^{-p_i} ; \; i = 0, 1, \ldots$$

Hence,

$$\Pi_{i=0}^{m} (1 - p_i) \le \exp\left(-\Sigma_{i=0}^{m} p_i \right) .$$

Letting $m \to \infty$ proves that (4.19) follows from (4.20).

The converse direction is proved indirectly: The assumption that (4.19) is true and (4.20) is wrong implies the existence of a positive integer k satisfying

$$0 < \Sigma_{i=k}^{m} p_i < 1 .$$

By induction,

$$\Pi_{i=k}^{m} (1 - p_i) > 1 - p_k - p_{k+1} - \cdots - p_m = 1 - \Sigma_{i=k}^{m} p_i .$$

Therefore,

$$\lim_{m \to \infty} \Pi_{i=k}^{m} (1 - p_i) > \lim_{m \to \infty} \left(1 - \Sigma_{i=k}^{m} p_i \right) > 0 .$$

This contradicts the assumption that condition (4.19) is true, and, hence, completes the proof of the proposition.

Thus, state 0 and with it the Markov chain are recurrent if and only if condition (4.20) is true. This is the case, for instance, if $p_i = p > 0 ; \; i = 0, 1, \ldots$ □

4.3 LIMIT THEOREMS AND STATIONARY DISTRIBUTION

Theorem 4.7 Let state i and j communicate, i.e. $i \Leftrightarrow j$. Then,

$$\lim_{n\to\infty} \frac{1}{n} \sum_{m=1}^{n} p_{ij}^{(m)} = \frac{1}{\mu_{jj}} . \tag{4.21}$$

Proof Analogously to the proof of theorem 4.5 it can be shown that, given the Markov chain is in state i at time $t = 0$, the sum

$$\Sigma_{m=1}^{n} \, p_{ij}^{(m)}$$

is equal to the mean number of transitions into state j in the time interval $(0, n]$. The theorem is, therefore, a direct consequence of the elementary renewal theorem (theorem 3.11). (If $i \neq j$, the corresponding renewal process is delayed.) ■

Theorem 4.7 even holds if the sequence $\left\{ p_{ij}^{(m)}; \; m = 1, 2, ... \right\}$ has no limit. This is, for instance, the case if

$$p_{ij}^{(1)} = 1, \;\; p_{ij}^{(2)} = 0, \;\; p_{ij}^{(3)} = 1, ...$$

However,

$$\lim_{n\to\infty} \frac{1}{n} \sum_{m=1}^{n} p_{ij}^{(m)} = \frac{1}{2} .$$

But, if the limits

$$\lim_{m\to\infty} p_{ij}^{(m)}$$

exist, then they coincide with the right hand side of (4.21) (indirect proof). Since it can be shown that in case of an irreducible Markov chain these limits exist for all $i, j \in \mathbf{Z}$, theorem 4.7 implies theorem 4.8:

Theorem 4.8 Let $p_{ij}^{(m)}$ be the m-step transition probabilities of an irreducible, aperiodic Markov chain. Then, for all $i,j \in \mathbf{Z}$,

$$\lim_{m\to\infty} p_{ij}^{(m)} = \frac{1}{\mu_{jj}} .$$

If state j is transient or null-recurrent, then

$$\lim_{m\to\infty} p_{ij}^{(m)} = 0 .$$

■

Corollary For an irreducible Markov chain with period d,

$$\lim_{m\to\infty} p_{ij}^{(md)} = \frac{d}{\mu_{jj}} .$$

Theorem 4.9 For any irreducible, aperiodic Markov chain, there are two possibilities:

(1) The Markov chain is transient or null-recurrent. Then a stationary distribution does not exist.

(2) The Markov chain is positive recurrent. Then there exists a unique stationary distribution $\{\pi_j, j \in \mathbf{Z}\}$, which for any $i \in \mathbf{Z}$ is given by

$$\pi_j = \lim_{m\to\infty} p_{ij}^{(m)} = \frac{1}{\mu_{jj}}.$$

Proof Without loss of generality, let $\mathbf{Z} = \{0, 1, ...\}$.

(1) By (4.10), a stationary distribution $\{p_j;\ j = 0, 1, ...\}$ satisfies for any $m = 1, 2, ...$ the system of linear algebraic equations

$$p_j = \Sigma_{i=0}^{\infty} p_i\, p_{ij}^{(m)}, \quad m = 1, 2, ... \tag{4.22}$$

If

$$\lim_{m\to\infty} p_{ij}^{(m)} = 0,$$

then there is no probability distribution $\{p_i;\ i = 0, 1, ...\}$, which is solution of (4.22).

(2) Next the existence of a stationary distribution is shown. For $M < \infty$, any $i \in \mathbf{Z}$, and any $m = 1, 2, ...,$

$$\Sigma_{j=0}^{M} p_{ij}^{(m)} < \Sigma_{j=0}^{\infty} p_{ij}^{(m)} = 1.$$

Passing to the limit as $m \to \infty$ yields for all M

$$\Sigma_{j=0}^{M} \pi_j < 1$$

Therefore,

$$\Sigma_{j=0}^{\infty} \pi_j \le 1. \tag{4.23}$$

Analogously, it follows from

$$p_{ij}^{(m+1)} = \Sigma_{k=0}^{\infty} p_{ik}^{(m)} p_{kj} > \Sigma_{k=0}^{M} p_{ik}^{(m)} p_{kj}$$

that

$$\pi_j \ge \Sigma_{j=0}^{\infty} \pi_k p_{kj}. \tag{4.24}$$

If there exists at least one state j for which (4.24) is a proper inequality, then, by summing up the inequalities (4.24) over all j,

$$\Sigma_{j=0}^{\infty} \pi_j > \Sigma_{j=0}^{\infty} \Sigma_{k=0}^{\infty} \pi_k p_{kj} = \Sigma_{k=0}^{\infty} \pi_k \Sigma_{j=0}^{\infty} p_{kj}$$
$$= \Sigma_{k=0}^{\infty} \pi_k.$$

But this is a contradiction to the fact that, by (4.23), the sum of the π_i is finite. Therefore

$$\pi_j = \Sigma_{k=0}^{\infty} \pi_k p_{kj}; \quad j = 0, 1, \ldots$$

Thus, at least one stationary distribution exists, namely $\{p_j;\ j = 0, 1, \ldots\}$ where

$$p_j = \frac{\pi_j}{\Sigma_{i=0}^{\infty} \pi_i}, \quad j \in \mathbf{Z}.$$

From theorem 4.8, letting $m \to \infty$ in (4.22) for any stationary distribution of the Markov chain $\{p_j;\ j = 0, 1, \ldots\}$

$$p_j = \Sigma_{i=0}^{\infty} p_i \pi_j = \pi_j \Sigma_{i=0}^{\infty} p_i = \pi_j, \quad j \in \mathbf{Z}.$$

Thus, $\{\pi_j\,; j = 0, 1, \ldots\}$ with $\pi_j = 1/\mu_{jj}$ is the only stationary distribution. ■

Example 4.15 A particle moves along the real axis. Starting from a position (state) i it jumps to state $i+1$ with probability p and to state $i-1$ with probability $q = 1-p$, $i = 1, 2, \ldots$ When the particle arrives at state 0, it remains there for a further time unit with probability q or jumps to state 1 with probability p. Let X_n denote the position of the particle after the nth jump (time unit). Under which condition has the Markov chain $\{X_0, X_1, \ldots\}$ a stationary distribution?

Since $p_{00} = q,\ p_{i\,i+1} = p$ and $p_{i\,i-1} = q = 1-p;\ i = 1, 2, \ldots,$ the system (4.9) is

$$\pi_0 = \pi_0\, q + \pi_1\, q$$
$$\pi_i = \pi_{i-1}\, p + \pi_{i+1}\, q; \quad i = 1, 2, \ldots$$

By recursively solving this system of equations,

$$\pi_i = \left(\frac{p}{q}\right)^i \pi_0; \quad i = 0, 1, \ldots$$

To ensure that $\Sigma_{i=0}^{\infty} \pi_i = 1$, condition $p < q$ or, equivalently, $p < 1/2$, must hold. In this case,

$$\pi_i = \frac{q-p}{q}\left(\frac{p}{q}\right)^i; \quad i = 0, 1, \ldots \tag{4.25}$$

The necessary condition $p < 1/2$ for the existence of a stationary distribution is intuitive, since otherwise the particle would tend to drift to infinity. But then no time-invariant behaviour of the Markov chain can be expected. □

Theorem 4.10 Let $\{X_0, X_1, \ldots\}$ be an irreducible, recurrent Markov chain with state space $\mathbf{Z}$ and stationary state probabilities $\pi_i,\ i \in \mathbf{Z}$. If g is any bounded function on $\mathbf{Z}$, then

$$\lim_{n\to\infty} \frac{1}{n} \sum_{j=0}^{n} g(X_j) = \sum_{i\in\mathbf{Z}} \pi_i\, g(i).$$ ■

For example, if $c_i = g(i)$ is the profit which accrues from the Markov chain by making a transition to state i, then

$$\Sigma_{i \in \mathbf{Z}} \pi_i c_i$$

is the mean profit resulting from a state change of the Markov chain. Thus, theorem 4.10 is the analogue to the renewal reward theorem (3.170) for compound stochastic processes. In particular, let

$$g(i) = \begin{cases} 1 & \text{for } i = k \\ 0 & \text{for } i \neq k \end{cases}.$$

If, as generally assumed in this chapter, changes of state of the Markov chain occur after unit time intervals, then the limit

$$\lim_{n \to \infty} \frac{1}{n} \Sigma_{j=0}^{n} g(X_j)$$

is equal to the mean percentage of time the system is in state k. By theorem 4.10, this percentage coincides with π_k. This property of the stationary state distribution illustrates once more that it refers to an equilibrium state of the Markov chain. A proof of theorem 4.10 under weaker assumptions can be found in [81].

Example 4.16 A system can be in one of the three states 1, 2, and 3: In state 1 it operates most efficiently. In state 2 it is still working but its efficiency is lower than in state 1. State 3 is the *down state*, the system is no longer operating and has to be maintained. State changes can only occur after a fixed time unit of length 1. Transitions into the same state are allowed. If X_n denotes the state of the system at time n, then $\{X_0, X_1, ...\}$ is assumed to be a Markov chain with transition matrix

$$\mathbf{P} = \begin{matrix} \\ 1 \\ 2 \\ 3 \end{matrix} \begin{matrix} \begin{matrix} 1 & \;\; 2 & \;\; 3 \end{matrix} \\ \begin{pmatrix} 0.8 & 0.1 & 0.1 \\ 0 & 0.6 & 0.4 \\ 0.8 & 0 & 0.2 \end{pmatrix} \end{matrix}.$$

Note that from state 3 the system most likely makes a transition to state 1, but it may also stay in state 3 for one or more time units (for example, if a maintenance action has not been successful). The corresponding stationary state probabilities satisfy the system of linear equations

$$\begin{aligned} \pi_1 &= 0.8\,\pi_1 \qquad\qquad + 0.8\,\pi_3 \\ \pi_2 &= 0.1\,\pi_1 + 0.6\,\pi_2 \\ \pi_3 &= 0.1\,\pi_1 + 0.4\,\pi_2 + 0.2\,\pi_3 \end{aligned}$$

Only two of these equations are linearly independent. Together with the normalizing constraint

$$\pi_1 + \pi_2 + \pi_3 = 1,$$

the unique solution is

$$\pi_1 = \frac{4}{6}, \quad \pi_2 = \pi_3 = \frac{1}{6}. \tag{4.26}$$

The profits the system makes per unit time in states 1 and 2 are

$$g(1) = \$\,1000, \;\; g(2) = \$\,600,$$

whereas, when in state 3, the system generates a loss of

$$g(3) = \; -\$100$$

per unit time. According to theorem 4.10, after an infinite (sufficiently long) running time, the mean profit per unit time is

$$\Sigma_{i=1}^{3} \pi_i \, g(i) = 1000 \cdot \frac{4}{6} + 600 \cdot \frac{1}{6} - 100 \cdot \frac{1}{6} \; = 250 \quad [\$ \text{ per unit time}].$$

Now, let Y be the random time in which the system is in the profitable states 1 and 2. According to the structure of the transition matrix, such a time period must begin with state 1. Further, let Z be the random time in which the system is in the unprofitable state 3. The mean values $E(Y)$ and $E(Z)$ are to be determined. The random vector (Y, Z) characterizes the typical cycle of an alternating renewal process. Therefore, by (3.163), the ratio

$$E(Y)/[E(Y) + E(Z)]$$

is equal to the mean percentage of time the system is in states 1 or 2. As pointed out after theorem 4.10, this percentage must be equal to $\pi_1 + \pi_2$:

$$\frac{E(Y)}{E(Y) + E(Z)} = \pi_1 + \pi_2. \tag{4.27}$$

Since the mean time between transitions into state 3 is equal to $E(Y) + E(Z)$, the ratio

$$1/[\,E(Y) + E(Z)]$$

is equal to the rate of transitions to state 3. On the other hand, this rate is

$$\pi_1 p_{13} + \pi_2 p_{23}.$$

Hence,

$$\frac{1}{E(Y) + E(Z)} = \pi_1 p_{13} + \pi_2 p_{23}. \tag{4.28}$$

From (4.27) and (4.28),

$$E(Y) = \frac{\pi_1 + \pi_2}{\pi_1 p_{13} + \pi_2 p_{23}},$$

$$E(Z) = \frac{\pi_3}{\pi_1 p_{13} + \pi_2 p_{23}}.$$

Substituting the numerical values (4.26) gives

$$E(Y) = 6.25 \quad \text{and} \quad E(Z) = 1.25.$$

□

4.4 BIRTH- AND DEATH PROCESSES

In some of the examples considered so far only direct transitions to 'neighbouring' states were possible. More exactly, if starting in state i and not staying there, only transitions to states $i-1$ or $i+1$ could be made in one step. In these cases, the positive one-step transition probabilities have structure (Figure 4.2)

$$p_{i\,i+1} = p_i, \quad p_{i\,i-1} = q_i, \quad p_{i\,i} = r_i \quad \text{with} \quad p_i + q_i + r_i = 1. \tag{4.29}$$

A discrete Markov chain with state space $\mathbf{Z} = \{0, 1, ..., n\}$, $n \le \infty$, and transition probabilities (4.29) is called a *birth- and death process.* (The state space implies that $q_0 = 0$.) The random walk considered in example 4.9 is a special birth- and death process with

$$p_i = p \text{ for } i = 0, 1, ...$$
$$q_i = q \text{ and } r_i = 0 \text{ for } i = 1, 2, ...,$$
$$q_0 = 0, \; r_0 = q = 1 - p$$

The unbounded random walk in example 4.7 also makes direct transitions only to neighbouring states, but its state space is $\mathbf{Z} = \{0, \pm 1, \pm 2, ...\}$.

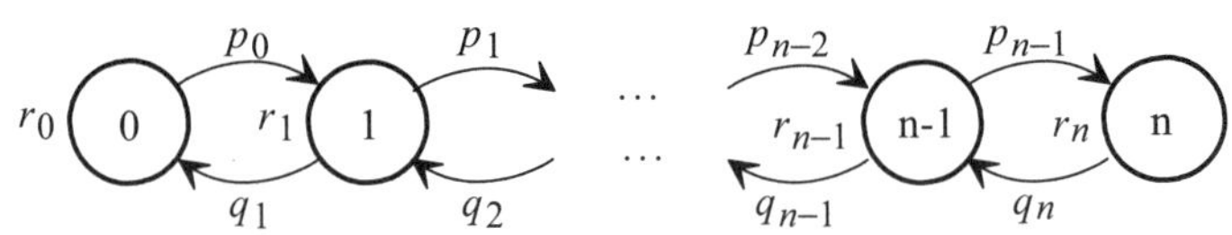

Figure 4.2 Transition graph of a birth- and death process with finite state space

Example 4.17 (***random walk with absorbing barriers***) A random walk with absorbing barriers 0 and s can be modeled by a birth- and death process. In addition to (4.29), its transition probabilities satisfy conditions

$$r_0 = r_s = 1, \; p_i > 0 \text{ and } q_i > 0 \text{ for } i = 1, 2, ..., s-1. \tag{4.30}$$

Let $p(k)$ be the probability that the random walk arrives at state 0 when starting from state k; $k = 1, 2, ..., s-1$. (Since s is absorbing, the Markov chain cannot have been in this state before arriving at 0.) In view of the total probability rule,

$$p(k) = p_k\, p(k+1) + q_k\, p(k-1) + r_k\, p(k),$$

or, replacing r_k with $r_k = 1 - p_k - q_k$,

$$p(k) - p(k+1) = \frac{q_k}{p_k}\,[p(k-1) - p(k)]; \quad k = 1, 2, ..., s-1.$$

Repeated application of this difference equation yields

$$p(j) - p(j+1) = Q_j\,[p(0) - p(1)]\,; \; j = 0, 1, ..., s-1, \tag{4.31}$$

where $p(0) = 1$, $p(s) = 0$ and

$$Q_j = \frac{q_j\, q_{j-1} \cdots q_1}{p_j\, p_{j-1} \cdots p_1}; \quad j = 1, 2, ..., s-1; \quad Q_0 = 1.$$

Summing the equations (4.31) from $j = k$ to $j = s-1$ yields

$$p(k) = \Sigma_{j=k}^{s-1} [p(j) - p(j+1)] = [p(0) - p(1)]\, \Sigma_{j=k}^{s-1} Q_j .$$

In particular, for $k = 0$,

$$1 = [p(0) - p(1)]\, \Sigma_{j=0}^{s-1} Q_j .$$

By combining the last two equations,

$$p(k) = \frac{\Sigma_{j=k}^{s-1} Q_j}{\Sigma_{j=0}^{s-1} Q_j}; \quad k = 0, 1, ..., s-1; \quad p(s) = 0. \tag{4.32}$$

Besides the interpretation of this birth- and death process as a random walk with absorbing barriers, the following application may be more interesting: Two gamblers begin a game with stakes of $\$k$ and $\$(s-k)$, respectively; k, s integers. After each move a gambler either wins or loses $\$1$ or the gambler's stake remains constant. These possibilities are governed by transition probabilities satisfying (4.29) and (4.30). The game is finished if a gambler has won the entire stake of the other one or, equivalently, if one gambler has lost her/his entire stake. Hence this birth- and death process is also called *gambler's ruin problem.* □

To ensure that a birth- and death process is irreducible, assumptions (4.29) have to be supplemented by

$$p_i > 0 \;\text{ for } i = 0, 1, ... \;\text{ and }\; q_i > 0 \;\text{ for } i = 1, 2, ... \tag{4.33}$$

Theorem 4.11 Under the additional assumptions (4.33) on its transition probabilities, a birth- and death process is recurrent if and only if

$$\sum_{j=1}^{\infty} \frac{q_j\, q_{j-1} \cdots q_1}{p_j\, p_{j-1} \cdots p_1} = \infty . \tag{4.34}$$

Proof It is sufficient to show that state 0 is recurrent. This can be established by using the result (4.32) of example 4.17, since

$$\lim_{s\to\infty} p(k) = f_{k0}; \quad k = 1, 2, ...,$$

where the first-passage time probabilities f_{k0} are given by (4.16). If state 0 is recurrent, then, from the irreducibility of the Markov chain

$$f_{00} = 1 \;\text{ and }\; f_{k0} = 1.$$

However, $f_{k0} = 1$ if and only if (4.34) is valid.

Conversely, let (4.34) be true. Then, by the total probability rule,

$$f_{00} = p_{00} + p_{01} f_{10} = r_0 + p_0 \cdot 1 = 1 .$$

This result completes the proof of the theorem. ■

The notation *birth- and death process* results from the application of these processes to describing the development in time of biological populations. In this context, X_n is the number of individuals of a population at time n assuming that the population does not increase or decrease by more than one individual per unit time. Correspondingly, the p_i are called *birth-* and the q_i *death probabilities*.

Discrete-time birth- and death processes may serve as approximations to continuous-time birth- and death processes, which are dealt with in section 5.6.

4.5 EXERCISES

4.1) A Markov chain $\{X_0, X_1, ...\}$ has state space $\mathbf{Z} = \{0, 1, 2\}$ and transition matrix

$$\mathbf{P} = \begin{pmatrix} 0.5 & 0 & 0.5 \\ 0.4 & 0.2 & 0.4 \\ 0 & 0.4 & 0.6 \end{pmatrix}.$$

(1) Determine $P(X_2 = 2 \,|\, X_1 = 0, X_0 = 1)$ and $P(X_2 = 2, X_1 = 0 \,|\, X_0 = 1)$

(2) Determine $P(X_2 = 2, X_1 = 0 \,|\, X_0 = 0)$ and, for $n > 1$,

$$P(X_{n+1} = 2, X_n = 0 \,|\, X_{n-1} = 0)$$

(3) Assuming the initial distribution

$$P(X_0 = 0) = 0.4; \; P(X_0 = 1) = P(X_0 = 2) = 0.3,$$

determine $P(X_1 = 2)$ and $P(X_1 = 1, X_2 = 2)$.

4.2) A Markov chain $\{X_0, X_1, ...\}$ has state space $\mathbf{Z} = \{0, 1, 2\}$ and transition matrix

$$\mathbf{P} = \begin{pmatrix} 0.2 & 0.3 & 0.5 \\ 0.8 & 0.2 & 0 \\ 0.6 & 0 & 0.4 \end{pmatrix}.$$

(1) Determine the matrix of the 2-step transition probabilities $\mathbf{P}^{(2)}$.

(2) Given the initial distribution

$$P(X_0 = i) = 1/3 ; \; i = 0, 1, 2 ;$$

determine the probabilities

$$P(X_2 = 0) \text{ and } P(X_0 = 0, X_1 = 1, X_2 = 2).$$

4.3) A Markov chain $\{X_0, X_1, ...\}$ has state space $\mathbf{Z} = \{0, 1, 2\}$ and transition matrix

$$\mathbf{P} = \begin{pmatrix} 0 & 0.4 & 0.6 \\ 0.8 & 0 & 0.2 \\ 0.5 & 0.5 & 0 \end{pmatrix}.$$

(1) Given the initial distribution

$$P(X_0 = 0) = P(X_0 = 1) = 0.4 \text{ and } P(X_0 = 2) = 0.2,$$

determine $P(X_3 = 2)$.

(2) Draw the corresponding transition graph.

(3) Determine the stationary distribution.

4.4) Let $\{Y_0, Y_1, ...\}$ be a sequence of independent, identically distributed binary random variables with

$$P(Y_i = 0) = P(Y_i = 1) = 1/2; \; i = 0, 1, ...$$

Define a sequence of random variables $\{X_1, X_2, ...\}$ by

$$X_n = \frac{1}{2}(Y_n - Y_{n-1}); \quad n = 1, 2, ...$$

Check whether the random sequence $\{X_1, X_2, ...\}$ has the Markov property.

4.5) A Markov chain $\{X_0, X_1, ...\}$ has state space $\mathbf{Z} = \{0, 1, 2, 3\}$ and transition matrix

$$\mathbf{P} = \begin{pmatrix} 0.1 & 0.2 & 0.4 & 0.3 \\ 0.2 & 0.3 & 0.1 & 0.4 \\ 0.4 & 0.1 & 0.3 & 0.2 \\ 0.3 & 0.4 & 0.2 & 0.1 \end{pmatrix}.$$

(1) Draw the corresponding transition graph.

(2) Determine the stationary distribution of this Markov chain.

4.6) Let $\{X_0, X_1, ...\}$ be an irreducible Markov chain with state space

$$\mathbf{Z} = \{1, 2, ..., n\}, \; n < \infty,$$

and with the doubly stochastic transition matrix $\mathbf{P} = ((p_{ij}))$, i.e.

$$\sum_{j \in \mathbf{Z}} p_{ij} = 1 \text{ for all } i \in \mathbf{Z} \text{ and } \sum_{i \in \mathbf{Z}} p_{ij} = 1 \text{ for all } j \in \mathbf{Z}.$$

(1) Prove that the stationary distribution of $\{X_0, X_1, ...\}$ is given by

$$\pi_j = \frac{1}{n}, \quad j \in \mathbf{Z}.$$

(2) Can $\{X_0, X_1, ...\}$ be a transient Markov chain?

4.7) A source emits symbols 0 and 1 for transmission to a sink. Random noises $S_1, S_2, \ldots$ successively and independently affect the transmission process of a symbol in the following way: if a '0' ('1') is to be transmitted, then S_i distorts it to a '1' ('0') with probability p (q); $i = 1, 2, \ldots$ Let $X_0 = 0$ or $X_0 = 1$ denote whether the source has emitted a '0' or a '1' for transmission. Further, let $X_i = 0$ or $X_i = 1$ denote whether the attack of noise S_i implies the transmission of a '0' or a '1'; $i = 1, 2, \ldots$ The random sequence $\{X_0, X_1, \ldots\}$ is an irreducible Markov chain with state space $\mathbf{Z} = \{0, 1\}$ and transition matrix

$$\mathbf{P} = \begin{pmatrix} 1-p & p \\ q & 1-q \end{pmatrix}.$$

(1) Verify: On condition $0 < p + q \le 1$, the m-step transition matrix is given by

$$\mathbf{P}^{(m)} = \frac{1}{p+q}\begin{pmatrix} q & p \\ q & p \end{pmatrix} + \frac{(1-p-q)^m}{p+q}\begin{pmatrix} p & -p \\ -q & q \end{pmatrix}.$$

(2) Let $p = q = 0.1$. The transmission of the symbols 0 and 1 is affected by the random noises $S_1, S_2, \ldots, S_5$.

Determine the probability that a '0' emitted by the source is actually received.

4.8) Weather is classified as (predominantly) sunny (S) and (predominantly) cloudy (C), where C includes rain. For the town of Musi, a fairly reliable prediction of tomorrow's weather can only be made on the basis of today's and yesterday's weather. Let (C,S) indicate that the weather yesterday was cloudy and today's weather is sunny and so on. Based on historical observations it is known that, given the constellation (S,S) today, the weather tomorrow will be sunny with probability 0.8 and cloudy with probability 0.2; given (S,C) today, the weather tomorrow will be sunny with probability 0.4 and cloudy with probability 0.6; given (C,S) today, the weather tomorrow will be sunny with probability 0.6 and cloudy with probability 0.4; given (C,C) today, the weather tomorrow will be cloudy with probability 0.8 and sunny with probability 0.2.

(1) Illustrate graphically the transitions between the states

1 = (S,S), 2 = (S,C), 3 = (C,S), and 4 = (C,C).

(2) Determine the matrix of the transition probabilities of the corresponding discrete-time Markov chain and its stationary state distribution.

4.9)* An area (e.g. a stiffy disc) is partitioned into n segments $S_1, S_2, \ldots, S_n$, and a collection of n objects $O_1, O_2, \ldots, O_n$ (e.g. pieces of information) are stored in these segments so that each segment contains exactly one object. At time points $t = 1, 2, \ldots$ one of the objects is needed. Since its location is assumed to be unknown, it has to be searched for. This is done in the following way: The segments are checked in increasing order of their indices. When the desired object O is found at segment S_k, then O

will be moved to segment S_1 and the objects originally located at $S_1, S_2, ..., S_{k-1}$ will be moved in this order to $S_2, S_3, ..., S_k$.

Let p_i be the probability that at a time point t object O_i is needed; $i = 1, 2, ..., n$. It is assumed that these probabilities do not depend on t.

(1) Describe the successive location of object O_1 by a homogeneous discrete-time Markov chain, i.e. determine the transition probabilities

$$p_{ij} = P(O_1 \text{ at segment } S_j \text{ at time } t+1 \,|\, O_1 \text{ at segment } S_i \text{ at time } t).$$

(2) What is the stationary distribution of the location of O_1 given that

$$p_1 = \alpha \ \text{ and } \ p_2 = p_3 = \cdots = p_n = \frac{1-\alpha}{n-1}\,?$$

4.10) A supplier of toner cartridges of a certain brand checks his stock every Monday. If the stock is less than or equal to s cartridges, he orders an amount of $S - s$ cartridges, which will be available the following Monday, $0 \le s < S$. The weekly demands of cartridges D are independent and identically distributed according to

$$p_i = P(D = i); \quad i = 0, 1, ...$$

Let X_n be the number of cartridges on stock on the nth Sunday (no business over weekends) given that the supplier starts his business on a Monday.

(1) Is $\{X_1, X_2, ...\}$ a Markov chain?

(2) If yes, obtain the matrix of the transition probabilities.

4.11) A Markov chain has state space $\mathbf{Z} = \{0, 1, 2, 3, 4\}$ and transition matrix

$$\mathbf{P} = \begin{pmatrix} 0.5 & 0.1 & 0.4 & 0 & 0 \\ 0.8 & 0.2 & 0 & 0 & 0 \\ 0 & 1 & 0 & 0 & 0 \\ 0 & 0 & 0 & 0.9 & 0.1 \\ 0 & 0 & 0 & 1 & 0 \end{pmatrix}.$$

(1) Determine the minimal closed sets.

(2) Check, whether inessential states exist.

4.12) A Markov chain has state space $\mathbf{Z} = \{0, 1, 2, 3\}$ and transition matrix

$$\mathbf{P} = \begin{pmatrix} 0 & 0 & 1 & 0 \\ 1 & 0 & 0 & 0 \\ 0.4 & 0.6 & 0 & 0 \\ 0.1 & 0.4 & 0.2 & 0.3 \end{pmatrix}.$$

Determine the classes of essential and inessential states.

4.13) A Markov chain has state space $\mathbf{Z} = \{0, 1, 2, 3, 4\}$ and transition matrix

$$\mathbf{P} = \begin{pmatrix} 0 & 0.2 & 0.8 & 0 & 0 \\ 0 & 0 & 0 & 0.9 & 0.1 \\ 0 & 0 & 0 & 0.1 & 0.9 \\ 1 & 0 & 0 & 0 & 0 \\ 1 & 0 & 0 & 0 & 0 \end{pmatrix}.$$

(1) Draw the transition graph.

(2) Verify that this Markov chain is irreducible with period 3.

(3) Determine the stationary distribution.

4.14) A Markov chain has state space $\mathbf{Z} = \{0, 1, 2, 3, 4\}$ and transition matrix

$$\mathbf{P} = \begin{pmatrix} 0 & 1 & 0 & 0 & 0 \\ 1 & 0 & 0 & 0 & 0 \\ 0.2 & 0.2 & 0.2 & 0.4 & 0 \\ 0.2 & 0.8 & 0 & 0 & 0 \\ 0.4 & 0.1 & 0.1 & 0 & 0.4 \end{pmatrix}.$$

(1) Find the essential and inessential states.

(2) Find the recurrent and transient states.

4.15) Determine the stationary distribution of the random walk considered in example 4.8 on condition $p_i = p, \quad 0 < p < 1$.

4.16) Let the transition probabilities of a birth- and death process be given by

$$p_i = \frac{1}{1 + [i/(i+1)]^2} \quad \text{and} \quad q_i = 1 - p_i; \;\; i = 1, 2, ...; \;\; p_0 = 1.$$

Show that the process is transient.

4.17) Let i and j be two different states with $f_{ij} = f_{ji} = 1$. Show that both i and j are recurrent.

4.18) The respective transition probabilities of two irreducible Markov chains (1) and (2) with common state space $\mathbf{Z} = \{0, 1, ...\}$ are

$$(1) \quad p_{i\,i+1} = \frac{1}{i+2}, \quad p_{i0} = \frac{i+1}{i+2}; \quad i = 0, 1, ...;$$

$$(2) \quad p_{i\,i+1} = \frac{i+1}{i+2}, \quad p_{i0} = \frac{1}{i+2}; \quad i = 0, 1, ...$$

Check whether these Markov chains are transient, null recurrent or positive recurrent.

4.19) Let N_i be the random number of time periods a discrete-time Markov chain stays in state *i* (sojourn time of the Markov chain in state *i*).

Determine $E(N_i)$ and $Var(N_i)$.

4.20) A haulier operates a fleet of trucks. His contract with an insurance company covers his whole fleet and has the following structure ('bonus malus system' in car insurance): The haulier has to pay his premium at the beginning of each year. There are 3 premium levels: λ_1, λ_2 and λ_3 with $\lambda_3 < \lambda_2 < \lambda_1$. If no claim had been made in the previous year and the premium level was λ_i, then the premium level in the current year is λ_{i+1} or λ_3 if $\lambda_i = \lambda_3$. If a claim had been made in the previous year, the premium level in the current year is λ_1. The haulier will claim only then if the total damage a year exceeds an amount of c_i given the premium level λ_i in that year; $i = 1, 2, 3$. In case of a claim, the insurance company will cover the full amount minus a profitincreasing amount of a_i, $0 \le a_i < c_i$. The total damages a year are independent random variables, identically distributed as *M*.

Given a vector of claim limits (c_1, c_2, c_3), determine the haulier's long-run mean loss cost a year.

Hint Introduce the Markov chain $\{X_1, X_2, ...\}$, where $X_n = i$ if the premium level at the beginning of year *n* is λ_i and make use of theorem 4.10.

(Loss cost = premium plus total damage not refunded by the insurance company.)

CHAPTER 5

Continuous-Time Markov Chains

5.1 BASIC CONCEPTS AND EXAMPLES

This chapter deals with Markov processes which have parameter set $\mathbf{T} = [0, \infty)$ and state space $\mathbf{Z} = \{0, \pm 1, \pm 2, ...\}$ or subsets of it. According to the terminology introduced in section 2.3, for having a discrete parameter space, this class of Markov processes are called *Markov chains.*

Definition 5.1 A stochastic process $\{X(t), t \geq 0\}$ with parameter set $\mathbf{T}$ and discrete state space $\mathbf{Z}$ is called a *continuous-time Markov chain* or a *Markov chain in continuous time* if, for any $n \geq 1$ and arbitrary sequences

$$\{t_0, t_1, ..., t_{n+1}\} \text{ with } t_0 < t_1 < \cdots < t_{n+1} \text{ and } \{i_0, i_1, \ldots, i_{n+1}\}, i_k \in \mathbf{Z},$$

the following relationship holds:

$$P(X(t_{n+1}) = i_{n+1} | X(t_n) = i_n, \ldots, X(t_1) = i_1, X(t_0) = i_0) \tag{5.1}$$
$$= P(X(t_{n+1}) = i_{n+1} | X(t_n) = i_n).$$
●

The intuitive interpretation of the *Markov property* (5.1) is the same as for discrete-time Markov chains:

> The future development of a continuous-time Markov chain depends only on its present state and not on its evolution in the past.

The conditional probabilities

$$p_{ij}(s,t) = P(X(t) = j | X(s) = i); \quad s < t; \; i,j \in \mathbf{Z};$$

are the *transition probabilities of the Markov chain.* A Markov chain is said to be *homogeneous* if for all $s, t \in \mathbf{T}$ and $i, j \in \mathbf{Z}$ the transition probabilities $p_{ij}(s,t)$ depend only on the difference $t - s$:

$$p_{ij}(s,t) = p_{ij}(0, t-s).$$

In this case the transition probabilities depend only on one variable:

$$p_{ij}(t) = p_{ij}(0,t).$$

Note This chapter only considers homogeneous Markov chains. Hence no confusion can arise if only *Markov chains* is referred to.

The transition probabilities are comprised in the *matrix of transition probabilities* **P** (simply: *transition matrix*):

$$\mathbf{P}(t) = ((p_{ij}(t))); \quad i, j \in \mathbf{Z}.$$

Besides the trivial property $p_{ij}(t) \geq 0$, transition probabilities are generally assumed to satisfy the conditions

$$\sum_{j \in \mathbf{Z}} p_{ij}(t) = 1; \quad t \geq 0, \ i \in \mathbf{Z}. \tag{5.2}$$

Comment It is theoretically possible that, for some $i \in \mathbf{Z}$,

$$\sum_{j \in \mathbf{Z}} p_{ij}(t) < 1; \quad t > 0, \ i \in \mathbf{Z}. \tag{5.3}$$

In this case, unboundedly many transitions between the states occur in any finite time interval $[0, t)$ with positive probability

$$1 - \sum_{j \in \mathbf{Z}} p_{ij}(t).$$

This situation approximately applies to nuclear chain reactions and population explosions of certain species of insects (e.g. locusts). In the sequel it is assumed that

$$\lim_{t \to +0} p_{ii}(t) = 1. \tag{5.4}$$

By (5.2), this assumption is equivalent to

$$p_{ij}(0) = \lim_{t \to +0} p_{ij}(t) = \delta_{ij}; \quad i, j \in \mathbf{Z}. \tag{5.5}$$

The *Kronecker symbol* δ_{ij} is defined by (4.4).

Analogously to (4.5), the ***Chapman-Kolmogorov equations*** are

$$p_{ij}(t + \tau) = \sum_{k \in \mathbf{Z}} p_{ik}(t) p_{kj}(\tau) \tag{5.6}$$

for any $t \geq 0$, $\tau \geq 0$, and $i, j \in \mathbf{Z}$. By making use of the total probability rule, the homogeneity and the Markov property, (5.6) is proved as follows:

$$p_{ij}(t+\tau) = P(X(t+\tau) = j | X(0) = i) = \frac{P(X(t+\tau) = j, X(0) = i)}{P(X(0) = i)}$$

$$= \sum_{k \in \mathbf{Z}} \frac{P(X(t+\tau) = j, X(t) = k, X(0) = i)}{P(X(0) = i)}$$

$$= \sum_{k \in \mathbf{Z}} \frac{P(X(t+\tau) = j | X(t) = k, X(0) = i) P(X(t) = k, X(0) = i)}{P(X(0) = i)}$$

$$= \sum_{k \in \mathbf{Z}} \frac{P(X(\tau+t) = j | X(t) = k) P(X(t) = k | X(0) = i) P(X(0) = i)}{P(X(0) = i)}$$

$$= \sum_{k \in \mathbf{Z}} P(X(\tau) = j | X(0) = k) P(X(t) = k | X(0) = i)$$

$$= \sum_{k \in \mathbf{Z}} p_{ik}(t) p_{kj}(\tau).$$

Absolute and Stationary Distributions Let $p_i(t) = P(X(t) = i)$ be the probability that the Markov chain is in state i at time t. $p_i(t)$ is called *absolute state probability* (of the Markov chain) at time t. Hence, $\{p_i(t), i \in \mathbf{Z}\}$ is said to be the *absolute (one-dimensional) probability distribution* of the Markov chain at time t. In particular, $\{p_i(0); i \in \mathbf{Z}\}$ is called an *initial (probability) distribution* of the Markov chain. By the total probability rule, given an initial distribution, the absolute probability distribution of the Markov chain at time t is

$$p_j(t) = \sum_{i \in \mathbf{Z}} p_i(0)\, p_{ij}(t), \quad j \in \mathbf{Z}. \tag{5.7}$$

For determining the *multidimensional distribution* of the Markov chain at time points $t_0, t_1, ..., t_n$ with $0 \le t_0 < t_1 < \cdots < t_n < \infty$, only its absolute probability distribution at time t_0 and its transition probabilities need to be known. This can be proved by repeated application of the formula of the conditional probability (1.6) and by making use of homogeneity of the Markov chain:

$$P(X(t_0) = i_0, X(t_1) = i_1, ..., X(t_n) = i_n)$$
$$= p_{i_0}(t_0) p_{i_0 i_1}(t_1 - t_0) p_{i_1 i_2}(t_2 - t_1) \cdots p_{i_{n-1} i_n}(t_n - t_{n-1}). \tag{5.8}$$

Definition 5.2 An initial distribution $\{\pi_i = p_i(0),\ i \in \mathbf{Z}\}$ is said to be *stationary* if

$$\pi_i = p_i(t) \quad \text{for all } t \ge 0 \text{ and } i \in \mathbf{Z}. \tag{5.9}$$

●

Thus, if at time $t = 0$ the initial state is determined by a stationary initial distribution, then the absolute state probabilities $p_j(t)$ do not depend on t and are equal to π_j. Consequently, the stationary initial probabilities π_j are the absolute state probabilities $p_j(t)$ for all $j \in \mathbf{Z}$ and $t \ge 0$. Moreover, it follows from (5.8) that in this case all n-dimensional distributions of the Markov chain, namely

$$\{P(X(t_1 + h) = i_1, X(t_2 + h) = i_2, ..., X(t_n + h) = i_n\},\ i_j \in \mathbf{Z} \tag{5.10}$$

do not depend on h, i.e. if the process starts with a stationary initial distribution, then the Markov chain is strictly stationary. (This result verifies the more general statement of theorem 2.1.) Moreover, it is justified to call $\{\pi_i, i \in \mathbf{Z}\}$ a *stationary (probability) distribution* of the Markov chain.

Example 5.1 A homogeneous Poisson process $\{N(t), t \ge 0\}$ with intensity λ is a homogeneous Markov chain with state space $\mathbf{Z} = \{0, 1, ...\}$ and transition probabilities

$$p_{ij}(t) = \frac{(\lambda t)^{j-i}}{(j-i)!} e^{-\lambda t}; \quad i \le j.$$

The sample paths of the process $\{N(t), t \ge 0\}$ are nondecreasing step-functions. Its trend function is linearly increasing:

$$m(t) = E(N(t)) = \lambda t.$$

Thus, a stationary initial distribution cannot exist. (However, according to the corollary following definition 3.1 in section 3.1, the homogeneous Poisson process is a stationary point process.) □

Example 5.2 At time $t = 0$, exactly n systems start operating. Their lifetimes are independent, identically distributed exponential random variables with parameter λ. If $X(t)$ denotes the number of systems still operating at time t, then $\{X(t), t \geq 0\}$ is a Markov chain with state space $\mathbf{Z} = \{0, 1, ..., n\}$, transition probabilities

$$p_{ij}(t) = \binom{i}{i-j} (1 - e^{-\lambda t})^{i-j} e^{-\lambda t j}, \quad n \geq i \geq j \geq 0.$$

and initial distribution $P(X(0) = n) = 1$. The structure of these transition probabilities is based on the memoryless property of the exponential distribution (example 1.14). Of course, this Markov chain cannot be stationary. □

Example 5.3 Let $\mathbf{Z} = \{0, 1)$ be the state space and

$$\mathbf{P}(t) = \begin{pmatrix} \frac{1}{t+1} & \frac{t}{t+1} \\ \frac{t}{t+1} & \frac{1}{t+1} \end{pmatrix}$$

the transition matrix of a stochastic process $\{X(t), t \geq 0\}$. It is to check whether this process is a Markov chain. Assuming the initial distribution

$$p_0(0) = P(X(0) = 0) = 1$$

and applying formula (5.7) yields the absolute probability of state 0 at time $t = 3$:

$$p_0(3) = p_0(0) p_{00}(3) = 1/4 .$$

On the other hand, applying (5.6) with $t = 2$ and $\tau = 1$ yields the (wrong) result

$$p_0(3) = p_{00}(2) p_{00}(1) + p_{01}(2) p_{10}(1) = 1/2 .$$

Therefore, Chapman-Kolmogorov's equations (5.6) are not valid so that $\{X(t), t \geq 0\}$ cannot be a Markov chain. □

Classification of States The classification concepts already introduced for discrete-time Markov chains can analogously be defined for continuous-time Markov chains. In what follows, some concepts are defined, but not discussed in detail.

A state set $\mathbf{C} \subseteq \mathbf{Z}$ is called *closed* if

$$p_{ij}(t) = 0 \text{ for all } t > 0, \, i \in \mathbf{C} \text{ and } j \notin \mathbf{C}.$$

If, in particular, $\{i\}$ is a closed set, then i is called an *absorbing state.* The state j is *accessible* from i if there exists a t with $p_{ij}(t) > 0$. If i and j are accessible from each other, then they are said to *communicate.* Thus, equivalence classes, essential and inessential states as well as irreducible and reducible Markov chains can be defined as in section 4.2 for discrete Markov chains.

State i is *recurrent* (*transient*) if

$$\int_0^\infty p_{ii}(t)\,dt = \infty \qquad \left(\int_0^\infty p_{ii}(t)\,dt < \infty\right).$$

A recurrent state i is *positive recurrent* if the mean value of its recurrence time (time between two successive occurences of state i) is finite. Since it can easily be shown that $p_{ij}(t_0) > 0$ implies $p_{ij}(t) > 0$ for all $t > t_0$, introducing the concept of a period analogously to section 4.3.3 makes no sense.

5.2 TRANSITION PROBABILITIES AND RATES

This section discusses some structural properties of continuous-time Markov chains which are fundamental to mathematically modeling real systems.

Theorem 5.1 On condition (5.4), the transition probabilities $p_{ij}(t)$ are differentiable in $[0, \infty)$ for all $i, j \in \mathbf{Z}$.

Proof For any $h > 0$, the Chapman-Kolmogorov equations (5.6) yield

$$\begin{aligned} p_{ij}(t+h) - p_{ij}(t) &= \sum_{k \in \mathbf{Z}} p_{ik}(h)\, p_{kj}(t) - p_{ij}(t) \\ &= -(1 - p_{ii}(h))\, p_{ij}(t) + \sum_{k \in \mathbf{Z},\, k \neq i} p_{ik}(h)\, p_{kj}(t). \end{aligned}$$

Thus,

$$\begin{aligned} -(1 - p_{ii}(h)) \le -(1 - p_{ii}(h))\, p_{ij}(t) &\le p_{ij}(t+h) - p_{ij}(t) \\ &\le \sum_{\substack{k \in \mathbf{Z} \\ k \neq i}} p_{ik}(h)\, p_{kj}(t) \le \sum_{\substack{k \in \mathbf{Z} \\ k \neq i}} p_{ik}(h) \\ &= 1 - p_{ii}(h). \end{aligned}$$

Hence,

$$\left| p_{ij}(t+h) - p_{ij}(t) \right| \le 1 - p_{ii}(h).$$

The uniform continuity of the transition probabilities and, therefore, their differentiability for all $t \ge 0$ is now a consequence of assumption (5.4). ■

Transition Rates The following limits play an important role in all future derivations. For any $i, j \in \mathbf{Z}$, let

$$q_i = \lim_{h \to 0} \frac{1 - p_{ii}(h)}{h}, \tag{5.11}$$

$$q_{ij} = \lim_{h \to 0} \frac{p_{ij}(h)}{h}, \quad i \neq j. \tag{5.12}$$

These limits exist, since, by (5.5),

$$p_{ii}(0)=1 \text{ and } p_{ij}(0)=0 \text{ for } i\neq j$$

so that, by theorem 5.1,

$$p'_{ii}(0)=\left.\frac{dp_{ii}(t)}{dt}\right|_{t=0}=-q_i, \tag{5.13}$$

$$p'_{ij}(0)=\left.\frac{dp_{ij}(t)}{dt}\right|_{t=0}=q_{ij}, \quad i\neq j. \tag{5.14}$$

For $h\to 0$, relationships (5.13) and (5.14) are equivalent to

$$p_{ii}(h)=1-q_i h+o(h) \tag{5.15}$$

$$p_{ij}(h)=q_{ij}h+o(h), \quad i\neq j, \tag{5.16}$$

respectively. The parameters q_i and q_{ij} are the *transition rates* of the Markov chain. More exactly, q_i is the *unconditional transition rate* of leaving state i for any other state, and q_{ij} is the *conditional transition rate* of making a transition from state i to state j. According to (5.2),

$$\sum_{\{j,j\neq i\}} q_{ij}=q_i, \quad i\in\mathbf{Z}. \tag{5.17}$$

Kolmogorov's Differential Equations In what follows, systems of differential equations for the transition probabilities and the absolute state probabilities of a Markov chain are derived. For this purpose, the system of Chapman-Kolmogorov equations is written in the form

$$p_{ij}(t+h)=\sum_{k\in\mathbf{Z}} p_{ik}(h)p_{kj}(t).$$

It follows that

$$\frac{p_{ij}(t+h)-p_{ij}(t)}{h}=\sum_{k\neq i}\frac{p_{ik}(h)}{h}p_{kj}(t)-\frac{1-p_{ii}(h)}{h}p_{ij}(t).$$

By (5.13) and (5.14), letting $h\to 0$ yields *Kolmogorov's backward equations* for the transition probabilities:

$$p'_{ij}(t)=\sum_{k\neq i} q_{ik}p_{kj}(t)-q_i p_{ij}(t), \quad t\geq 0. \tag{5.18}$$

Analogously, starting with

$$p_{ij}(t+h)=\sum_{k\in\mathbf{Z}} p_{ik}(t)p_{kj}(h),$$

yields *Kolmogorov's forward equations* for the transition probabilities:

$$p'_{ij}(t)=\sum_{k\neq j} p_{ik}(t)q_{kj}-q_j p_{ij}(t), \quad t\geq 0. \tag{5.19}$$

Let $\{p_i(0),\ i \in \mathbf{Z}\}$ be any initial distribution. Multiplying Kolmogorov's forward equations (5.19) by $p_i(0)$ and summing with respect to i yields

$$\sum_{i\in\mathbf{Z}} p_i(0)p'_{ij}(t) = \sum_{i\in\mathbf{Z}} p_i(0)\sum_{k\neq j} p_{ik}(t)q_{kj} - \sum_{i\in\mathbf{Z}} p_i(0)q_j p_{ij}(t)$$

$$= \sum_{k\neq j} q_{kj}\sum_{i\in\mathbf{Z}} p_i(0)p_{ik}(t) - q_j\sum_{i\in\mathbf{Z}} p_i(0)p_{ij}(t).$$

Thus, in view of (5.7), the absolute state probabilities satisfy the system of linear differential equations

$$p'_j(t) = \sum_{k\neq j} q_{kj}p_k(t) - q_j p_j(t), \quad t \geq 0, \quad j \in \mathbf{Z}. \tag{5.20}$$

In future, the absolute state probabilities are assumed to satisfy

$$\sum_{i\in\mathbf{Z}} p_i(t) = 1. \tag{5.21}$$

This *normalizing condition* is always fulfilled if $\mathbf{Z}$ is finite.

Note If the initial distribution has structure

$$p_i(0) = 1,\ p_j(0) = 0 \ \text{ for } j \neq i,$$

then the absolute state probabilities are equal to the transition probabilities

$$p_j(t) = p_{ij}(t), \ \ j \in \mathbf{Z}.$$

Transition Times and Transition Rates It is only possible to exactly model real systems by continuous-time Markov chains if the lengths of the time periods between changes of states are exponentially distributed, since in this case the 'memoryless property' of the exponential distribution (example 1.14) implies the Markov property. If the times between transitions have known exponential distributions, then it is no problem to determine the transition rates. For instance, if the sojourn time of a Markov chain in state 0 has an exponential distribution with parameter λ_0, then, according to (5.11), the unconditional rate of leaving this state is given by

$$q_0 = \lim_{h\to 0}\frac{1 - p_{00}(h)}{h} = \lim_{h\to 0}\frac{1 - e^{-\lambda_0 h}}{h}$$

$$= \lim_{h\to 0}\frac{\lambda_0 h + o(h)}{h} = \lambda_0 + \lim_{h\to 0}\frac{o(h)}{h}.$$

Hence,

$$q_0 = \lambda_0. \tag{5.22}$$

Now let the sojourn time of a Markov chain in state 0 have structure

$$Y_0 = \min(Y_{01}, Y_{02}),$$

where Y_{01} and Y_{02} are independent exponential random variables with respective parameters λ_1 and λ_2. If $Y_{01} < Y_{02}$, the Markov chain makes a transition to state 1 and if $Y_{01} > Y_{02}$ to state 2. Thus, by (5.12), the conditional transition rate from state 0 to state 1 is,

$$\begin{aligned} q_{01} &= \lim_{h\to 0} \frac{p_{01}(h)}{h} = \lim_{h\to 0} \frac{(1-e^{-\lambda_1 h})\, e^{-\lambda_2 h} + o(h)}{h} \\ &= \lim_{h\to 0} \frac{\lambda_1 h(1-\lambda_2 h)}{h} + \lim_{h\to 0} \frac{o(h)}{h} \\ &= \lim_{h\to 0} (\lambda_1 - \lambda_1\lambda_2 h) = \lambda_1 . \end{aligned}$$

Hence, since the roles of Y_{01} and Y_{02} can be interchanged,

$$q_{01} = \lambda_1, \quad q_{02} = \lambda_2, \quad q_0 = \lambda_1 + \lambda_2 . \tag{5.23}$$

The results (5.22) and (5.23) will be generalized in section 5.4.

Transition Graphs Establishing the Kolmogorov equations can be facilitated by *transition graphs.* These graphs are constructed analogously to the transition graphs for discrete-time Markov chains: The nodes of a transition graph represent the states of the Markov chain. A (directed) edge from node i to node j exists if and only if $q_{ij} > 0$. The edges are weighted by their corresponding transition rates. Thus, two sets of states (possibly empty ones) can be assigned to each node i: firstly edges with initial node i and secondly edges with end node i, that is, edges which leave node i and edges which end in node i. The unconditional transition rate q_i equals the sum of the weights of all those edges leaving node i. If there is an edge ending in state i and no edge leaving state i, then i is an absorbing state.

Example 5.4 (*system with renewal*) The lifetime L of a system has an exponential distribution with parameter λ. After a failure the system is replaced by an equivalent new one. A replacement takes a random time Z, which is exponentially distributed with parameter μ. All life- and replacement times are assumed to be independent. Thus, the operation of the system can be described by an alternating renewal process (section 3.3.6) with 'typical renewal cycle' (L, Z). Consider the Markov chain $\{X(t), t \geq 0\}$ defined by

$$X(t) = \begin{cases} 1 & \text{if the system is operating} \\ 0 & \text{if the system is being replaced} \end{cases} .$$

Its state space is $\mathbf{Z} = \{0, 1\}$. The absolute state probability

$$p_1(t) = P(X(t) = 1)$$

of this Markov chain is the *point availability* of the system. In this simple example, only state changes from 0 to 1 and from 1 to 0 are possible. Hence, by (5.22),

$$q_0 = q_{01} = \mu \text{ and } q_1 = q_{10} = \lambda.$$

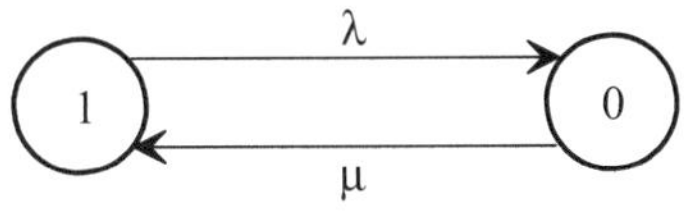

Figure 5.1 Transition graph of an alternating renewal process (example 5.4)

The corresponding Kolmogorov differential equations (5.20) are

$$p_0'(t) = -\mu p_0(t) + \lambda p_1(t),$$
$$p_1'(t) = +\mu p_0(t) - \lambda p_1(t).$$

These two equations are linearly dependent. (The sums of the left hand sides and the right hand sides are equal to 0.) Replacing $p_0(t)$ in the second equation by $1 - p_1(t)$ yields a first-order nonhomogeneous differential equation with constant coefficients for $p_1(t)$:

$$p_1'(t) + (\lambda + \mu) p_1(t) = \mu .$$

Given the initial condition $p_1(0) = 1$, the solution is

$$p_1(t) = \frac{\mu}{\lambda + \mu} + \frac{\lambda}{\lambda + \mu} e^{-(\lambda+\mu)t}, \quad t \geq 0 .$$

The corresponding stationary availability is

$$\pi_1 = \lim_{t \to \infty} p_1(t) = \frac{\mu}{\lambda + \mu} .$$

In example 3.17, the same results have been obtained by applying the Laplace transform. (There the notation $L = Y$, $\lambda = \lambda_1$ and $\mu = \lambda_0$ had been used.) □

Example 5.5 ***(two-unit redundant system, standby redundancy)*** A system consists of two identical units. The system is available if and only if at least one of its units is available. If both units are available, then one of them is in standby redundancy (cold redundancy), that is, in this state it does not age and cannot fail. After the failure of a unit, the other one (if available) is immediately switched from the redundancy state to the operating state and the replacement of the failed unit begins. The replaced unit becomes the standby unit if the other unit is still operating. Otherwise it immediately resumes its work. The lifetimes and replacement times of the units are independent random variables, identically distributed as L and Z, respectively. L and Z are assumed to be exponentially distributed with respective parameters λ and μ. Let L_s denote the system lifetime, i.e. the random time to a system failure. A system failure occurs, when a unit fails whilst the other unit is being replaced. A Markov chain $\{X(t), t \geq 0\}$ with state space $\mathbf{Z} = \{0, 1, 2\}$ is introduced in the following way: $X(t) = i$ if i units are unavailable at time t. Let Y_i be the unconditional sojourn time of the system in state i and Y_{ij} be the conditional sojourn time of the system in state i given that the system makes a transition from state i into state j. From state 0, the

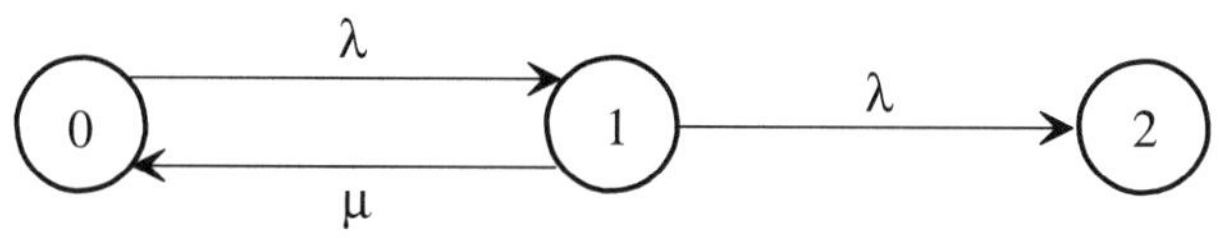

Figure 5.2 Transition graph for example 5.5 a)

system can only make a transition to state 1. Hence, $Y_0 = Y_{01} = L$. According to (5.22), the corresponding transition rate is given by

$$q_0 = q_{01} = \lambda.$$

If the system makes a transition from state 1 to state 2, then its conditional sojourn time in state 1 is $Y_{12} = L$, whereas in case of a transition to state 0, it stays a time $Y_{10} = Z$ in state 1. The unconditional sojourn time of the system in state 1 is

$$Y_1 = \min(L, Z).$$

Hence, according to (5.23), the corresponding transition rates are

$$q_{12} = \lambda, \; q_{10} = \mu \quad \text{and} \quad q_1 = \lambda + \mu.$$

When the system returns from state 1 to state 0, then it again spends time L in state 0, since the operating unit is 'as good as new' in view of the memoryless property of the exponential distribution.

a) *Survival probability* In this case, only the time to entering state 2 (system failure) is of interest. Hence, state 2 must be considered absorbing (Figure 5.2) so that

$$q_{20} = q_{21} = 0.$$

The survival probability of the system has the structure

$$\overline{F}_s(t) = P(L_s > t) = p_0(t) + p_1(t).$$

The corresponding system of differential equations (5.20) is

$$\begin{aligned} p_0'(t) &= -\lambda p_0(t) + \mu p_1(t), \\ p_1'(t) &= +\lambda p_0(t) - (\lambda + \mu) p_1(t), \\ p_2'(t) &= +\lambda p_1(t). \end{aligned} \tag{5.24}$$

This system of differential equations will be solved on condition that both units are available at time $t = 0$. Combining the first two differential equations in (5.24) yields a homogeneous differential equation of the second order with constant coefficients for $p_0(t)$:

$$p_0''(t) + (2\lambda + \mu) p_0'(t) + \lambda^2 p_0(t) = 0.$$

The corresponding characteristic equation is

$$x^2 + (2\lambda + \mu)x + \lambda^2 = 0.$$

Its solutions are

$$x_{1,2} = -\left(\lambda + \frac{\mu}{2}\right) \pm \sqrt{\lambda\mu + \mu^2/4}\ .$$

Hence, since $p_0(0) = 1$, for $t \ge 0$,

$$p_0(t) = a \sinh \frac{c}{2} t \quad \text{with} \quad c = \sqrt{4\lambda\mu + \mu^2}\ .$$

Since $p_1(0) = 0$, the first differential equation in (5.24) yields $a = 2\lambda/c$ and

$$p_1(t) = e^{-\frac{2\lambda+\mu}{2}t}\left(\frac{\mu}{c}\sinh\frac{c}{2}t + \cosh\frac{c}{2}t\right), \quad t \ge 0\,.$$

Thus, the survival probability of the system is

$$\overline{F}_s(t) = e^{-\frac{2\lambda+\mu}{2}}\left[\cosh\frac{c}{2}t + \frac{2\lambda+\mu}{c}\sinh\frac{c}{2}t\right], \quad t \ge 0\,.$$

(For a definition of the hyperbolic functions sinh and cosh, see section 3.2.1.) The mean value of the system lifetime L_s is most easily obtained from formula (1.12):

$$E(L_s) = \frac{2}{\lambda} + \frac{\mu}{\lambda^2}\,. \tag{5.25}$$

For the sake of comparison, in case of no replacement $(\mu = 0)$, the system lifetime L_s has an Erlang distribution with parameters 2 and λ:

$$\overline{F}_s(t) = (1 + \lambda t)\, e^{-\lambda t}, \quad E(L_s) = 2/\lambda\,.$$

b) *Availability* If the replacement of failed units is continued after system failures, then the point availability

$$A(t) = p_0(t) + p_1(t)$$

of the system is of particular interest. In this case, the transition rate q_{21} from state 2 to state 1 is positive. However, q_{21} depends on the number $r = 1$ or $r = 2$ of mechanics which are in charge of the replacement of failed units. Assuming that a mechanic cannot replace two failed units at the same time, then (Figure 5.3)

$$q_{21} = q_2 = r\mu.$$

For $r = 2$, the sojourn time of the system in state 2 is given by $Y_2 = \min(Z_1, Z_2)$, where Z_1 and Z_2 are independent and identically as Z distributed. Analogously, the sojourn time in state 1 is given by $Y_1 = \min(L, Z)$.

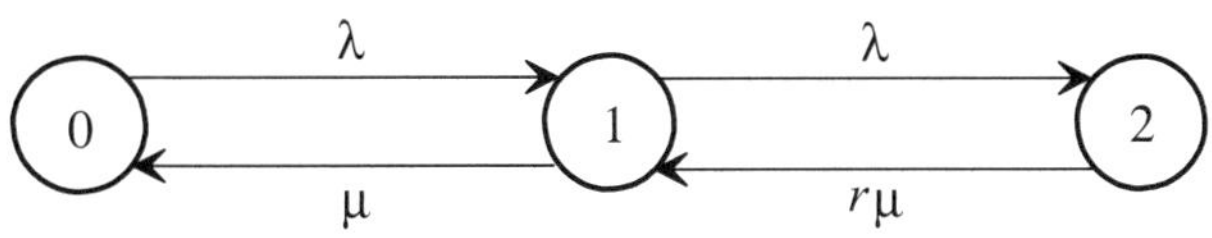

Figure 5.3 Transition graph for example 5.5 b)

Hence, the transition rates q_{10} and q_{12} have the same values as under a). The corresponding system of differential equations (5.20) becomes, when replacing the last differential equation by the normalizing condition (5.21),

$$p_0'(t) = -\lambda p_0(t) + \mu p_1(t),$$
$$p_1'(t) = +\lambda p_0(t) - (\lambda + \mu) p_1(t) + r\mu p_2(t),$$
$$1 = p_0(t) + p_1(t) + p_2(t).$$

The solution is left as an exercise to the reader. □

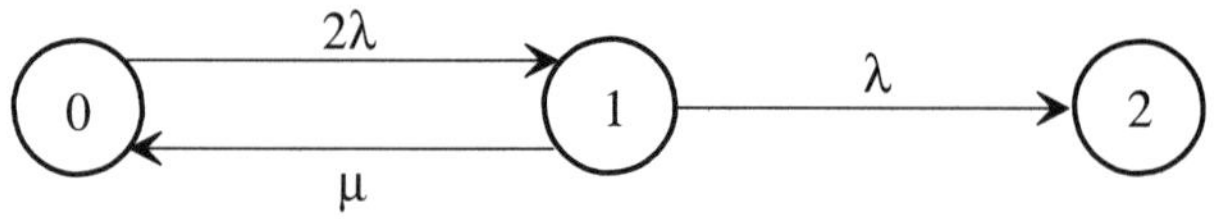

Figure 5.4 Transition graph for example 5.6 a)

Example 5.6 (*two-unit system, parallel redundancy*) Now assume that both units of the system operate at the same time when they are available. All other assumptions and the notation of the previous example are retained. In particular, the system is available if and only if at least one unit is available. In view of the initial condition $p_0(0) = 1$, the system spends

$$Y_0 = \min\,(L_1, L_2)$$

time units in state 0. Y_0 has an exponential distribution with parameter 2λ and from state 0 only a transition to state 1 is possible. Therefore, $Y_0 = Y_{01}$ and

$$q_0 = q_{01} = 2\lambda.$$

When the system is in state 1, then it behaves as in example 5.5:

$$q_{10} = \mu, \quad q_{12} = \lambda, \quad q_1 = \lambda + \mu.$$

a) *Survival probability* As in the previous example, state 2 has to be thought of as absorbing: $q_{20} = q_{21} = 0$ (Figure 5.4). Hence, from (5.20) and (5.21),

$$p_0'(t) = -2\lambda p_0(t) + \mu p_1(t),$$
$$p_1'(t) = +2\lambda p_0(t) - (\lambda + \mu) p_1(t),$$
$$1 = p_0(t) + p_1(t) + p_2(t).$$

Combining the first two differential equations yields a homogeneous differential equation of the second order with constant coefficients for $p_0(t)$:

$$p_0''(t) + (3\lambda + \mu) p_0'(t) + 2\lambda^2 p_0(t) = 0.$$

The solution is

$$p_0(t) = e^{-\left(\frac{3\lambda+\mu}{2}\right)t}\left[\cosh\frac{c}{2}t + \frac{\mu-\lambda}{c}\sinh\frac{c}{2}t\right]$$

where

$$c = \sqrt{\lambda^2 + 6\lambda\mu + \mu^2}\ .$$

Furthermore,

$$p_1(t) = \frac{4\lambda}{c}e^{-\left(\frac{3\lambda+\mu}{2}\right)t}\sinh\frac{c}{2}t\ .$$

The survival probability of the system is

$$\overline{F}_s(t) = P(L_s > t) = p_0(t) + p_1(t)$$

Hence,

$$\overline{F}_s(t) = e^{-\left(\frac{3\lambda+\mu}{2}\right)t}\left[\cosh\frac{c}{2}t + \frac{3\lambda+\mu}{c}\sinh\frac{c}{2}t\right], \quad t \ge 0. \tag{5.26}$$

The mean system lifetime is

$$E(L_s) = \frac{3}{2\lambda} + \frac{\mu}{2\lambda^2}\,.$$

For the sake of comparison, in the case without replacement ($\mu = 0$)),

$$\overline{F}(t) = 2e^{-\lambda t} - e^{-2\lambda t}, \quad E(L_s) = \frac{3}{2\lambda}\,.$$

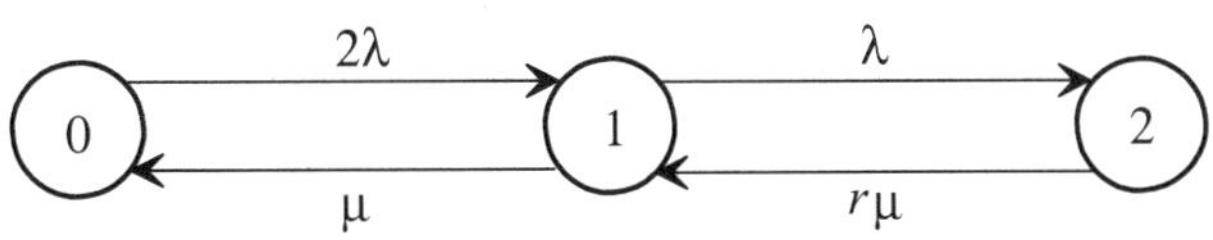

Figure 5.5 Transition graph for example 5.6 b)

b) *Availability* If r ($r = 1$ or $r = 2$) mechanics replace failed units, then

$$q_2 = q_{21} = r\mu\,.$$

The other transition rates are the same as those under a) (Figure 5.5 b). The absolute state probabilities satisfy the system of differential equations

$$p_0'(t) = -2\lambda p_0(t) + \mu p_1(t),$$

$$p_1'(t) = +2\lambda p_0(t) - (\lambda+\mu)p_1(t) + r p_2(t),$$

$$1 = p_0(t) + p_1(t) + p_2(t).$$

Solving this system of linear differential equations is left to the reader. □

5.3 STATIONARY STATE PROBABILITIES

If $\{\pi_j, j \in \mathbf{Z}\}$ is a stationary distribution of the Markov chain $\{X(t), t \geq 0\}$, then this special absolute distribution must satisfy Kolmogorov's equations (5.20). Since the π_j are constant, all the left-hand sides of these equations are equal to 0. Therefore, the system of linear differential equations (5.20) simplifies to a system of linear algebraic equations in the unknowns π_j:

$$0 = \sum_{k\in\mathbf{Z}, k\neq j} q_{kj}\pi_k - q_j\pi_j, \quad j \in \mathbf{Z}. \tag{5.27}$$

This system of equations is frequently written in the form

$$q_j\pi_j = \sum_{k\in\mathbf{Z}, k\neq j} q_{kj}\pi_k, \quad j \in \mathbf{Z}. \tag{5.28}$$

This form clearly illustrates that the stationary state probabilities refer to an equilibrium state of the Markov chain:

> The mean intensity per unit time of leaving state j, which is $q_j\pi_j$, is equal to the mean intensity per unit time of arriving in state j.

According to assumption (5.21), only those solutions $\{\pi_j, j \in \mathbf{Z}\}$ of (5.27) which satisfy the normalizing condition are of interest:

$$\sum_{j\in\mathbf{Z}} \pi_j = 1. \tag{5.29}$$

It is now assumed that the Markov chain is irreducible and positive recurrent. (Recall that an irreducible Markov chain with finite state space $\mathbf{Z}$ is always positive recurrent.) Then it can be shown that a unique stationary distribution $\{\pi_j, j \in \mathbf{Z}\}$ exists which satisfies (5.27) and (5.29). Moreover, in this case the limits

$$p_j = \lim_{t\to\infty} p_{ij}(t)$$

exist and are independent of i. Hence, for any initial distribution, there exist the limits of the absolute state probabilities $\lim_{t\to\infty} p_j(t)$ and they are equal to p_j:

$$p_j = \lim_{t\to\infty} p_j(t), \quad j \in \mathbf{Z}. \tag{5.30}$$

Furthermore, for all $j \in \mathbf{Z}$,

$$\lim_{t\to\infty} p_j'(t) = 0.$$

Otherwise, $p_j(t)$ would unboundedly increase as $t \to \infty$, contradictory to $p_j(t) \leq 1$. Hence, when passing to the limit as $t \to \infty$ in (5.20) and (5.21), the limits (5.30) are seen to satisfy the system of equations (5.27) and (5.29). Since this system has a unique solution, the limits p_j and the stationary probabilities π_j must coincide:

$$p_j = \pi_j, \quad j \in \mathbf{Z}.$$

For a detailed discussion of the relationship between the solvability of (5.27) and the existence of a stationary distribution, see Feller [27].

Continuation of example 5.5 (*two-unit system, standby redundancy*) Since the system is available if at least one unit is available, its stationary availability is

$$A = \pi_0 + \pi_1 .$$

Substituting the transition rates from Figure 5.3 into (5.27) and (5.29), the π_j are seen to satisfy the following system of algebraic equations:

$$\begin{aligned} -\lambda\pi_0 + \quad \mu\pi_1 \qquad &= 0, \\ +\lambda\pi_0 - (\lambda+\mu)\pi_1 + r\pi_2 &= 0, \\ \pi_0 + \qquad \pi_1 + \pi_2 &= 1. \end{aligned}$$

Case $r = 1$

$$\pi_0 = \frac{\mu^2}{(\lambda+\mu)^2 - \lambda\mu}, \quad \pi_1 = \frac{\lambda\mu}{(\lambda+\mu)^2 - \lambda\mu}, \quad \pi_2 = \frac{\lambda^2}{(\lambda+\mu)^2 - \lambda\mu},$$

$$A = \pi_0 + \pi_1 = \frac{\mu^2 + \lambda\mu}{(\lambda+\mu)^2 - \lambda\mu}.$$

Case $r = 2$

$$\pi_0 = \frac{2\mu^2}{(\lambda+\mu)^2 + \mu^2}, \quad \pi_1 = \frac{2\lambda\mu}{(\lambda+\mu)^2 + \mu^2}, \quad \pi_2 = \frac{\lambda^2}{(\lambda+\mu)^2 + \mu^2},$$

$$A = \pi_0 + \pi_1 = \frac{2\mu^2 + 2\lambda\mu}{(\lambda+\mu)^2 + \mu^2}.$$

Continuation of example 5.6 (*two-unit system, parallel redundancy*) Given the transition rates in Figure 5.5, the π_j are solutions of

$$\begin{aligned} -2\lambda\pi_0 + \quad \mu\pi_1 \qquad &= 0, \\ +2\lambda\pi_0 - (\lambda+\mu)\pi_1 + r\mu\pi_2 &= 0, \\ \pi_0 + \qquad \pi_1 + \pi_2 &= 1. \end{aligned}$$

Case $r = 1$

$$\pi_0 = \frac{\mu^2}{(\lambda+\mu)^2 + \lambda^2}, \quad \pi_1 = \frac{2\lambda\mu}{(\lambda+\mu)^2 + \lambda^2}, \quad \pi_2 = \frac{2\lambda^2}{(\lambda+\mu)^2 + \lambda^2},$$

$$A = \pi_0 + \pi_1 = \frac{\mu^2 + 2\lambda\mu}{(\lambda+\mu)^2 + \lambda^2}.$$

Case r = 2

$$\pi_0 = \frac{\mu^2}{(\lambda+\mu)^2}, \quad \pi_1 = \frac{2\lambda\mu}{(\lambda+\mu)^2}, \quad \pi_2 = \frac{\mu^2}{(\lambda+\mu)^2},$$

$$A = \pi_0 + \pi_1 = 1 - \left(\frac{\lambda}{\lambda+\mu}\right)^2.$$

Figure 5.6 shows a) the mean lifetimes and b) the stationary availabilities of the two-unit system for $r = 1$ as functions of $\rho = \lambda/\mu$. As anticipated, standby redundancy yields better results if switching a unit from a standby redundancy state to the operating state is absolutely reliable. With parallel redundancy, this switching problem does not exist since an available spare unit is also operating. □

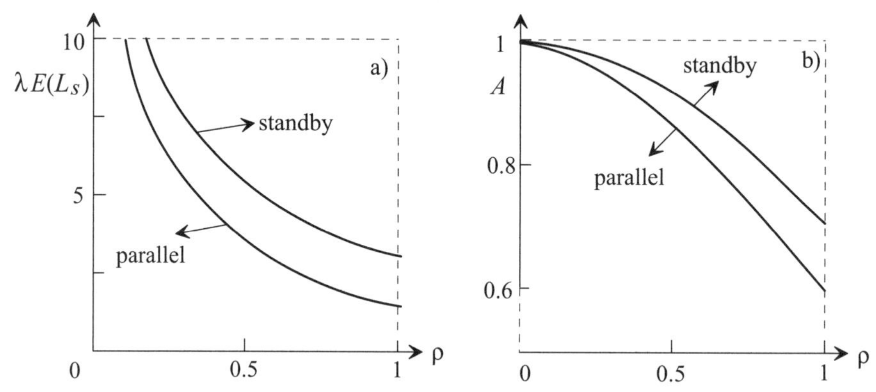

Figure 5.6 Mean lifetime a) and stationary availability b

Example 5.7 A system has two different failure types: type 1 and type 2. After a type i-failure the system is said to be in failure state i; $i = 1, 2$. The time L_i to a type i-failure is assumed to have an exponential distribution with parameter λ_i and the random variables L_1 and L_2 are assumed to be independent. Thus, if at time $t = 0$ a new system starts working, the time to its first failure is $Y_0 = \min(L_1, L_2)$. After a type 1-failure, the system is switched from failure state 1 into failure state 2. The time required for this is exponentially distributed with parameter ν. After entering failure state 2, the renewal of the system begins. A renewed system immediately starts working. The renewal time is exponentially distributed with parameter μ. This process continues to infinity. All life- and renewal times as well as switching times are assumed to be independent. This model is, for example, of importance in traffic safety engineering: When the red signal in a traffic light fails (type 1-failure), then the whole traffic light is switched off (type 2-failure). That is, a *dangerous failure state* is removed by inducing a *blocking failure state*.

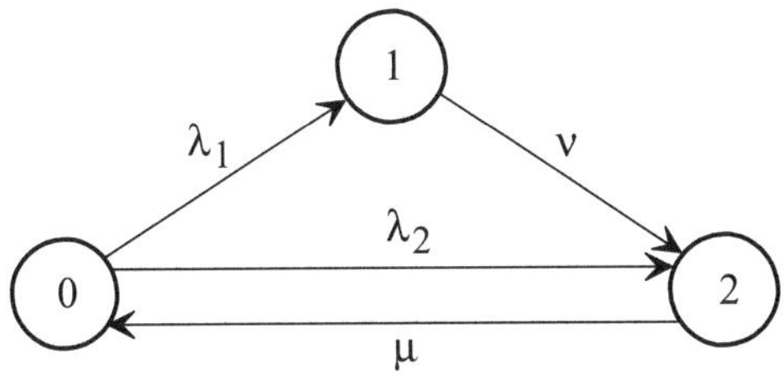

Figure 5.7 Transition graph for example 5.7

Consider the following system states:

0 system is operating
1 type 1-failure state
2 type 2-failure state

If $X(t)$ denotes the state of the system at time t, then $\{X(t), t \geq 0\}$ is a homogeneous Markov chain with state space $\mathbf{Z} = \{0, 1, 2\}$. Its transition rates are (Figure 5.7)

$$q_{01} = \lambda_1, \ q_{02} = \lambda_2, \ q_0 = \lambda_1 + \lambda_2, \ q_{12} = q_1 = \nu, \ q_{20} = q_2 = \mu .$$

Hence, the stationary state probabilities satisfy the system of algebraic equations

$$-(\lambda_1 + \lambda_2)\pi_0 + \mu\pi_2 = 0,$$
$$\lambda_1\pi_0 - \nu\pi_1 = 0,$$
$$\pi_0 + \pi_1 + \pi_2 = 1.$$

The solution is

$$\pi_0 = \frac{\mu\nu}{(\lambda_1 + \lambda_2)\nu + (\lambda_1 + \nu)\mu},$$
$$\pi_1 = \frac{\lambda_1\mu}{(\lambda_1 + \lambda_2)\nu + (\lambda_1 + \nu)\mu},$$
$$\pi_2 = \frac{(\lambda_1 + \lambda_2)\nu}{(\lambda_1 + \lambda_2)\nu + (\lambda_1 + \nu)\mu}.$$ □

5.4 SOJOURN TIMES IN PROCESS STATES

So far the fact has been used that independent, exponentially distributed times between changes of system states allow for modeling system behaviour by homogeneous Markov chains. Conversely, it can be shown that for any $i \in \mathbf{Z}$ the sojourn time Y_i of a homogeneous Markov chain $\{X(t), t \geq 0\}$ in state i also has an exponential distribution: By properties (5.8) and (5.15) of a homogeneous Markov chain,

$$P(Y_i > t|X(0) = i) = P(X(s) = i,\ 0 < s \le t|X(0) = i)$$
$$= \lim_{n\to\infty} P\left(X\left(\frac{k}{n}t\right) = i;\ k = 1, 2, ..., n \middle| X(0) = i\right)$$
$$= \lim_{n\to\infty} \left[p_{ii}\left(\frac{1}{n}t\right)\right]^n$$
$$= \lim_{n\to\infty} \left[1 - q_i \frac{t}{n} + o\left(\frac{1}{n}\right)\right]^n.$$

It follows that

$$P(Y_i > t|X(0) = i) = e^{-q_i t}, \quad t \ge 0, \tag{5.31}$$

since e can be represented by the limit

$$e = \lim_{x\to\infty} \left(1 + \frac{1}{x}\right)^x. \tag{5.32}$$

Thus, Y_i has an exponential distribution with parameter q_i.

Given $X(0) = i$, $X(Y_i) = X(Y_i + 0)$ is the state to which the Markov chain makes a transition on leaving state i. Let $m(nt)$ denote the greatest integer m satisfying the inequality $m/n \le t$ or, equivalently,

$$nt - 1 < m(nt) \le nt.$$

By making use of the geometric series, the joint probability distribution of the random vector $(Y_i, X(Y_i))$, $i \ne j$, can be obtained as follows:

$$P(X(Y_i) = j,\ Y_i > t|X(0) = i)$$
$$= P(X(Y_i) = j,\ X(s) = i \text{ for } 0 < s \le t|X(0) = i)$$
$$= \lim_{n\to\infty} \sum_{m=m(nt)}^{\infty} P\left(\left(X\left(\frac{m+1}{n}\right) = j,\ Y_i \in \left[\frac{m}{n}, \frac{m+1}{n}\right)\right| X(0) = i\right)$$
$$= \lim_{n\to\infty} \sum_{m=m(nt)}^{\infty} P\left(\left(X\left(\frac{m+1}{n}\right) = j,\ X\left(\frac{k}{n}\right) = i \text{ for } 1 \le k \le m\right| X(0) = i\right)$$
$$= \lim_{n\to\infty} \sum_{m=m(nt)}^{\infty} \left[q_{ij}\frac{1}{n} + o\left(\frac{1}{n}\right)\right]\left[1 - q_i\frac{1}{n} + o\left(\frac{1}{n}\right)\right]^m$$
$$= \lim_{n\to\infty} \frac{\left[q_{ij}\frac{1}{n} + o\left(\frac{1}{n}\right)\right]}{q_i\frac{1}{n} + o\left(\frac{1}{n}\right)}\left[1 - q_i\frac{1}{n} + o\left(\frac{1}{n}\right)\right]^{m(nt)}.$$

Hence, by (5.32),

$$P(X(Y_i) = j,\ Y_i > t|X(0) = i) = \frac{q_{ij}}{q_i} e^{-q_i t}; \quad i \ne j;\ i,j \in \mathbf{Z}. \tag{5.33}$$

Passing to the marginal distribution of Y_i (i.e. summing the equations (5.33) with respect to $j \in \mathbf{Z}$) verifies (5.31). Two other important conclusions are:

1) Letting in (5.33) $t=0$ yields the one-step transition probability from state i into state j:

$$p_{ij} = P(X(Y_i + 0) = j | X(0) = i) = \frac{q_{ij}}{q_i}, \quad j \in \mathbf{Z}. \tag{5.34}$$

2) The state following state i is independent of Y_i (and, of course, independent of the history of the Markov chain before arriving at state i).

Knowledge of the transition probabilities p_{ij} suggests to observe a continuous-time Markov chain $\{X(t), t \geq 0\}$ only at those discrete time points at which state changes take place. Let X_n be the state of the Markov chain immediately after the nth change of state and $X_0 = X(0)$. Then $\{X_0, X_1, ...\}$ is a discrete-time homogeneous Markov chain with transition probabilities given by (5.34):

$$p_{ij} = P(X_n = j | X_{n-1} = i) = \frac{q_{ij}}{q_i}, \quad i, j \in \mathbf{Z}.; \; n = 1, 2, ... \tag{5.35}$$

In this sense, the discrete-time Markov chain $\{X_0, X_1, ...\}$ is *embedded* in the continuous-time Markov chain $\{X(t), t \geq 0\}$. Embedded Markov chains can also be found in non-Markov processes. In these cases, they may facilitate the investigation of non-Markov processes. Actually, discrete-time Markov chains, which are embedded in arbitrary continuous-time stochastic processes, are frequently an efficient (if not the only) tool for analyzing these processes. Examples for the application of the *method of embedded Markov chains* to analyzing queueing systems are given in sections 5.7.3.2 and 5.7.3.3. Section 5.8 deals with semi-Markov chains, the framework of which is an embedded Markov chain.

5.5 CONSTRUCTION OF MARKOV SYSTEMS

In a *Markov system*, state changes are controlled by a Markov process. Markov systems, in which the underlying Markov process is a homogeneous, continuous-time Markov chain with state space **Z**, are frequently special cases of the following basic model: The sojourn time of the system in state i is given by

$$Y_i = \min\,(Y_{i1}, Y_{i2}, ..., Y_{in_i}),$$

where the Y_{ij} are independent, exponentially distributed random variables with parameters λ_{ij}; $j = 1, 2, ..., n_i$; $i, j \in \mathbf{Z}$. A transition from state i to state j is made if and only if $Y_i = Y_{ij}$. If $X(t)$ as usual denotes the state of the system at time t, then, by the memoryless property of the exponential distribution, $\{X(t), t \geq 0\}$ is a homogeneous Markov chain with transition rates

$$q_{ij} = \lim_{h \to 0} \frac{p_{ij}(h)}{h} = \lambda_{ij}, \quad q_i = \Sigma_{j=1}^{n_i} \lambda_{ij}.$$

This representation of q_i results from (5.12) and (5.17). It reflects the fact that Y_i as the minimum of independent, exponentially distributed random variables Y_{ij} also has an exponential distribution, the parameter of which is obtained by summing the parameters of the Y_{ij}.

Example 5.8 (*repairman problem*) n machines with lifetimes $L_1, L_2, ..., L_n$ start operating at time $t = 0$. The L_i are assumed to be independent, exponential random variables with parameter λ. Failed machines are repaired. A repaired machine is 'as good as new'. There is one mechanic who can only handle one failed machine at a time. Thus, when there are $k \geq 1$ failed machines, $k-1$ have to wait for repair. The repair times are assumed to be mutually independent and identically distributed as an exponential random variable Z with parameter μ. Life- and repair times are independent. Immediately after completion of its repair, a machine resumes its work.

Let $X(t)$ denote the number of machines which are in the failed state at time t. Then $\{X(t), t \geq 0\}$ is a Markov chain with state space $\mathbf{Z} = \{0, 1, ..., n\}$. The system stays in state 0 for a random time

$$Y_0 = \min\,(L_1, L_2, \dots, L_n)$$

and then it makes a transition to state 1. The corresponding transition rate is

$$q_0 = q_{01} = n\lambda.$$

The system stays in state 1 for a random time

$$Y_1 = \min\,(L_1, L_2, \dots, L_{n-1}, Z).$$

From state 1 it makes a transition to state 2 if $Y_1 = L_k$ for $k \in \{1, 2, ..., n-1\}$, and a transition to state 0 if $Y_1 = Z$. Hence,

$$q_{10} = \mu,\ q_{12} = (n-1)\lambda \quad \text{and} \quad q_1 = (n-1)\lambda + \mu.$$

In general (Figure 5.8),

$$q_{j-1,j} = (n-j+1)\lambda; \quad j = 1, 2, ..., n,$$

$$q_{j+1,j} = \mu; \quad j = 0, 1, ..., n-1,$$

$$q_{ij} = 0; \quad |i-j| \geq 2,$$

$$q_j = (n-j)\lambda + \mu; \quad j = 1, 2, ..., n,$$

$$q_0 = n\lambda.$$

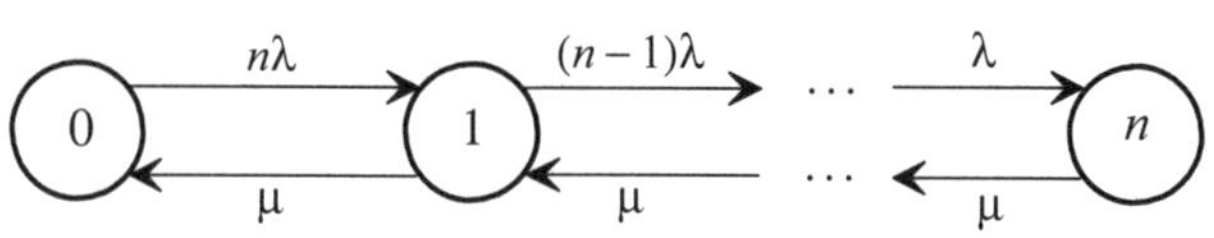

Figure 5.8 Transition graph for the repairman problem (example 5.8)

The corresponding system of equations (5.28) is

$$\mu\pi_1 = n\lambda\pi_0$$
$$(n-j+1)\lambda\pi_{j-1} + \mu\pi_{j+1} = ((n-j)\lambda + \mu)\pi_j; \quad j = 1, 2, ..., n-1$$
$$\mu\pi_n = \lambda\pi_{n-1}$$

Beginning with the first equation, the stationary state probabilities are obtained by successively solving for the π_i:

$$\pi_j = \frac{n!}{(n-j)!}\rho^j\pi_0; \quad j = 0, 1, ..., n;$$

where $\rho = \lambda/\mu$. From the normalizing condition (5.29),

$$\pi_0 = \left[\sum_{i=0}^{n} \frac{n!}{(n-i)!}\rho^i\right]^{-1}.$$ □

Erlang's Phase Method Systems with Erlang distributed sojourn times in their states can be transformed into Markov systems by introducing dummy states. This is due to the fact that a random variable, which is Erlang distributed with parameters n and μ, can be represented as a sum of n independent exponential random variables with parameter μ (example 1.23, section 1.7.2). Hence, if the time interval, which the system stays in state i, is Erlang distributed with parameters n_i and μ_i, then this interval is partitioned into n_i disjoint subintervals (*phases*), the lengths of which are independent, identically distributed exponential random variables with parameter μ_i. By introducing the new states $j_1, j_2, ..., j_{n_i}$ to label these phases, the original non-Markov system becomes a Markov system. In what follows, instead of presenting a general treatment of this approach, the application of *Erlang's phase* method is demonstrated by an example.

Example 5.9 (*two-unit system, parallel redundancy*) As in example 5.6, a two-unit system with parallel redundancy is considered. The lifetimes of the units are identically distributed as an exponential random variable L with parameter λ. The replacement times of the units are identically distributed as Z, where Z has an Erlang distribution with parameters $n = 2$ and μ. There is only one mechanic in charge of the replacement of failed units. All other assumptions and model specifications are as in example 5.6. The following system states are introduced:

0 both units are operating

1 one unit is operating, the replacement of the other one is in phase 1

2 one unit is operating, the replacement of the other one is in phase 2

3 no unit is operating, the replacement of the one being maintained is in phase 1

4 no unit is operating, the replacement of the one being maintained is in phase 2

The transition rates are (Figure 5.9):

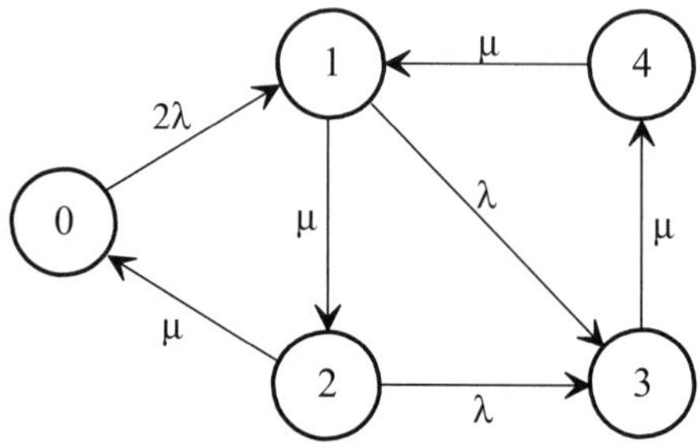

Figure 5.9 Transition graph for example 5.9

$$q_{01} = 2\lambda, \quad q_0 = 2\lambda,$$
$$q_{12} = \mu, \quad q_{13} = \lambda, \quad q_1 = \lambda + \mu$$
$$q_{20} = \mu, \quad q_{23} = \lambda, \quad q_2 = \lambda + \mu$$
$$q_{34} = \mu, \quad q_3 = \mu$$
$$q_{41} = \mu, \quad q_4 = \mu$$

Hence the stationary state probabilities satisfy the following system of equations:

$$\mu \pi_2 = 2\lambda \pi_0$$
$$2\lambda\pi_0 + \mu \pi_4 = (\lambda + \mu)\pi_1$$
$$\mu \pi_1 = (\lambda + \mu)\pi_2$$
$$\lambda \pi_1 + \lambda \pi_2 = \mu \pi_3$$
$$\mu \pi_3 = \mu \pi_4$$
$$1 = \pi_0 + \pi_1 + \pi_2 + \pi_3 + \pi_4$$

Let π_i^* denote the stationary probability that i units are failed. Then,

$$\pi_0^* = \pi_0, \quad \pi_1^* = \pi_1 + \pi_2, \quad \pi_2^* = \pi_3 + \pi_4 .$$

The probabilities π_i^* are the ones of interest. Letting $\rho = E(Z)/E(L) = 2\lambda/\mu$, they are

$$\pi_0^* = \left[1 + 2\rho + \tfrac{3}{2}\rho^2 + \tfrac{1}{4}\rho^3\right]^{-1},$$

$$\pi_1^* = \left[2\rho + \tfrac{1}{2}\rho^2\right]^{-1}\pi_0^*, \quad \pi_2^* = \left[\rho^2 + \tfrac{1}{4}\rho^3\right]^{-1}\pi_0^* .$$

The stationary system availability is given by $A = \pi_0^* + \pi_1^*$. □

Unfortunately, applying Erlang's phase method to structurally complicated systems leads to rather complex Markov systems.

5.6 BIRTH- AND DEATH PROCESSES

In this section, continuous-time Markov chains with property that only transitions to 'neighbouring' states are possible, are discussed in more detail. These processes, called (continuous-time) birth- and death processes, have proved to be an important tool for modeling queueing, reliability and inventory systems. In the economical sciences, birth- and death processes are among other things used for describing the development of the number of enterprises in a particular area and manpower fluctuations. In physics, flows of radioactive, cosmic and other particles are modeled by birth- and death processes. Their name, however, comes from applications in biology, where they have been used to stochastically model the development in time of the number of individuals in populations of organisms.

5.6.1 Birth Processes

A continuous-time Markov chain with state space $\mathbf{Z} = \{0, 1, ..., n\}$ is called a *(pure) birth process* if, for all $i = 0, 1, ..., n-1$, only a transition from state i to $i+1$ is possible. State n is absorbing if $n < \infty$.

Thus, the positive transition rates of a birth process are given by $q_{i,\,i+1}$. In what follows, they will be called *birth rates* and denoted as

$$\lambda_i = q_{i,i+1}, \quad i = 0, 1, ..., n-1,$$
$$\lambda_n = 0 \quad \text{for } n < \infty.$$

The sample paths of birth processes are nondecreasing step functions with jump height 1. The homogeneous Poisson process with intensity λ is the simplest example of a birth process. In this case, $\lambda_i = \lambda,\ i = 0, 1, ...$ Given the initial distribution

$$p_m(0) = P(X(0) = m) = 1$$

(i.e. in the beginning the 'population' consists of m individuals), the absolute state probabilities $p_j(t)$ are equal to the transition probabilities $p_{mj}(t)$. The $p_j(t)$ are identically equal to 0 for $j < m$ and, according to (5.20), for $j \geq m$ they satisfy the system of linear differential equations

$$\begin{aligned} p'_m(t) &= -\lambda_m\, p_m(t), \\ p'_j(t) &= +\lambda_{j-1}\, p_{j-1}(t) - \lambda_j\, p_j(t); \quad j = m+1, m+2, ... \\ p'_n(t) &= +\lambda_{n-1}\, p_{n-1}(t), \quad n < \infty. \end{aligned} \tag{5.36}$$

From the first differential equation,

$$p_m(t) = e^{-\lambda_m t}, \quad t \geq 0. \tag{5.37}$$

For $j = m+1, m+2, ...$, the differential equations (5.36) are equivalent to

$$e^{\lambda_j t}\left(p_j'(t)+\lambda_j p_j(t)\right)=\lambda_{j-1}e^{\lambda_j t}p_{j-1}(t)$$

or

$$\frac{d}{dt}\left(e^{\lambda_j t}p_j(t)\right)=\lambda_{j-1}e^{\lambda_j t}p_{j-1}(t).$$

By integration,

$$p_j(t)=\lambda_{j-1}e^{-\lambda_j t}\int_0^t e^{\lambda_j x}p_{j-1}(x)\,dx. \tag{5.38}$$

Formulas (5.37) and (5.38) allow the successive calculation of the probabilities $p_j(t)$ for $j=m+1,m+2,\ldots$ For instance, on conditions $p_0(0)=1$ and $\lambda_0\neq\lambda_1$,

$$\begin{aligned}p_1(t)&=\lambda_0\,e^{-\lambda_1 t}\int_0^t e^{\lambda_1 x}\,e^{-\lambda_0 x}\,dx\\&=\lambda_0\,e^{-\lambda_1 t}\int_0^t e^{-(\lambda_0-\lambda_1)x}dx\\&=\frac{\lambda_0}{\lambda_0-\lambda_1}\left(e^{-\lambda_1 t}-e^{-\lambda_0 t}\right),\quad t\geq 0.\end{aligned}$$

If all the birth rates are different from each other, then this result and (5.38) yields by induction:

$$p_j(t)=\sum_{i=0}^{j}C_{ij}\lambda_i e^{-\lambda_i t},\quad j=0,1,\ldots,$$

$$C_{ij}=\frac{1}{\lambda_j}\prod_{k=0,\,k\neq i}^{j}\frac{\lambda_k}{\lambda_k-\lambda_i},\quad 0\leq i\leq j,\quad C_{00}=\frac{1}{\lambda_0}.$$

Linear Birth Process A birth process is called a *linear birth process* or *a Yule-Furry process* if its birth rates are given by

$$\lambda_i=i\lambda;\quad i=0,1,2,\ldots$$

Since state 0 is absorbing, an initial distribution should not concentrate probability 1 on state 0. Linear birth processes occur, for instance, if in the interval $[t,\,t+h]$ each member of a population (bacterium, physical particle) independently of each other splits with probability $\lambda h+o(h)$ as $h\to 0$.

Assuming $p_1=P(X(0)=1)=1$, the system of differential equations (5.36) becomes

$$p_j'(t)=-\lambda\,[j\,p_j(t)-(j-1)\,p_{j-1}(t)];\quad j=1,2,\ldots \tag{5.39}$$

with

$$p_1(0)=1,\quad p_j(0)=0;\quad j=2,3,\ldots \tag{5.40}$$

The solution of (5.39) under the initial distribution (5.40) is

$$p_i(t)=e^{-\lambda t}(1-e^{-\lambda t})^{i-1};\quad i=1,2,\ldots$$

Thus, $X(t)$ has a geometric distribution with parameter $p = e^{-\lambda t}$. Hence, the trend function of the linear birth process is

$$m(t) = e^{\lambda t}, \quad t \ge 0.$$

If $\mathbf{Z}$ is finite, then there always exists a solution of (5.36) which satisfies

$$\sum_{i \in \mathbf{Z}} p_i(t) = 1. \tag{5.41}$$

In case of an infinite state space $\mathbf{Z} = \{0, 1, ...\}$, the following theorem gives a necessary and sufficient condition for the existence of a solution of (5.36) with property (5.41). Without loss of generality, the theorem is proved on condition (5.40).

Theorem 5.2 (*Feller-Lundberg*) A solution $\{p_0(t), p_1(t), ...\}$ of the system of differential equations (5.36) satisfies condition (5.41) if and only if the series

$$\sum_{i=0}^{\infty} \frac{1}{\lambda_i} \tag{5.42}$$

diverges.

Proof Let

$$s_k(t) = p_0(t) + p_1(t) + \cdots + p_k(t).$$

Summing the middle equation of (5.36) from $j = 1$ to k yields

$$s_k'(t) = -\lambda_k p_k(t).$$

By integration, taking into account $s_k(0) = 1$,

$$1 - s_k(t) = \lambda_k \int_0^t p_k(x)\, dx. \tag{5.43}$$

Since $s_k(t)$ is monotonically increasing as $k \to \infty$, the following limit exists:

$$r(t) = \lim_{k \to \infty} (1 - s_k(t)).$$

From (5.43),

$$\lambda_k \int_0^t p_k(x)\, dx \ge r(t).$$

Dividing by λ_k and summing the arising inequalities from 0 to k,

$$\int_0^t s_k(x)\, dx \ge r(t)\left(\frac{1}{\lambda_0} + \frac{1}{\lambda_1} + \cdots + \frac{1}{\lambda_k}\right).$$

Since $s_k(t) \le 1$ for all $t \ge 0$,

$$t \ge r(t)\left(\frac{1}{\lambda_0} + \frac{1}{\lambda_1} + \cdots + \frac{1}{\lambda_k}\right).$$

If the series (5.42) diverges, then this inequality implies that $r(t) = 0$ for all $t > 0$. But this result is equivalent to (5.41).

Conversely, from (5.43),

$$\lambda_k \int_0^t p_k(x)\,dx \le 1$$

so that

$$\int_0^t s_k(x)\,dx \le \frac{1}{\lambda_0} + \frac{1}{\lambda_1} + \cdots + \frac{1}{\lambda_k}.$$

By passing to the limit as $k \to \infty$,

$$\int_0^t (1 - r(t))\,dt \le \sum_{i=0}^{\infty} \frac{1}{\lambda_i}.$$

If $r(t) \equiv 0$, the left-hand side of this inequality is equal to t. Since t can be arbitrarily large, the series (5.42) must diverge. This result completes the proof. ■

According to this theorem, it is theoretically possible that within a finite interval $[0, t]$ the population grows beyond all finite bounds. The probability of such an *explosive growth* is

$$1 - \sum_{i=0}^{\infty} p_i(t).$$

This probability is positive if the birth rates grow so fast that the series (5.42) converges. For example, an explosive growth would occur if

$$\lambda_i = i^2 \lambda; \quad i = 1, 2, \ldots$$

since

$$\sum_{i=1}^{\infty} \frac{1}{\lambda_i} = \frac{1}{\lambda} \sum_{i=1}^{\infty} \frac{1}{i^2} = \frac{\pi^2}{6\lambda} < \infty.$$

It is remarkable that an explosive growth occurs in an arbitrarily small time interval, since the convergence of the series (5.42) does not depend on t.

5.6.2 Death Processes

A continuous-time Markov chain with state space $\mathbf{Z} = \{0, 1, \ldots\}$ is called a (*pure*) *death process* if, for all $i = 1, 2, \ldots$ only transitions from state i to $i - 1$ are possible. State 0 is absorbing.

Thus, the positive transition rates of pure death processes are given by $q_{i,i-1}$, $i \ge 1$. In what follows, they will be called *death rates* and denoted as

$$\mu_0 = 0, \ \ \mu_i = q_{i,i-1}; \ \ i = 1, 2, \ldots$$

The sample paths of such processes are non-increasing step functions. For pure death processes, on condition

$$p_n(0) = P(X(0) = n) = 1,$$

the system of differential equations (5.20) becomes

$$p'_n(t) = -\mu_n p_n(t)$$

$$p'_j(t) = -\mu_j p_j(t) + \mu_{j+1} p_{j+1}(t); \quad j = 0, 1, ..., n-1. \tag{5.44}$$

The solution of the first differential equation is

$$p_n(t) = e^{-\mu_n t}, \quad t \ge 0.$$

Integrating (5.44) yields

$$p_j(t) = \mu_{j+1} e^{-\mu_j t} \int_0^t e^{\mu_j x} p_{j+1}(x)\, dx; \quad j = n-1, ..., 1, 0. \tag{5.45}$$

Starting with $p_n(t)$, the probabilities

$$p_j(t), \quad j = n-1, n-2, ..., 0,$$

can be recursively determined from (5.45). For instance, assuming $\mu_n \ne \mu_{n-1}$,

$$p_{n-1}(t) = \mu_n e^{-\mu_{n-1} t} \int_0^t e^{-(\mu_n - \mu_{n-1})x} dx$$

$$= \frac{\mu_n}{\mu_n - \mu_{n-1}} \left(e^{-\mu_{n-1} t} - e^{-\mu_n t} \right).$$

More generally, if all the death rates are different from each other, then

$$p_j(t) = \sum_{i=j}^{n} D_{ij} \mu_i e^{-\mu_i t}, \quad 0 \le j \le n, \tag{5.46}$$

where

$$D_{ij} = \frac{1}{\mu_j} \prod_{\substack{k=j \\ k \ne i}}^{n} \frac{\mu_k}{\mu_k - \mu_i}, \quad j \le i \le n, \quad D_{nn} = \frac{1}{\mu_n}.$$

Linear Death Process A death process $\{X(t), t \ge 0\}$ is called a *linear death process* if it has death rates

$$\mu_i = i\lambda; \quad i = 0, 1, ...$$

Under the initial distribution

$$p_n(0) = P(X(0) = n) = 1$$

the process stays in state n an exponentially with parameter $n\lambda$ distributed time:

$$p_n(t) = e^{-n\lambda t}, \quad t \ge 0.$$

Starting with $p_n(t)$, one obtains inductively from (5.45) or simply from (5.46):

$$p_i(t) = \binom{n}{i} e^{-i\lambda t} (1 - e^{-\lambda t})^{n-i}; \quad i = 0, 1, ..., n.$$

Hence, $X(t)$ has a binomial distribution with parameters n and $p = e^{-\lambda t}$ so that the trend function of a linear death process is

$$m(t) = n e^{-\lambda t}, \quad t \ge 0.$$

Example 5.10 A system consisting of n subsystems starts operating at time $t=0$. The lifetimes of the subsystems are independent, exponentially with parameter λ distributed random variables. If $X(t)$ denotes the number of subsystems still working at time t, then $\{X(t), t \ge 0\}$ is a linear death process with death rates

$$\mu_i = i\lambda; \quad i = 0, 1, \ldots$$ □

5.6.3 Birth- and Death Processes

5.6.3.1 Time-Dependent State Probabilities

A continuous-time Markov chain $\{X(t), t \ge 0\}$ with state space

$$\mathbf{Z} = \{0, 1, \ldots, n\}, \; n \le \infty,$$

is called a *birth- and death process* if from any state i only a transition to $i-1$ or $i+1$ is possible, provided that $i-1 \in \mathbf{Z}$ and $i+1 \in \mathbf{Z}$, respectively.

Therefore, the transition rates of a birth- and death process have property

$$q_{i,j} = 0 \quad \text{for } |i-j| > 1.$$

The transition rates $\lambda_i = q_{i,i+1}$ and $\mu_i = q_{i,i-1}$ are called *birth rates* and *death rates*, respectively. According to the restrictions given by the state space, $\lambda_n = 0$ for $n < \infty$ and $\mu_0 = 0$ (Figure 5.10). Hence, a birth process (death process) is a birth- and death process the death rates (birth rates) of which are equal to 0. If a birth- and death process describes the number of individuals in a population of organisms, then, when arriving in state 0, the population is extinguished. Thus, without the possibility of immigration, state 0 is absorbing ($\lambda_0 = 0$).

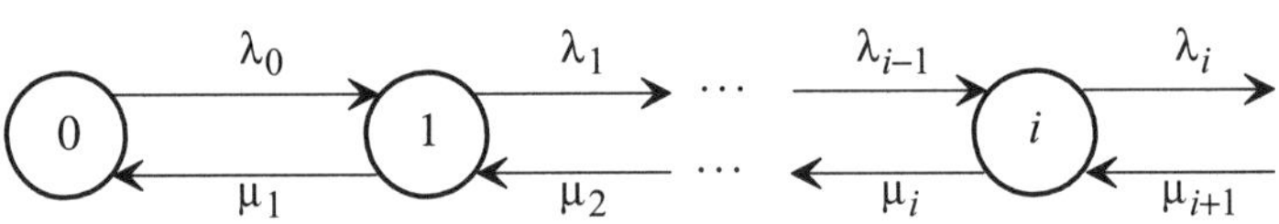

Figure 5.10 Transition graph of the birth- and death process

According to (5.20), the absolute state probabilities $p_j(t) = P(X(t) = j), \; j \in \mathbf{Z}$, of a birth- and death process satisfy the system of linear differential equations

$$\begin{aligned} p'_0(t) &= -\lambda_0 p_0(t) + \mu_1 p_1(t), \\ p'_j(t) &= +\lambda_{j-1} p_{j-1}(t) - (\lambda_j + \mu_j) p_j(t) + \mu_{j+1} p_{j+1}(t), \quad j = 1, 2, \ldots, \\ p'_n(t) &= +\lambda_{n-1} p_{n-1}(t) - \mu_n p_n(t), \quad n < \infty. \end{aligned} \tag{5.47}$$

In the following two examples, the state probabilities $(p_0(t), p_1(t), ...\}$ of two important birth- and death processes are determined via their respective z-transforms

$$M(t,z) = \Sigma_{i=0}^{\infty} p_i(t) z^i$$

under initial conditions of type

$$p_n(0) = P(X(0) = n) = 1.$$

In terms of the z-transform, this condition is equivalent to

$$M(0,z) \equiv z^n, \quad n = 0, 1, ...$$

Furthermore, partial derivatives of the z-transforms will be needed:

$$\frac{\partial M(t,z)}{\partial t} = \sum_{i=0}^{\infty} p_i'(t) z^i \quad \text{and} \quad \frac{\partial M(t,z)}{\partial z} = \sum_{i=1}^{\infty} i p_i(t) z^{i-1}. \tag{5.48}$$

Partial differential equations for $M(t,z)$ will be established and solved by applying the characteristic method.

Example 5.11 (***linear birth- and death process***) $\{X(t), t \geq 0\}$ is called a *linear birth- and death process* if it has transition rates

$$\lambda_i = i\lambda, \quad \mu_i = i\mu, \quad i = 0, 1, ...$$

In what follows, this process is analyzed on condition that

$$p_1(0) = P(X(0) = 1) = 1.$$

Assuming $p_0(0) = 1$ would make no sense since state 0 is absorbing. The system of differential equations (5.20) becomes

$$p_0'(t) = \mu p_1(t),$$

$$p_j'(t) = (j-1)\lambda p_{j-1}(t) - j(\lambda + \mu) p_j(t) + (j+1)\mu p_{j+1}(t); \quad j = 1, 2, ... \tag{5.49}$$

Multiplying the jth differential equation by z^j and summing from $j = 0$ to $j = \infty$, taking into account (5.48), yields the following linear homogeneous partial differential equation in $M(t, z)$:

$$\frac{\partial M(t,z)}{\partial t} - (z-1)(\lambda z - \mu) \frac{\partial M(t,z)}{\partial z} = 0. \tag{5.50}$$

The corresponding (ordinary) *characteristic differential equation* is a *Riccati differential equation* with constant coefficients:

$$\frac{dz}{dt} = -(z-1)(\lambda z - \mu) = -\lambda z^2 + (\lambda + \mu) z - \mu. \tag{5.51}$$

a) $\lambda \neq \mu$ By separation of variables, (5.51) can be written in the form

$$\frac{dz}{(z-1)(\lambda z - \mu)} = -dt.$$

Integration on both sides of this relationship yields

$$-\frac{1}{\lambda-\mu}\ln\left(\frac{\lambda z-\mu}{z-1}\right)=-t+C.$$

The general solution $z=z(t)$ of the characteristic differential equation in implicit form is, therefore, given by

$$c=(\lambda-\mu)\,t-\ln\left(\frac{\lambda z-\mu}{z-1}\right),$$

where c is an arbitrary constant. Thus, the general solution $M(t,z)$ of (5.50) has structure

$$M(t,z)=f\left((\lambda-\mu)t-\ln\left(\frac{\lambda z-\mu}{z-1}\right)\right),$$

where f can be any function with a continuous derivative. f can be determined by making use of the initial condition $p_1(0)=1$ or, equivalently, $M(0,z)=z$. Since

$$M(0,z)=f\left(\ln\left(\frac{z-1}{\lambda z-\mu}\right)\right)=z,$$

f must have structure

$$f(x)=\frac{\mu e^x-1}{\lambda e^x-1}.$$

Thus, $M(t,z)$ is

$$M(t,z)=\frac{\mu\exp\left\{(\lambda-\mu)t-\ln\left(\frac{\lambda z-\mu}{z-1}\right)\right\}-1}{\lambda\exp\left\{(\lambda-\mu)t-\ln\left(\frac{\lambda z-\mu}{z-1}\right)\right\}-1}.$$

After simplification, $M(t,z)$ becomes

$$M(t,z)=\frac{\mu\left[1-e^{(\lambda-\mu)t}\right]-\left[\lambda-\mu e^{(\lambda-\mu)t}\right]z}{\left[\mu-\lambda e^{(\lambda-\mu)t}\right]-\lambda\left[1-\mu e^{(\lambda-\mu)t}\right]z}.$$

This representation of $M(t,z)$ allows its expansion as a power series in z. The coefficient of z^j is the desired absolute state probability $p_j(t)$. Letting $\rho=\lambda/\mu$ yields

$$p_0(t)=\frac{1-e^{(\lambda-\mu)t}}{1-\rho e^{(\lambda-\mu)t}},$$

$$p_j(t)=(1-\rho)\rho^{j-1}\frac{\left[1-e^{(\lambda-\mu)t}\right]^{j-1}}{\left[1-\rho\, e^{(\lambda-\mu)t}\right]^{j+1}}e^{(\lambda-\mu)t},\quad j=1,2,...$$

Since state 0 is absorbing, $p_0(t)$ is the probability that the population is extinguished at time t. Moreover,

$$\lim_{t\to\infty} p_0(t) = \begin{cases} 1 & \text{for } \lambda < \mu \\ \frac{\mu}{\lambda} & \text{for } \lambda > \mu \end{cases}.$$

Thus, if $\lambda > \mu$, the population will survive to infinity with positive probability μ/λ. If $\lambda < \mu$, the population sooner or later will disappear with probability 1. In the latter case, the distribution function of the lifetime L of the population is

$$P(L \le t) = p_0(t) = \frac{1 - e^{(\lambda-\mu)t}}{1 - \rho e^{(\lambda-\mu)t}}, \quad t \ge 0.$$

Hence, the population will survive interval $[0, t]$ with probability

$$P(L > t) = 1 - p_0(t).$$

From this, applying (1.17),

$$E(L) = \frac{1}{\mu - \lambda} \ln\left(2 - \frac{\lambda}{\mu}\right).$$

The trend function $m(t) = E(X(t))$ is principally given by

$$m(t) = \Sigma_{j=0}^{\infty}\, j p_j(t).$$

By (1.23), $m(t)$ can also be obtained from the z-transform:

$$m(t) = \left.\frac{\partial M(t,z)}{\partial z}\right|_{z=1}.$$

If only the trend function of the process is of interest, then here as in many other cases knowledge of the z-transform or the absolute state distribution is not necessary, since $m(t)$ can be determined from the respective system of differential equations (5.47). In this example, multiplying the jth differential equation of (5.49) by j and summing from $j = 0$ to ∞ yields the following first-order differential equation:

$$m'(t) = (\lambda - \mu) m(t). \tag{5.52}$$

Taking into account the initial condition $p_1(0) = 1$, its solution is

$$m(t) = e^{(\lambda-\mu)t}.$$

By multiplying the j-th differential equation of (5.47) by j^2 and summing from $j = 0$ to ∞, a second order differential equation in $Var(X(t))$ is obtained. Its solution is

$$Var(X(t)) = \frac{\lambda+\mu}{\lambda-\mu}\left[1 - e^{-(\lambda-\mu)\,t}\right] e^{2(\lambda-\mu)t}.$$

Of course, since $M(t,z)$ is known, $Var(X(t))$ can be obtained from (1.23), too.

If the linear birth- and death process starts in states $s = 2, 3, \ldots,$ no principal additional problems arise up to the determination of $M(t,z)$. But it will be more complicated to expand $M(t,z)$ as a power series in z. The corresponding trend function, however, is easily obtained as solution of (5.52) with the initial condition $p_s(0) = 1$:

$$m(t) = s\,e^{(\lambda-\mu)t}, \quad t \geq 0.$$

b) $\boldsymbol{\lambda = \mu}$ In this case, the characteristic differential equation (5.51) simplifies to

$$\frac{dz}{\lambda(z-1)^2} = -dt.$$

Integration yields

$$c = \lambda t - \frac{1}{z-1},$$

where c is an arbitrary constant. Therefore, $M(t,z)$ has structure

$$M(t,z) = f\left(\lambda t - \frac{1}{z-1}\right),$$

where f is a continuously differentiable function. Since $p_1(0) = 1$, f satisfies

$$f\left(-\frac{1}{z-1}\right) = z.$$

Hence, the desired function f is given by

$$f(x) = 1 - \frac{1}{x}, \quad x \neq 0.$$

The corresponding z-transform is

$$M(t,z) = \frac{\lambda t + (1-\lambda t)z}{1 + \lambda t - \lambda t z}.$$

Expanding $M(t,z)$ as a power series in z yields the absolute state probabilities:

$$p_0(t) = \frac{\lambda t}{1+\lambda t}, \quad p_j(t) = \frac{(\lambda t)^{j-1}}{(1+\lambda t)^{j+1}}; \quad j = 1, 2, \ldots, \; t \geq 0.$$

An equivalent form of the absolute state probabilities is

$$p_0(t) = \frac{\lambda t}{1+\lambda t}, \quad p_j(t) = \left[1 - p_0(t)\right]^2 \left[p_0(t)\right]^{j-1}; \quad j = 1, 2, \ldots, \; t \geq 0.$$

Mean value and variance of $X(t)$ are

$$E(X(t)) = 1, \quad Var(X(t)) = 2\lambda t.$$

This example shows that the analysis of apparently simple birth- and death processes requires some effort. □

Example 5.12 Consider a birth- and death process with transition rates

$$\lambda_i = \lambda, \; \mu_i = i\mu; \quad i = 0, 1, \ldots$$

and initial distribution and $p_0(0) = P(X(0) = 0) = 1$.

The corresponding system of linear differential equations (5.47) is

$$p_0'(t) = \mu p_1(t) - \lambda p_0(t),$$
$$p_j'(t) = \lambda p_{j-1}(t) - (\lambda + \mu j) p_j(t) + (j+1)\mu p_{j+1}(t); \quad j = 1, 2, ... \tag{5.53}$$

Multiplying the jth equation by z^j and summing from $j = 0$ to ∞ yields a homogeneous linear partial differential equation for the moment generating function:

$$\frac{\partial M(t,z)}{\partial t} + \mu(z-1)\frac{\partial M(t,z)}{\partial z} = \lambda(z-1)\, M(t,z). \tag{5.54}$$

The corresponding system of characteristic differential equations is

$$\frac{dz}{dt} = \mu\,(z-1), \quad \frac{dM(t,z)}{dt} = \lambda(z-1)\, M(t,z).$$

After separation of variables and subsequent integration, the first differential equation yields

$$c_1 = \ln\,(z-1) - \mu\, t$$

with an arbitrary constant c_1. By combining both differential equations and letting $\rho = \lambda/\mu$,

$$\frac{dM(t,z)}{M(t,z)} = \rho\, dz.$$

Integration yields

$$c_2 = \ln M(t,z) - \rho\, z,$$

where c_2 is an arbitrary constant. As a solution of (5.54), $M(t,z)$ satisfies

$$c_2 = f(c_1)$$

with an arbitrary continuous function f, i.e. $M(t,z)$ satisfies

$$\ln M(t,z) - \rho\, z = f(\ln(z-1) - \mu t).$$

Therefore,

$$M(t,z) = \exp\{f(\ln(z-1) - \mu t) + \rho\, z\}.$$

Since condition $p_0(0) = 1$ is equivalent to $M(0,z) \equiv 1$, f is implicitely given by

$$f(\ln(z-1)) = -\rho\, z.$$

Hence, the explicit representation of f is

$$f(x) = -\rho\,(e^x + 1).$$

Thus,

$$M(t,z) = \exp\left\{-\rho\left(e^{\ln(z-1)-\mu\, t} + 1\right) + \rho\, z\right\}.$$

Equivalently,

$$M(t,z) = e^{-\rho(1-e^{-\mu t})} \cdot e^{+\rho(1-e^{-\mu t})z}.$$

Now it is easy to expand $M(t, z)$ as a power series in z. The coefficients of z^j are

$$p_j(t) = \frac{\left(\rho(1-e^{-\mu t})\right)^j}{j!} e^{-\rho(1-e^{-\mu t})}; \quad j = 0, 1, ... \tag{5.55}$$

This is a Poisson distribution with intensity function $\rho(1-e^{-\mu t})$. Therefore, this birth- and death process has trend function

$$m(t) = \rho(1-e^{-\mu t}).$$

For $t \to \infty$ the absolute state probabilities $p_j(t)$ converge to the stationary state probabilities:

$$\pi_j = \lim_{t\to\infty} p_j(t) = \frac{\rho^j}{j!} e^{-\rho}; \quad j = 0, 1, ...$$

If the process starts in a state $s > 0$, the absolute state probability distribution is not Poisson. In this case this distribution has a rather complicated structure, which will not be presented here. Instead, the system of linear differential equations (5.53) can be used to establish ordinary differential equations for the trend function $m(t)$ and the variance of $X(t)$. Given the initial distribution $p_s(0) = 1$, $s = 1, 2, ...,$ their respective solutions are

$$m(t) = \rho(1-e^{-\mu t}) + s\,e^{-\mu t},$$

$$Var(X(t)) = (1-e^{-\mu t})\left(\rho + s\,e^{-\mu t}\right).$$

The birth- and death process considered in this example is of some importance in queueing theory (section 5.7). □

Example 5.13 (*birth- and death process with immigration*) For positive parameters λ, μ and ν, let transition rates be given by

$$\lambda_i = i\lambda + \nu, \quad \mu_i = i\mu; \quad i = 0, 1, ...$$

If this model is used to describe the development in time of a population, then each individual will produce a new individual in $[t, t+\Delta t]$ with probability $\lambda\,\Delta t + o(\Delta t)$ or leave the population in this interval with probability $\mu\,\Delta t + o(\Delta t)$. Moreover, due to immigration from outside, the population will increase by one individual in $[t, t+\Delta t]$ with probability $\nu\,t + o(\Delta t)$. Thus, if $X(t) = i$, the probability that the population will increase or decrease by one individual in the interval $[t, t+\Delta t]$ is

$$(i\lambda + \nu)\,\Delta t + o(\Delta t) \quad \text{or} \quad i\mu\,\Delta t + o(\Delta t),$$

respectively. These probabilities do not depend on t and refer to $\Delta t \to 0$. As in the previous example, state 0 is not absorbing. The differential equations (5.47) become

$$p_0'(t) = \mu p_1(t) - \nu p_0(t),$$

$$p_j'(t) = (\lambda(j-1)+\nu)p_{j-1}(t) + \mu(j+1)p_{j+1}(t) - (\lambda j + \nu + \mu j)p_j(t).$$

Analogously to the previous examples, the z-transformation $M(t,z)$ of the probability distribution $\{p_0(t), p_1(t), ...\}$ is seen to satisfy the partial differential equation

$$\frac{\partial M(t,z)}{\partial t} = (\lambda z - \mu)(z-1)\frac{\partial M(t,z)}{\partial z} + \nu(z-1)M(t,z). \tag{5.56}$$

The system of the characteristic differential equations belonging to (5.56) is

$$\frac{dz}{dt} = -(\lambda z - \mu)(z-1),$$

$$\frac{dM(t,z)}{dt} = \nu(z-1)M(t,z).$$

From this, with the initial condition $p_0(0) = 1$ or, equivalently, $M(0,z) \equiv 1$, the solution is obtained analogously to the previous example

$$M(t,z) = \left\{\frac{\lambda - \mu}{\lambda z + \lambda(1-z)e^{(\lambda-\mu)t} - \mu}\right\}^{\nu/\lambda} \quad \text{for} \quad \lambda \neq \mu,$$

$$M(t,z) = (1+\lambda t)^{\nu/\lambda}\left\{1 - \frac{\lambda t z}{1+\lambda t}\right\}^{-\nu/\lambda} \quad \text{for} \quad \lambda = \mu.$$

Generally it is not possible to expand $M(t,z)$ as a power series in z. But the absolute state probabilities $p_i(t)$ can be obtained by differentiation of $M(t,z)$:

$$p_i(t) = \left.\frac{\partial^i M(t,z)}{\partial z^i}\right|_{z=0} \quad \text{for } i = 1, 2, ...$$

The trend function

$$m(t) = E(X(t)) = \left.\frac{\partial M(t,z)}{\partial z}\right|_{z=1}$$

of this birth- and death process is

$$m(t) = \frac{\nu}{\lambda-\mu}\left[e^{(\lambda-\mu)t} - 1\right] \quad \text{for} \quad \lambda \neq \mu, \tag{5.57}$$

$$m(t) = \nu t \quad \text{for} \quad \lambda = \mu.$$

If $\lambda < \mu$, the limit as $t \to \infty$ of the z-transform exists:

$$\lim_{t\to\infty} M(t,z) = \left(1 - \frac{\lambda}{\mu}\right)^{\nu/\lambda}\left(1 - \frac{\lambda}{\mu z}\right)^{-\nu/\lambda}.$$

For $\lambda < \mu$, the trend function (5.57) tends to a positive limit as $t \to \infty$:

$$\lim_{t\to\infty} m(t) = \frac{\nu}{\mu - \lambda} \quad \text{for} \quad \lambda < \mu.$$

□

5.6.3.2 Stationary State Probabilities

By (5.27), in case of their existence the stationary distribution $\{\pi_0, \pi_1, ...\}$ of a birth- and death process satisfies the following system of linear algebraic equations

$$\lambda_0\pi_0 - \mu_1\pi_1 = 0$$

$$\lambda_{j-1}\pi_{j-1} - (\lambda_j + \mu_j)\pi_j + \mu_{j+1}\pi_{j+1} = 0, \quad j = 1, 2, ... \tag{5.58}$$

$$\lambda_{n-1}\pi_{n-1} - \mu_n\pi_n = 0, \quad n < \infty.$$

This system is equivalent to the following one:

$$\mu_1\pi_1 = \lambda_0\pi_0$$

$$\mu_{j+1}\pi_{j+1} + \lambda_{j-1}\pi_{j-1} = (\lambda_j + \mu_j)\pi_j; \quad j = 1, 2, ... \tag{5.59}$$

$$\mu_n\pi_n = \lambda_{n-1}\pi_{n-1}, \quad n < \infty.$$

Provided its existence, it is possible to obtain the general solution of (5.58): Let

$$h_j = -\lambda_j\pi_j + \mu_{j+1}\pi_{j+1}; \; j = 0, 1, ...$$

Then the system (5.58) simplifies to

$$h_0 = 0,$$

$$h_j - h_{j-1} = 0, \; j = 1, 2, ...$$

$$h_{n-1} = 0, \quad n < \infty.$$

Starting with $j = 0$, one successively obtains

$$\pi_j = \prod_{i=1}^{j} \frac{\lambda_{i-1}}{\mu_i} \pi_0; \quad j = 1, 2, ..., n. \tag{5.60}$$

1) If $n < \infty$, then the stationary state probabilities satisfy the normalizing condition

$$\Sigma_{i=0}^{n} \pi_i = 1.$$

Solving for π_0 yields

$$\pi_0 = \left[1 + \sum_{j=1}^{n} \prod_{i=1}^{j} \frac{\lambda_{i-1}}{\mu_i} \right]^{-1}. \tag{5.61}$$

2) If $n = \infty$, then equation (5.61) shows that the convergence of the series

$$\sum_{j=1}^{\infty} \prod_{i=1}^{j} \frac{\lambda_{i-1}}{\mu_i} \tag{5.62}$$

is necessary for the existence of a stationary distribution. A sufficient condition for the convergence of this series is the existence of a positive integer N such that

$$\frac{\lambda_{i-1}}{\mu_i} \le \alpha < 1 \quad \text{for all } i > N. \tag{5.63}$$

Intuitively, this condition is not surprising: If the birth rates are greater than the corresponding death rates, the process will drift to infinity with probability 1. But this exludes the existence of a stationary distribution of the process. For a proof of the following theorem see Karlin and Taylor [45].

Theorem 5.3 The convergence of the series (5.62) and the divergence of the series

$$\sum_{j=1}^{\infty} \prod_{i=1}^{j} \frac{\mu_i}{\lambda_i} \tag{5.64}$$

is sufficient for the existence of a stationary state distribution. The divergence of (5.64) is, moreover, sufficient for the existence of such a time-dependent solution $\{p_0(t), p_1(t), ...\}$ of (5.47) which satisfies the normalizing condition (5.21). ■

Example 5.14 (*repairman problem*) The repairman problem introduced in example 5.8 is considered once more. However, it is now assumed that there are r mechanics for repairing failed n machines, $1 \le r \le n$. A failed machine can be attended only by one mechanic. (For a modification of this assumption see example 5.14.) All the other assumptions as well as the notation are as in example 5.8.

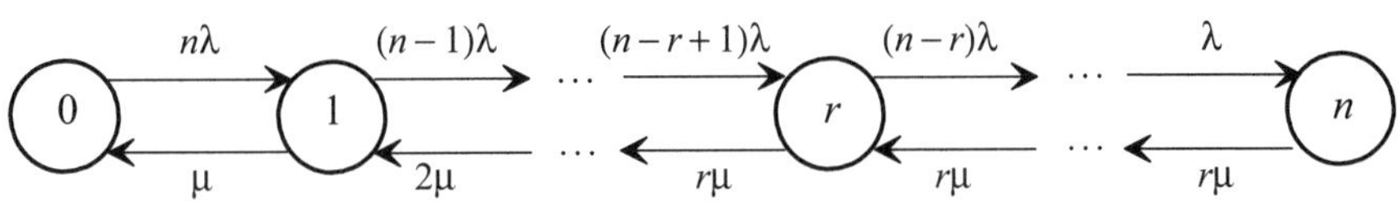

Figure 5.11 Transition graph of the general repairman problem

Let $X(t)$ denote the number of failed machines at time t. Then $\{X(t), t \ge 0\}$ is a birth- and death process with state space $\mathbf{Z} = \{0, 1, ..., n\}$. Its transition rates are

$$\lambda_j = (n-j)\lambda, \quad 0 \le j \le n,$$

$$\mu_j = \begin{cases} j\mu, & 0 \le j \le r \\ r\mu, & r < j \le n \end{cases}$$

(Figure 5.11). Note that in this example the terminology 'birth- and death rates' does not reflect the technological situation. If the *service rate* $\rho = \lambda/\mu$ is introduced, formulas (5.57) and (5.58) yield the stationary state probabilities

$$\pi_j = \begin{cases} \binom{n}{j} \rho^j \pi_0 ; & 1 \le j \le r \\ \dfrac{n!}{r^{j-r}\, r!\, (n-j)!} \rho^j \pi_0 ; & r \le j \le n \end{cases}, \tag{5.65}$$

$$\pi_0 = \left[\sum_{j=0}^{r} \binom{n}{j} \rho^j + \sum_{j=r+1}^{n} \frac{n!}{r^{j-r}\, r!\, (n-j)!} \rho^j \right]^{-1}.$$

Table 5.1 Stationary state probabilities for example 5.14

Policy 1:	n=10, r = 2	Policy 2:	n=5, r = 1
j	$\pi_{j,1}$	j	$\pi_{j,2}$
0	0.0341	0	0.1450
1	0.1022	1	0.2175
2	0.1379	2	0.2611
3	0.1655	3	0.2350
4	0.1737	4	0.1410
5	0.1564	5	0.0004
6	0.1173		
7	0.0704		
8	0.0316		
9	0.0095		
10	0.0014		

A practical application of the stationary state probabilities (5.65) is illustrated by a numerical example: Let $n = 10$, $\rho = 0.3$, and $r = 2$. The efficiencies of the following two maintenance policies will be compared:

1) Both mechanics are in charge of the repair of any of the 10 machines.

2) The mechanics are assigned 5 machines each for the repair of which they alone are responsible.

Let $X_{n,r}$ be the random number of failed machines and $Z_{n,r}$ the random number of mechanics which are busy with repairing failed machines, dependent on the number n of machines and the number r of available mechanics. From table 5.1, for policy 1,

$$E(X_{10,2}) = \Sigma_{j=1}^{10} j\,\pi_{j,1} = 3.902$$

$$E(Z_{10,2}) = 1 \cdot \pi_{1,1} + 2\,\Sigma_{j=2}^{10}\,\pi_{j,1} = 1.8296\,.$$

For policy 2,

$$E(X_{5,1}) = \Sigma_{j=1}^{5} j\,\pi_{j,2} = 2.011$$

$$E(Z_{5,1}) = 1 \cdot \pi_{1,2} + \Sigma_{j=2}^{5}\,\pi_{j,2} = 0.855.$$

Hence, when applying policy 2, the average number of failed machines out of 10 and the average number of busy mechanics out of 2 are

$$2\,E(X_{5,1}) = 4.022 \quad \text{and} \quad 2\,E(Z_{5,1}) = 1.710.$$

Thus, on the one hand, the mean number of failed machines under policy 1 is smaller than under policy 2, and, on the other hand, the mechanics are less busy under policy 2 than under policy 1. Hence, policy 1 should be preferred if there are no other relevant performance criteria. □

Example 5.15 The repairman problem of example 5.14 is modified in the following way: The available maintenance capacity of r units (which need not necessarily be human) is always fully used for repairing failed machines. Thus, if only one machine has failed, then all r units are busy with repairing this machine. If several machines are down, the full maintenance capacity of r units is uniformly distributed to the failed machines. This adaptation is repeated after each failure of a machine and after each completion of a repair. In this case, no machines have to wait for repair.

If j machines have failed, then the repair rate of each failed machine is

$$r\mu/j.$$

Therefore, the death rates of the corresponding birth- and death process are constant, i.e. they do not depend on the system state:

$$\mu_j = j \cdot \frac{r}{j}\mu = r\mu; \quad j = 1, 2, ...$$

The birth rates are the same as in example 5.14:

$$\lambda_j = (n-j)\lambda; \quad j = 0, 1, ...$$

Thus, the stationary state probabilities are according to (5.60) and (5.61):

$$\pi_0 = \left[\sum_{j=1}^{n} \frac{n!}{(n-j)!}\left(\frac{\lambda}{r\mu}\right)^j\right]^{-1},$$

$$\pi_j = \frac{n!}{(n-j)!}\left(\frac{\lambda}{r\mu}\right)^j \pi_0; \quad j = 1, 2, ...$$

Comparing this result with the stationary state probabilities (5.65), it is apparent that in case $r = 1$ the uniform distribution of the repair capacity to the failed machines has no influence on the stationary state probabilities. This fact is not surprising, since in this case the available maintenance capacity of one unit (if required) is always fully used. □

Many of the results presented so far in section 5.6 are due to Kendall [47].

5.6.3.3 Nonhomogeneous Birth- and Death Processes

Up till now, chapter 5 has been restricted to homogeneous Markov chains. They are characterized by transition rates which do not depend on time.

Nonhomogeneous Birth Processes 1) *Nonhomogeneous Poisson process* The most simple representative of a nonhomogeneous birth process is the nonhomogeneous Poisson process (section 3.2.2). Its birth rates are

$$\lambda_i(t) = \lambda(t); \quad i = 0, 1, ...$$

Thus, the process makes a transition from state i at time t to state $i+1$ in $[t, t+\Delta t]$ with probability $\lambda(t)\,\Delta t + o(\Delta t)$.

2) *Mixed Poisson process* If certain conditions are fulfilled, mixed Poisson processes (section 3.2.3) belong to the class of nonhomogeneous birth processes.

Lundberg [56] proved that a birth process is a mixed Poisson process if and only if its birth rates $\lambda_i(t)$ have properties

$$\lambda_{i+1}(t) = \lambda_i(t) - \frac{d\ln\lambda_i(t)}{dt}; \quad i = 0, 1, \ldots$$

Equivalently, a pure birth process $\{X(t), t \geq 0\}$ with transition rates $\lambda_i(t)$ and with absolute state distribution

$$\{p_i(t) = P(X(t) = i);\ i = 0, 1, \ldots\}$$

is a mixed Poisson process if and only if

$$p_i(t) = \frac{t}{i}\lambda_{i-1}(t)\,p_{i-1}(t); \quad i = 1, 2, \ldots$$

(see also Grandel [35]).

Nonhomogeneous Linear Birth- and Death Process In generalizing the birth- and death process of example 5.11, now a birth- and death process $\{X(t), t \geq 0\}$ is considered which has transition rates

$$\lambda_i(t) = \lambda(t)\, i, \quad \mu_i(t) = \mu(t)\, i; \quad i = 0, 1, \ldots$$

and initial distribution

$$p_1(0) = P(X(0) = 1) = 1.$$

Thus, $\lambda(t)$ can be interpreted as the transition rate from state 1 into state 2 at time t, and $\mu(t)$ is the transition rate from state 1 into the absorbing state 0 at time t. According to (5.47), the absolute state probabilities $p_j(t)$ satisfy

$$p_0'(t) = \mu(t)\,p_1(t),$$

$$p_j'(t) = (j-1)\lambda(t)\,p_{j-1}(t) - j\,(\lambda(t) + \mu(t))\,p_j(t) + (j+1)\mu(t)\,p_{j+1}(t); \quad j = 1, 2, \ldots$$

Hence, the corresponding z-transform $M(t, z)$ of

$$\{p_i(t) = P(X(t) = i);\ i = 0, 1, \ldots\}$$

is given by the partial differential equation (5.50) with time-dependent λ and μ :

$$\frac{\partial M(t,z)}{\partial t} - (z-1)\,[\lambda(t)\,z - \mu(t)]\,\frac{\partial M(t,z)}{\partial z} = 0\,. \tag{5.66}$$

The corresponding characteristic differential equation is a differential equation of Riccati type with time-dependent coefficients (compare with (5.51)):

$$\frac{dz}{dt} = -\lambda(t)\,z^2 + [\lambda(t) + \mu(t)]\,z - \mu\,.$$

A property of this differential equation is that there exist functions

$$\varphi_i(x); \quad i = 1, 2, 3, 4$$

so that its general solution $z = z(t)$ can be implicitely written in the form

$$c = \frac{z\,\varphi_1(t) - \varphi_2(t)}{\varphi_3(t) - z\,\varphi_4(t)}.$$

Hence, for all differentiable functions $g(\cdot)$, the general solution of (5.66) has the form

$$M(t,z) = g\left(\frac{z\,\varphi_1(t) - \varphi_2(t)}{\varphi_3(t) - z\,\varphi_4(t)}\right).$$

From this and the initial condition $M(0,z) = z$ it follows that there exist two functions $a(t)$ and $b(t)$ so that

$$M(t,z) = \frac{a(t) + [1 - a(t) - b(t)]\,z}{1 - b(t)\,z}. \tag{5.67}$$

By expanding $M(t,z)$ as a powers series in z,

$$p_0(t) = a(t),$$

$$p_i(t) = [1 - a(t)][1 - b(t)][\,b(t)]^{i-1}; \quad i = 1, 2, ... \tag{5.68}$$

Inserting (5.67) in (5.66) and comparing the coefficients of z yields a system of differential equations for $a(t)$ and $b(t)$:

$$(a'b - ab') + b' = \lambda\,(1 - a)\,(1 - b)$$

$$a' = \mu\,(1 - a)\,(1 - b).$$

The transformations $A = 1 - a$ and $B = 1 - b$ simplify this system to

$$B' = (\mu - \lambda)\,B - \mu\,B^2 \tag{5.69}$$

$$A' = -\mu\,A\,B. \tag{5.70}$$

The first differential equation is of Bernoulli type. Substituting in (5.69)

$$y(t) = 1/B(t)$$

gives a linear differential equation in y:

$$y' + (\mu - \lambda)\,y = \mu. \tag{5.71}$$

Since

$$a(0) = b(0) = 0,$$

y satisfies $y(0) = 1$. Hence the solution of (5.71) is

$$y(t) = e^{-\omega(t)}\left[\int_0^t e^{\omega(x)}\mu(x)\,dx + 1\right],$$

where

$$\omega(t) = \int_0^t [\mu(x) - \lambda(x)]\,dx.$$

From (5.70) and (5.71),

$$\frac{A'}{A} = -\mu B = -\frac{\mu}{y} = -\frac{y'}{y} - \omega' .$$

Therefore, the desired functions a and b are

$$a(t) = 1 - \frac{1}{y(t)}\, e^{-\omega(t)}$$

$$b(t) = 1 - \frac{1}{y(t)}, \quad t \geq 0.$$

With $a(t)$ and $b(t)$ known, the one-dimensional probability distribution (5.68) of the nonhomogeneous birth- and death process $\{X(t), t \geq 0\}$ is completely characterized. In particular, the probability that the process is in the absorbing state 0 at time t is

$$p_0(t) = \frac{\int_0^t e^{\omega(x)} \mu(x)\, dx}{\int_0^t e^{\omega(x)} \mu(x)\, dx + 1} .$$

Hence, the process $\{X(t), t \geq 0\}$ will reach state 0 with probability 1 if the integral

$$\int_0^t e^{\omega(x)} \mu(x)\, dx. \tag{5.72}$$

diverges as $t \to \infty$.

Let L denote the first passage time of the process with regard to state 0, i.e.

$$L = \inf_t \{t,\ X(t) = 0\}.$$

Since state 0 is absorbing, it is justified to call L the *lifetime* of the process. On condition that the integral (5.72) diverges as $t \to \infty$, L has distribution function

$$F_L(t) = P(L \leq t) = p_0(t), \quad t \geq 0.$$

Mean value and variance of $X(t)$ are

$$E(X(t)) = e^{-\omega(t)}, \tag{5.73}$$

$$Var(X(t)) = e^{-2\omega(t)} \int_0^t e^{\omega(x)} [\lambda(x) + \mu(x)]\, dx . \tag{5.74}$$

If the process $\{X(t), t \geq 0\}$ starts at $s = 2, 3, \ldots$ i.e. it has the initial distribution

$$p_s(0) = P(X(0) = s) = 1 \quad \text{for an } s = 2, 3, \ldots$$

then the corresponding z-transform is

$$M(t, z) = \left(\frac{a(t) + [1 - a(t) - b(t)]\, z}{1 - b(t)\, z} \right)^s .$$

In this case, mean value and variance of $X(t)$ are obtained by multiplying (5.73) and (5.74), respectively, by s.

5.7 APPLICATIONS TO QUEUEING MODELS

5.7.1 Basic Concepts

One of the most important applications of continuous-time Markov chains is stochastic modeling of service facilities. The basic situation is the following: Customers arrive at a service system (queueing system) according to a random point process. If all servers are busy, an arriving customer either waits for service or leaves the system without having been served. Otherwise, an available server takes care of the customer. After random service times customers leave the system. The arriving customers constitute the *input* (*input flow*, *traffic*, *flow of demands*) and the leaving customers the *output* (*output flow*) of the queueing system. A queueing system is called a *loss system* if it has no waiting capacity for customers which do not find an available server on arriving at the system. These customers leave the system immediately after arrival and are said to be *lost.* A *waiting system* has unlimited waiting capacity for those customers who do not immediately find an available server and are willing to wait any length of time for service. A *waiting-loss system* has only limited waiting capacity for customers. An arriving customer is lost if it finds all servers busy and the waiting capacity fully occupied. A *multi-server queueing system* has $s > 1$ servers. A *single-server queueing system* has only one server. Of course, 'customers' or 'servers' need not be persons.

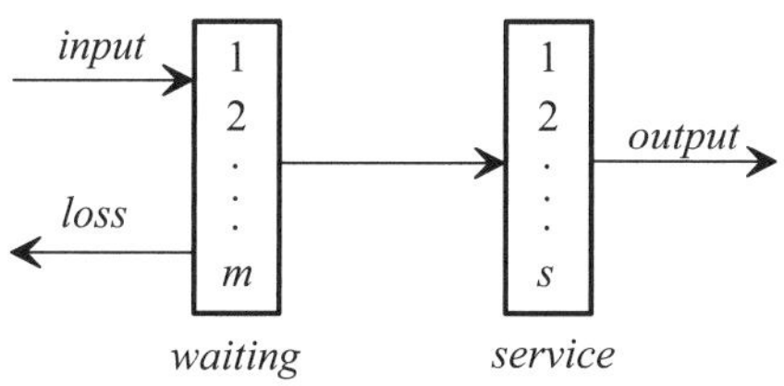

Figure 5.12 Scheme of a standard queueing system

Supermarkets are simple examples of queueing systems. Their customers are served at checkout counters. Filling stations also can be thought of as queueing systems with petrol pumps being the servers. Even a car park has the typical features of a waiting system. In this case, the parking lots are the 'servers' and the 'service times' are generated by the customers themselves. An anti-aircraft battery is a queueing system in the sense that it 'serves' the enemy aircraft. During recent years the stochastic modeling of communication systems, in particular computer networks, has stimulated the application of standard queueing models and the creation of new, more sophisticated ones. But the investigation of queueing systems goes back to the Danish engineer *A. K. Erlang* in the early 1900s, when he was in charge of designing telephone exchanges to meet criteria such as 'what is the mean waiting time of a customer before being connected' or 'how many lines (servers) are necessary to guarantee that with a given probability a customer can immediately be connected'? The repairman problem considered in example 5.14 also fits into the framework of a queueing system. The failed machines constitute the input and the mechanics are the servers. This example is distinguished by a particular feature: each demand (customer) is produced by one of a finite number n of different sources 'inside the system', namely by one of the n machines. Classes of queueing systems having this particular feature are called *closed queueing systems*.

The global objective of queueing theory is to provide theoretical tools for the design and the quantitative analysis of service systems. Designers of service systems need to make sure that the required service can be reliably delivered at minimal expense. Managers of service systems do not want to 'employ' more servers than necessary for meeting given performance criteria. Important criteria are:

1) The probability that an arriving customer finds an available server.

2) The mean waiting time of a customer for service.

It is common practice to characterize the structure of standard queueing systems by *Kendall's notation* $A/B/s/m$. In this code, A characterizes the input and B the ser-vice, s is the number of servers, and waiting capacity is available for m customers. Using this notation, standard classes of queueing systems are:

$A = M$ (*Markov*): Customers arrive in accordance with a homogeneous Poisson process (*Poisson input*).

$A = GI$ (*general independent*): Customers arrive in accordance with an ordinary renewal process (*recurrent input*).

$A = D$ (*deterministic*): The distances between the arrivals of neighbouring customers are constant (*deterministic input*).

$B = M$ (*Markov*) The service times are independent, identically distributed exponential random variables.

$B = G$ (*general*) The service times are independent, identically distributed random variables with arbitrary probability distribution.

For instance, $M/M/1/0$ is a loss system with Poisson input, one server, and exponential service times. $GI/M/3/\infty$ is a waiting system with recurrent input, exponential service times, and 3 servers. For queueing systems with an infinite number of servers, no waiting capacity is necessary. Hence their code is $A/B/\infty$.

In waiting systems and waiting-loss systems there are several ways of choosing waiting customers for service. These possibilities are called *service disciplines* (*queueing disciplines*). The most important ones are:

1) *FCFS* (*first come-first served*) Waiting customers are served in accordance with their order of arrival. This discipline is also called *FIFO* (*first in-first out*), although 'first in' does not necessarily imply 'first out'.

2) *LCFS* (*last come-first served*) The customer which arrived last is served first. This discipline is also called *LIFO* (*last in-first out*).

3) *SIRO* (*service in random order*) A server, when having finished with a customer, randomly picks one of the waiting customers for service.

There is a close relationship between service disciplines and priority (queueing) systems: In a *priority system* arriving customers have different *priorities* of being served. A customer with higher priority is served before a customer with lower priority, but no interruption of service takes place (*head of the line priority discipline*). When a customer with *absolute priority* arrives and finds all servers busy, then the service of a customer with lower priority has to be interrupted (*preemptive priority discipline*).

System Parameter The intensity of the input flow (mean number of arriving customers per unit time) is denoted as λ and referred to as *arrival rate* or *arrival intensity*. The service times of all servers are assumed to be independent and identically distributed. The *service intensity* or *service rate* of the servers is denoted as μ. Thus, if Y denotes the random times between the arrival of two neighbouring customers and Z the random service time of a customer, then

$$E(Y) = 1/\lambda \text{ and } E(Z) = 1/\mu.$$

The *traffic intensity* of a queueing system is defined as the ratio

$$\rho = \lambda/\mu,$$

and the *degree of server utilisation* is $\eta = E(S)/s$, where S is the random number of busy servers in the steady state. Thus, in the steady state, the coefficient η can be interpreted as the proportion of time a server is busy. Note that here and in what follows *in the steady state* refers to stationarity. More precisely, a (queueing) system is in the steady state if the underlying stochastic process $\{X(t), t \ge 0\}$ is stationary. In what follows, if not stated otherwise, $X(t)$ denotes the total number of customers at a service station (either wait- ing or being served) at time t. If X is the corresponding number in the steady state and π_j the stationary probability of state j, then

$$\pi_j = \lim_{t\to\infty} p_j(t) = \lim_{t\to\infty} P(X(t) = j) = P(X = j)$$

with $j = 0, 1, ..., s+m;\ s, m \le \infty$.

5.7.2 Loss Systems

5.7.2.1 *M/M/∞*- System

Strictly speaking, this system is neither a loss nor a waiting system. In this model, $\{X(t), t \ge 0\}$ is a homogeneous birth-and death process with state space $\mathbf{Z} = \{0, 1, ...\}$ and transition rates (example 5.12)

$$\lambda_i = \lambda; \quad \mu_i = i\mu; \quad i = 0, 1, ...$$

The corresponding time-dependent state probabilities $p_j(t)$ of this queueing system are given by (5.55). The stationary state probabilities are obtained by passing to the limit as $t \to \infty$ in these $p_j(t)$ or by inserting the transition rates $\lambda_i = \lambda$ and $\mu_i = i\mu$ with $n = \infty$ into (5.60) and (5.61):

$$\pi_j = \frac{\rho^j}{j!} e^{-\rho}; \quad j = 0, 1, ... \tag{5.75}$$

This is a Poisson distribution with parameter ρ. Hence, in the steady state the mean number of busy servers is equal to the traffic intensity of the system: $E(X) = \rho$.

5.7.2.2 *M/M/s/0* - System

In this case, $\{X(t), t \geq 0\}$ is a birth- and death process with $\mathbf{Z} = \{0, 1, ..., s\}$ and

$$\lambda_i = \lambda; \quad i = 0, 1, ..., s-1,$$
$$\lambda_i = 0 \text{ for } i \geq s,$$
$$\mu_i = i\mu; \quad i = 0, 1, ..., s.$$

Inserting these transition rates into the stationary state probabilities (5.60) and (5.61) with $n = s$ yields

$$\pi_0 = \left[\sum_{i=0}^{s} \frac{1}{i!}\rho^i\right]^{-1}; \quad \pi_j = \frac{1}{j!}\rho^j \pi_0; \quad j = 0, 1, ..., s. \tag{5.76}$$

The probability π_0 is called *vacant probability*. The *loss probability*, i.e. the probability that an arriving customer does not find an idle server, and, hence, leaves the system immediately, is

$$\pi_s = \frac{\frac{1}{s!}\rho^s}{\sum_{i=0}^{s} \frac{1}{i!}\rho^i}. \tag{5.77}$$

This is the famous *Erlang loss formula*. The following recursive formula for the loss probability as a function of s can easily be verified:

$$\pi_0 = 1 \text{ for } s = 0; \ \frac{1}{\pi_s} = \frac{s}{\rho}\frac{1}{\pi_{s-1}} + 1; \quad s = 1, 2, ...$$

The mean number of busy servers is

$$E(X) = \sum_{i=1}^{s} i\pi_i = \sum_{i=1}^{s} i\frac{\rho^i}{i!}\pi_0 = \rho\sum_{i=0}^{s-1}\frac{\rho^i}{i!}\pi_0.$$

By comparing to (5.77),

$$E(X) = \rho(1 - \pi_s).$$

Hence, the *degree of server utilization* is

$$\eta = \frac{\rho}{s}(1 - \pi_s).$$

Single-Server Loss System In case $s = 1$, vacant and loss probability are

$$\pi_0 = \frac{1}{1+\rho} \quad \text{and} \quad \pi_1 = \frac{\rho}{1+\rho}. \tag{5.78}$$

Since $\rho = E(Z)/E(Y)$,

$$\pi_0 = \frac{E(Y)}{E(Y)+E(Z)} \quad \text{and} \quad \pi_1 = \frac{E(Z)}{E(Y)+E(Z)}.$$

Hence, π_0 (π_1) is formally equal to the stationary availability (nonavailability) of a system with mean lifetime $E(Y)$ and mean renewal time $E(Z)$ the operation of which is governed by an alternating renewal process (section 3.3.6, formula (3.123)).

Example 5.16 A 'classical' application of loss models of type *M/M/s/0* is a telephone exchange. Assume that the input (calls of subscribers wishing to be connected) has intensity $\lambda = 2\ [min^{-1}]$. Thus, the mean time between successive calls is

$$E(Y) = 1/\lambda = 0.5\ [min].$$

On average, each subscriber occupies a line for $E(Z) = 1/\mu = 3\ [min]$.

1) What is the loss probability in case of $s = 7$ lines? The corresponding traffic intensity is $\rho = \lambda/\mu = 6$. Thus, the loss probability equals

$$\pi_7 = \frac{\frac{1}{7!}6^7}{1+6+\frac{6^2}{2!}+\frac{63}{3!}+\frac{6^4}{4!}+\frac{6^5}{5!}+\frac{6^6}{6!}+\frac{6^7}{7!}} = 0.185.$$

Hence, the mean number of occupied lines is

$$E(X) = \rho(1-\pi_7) = 6\,(1-0.185) = 4.89$$

and the degree of server (line) utilization is

$$\eta = \eta(7) = 4.89/7 = 0.698.$$

2) What is the minimal number of lines which have to be provided in order to make sure that at least 95% of the desired connections can be made? The respective loss probabilities for $s = 9$ and $s = 10$ are

$$\pi_9 = 0.075 \text{ and } \pi_{10} = 0.043.$$

Hence, the minimal number of lines required is

$$s_{\min} = 10.$$

However, in this case the degree of server utilization is smaller than with $s = 7$ lines:

$$\eta = \eta(10) = 0.574.$$ □

It is interesting and practically important that the stationary state probabilities of the queueing system *M/G/s/0* also have the structure (5.76). That is, if the respective traffic intensities ρ of the systems *M/M/s/0* and *M/G/s/0* are equal, then their stationary state probabilities coincide: for both systems they are given by (5.76). A corresponding result holds for the queueing systems $M/M/\infty$ and $M/G/\infty$. (Compare the stationary state probabilities (5.75) with the stationary state probabilities (3.37) deriv- ed in example 3.5 for the $M/G/\infty$-system.) Queueing systems having this property are said to be *insensitive* with respect to the probability distribution of the servicetimes. An analogous property can be defined with regard to the input. In view of (5.78), the $M/M/1/0$ - system is insensitive both with regard to arrival and service time distributions (*full insensitivity*). A comprehensive treatment of the *insensitivity* of queueing systems and other stochastic models is given in [33].

5.7.2.3 Engset's Loss System

Assume that n sources generate n independent Poisson inputs with common intensity λ which are served by s servers, $s \le n$. The service times are independent, exponentially distributed random variables with parameter μ. As long as a customer from a particular source is being served, this source cannot produce another customer (Compare to the repairman problem, example 5.14: during the repair of a machine, this machine cannot produce another demand for repair.) A customer which does not find an available server is lost. Let $X(t)$ denote the number of customers being served at time t. Then $\{X(t), t \ge 0\}$ is a birth- and death process with state space

$$\mathbf{Z} = \{0, 1, ..., s\}.$$

In case $X(t) = j$, only $n-j$ sources are active, that is they are able to generate customers. Therefore, the transition rates of this birth- and death process are

$$\lambda_j = (n-j)\lambda; \quad j = 0, 1, 2, \dots, s-1;$$
$$\mu_j = j\mu; \quad j = 1, 2, \dots, s.$$

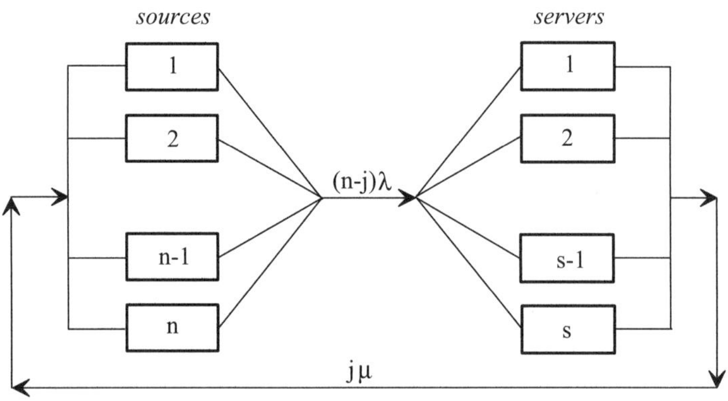

Figure 5.13 Engset's loss system in state $X(t)=j$

Inserting these transition rates into (5.60) and (5.61) with $n = s$ yields the stationary state distribution for Engset's loss system:

$$\pi_j = \frac{\binom{n}{j}\rho^j}{\sum_{i=0}^{s}\binom{n}{i}\rho^i}; \quad j = 0, 1, \ldots, s.$$

In particular, π_0 and the loss probability π_s are

$$\pi_0 = \frac{1}{\sum_{i=0}^{s}\binom{n}{i}\rho^i}, \qquad \pi_s = \frac{\binom{n}{s}\rho^s}{\sum_{i=0}^{s}\binom{n}{i}\rho^i}.$$

Engset's loss system is, just as the repairman problem considered in example 5.14, a closed queueing system. □

5.7.3 Waiting Systems

5.7.3.1 *M/M/s/∞* - System

The Markov chain $\{X(t), t \geq 0\}$ which models this system is defined as follows: If $X(t) = j$ with $0 \leq j \leq s$, then j servers are busy at time t. If $X(t) = j$ with $s > j$, then s servers are busy and $j - s$ customers are waiting for service. In either case, $X(t)$ is the total number of customers in the queueing system at time t. $\{X(t), t \geq 0\}$ is a birth- and death process with state space $\mathbf{Z} = \{0, 1, \ldots\}$ and transition rates

$$\lambda_j = \lambda; \quad j = 0, 1, \ldots,$$
$$\mu_j = j\mu \text{ for } j = 0, 1, \ldots, s; \quad \mu_j = s\mu \text{ for } j > s. \tag{5.79}$$

In what follows it is assumed that

$$\rho = \lambda/\mu < s.$$

If $\rho > s$, then the arrival intensity λ of customers is greater than the maximum service rate μs of the system so that, at least in the long-run, the system cannot cope with the input and the length of the waiting queue will tend to infinity as $t \to \infty$. Hence, no equilibrium state between arriving and leaving customers is possible. On the other hand, the condition $\rho < s$ is necessary and sufficient for the existence of a stationary state distribution, since in this case the corresponding series (5.62) converges and condition (5.63) is fulfilled. Inserting the transition rates (5.79) into (5.60) yields

$$\pi_j = \frac{\rho^j}{j!}\pi_0 \quad \text{for } j = 0, 1, \ldots, s-1$$

$$\pi_j = \frac{\rho^j}{s!\, s^{j-s}}\pi_0 \quad \text{for } j \geq s. \tag{5.80}$$

The normalizing condition and the geometric series yields the vacant probability π_0:

$$\pi_0 = \left[\sum_{i=0}^{s-1} \frac{1}{i!} \rho^i + \frac{\rho^s}{(s-1)!\,(s-\rho)} \right]^{-1}.$$

The probability π_w that an arriving customer finds all servers busy is

$$\pi_w = \sum_{i=s}^{\infty} \pi_i .$$

π_w is called *waiting probability*, since it is the probability that an arriving customer must wait for service. Making again use of the geometrical series yields a simple formula for π_w:

$$\pi_w = \frac{\pi_s}{1-\rho/s}. \tag{5.81}$$

In what follows, all derivations refer to the system in the steady state.
If S denotes the random number of busy servers, then its mean value is

$$E(S) = \sum_{i=0}^{s-1} i\,\pi_i + s\,\pi_w. \tag{5.82}$$

From this,

$$E(S) = \rho. \tag{5.83}$$

(The details of the derivation of (5.83) are left as an exercise to the reader.) Also without proof: Formula (5.83) holds for any $GI/G/s/\infty$-system. Hence the degree of server utilization in the $M/M/s/\infty$-system is $\eta = \rho/s$. By making use of (5.83), the mean value of the total number X of customers in the system is seen to be

$$E(X) = \sum_{i=1}^{\infty} i\,\pi_i = \rho \left[1 + \frac{s}{(s-\rho)^2}\,\pi_s \right]. \tag{5.84}$$

Let L denote the random number of customers waiting for service (queue length). Then the mean queue length is

$$E(L) = \sum_{i=s}^{\infty} (i-s)\,\pi_i = \sum_{i=s}^{\infty} i\,\pi_i - s\,\pi_w.$$

Combining this formula with (5.82)-(5.84) yields

$$E(L) = \frac{\rho\, s}{(s-\rho)^2}\,\pi_s . \tag{5.85}$$

Waiting Time Distribution Let W be the random time a customer has to wait for service if the service discipline *FCFS* is in effect. By the total probability rule,

$$P(W > t) = \sum_{i=s}^{\infty} P(W > t \,|\, X = i)\,\pi_i . \tag{5.86}$$

If a customer enters the system when it is in state $X = i \geq s$, then all servers are busy so that the current output is a Poisson process with intensity $s\mu$. The random event "$W > t$" occurs if within t time units after the arrival of a customer the service of at most $i-s$ customers has been finished. Therefore, the probability that the service of precisely k customers, $0 \le k \le i-s$, will be finished in this interval of length t is

$$\frac{(s\mu t)^k}{k!} e^{-s\mu t}.$$

Hence,
$$P(W > t | X = i) = e^{-s\mu t} \sum_{k=0}^{i-s} \frac{(s\mu t)^k}{k!}$$

and, by (5.86),

$$P(W > t) = e^{-s\mu t} \sum_{i=s}^{\infty} \pi_i \sum_{k=0}^{i-s} \frac{(s\mu t)^k}{k!} = \pi_0 e^{-s\mu t} \sum_{i=s}^{\infty} \frac{\rho^i}{s!s^{i-s}} \sum_{k=0}^{i-s} \frac{(s\mu t)^k}{k!}.$$

By performing the index transformation $j = i-s$, changing the order of summation according to formula (1.25), and making use of both the power series of e^x and the geometrical series yields

$$P(W > t) = \pi_0 \frac{\rho^s}{s!} e^{-s\mu t} \sum_{j=0}^{\infty} \left(\frac{\rho}{s}\right)^j \sum_{k=0}^{j} \frac{(s\mu t)^k}{k!}$$

$$= \pi_s e^{-s\mu t} \sum_{k=0}^{\infty} \frac{(s\mu t)^k}{k!} \sum_{j=k}^{\infty} \left(\frac{\rho}{s}\right)^j$$

$$= \pi_s e^{-s\mu t} \sum_{k=0}^{\infty} \frac{(\lambda t)^k}{k!} \sum_{i=0}^{\infty} \left(\frac{\rho}{s}\right)^i = \pi_s e^{-s\mu t} e^{\lambda t} \frac{1}{1-\rho/s}.$$

Hence, the distribution function of W is

$$F_W(t) = 1 - \frac{s}{s-\rho} \pi_s e^{-\mu(s-\rho)t}, \quad t \ge 0.$$

Note that $P(W > 0)$ is the waiting probability (5.81):

$$\pi_w = P(W > 0) = 1 - F_W(0) = \frac{s}{s-\rho} \pi_s.$$

The mean waiting time of a customer is

$$E(W) = \int_0^{\infty} P(W > t)\, dt = \frac{s}{\mu(s-\rho)^2} \pi_s. \tag{5.87}$$

A comparison of (5.85) and (5.87) yields *Little's formula* or *Little's law*:

$$E(L) = \lambda E(W). \tag{5.88}$$

Little's formula can be motivated as follows: The mean value of the sum of the waiting times arising in an interval of length τ is $\tau E(L)$. On the other hand, the same mean value is given by $\lambda\tau E(W)$, since the mean number of customers arriving in an interval of length τ is $\lambda\tau$. Hence,

$$\tau E(L) = \lambda\tau E(W),$$

which is Little's formula.

With $E(X)$ given by (5.84), an equivalent representation of Little's formula is

$$E(X) = \lambda\, E(T), \tag{5.89}$$

where T is the total sojourn time of a customer in the system, i.e. waiting plus service time: $T = W + Z$. Hence, the mean value of T is

$$E(T) = E(W) + 1/\mu .$$

Little's formula holds for any $GI/G/s/\infty$-system. For a proof of this proposition and other 'Little type formulas' see Franken et al. [29].

5.7.3.2 *M/G/1/∞* - System

In this single-server system, the service time Z is assumed to have an arbitrary probability density $g(t)$ and a finite mean $E(Z) = 1/\mu$. Hence, the corresponding stochastic process $\{X(t), t \geq 0\}$ describing the development in time of the number of customers in the system need no longer be a homogeneous Markov chain as in the previous queuing models. However, there exists an embedded homogeneous discrete-time Markov chain, which can be used to analyze this system (see section 5.4).

The system starts operating at time $t = 0$. Customers arrive according to a homogeneous Poisson process with positive intensity λ. Let A be the random number of customers, which arrive whilst a customer is being served, and

$$\{a_i = P(A = i);\ i = 0, 1, ...\}$$

be its probability distribution. To determine the a_i, note that the conditional probability that during a service time of length $Z = t$ exactly i new customers arrive is

$$\frac{(\lambda t)^i}{i!}\, e^{-\lambda t}.$$

Hence,

$$a_i = \int_0^\infty \frac{(\lambda t)^i}{i!}\, e^{-\lambda t} g(t)\, dt, \quad i = 0, 1, ...$$

This and the power series representation of e^x yield the z-transform $M_A(z)$ of A:

$$M_A(z) = \Sigma_{i=0}^{\infty} a_i z^i = \int_0^\infty e^{-(\lambda-\lambda z)\, t}\, g(t)\, dt .$$

Consequently, if $\hat{g}(\cdot)$ denotes the Laplace transform of $g(t)$, then

$$M_A(z) = \hat{g}(\lambda - \lambda z) . \tag{5.90}$$

By (1.23), letting as usual $\rho = \lambda/\mu$, the mean value of A is

$$E(A) = \frac{dM_A(z)}{dz}\Big|_{z=1} = -\lambda \frac{d\hat{g}(r)}{dr}\Big|_{r=0} = \rho. \tag{5.91}$$

Embedded Markov chain Let T_n be the random time point at which the nth customer leaves the system. If X_n denotes the number of customers in the system immediately after T_n, then $\{X_1, X_2, ...\}$ is a homogeneous, discrete-time Markov chain with state space $\mathbf{Z} = \{0, 1, ...\}$ and one-step transition probabilities

$$p_{ij} = P(X_{n+1} = j | X_n = i) = \begin{cases} a_j & \text{if } i = 0 \text{ and } j = 0, 1, 2, ... \\ a_{j-i+1} & \text{if } i - 1 \le j \text{ and } i = 1, 2, ... \\ 0 & \text{otherwise} \end{cases} \tag{5.92}$$

for all $n = 0, 1, ...;$ $X_0 = 0$. This Markov chain is embedded in $\{X(t), t \ge 0\}$ since

$$X_n = X(T_n + 0); \quad n = 0, 1, ...$$

The discrete-time Markov chain $\{X_0, X_1, ...\}$ is irreducible and aperiodic. Hence, on condition $\rho = \lambda/\mu < 1$ it has a stationary state distribution $\{\pi_0, \pi_1, ...\}$ which can be obtained by solving the corresponding system of algebraic equations (4.9): Inserting the transition probabilities p_{ij} given by (5.92) into (4.9) gives

$$\pi_0 = a_0(\pi_0 + \pi_1),$$

$$\pi_j = \pi_0 a_j + \sum_{i=1}^{j+1} \pi_i a_{j-i+1}; \quad j = 1, 2, ... \tag{5.93}$$

Let $M_X(z)$ be the z-transform of the state X of the system in the steady state:

$$M_X(z) = \sum_{j=0}^{\infty} \pi_j z^j.$$

Then, multiplying (5.93) by z^j and summing up from $j = 0$ to ∞ yields

$$\begin{aligned} M_X(z) &= \pi_0 \sum_{j=0}^{\infty} a_j z^j + \sum_{j=0}^{\infty} z^j \sum_{i=1}^{j+1} \pi_i a_{j-i+1} \\ &= \pi_0 M_A(z) + M_A(z) \sum_{i=1}^{\infty} \pi_i z^{i-1} a_{j-i+1} \\ &= \pi_0 M_A(z) + M_A(z) \frac{M_X(z) - \pi_0}{z}. \end{aligned}$$

Solving this equation for $M_X(z)$ yields

$$M_X(z) = \pi_0 M_A(z) \frac{1 - z}{M_A(z) - z}, \quad |z| < 1. \tag{5.94}$$

To determine π_0, note that

$$M_A(1) = M_X(z) = 1$$

and

$$\lim_{z \uparrow 1} \frac{M_A(z) - z}{1 - z} = \lim_{z \uparrow 1} \left(1 + \frac{M_A(z) - 1}{1 - z} \right) = 1 - \frac{dM_A(z)}{dz} \Big|_{z=1} = 1 - \rho.$$

Therefore, by letting $z \uparrow 1$ in (5.94),

$$\pi_0 = 1 - \rho. \tag{5.95}$$

Combining (5.90), (5.94) and (5.95) yields the *Formula of Pollaczek-Khinchin*:

$$M_X(z) = (1-\rho)\frac{1-z}{1-\dfrac{z}{\hat{g}(\lambda-\lambda z)}}, \quad |z|<1. \tag{5.96}$$

According to its derivation, this formula gives the z-transform of the stationary distribution of the random number X of customers in the system immediately after the completion of a customer's service. However, in view of the homogeneous Poisson input, it is even the stationary probability distribution of the 'original' Markov chain $\{X(t), t \geq 0\}$ itself. Thus, X is the random number of customers at the system in its steady state. Its probability distribution $\{\pi_0, \pi_1, ...\}$ exists and is a solution of (5.93). Hence, numerical parameters as mean value and variance of the number of customers in the system in the steady state can be determined by (5.96) via (1.23). For instance, the mean number of customers in the system is

$$E(X) = \frac{dM_X(z)}{dz}\Big|_{z=1} = \rho + \frac{\lambda^2[E(Z))^2 + Var(Z)]}{2(1-\rho)}. \tag{5.97}$$

Sojourn time Let T be the time a customer spends in the system (sojourn time) if the *FCFS*-queueing discipline is in effect. Then T has structure

$$T = W + Z,$$

where W is the time a customer has to wait for service (waiting time). Let $F_T(t)$ and $F_W(t)$ be the respective distribution functions of T and W and $f_T(t)$ and $f_W(t)$ the corresponding densities with Laplace transforms $\hat{f}_T(r)$ and $\hat{f}_W(r)$. Since W and Z are independent,

$$\hat{f}_T(r) = \hat{f}_W(r)\,\hat{g}(r). \tag{5.98}$$

The number of customers in the system after the departure of a served one is equal to the number of customers which arrived during the sojourn time of this customer. Hence, analogously to the structure of the a_i, the probabilities π_i are given by

$$\pi_i = \int_0^\infty \frac{(\lambda t)^i}{i!} e^{-\lambda t} f_T(t)\,dt; \quad i = 0, 1, ...$$

The corresponding z-transform $M_X(z)$ of X or, equivalently, the z-transform of the stationary distribution $\{\pi_0, \pi_1, ...\}$ is (compare to the derivation of (5.90))

$$M_X(z) = \hat{f}_T(\lambda - \lambda z).$$

Thus, by (5.98),

$$M_X(z) = \hat{f}_W(\lambda - \lambda z)\,\hat{g}(\lambda - \lambda z).$$

This formula and (5.96) yields the Laplace transform of $f_W(r)$:

$$\hat{f}_W(r) = (1-\rho)\frac{r}{\lambda\hat{g}(r)+r-\lambda}.$$

By (1.19) and (1.28), $E(W)$ and $Var(W)$ can be determined from $\hat{f}_W(s)$:

$$E(W) = \frac{\lambda[(E(Z))^2 + Var(Z)]}{2(1-\rho)}, \tag{5.99}$$

$$Var(W) = \frac{\lambda^2[(E(Z))^2 + Var(Z)]^2}{4(1-\rho)^2} + \frac{\lambda E(Z^3)}{3(1-\rho)}.$$

The random number of busy servers S has the stationary distribution

$$P(S=0) = \pi_0 = 1-\rho, \quad P(S=1) = 1-\pi_0 = \rho.$$

Thus,

$$E(S) = \rho.$$

The queue length is $L = X - S$. Hence, by (5.97),

$$E(L) = \frac{\lambda^2[E(Z))^2 + Var(Z)]}{2(1-\rho)}. \tag{5.100}$$

Comparing (5.97) and (5.100) verifies Little's formula (5.88):

$$E(L) = \lambda E(W).$$

5.7.3.3 *GI/M/1/∞* - System

In this single-server system, the interarrival times are given by an ordinary renewal process $\{Y_1, Y_2, ...\}$, where the Y_i are identically distributed as Y with probability density $f_Y(t)$ and finite mean value $E(Y) = 1/\lambda$. The service times are identically exponential distributed with parameter μ. A customer leaves the system immediately after completion of its service. If an arriving customer finds the server busy, it joins the queue. The stochastic process $\{X(t), t \geq 0\}$ describing the development of the number of customers in the system in time, need not be a homogeneous Markov chain. However, as in the previous section, an embedded homogeneous discrete-time Markov chain can be identified: The nth customer arrives at time

$$T_n = \Sigma_{i=1}^n Y_i; \quad n = 1, 2, ...$$

Let X_n denote the number of customers in the station immediately before arrival of the $(n+1)$th customer (being served or waiting). Then, $0 \leq X_n \leq n;\ n = 0, 1, ...$ The discrete-time stochastic process $\{X_0, X_1, ...\}$ is a Markov chain with parameter space $\mathbf{T} = \{0, 1, ...\}$ and state space $\mathbf{Z} = \{0, 1, ...\}$. Given that the system starts operating at time $t = 0$, the initial distribution of this discrete-time Markov chain is

$$P(X_0 = 0) = 1.$$

For obtaining the transition probabilities of $\{X_0, X_1, ...\}$, let D_n be the number of customers leaving the station in the interval $[T_n, T_{n+1})$ of length Y_{n+1}. Then,

$$X_n = X_{n-1} - D_n + 1 \text{ with } 0 \le D_n \le X_n\,; \quad n = 1, 2, ...,$$

By theorem 3.2, on condition $Y_{n+1} = t$ the random variable D_n has a Poisson distribution with parameter μt if the server is busy throughout the interval $[T_n, T_{n+1})$. Hence, for $i \ge 0$ and $1 \le j \le i+1$,

$$P\left(X_n = j \,\middle|\, X_{n-1} = i, Y_{n+1} = t\right) = \frac{(\mu\, t)^{i+1-j}}{(i+1-j)!}\, e^{-\mu t}; \quad n = 1, 2, ...$$

Consequently, the one-step transition probabilities

$$p_{ij} = P(X_n = j \,|\, X_{n-1} = i); \quad i, j \in \mathbf{Z}; \quad n = 1, 2, ...$$

of the Markov chain $\{X_0, X_1, ...\}$ are

$$p_{ij} = \int_0^\infty \frac{(\mu\, t)^{i+1-j}}{(i+1-j)!}\, e^{-\mu t} f_Y(t)\, dt; \quad 1 \le j \le i+1.$$

The normalizing condition yields p_{i0} :

$$p_{i0} = 1 - \Sigma_{j=1}^{i+1}\, p_{ij}.$$

The transition probabilities p_{ij} do not depend on n so that $\{X_0, X_1, ...\}$ is a homogeneous Markov chain. It is *embedded* in the original state process $\{X(t), t \ge 0\}$ since

$$X_n = X(T_{n+1} - 0)\,; \quad n = 0, 1, ...$$

Based on the embedded Markov chain $\{X_0, X_1, ...\}$, a detailed analysis of the queueing system *GI/M/1/∞* can be carried out analogously to the one of system *M/G/1/∞*.

5.7.4 Waiting-Loss Systems

5.7.4.1 *M/M/s/m* - System

This system has s servers and waiting capacity for m customers, $m \ge 1$. A customer which at arrival finds no idle server and the waiting capacity occupied is lost, that is it leaves the system immediately after arrival. The number of customers $X(t)$ in the system at time t generates a birth- and death process $\{X(t), t \ge 0\}$ with state space $\mathbf{Z} = \{0, 1, ..., s+m\}$ and transition rates

$$\lambda_j = \lambda, \qquad 0 \le j \le s+m-1,$$

$$\mu_j = \begin{cases} j\mu & \text{for} \quad 1 \le j \le s \\ s\mu & \text{for} \quad s < j \le s+m \end{cases}.$$

According to (5.60) and (5.61), the stationary state probabilities are

$$\pi_j = \begin{cases} \frac{1}{j!}\rho^j \pi_0 & \text{for } 1 \le j \le s-1 \\ \frac{1}{s!\, s^{j-s}}\rho^j \pi_0 & \text{for } s \le j \le s+m \end{cases},$$

$$\pi_0 = \left[\sum_{j=0}^{s-1} \frac{1}{j!}\rho^j + \sum_{j=s}^{s+m} \frac{1}{s!\, s^{j-s}}\rho^j \right]^{-1}.$$

The second series in π_0 can be summed up to obtain

$$\pi_0 = \begin{cases} \left[\sum_{j=0}^{s-1} \frac{1}{j!}\rho^j + \frac{1}{s!}\rho^s \frac{1-(\rho/s)^{m+1}}{1-\rho/s} \right]^{-1} & \text{for } \rho \ne s \\ \left[\sum_{j=0}^{s-1} \frac{1}{j!}\rho^j + (m+1)\frac{s^s}{s!} \right]^{-1} & \text{for } \rho = s \end{cases}.$$

The *vacant probability* π_0 is the probability that there is no customer in the system and π_{s+m} is the *loss probability*, i.e. the probability that an arriving customer is lost (rejected). The respective probabilities π_f and π_w that an arriving customer finds a free (idle) server or waits for service are

$$\pi_f = \sum_{i=0}^{s-1} \pi_i, \quad \pi_w = \sum_{i=s}^{s+m-1} \pi_i.$$

Analogously to the loss system *M/M/s/*0, the mean number of busy servers is

$$E(S) = \rho\,(1 - \pi_{s+m}).$$

Thus, the degree of server utilisation is

$$\eta = \rho\,(1 - \pi_{s+m})/s.$$

In the following example, the probabilities π_0 and π_{s+m} which refer to a queueing system with s servers and waiting capacity for m customers are denoted as $\pi_0(s, m)$ and $\pi_{s+m}(s, m)$, respectively.

Example 5.17 A filling station has $s = 8$ petrol pumps and waiting capacity for $m = 6$ cars. On average, 1.2 cars arrive at the filling station per minute. The mean time a car occupies a petrol pump is 5 minutes. It is assumed that the filling station behaves like an *M/M/s/m*-queueing system. Since $\lambda = 1.2$ and $\mu = 0.2$, the traffic intensity is $\rho = 6$. The corresponding loss probability $\pi_{16} = \pi_{16}(6, 10)$ is

$$\pi_{14}(8, 6) = \frac{1}{8!\, 8^6} 6^{14} \pi_0(8, 6) = 0.0167.$$

with

$$\pi_0(8,6) = \left[\sum_{j=0}^{7} \frac{1}{j!} 6^j + \frac{1}{8!} 6^8 \frac{1-(6/8)^7}{1-6/8} \right]^{-1} = 0.00225.$$

Consequently, the average number of occupied petrol pumps is

$$E(S) = 6 \cdot (1 - 0.0167) = 5.9.$$

After having obtained these figures, the owner of the filling station considers 2 out of the 8 petrol pumps superfluous and has them pulled down. It is assumed that this change does not influence the input flow so that cars continue to arrive with traffic intensity of $\rho = 6$. The corresponding loss probability $\pi_{12} = \pi_{12}(6,6)$ becomes

$$\pi_{12}(6,6) = \frac{6^6}{6!} \pi_0(6,6) = 0.1023 .$$

Thus, about 10% of all arriving cars leave the station without having filled up. To counter this drop, the owner provides waiting capacity for another 4 cars so that $m = 10$. The corresponding loss probability $\pi_{16} = \pi_{16}(6,10)$ is

$$\pi_{16}(6,10) = \frac{6^6}{6!} \pi_0(6,10) = 0.0726.$$

Formula

$$\pi_{6+m}(6,m) = \frac{6^6}{6!} \left[\sum_{j=0}^{5} \frac{1}{j!} 6^j + (m+1) \frac{6^6}{6!} \right]^{-1}$$

yields that additional waiting capacity for 51 cars has to be provided to equalize the loss caused by reducing the number of pumps from 8 to 6. □

5.7.4.2 *M/M/s/∞*-System with Impatient Customers

Even if there is waiting capacity for arbitrarily many customers, some customers might leave the system without having been served. This happens when customers can only spend a finite time, their *patience time*, in the queue. If the service of a customer does not begin before its patience time expires, the customer leaves system. For example, if somebody, whose long-distance train will depart in 10 minutes, has to wait 15 minutes to buy a ticket, then this person will leave the counter without a ticket. Real time monitoring and control systems have memories for data to be processed. But these data 'wait' only as long as they are up to date. Bounded waiting times are also typical for packed switching systems, for instance in computer-aided booking systems. Generally one expects that 'intelligent' customers adapt their behaviour to the actual state of the queueing system. Of the many available models dealing with such situations, the following one is considered in some detail: Customers arriving at an *M/M/s/∞*-system have independent, exponentially with parameter ν distributed patience times. If $X(t)$ as usual denotes the number of customers in the system at time t, then $\{X(t), t \geq 0\}$ is a birth- and death process with transition rates

$$\lambda_j = \lambda; \quad j = 0, 1, \ldots,$$

$$\mu_j = \begin{cases} j\mu & \text{for } j = 1, 2, \ldots, s \\ s\mu + (j-s)\nu & \text{for } j = s, s+1, \ldots \end{cases}.$$

If $j \to \infty$, then $\mu_j \to \infty$, whereas the birth rate remains constant. Hence the sufficient condition for the existence of a stationary distribution stated in theorem 5.3 (section 5.6.3.2) is fulfilled. Once the queue length exceeds a certain level, the number of customers leaving the system is on average greater than the number of arriving customers per unit time. That is, the system is self-regulating, aiming at reaching the equilibrium state. Now formulas (5.60) and (5.61) yield the corresponding stationary state probabilities:

$$\pi_j = \begin{cases} \dfrac{1}{j!}\rho^j \pi_0 & \text{for } j = 1, 2, \ldots, s \\ \dfrac{\rho^s}{s!} \dfrac{\lambda^{j-s}}{\prod\limits_{i=1}^{j-s}(s\mu + i\nu)} \pi_0 & \text{for } j = s+1, s+2, \ldots \end{cases}$$

$$\pi_0 = \left[\sum_{j=0}^{s} \frac{1}{j!}\rho^j + \frac{\rho^s}{s!} \sum_{j=s+1}^{\infty} \frac{\lambda^{j-s}}{\prod\limits_{i=1}^{j-s}(s\mu + i\nu)} \right]^{-1}.$$

Let L denote the random length of the queue in the steady state. Then,

$$E(L) = \sum_{j=s+1}^{\infty} (j-s)\,\pi_j .$$

Inserting the π_j yields after some algebra

$$E(L) = \pi_s \sum_{j=1}^{\infty} j\lambda^j \left[\prod_{i=1}^{j} (s\mu + i\nu) \right]^{-1}.$$

In this model, the *loss probability* π_v is not strictly associated with the number of customers in the system. It is the probability that a customer leaves the system without having been served, because its patience time has expired. Therefore, $1 - \pi_v$ is the probability that a customer leaves the system after having been served. By applying the total probability rule with the exhaustive and mutually exclusive set of events $\{"X = j"; \ j = s, s+1, \ldots\}$ one obtains

$$E(L) = \frac{\lambda}{\nu}\pi_v .$$

Thus, the mean queue length is directly proportional to the loss probability. (Compare to Little's formula (5.88).)

Variable Arrival Intensity Finite waiting capacities and patience times imply that in the end only a 'thinned flow' of potential customers will be served. Thus, it seems to be appropriate to investigate queueing systems whose arrival (input) intensities depend on the state of the system. However, those customers which actually enter the system do not leave it without service. Since the tendency of customers to leave the system immediately after arrival increases with the number of customers in the system, the birth rates should decrease for $j \geq s$ as j tends to infinity. For example, the following birth rates have this property:

$$\lambda_j = \begin{cases} \lambda & \text{for } j = 0, 1, \ldots, s-1 \\ \frac{s}{j+\alpha}\lambda & \text{for } j = s, s+1, \ldots \end{cases}, \quad \alpha \geq 0.$$

5.7.5 Special Single-Server Queueing Systems

5.7.5.1 System with Priorities

A single-server queueing system with waiting capacity for $m = 1$ customer is subject to two independent Poisson inputs 1 and 2 with respective intensities λ_1 and λ_2. The corresponding customers are called type 1 and type 2-customers. Type 1-customers have absolute (preemptive) priority, i.e. when a type 1 and a type 2-customer are in the system, the type 1-customer is being served. Thus, the service of a type 2-customer is interrupted as soon as a type 1-customer arrives. The displaced customer will occupy the waiting facility if it is empty. Otherwise it leaves the system. A waiting type 2-customer also has to leave the system when a type 1-customer arrives, since the newcomer will occupy the waiting facility. (Such a situation can only happen when a type 1-customer is being served.) An arriving type 1-customer is lost only when both server and waiting facility are occupied by other type 1-customers. Thus, if only the number of type 1-customers in the system is of interest, then this priority queueing system becomes the waiting-loss-system $M/M/s/1$ with $s = 1$, since type 2-customers have no impact on the service of type 1-customers at all.

The service times of type 1- and type 2- customers are assumed to have exponential distributions with respective parameters μ_1 and μ_2. The state space of the system is represented in the form

$$\mathbf{Z} = \{(i,j);\ i,j = 0, 1, 2\},$$

where i denotes the number of type 1-customers and j the number of type 2-customers in the system. Note that if $X(t)$ denotes the system state at time t, the stochastic process $\{X(t), t \geq 0\}$ can be treated as a one-dimensional Markov chain, since scalars can be assigned to the six possible system states, which are given as two-component vectors. However, $\{X(t), t \geq 0\}$ is not a birth- and death process. Figure 5.14 shows the transition graph of this Markov chain.

According to (5.28), the stationary state probabilities satisfy the system of equations

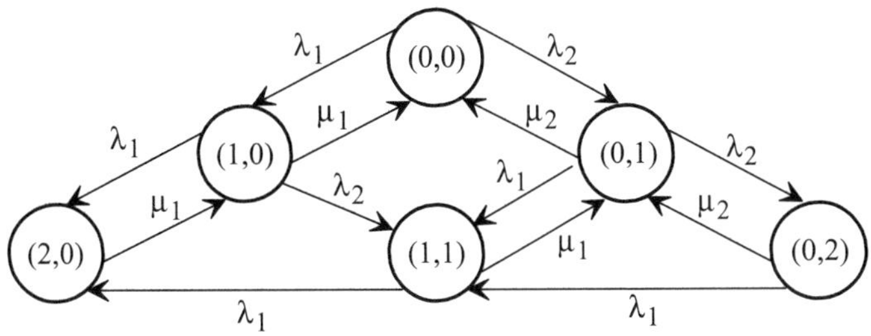

Figure 5.14 Transition graph for a single-server priority queueing system with $m = 1$

$$(\lambda_1 + \lambda_2)\pi_{(0,0)} = \mu_1\pi_{(1,0)} + \mu_2\pi_{(0,1)}$$
$$(\lambda_1 + \lambda_2 + \mu_1)\pi_{(1,0)} = \lambda_1\pi_{(0,0)} + \mu_1\pi_{(2,0)}$$
$$(\lambda_1 + \lambda_2 + \mu_2)\pi_{(0,1)} = \lambda_2\pi_{(0,0)} + \mu_1\pi_{(1,1)} + \mu_2\pi_{(0,2)}$$
$$(\lambda_1 + \mu_1)\pi_{(1,1)} = \lambda_2\pi_{(1,0)} + \lambda_1\pi_{(0,1)} + \lambda_1\pi_{(0,2)}$$
$$\mu_1\pi_{(2,0)} = \lambda_1\pi_{(1,0)} + \lambda_1\pi_{(1,1)}$$
$$(\lambda_1 + \mu_2)\pi_{(0,2)} = \lambda_2\pi_{(0,1)}$$
$$\pi_{(0,0)} + \pi_{(1,0)} + \pi_{(0,1)} + \pi_{(1,1)} + \pi_{(2,0)} + \pi_{(0,2)} = 1$$

$\boldsymbol{m = 0}$ Since there is no waiting capacity, each customer, notwithstanding its type, is lost if the server is busy with a type 1-customer. In addition, a type 2-customer is lost if, while being served, a type 1-customer arrives. The state space is

$$\mathbf{Z} = \{(0,0), (0,1), (1,0)\}.$$

Figure 5.15 shows the transition rates. The corresponding system (4.9) for the stationary state probabilities is

$$(\lambda_1 + \lambda_2)\pi_{(0,0)} = \mu_1\pi_{(1,0)} + \mu_2\pi_{(0,1)}$$
$$\mu_1\pi_{(1,0)} = \lambda_1\pi_{(0,0)} + \lambda_1\pi_{(0,1)}$$
$$1 = \pi_{(0,0)} + \pi_{(1,0)} + \pi_{(0,1)}$$

The solution is

$$\pi_{(0,0)} = \frac{\mu_1(\lambda_1 + \mu_2)}{(\lambda_1 + \mu_1)(\lambda_1 + \lambda_2 + \mu_2)},$$

$$\pi_{(0,1)} = \frac{\lambda_2\mu_1}{(\lambda_1 + \mu_1)(\lambda_1 + \lambda_2 + \mu_2)}, \qquad \pi_{(1,0)} = \frac{\lambda_1}{\lambda_1 + \mu_1}.$$

$\pi_{(1,0)}$ is the loss probability for type 1-customers. It is simply the probability that the service time of type 1-customers is greater than their interarrival time. On condition

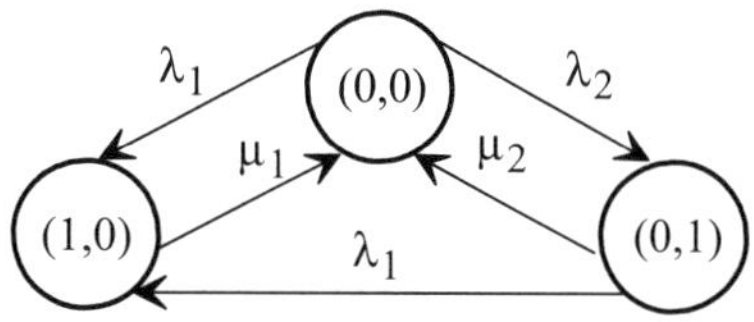

Figure 5.15 Transition graph for a 1-server priority loss system

that at the arrival time of a type 2-customer the server is idle, this customer is lost if and only if during its service a type 1-customer arrives. The conditional probability of this event is

$$\int_0^\infty e^{-\mu_2 t}\lambda_1 e^{-\lambda_1 t}\,dt = \lambda_1 \int_0^\infty e^{-(\lambda_1+\mu_2)t}dt = \frac{\lambda_1}{\lambda_1+\mu_2}.$$

Therefore, the (total) loss probability for type 2-customers is

$$\pi_l = \frac{\lambda_1}{\lambda_1+\mu_2}\,\pi_{(0,0)} + \pi_{(0,1)} + \pi_{(1,0)}.$$

Example 5.18 Let $\lambda_1 = 0.1$, $\lambda_2 = 0.2$, and $\mu_1 = \mu_2 = 0.2$. Then the stationary state probabilities are

$$\pi_{(0,0)} = 0.2105, \quad \pi_{(0,1)} = 0.3073, \quad \pi_{(1,0)} = 0.0085,$$
$$\pi_{(1,1)} = 0.1765, \quad \pi_{(0,2)} = 0.2048, \quad \pi_{(2,0)} = 0.0924.$$

In case $m = 0$, with the same numerical values for the transition rates,

$$\pi_{(0,0)} = 0.4000, \quad \pi_{(1,0)} = 0.3333, \quad \pi_{(0,1)} = 0.2667.$$

The loss probability for type 2-customers is $\pi_l = 0.7333$. □

5.7.5.2 *M/M/1/m* - System with Unreliable Server

If the implications of server failures on the system performance are not negligible, server failures have to be taken into account when building up a mathematical model. In what follows, the principal approach is illustrated by a single-server queuing system with waiting capacity for m customers, Poisson input, and independent, identically distributed exponential service times with parameter μ. The lifetime of the server is assumed to have an exponential distribution with parameter α, both in its busy phase and in its idle phase, and the subsequent renewal time of the server is assumed to be exponentially distributed with parameter β. It is further assumed that the sequence of life and renewal times of the server can be described by an alternating renewal process. When the server fails, all customers leave the system, i.e., the custom- er being served and the waiting customers if there are any are lost. Customers arriving during a renewal phase of the server are rejected, i.e. they are lost, too.

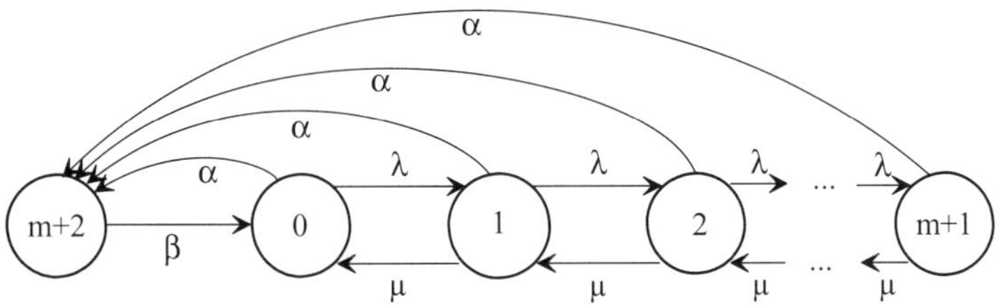

Figure 5.16 Transition graph of a queueing system with unreliable server

The stochastic process $\{X(t), t \geq 0\}$ describing the behaviour of the system is characterized as follows:

$$X(t) = \begin{cases} j & \text{if there are } j \text{ customers in the system at time } t;\ j = 0, 1, \dots, m+1 \\ m+2 & \text{if the server is being renewed at time } t \end{cases}.$$

Its transition rates are (Figure 5.16):

$$\begin{aligned} q_{j,j+1} &= \lambda; \quad j = 0, 1, \dots, m \\ q_{j,j-1} &= \mu; \quad j = 1, 2, \dots, m+1 \\ q_{j,m+2} &= \alpha; \quad j = 0, 1, \dots m+1 \\ q_{m+2,0} &= \beta \end{aligned} \tag{5.101}$$

According to (5.28), the stationary state probabilities satisfy the system of equations

$$\begin{aligned} (\alpha + \lambda)\,\pi_0 &= \mu\,\pi_1 + \beta\,\pi_{m+2} \\ (\alpha + \lambda + \mu)\,\pi_j &= \lambda\,\pi_{j-1} + \mu\,\pi_{j+1}\,; \quad j = 1, 2, \dots, m \\ (\alpha + \mu)\,\pi_{m+1} &= \lambda\,\pi_m \\ \beta\,\pi_{m+2} &= \alpha\,\pi_0 + \alpha\,\pi_1 + \cdots + \alpha\,\pi_{m+1} \end{aligned} \tag{5.102}$$

The last equation is equivalent to

$$\beta\,\pi_{m+2} = \alpha\,(1 - \pi_{m+2})\,.$$

Hence,

$$\pi_{m+2} = \frac{\alpha}{\alpha + \beta}\,.$$

Now, starting with the first equation in (5.102), the stationary state probabilities of the system $\pi_1, \pi_2, \dots, \pi_{m+1}$ can be successively determined. The probability π_0 is as usual obtained from the normalizing condition

$$\Sigma_{i=0}^{m+2}\,\pi_i = 1\,. \tag{5.103}$$

For the corresponding loss system ($m = 0$), the stationary state probabilities are

$$\pi_0 = \frac{\beta(\alpha+\mu)}{(\alpha+\beta)(\alpha+\lambda+\mu)}, \quad \pi_1 = \frac{\beta\lambda}{(\alpha+\beta)(\alpha+\lambda+\mu)},$$

$$\pi_2 = \frac{\alpha}{\alpha+\beta}.$$

Modification of the Model It makes sense to assume that the server can only fail when it is busy. In this case,

$$q_{j,m+2} = \alpha \quad \text{for } j = 1, 2, \ldots, m+1.$$

The other transition rates given by (5.101) remain valid. Thus, the corresponding transition graph is again given by Figure 5.16 with the arrow from node 0 to node $m+2$ deleted. The stationary state probabilities satisfy the system of equations

$$\begin{aligned} \lambda\pi_0 &= \mu\pi_1 + \beta\pi_{m+2} \\ (\alpha+\lambda+\mu)\pi_j &= \lambda\pi_{j-1} + \mu\pi_{j+1}; \quad j = 1, 2, \ldots, m \\ (\alpha+\mu)\pi_{m+1} &= \lambda\pi_m \\ \beta\pi_{m+2} &= \alpha\pi_1 + \alpha\pi_2 + \cdots + \alpha\pi_{m+1} \end{aligned} \tag{5.104}$$

The last equation is equivalent to

$$\beta\pi_{m+2} = \alpha(1 - \pi_0 - \pi_{m+2}).$$

It follows

$$\pi_{m+2} = \frac{\alpha}{\alpha+\beta}(1 - \pi_0).$$

Starting with the first equation in (5.104), the solution $\pi_0, \pi_1, \pi_2, \ldots, \pi_{m+1}$ can be obtained as above. In case $m = 0$ the stationary state probabilities are

$$\pi_0 = \frac{\beta(\alpha+\mu)}{\beta(\alpha+\mu)+\lambda(\alpha+\beta)},$$

$$\pi_1 = \frac{\lambda\beta}{\beta(\alpha+\mu)+\lambda(\alpha+\beta)},$$

$$\pi_2 = \frac{\alpha\lambda}{\beta(\alpha+\mu)+\lambda(\alpha+\beta)}.$$

Comment It is interesting that this queueing system with unreliable server can be interpreted as a queueing system with priorities and absolutely reliable server. To see this, a failure of the server has to be declared as the arrival of a 'customer' with absolute priority. The service provided to this 'customer' consists in the renewal of the server. Such a 'customer' pushes away any other customer from the server, in this model even from the waiting facility. Hence it is not surprising that the theory of queueing systems with priorities also provides solutions for more complicated queueing systems with unreliable servers than the one considered in this section.

5.7.6 Networks of Queueing Systems

5.7.6.1 Introduction

Customers frequently need several kinds of service so that, after leaving one service station, they have to visit one or more other service stations in a fixed or random order. Each of these service stations is assumed to behave like the basic queueing sys- tem sketched in Figure 5.12. A set of queueing systems together with rules of their interactions is called a *network of queueing systems* or a *queueing network.* Typical examples are technological processes for manufacturing (semi-) finished products. In such a case the order of service by different queueing systems is usually fixed. Queuing systems are frequently subject to several inputs, i.e. customers with different service requirements have to be attended. In this case they may visit the service stations in different orders. Examples of such situations are computer- and communication networks. Depending on whether and how data are to be provided, processed, or transmitted, the terminals (service stations) will be used in differen orders. If technical systems have to be repaired, then, depending on the nature and the extent of the damage, service of different production departments in a workshop is needed. Transport and loading systems also fit into the scheme of queueing networks.

Using a concept from graph theory, the service stations of a queueing network are called *nodes.* In an *open queueing network* customers arrive from 'outside' at the system (external input). Each node may have its own external input. Once in the system, customers visit other nodes in a deterministic or random order before leaving the network. Thus, in an open network, each node may have to serve *external* and *internal customers*, where internal customers are the ones which arrive from other nodes. In *closed queueing networks* there are no external inputs into the nodes and the total number of customers in the network is constant. Consequently, no customer departs from the network. Queueing networks can be represented by directed graphs. The directed edges between the nodes symbolize the possible transitions of customers from one node to another. The nodes in the network are denoted by $1, 2, ..., n$. Node i is assumed to have s_i servers; $1 \le s_i \le \infty$.

5.7.6.2 Open Queueing Networks

A mathematically exact analysis of queueing systems becomes extremely difficult or even impossible when dropping the assumptions of Poisson input and/or exponentially distributed service times. Hence, this section is restricted to a rather simple class of queueing networks, the *Jackson queueing networks.* They are characterized by four properties:

1) Each node has an unbounded waiting capacity.

2) The service times of all servers at node i are independent, identically distributed exponential random variables with parameter (intensity) μ_i. They are also independent of the service times at other nodes.

3) External customers arrive at node i in accordance with a homogeneous Poisson process with intensity λ_i. All external inputs are independent of each other and of all service times.

4) When the service of a customer at node i has been finished, the customer makes a transition to node j with probability p_{ij} or leaves the network with probability a_i.

The *transition* or *routing matrix* $\mathbf{P} = ((p_{ij}))$ is independent of the current state of the network and of its past. Let $\mathbf{I}$ be the identity matrix. The matrix $\mathbf{I} - \mathbf{P}$ is assumed to be nonsingular so that the inverse matrix $(\mathbf{I} - \mathbf{P})^{-1}$ exists.

According to the definition of the a_i and p_{ij},

$$a_i + \sum_{j=1}^{n} p_{ij} = 1. \tag{5.105}$$

In a Jackson queueing network, each node is principally subjected to both external and internal input. Let α_j be the total input (arrival) intensity at node j. In the steady state, α_j must be equal to the total output intensity from node j. The portion of internal input intensity to node j, which is due to customers from node i, is $\alpha_i p_{ij}$. Thus,

$$\sum_{i=1}^{n} \alpha_i p_{ij}$$

is the total internal input intensity to node j. Consequently, in the steady state,

$$\alpha_j = \lambda_j + \sum_{i=1}^{n} \alpha_i p_{ij}; \quad j = 1, 2, \ldots, n. \tag{5.106}$$

By introducing vectors the $\boldsymbol{\alpha} = (\alpha_1, \alpha_2, \ldots, \alpha_n)$ and $\boldsymbol{\lambda} = (\lambda_1, \lambda_2, \ldots, \lambda_n)$, the relationship (5.106) can be written as

$$\boldsymbol{\alpha}(\mathbf{I} - \mathbf{P}) = \boldsymbol{\lambda}.$$

Since $\mathbf{I} - \mathbf{P}$ is assumed to be nonsingular, the vector of the total input intensities $\boldsymbol{\alpha}$ is

$$\boldsymbol{\alpha} = \boldsymbol{\lambda}(\mathbf{I} - \mathbf{P})^{-1}. \tag{5.107}$$

Even under the assumptions stated, the total inputs at the nodes and the outputs from the nodes are generally nonhomogeneous Poisson processes.

Let $X_i(t)$ be the random number of customers at node i at time t. Its realizations are denoted as x_i; $x_i = 0, 1, \ldots$ The random state of the network at time t is characterized by the vector

$$\mathbf{X}(t) = (X_1(t), X_2(t), \ldots, X_n(t))$$

with realizations $\mathbf{x} = (x_1, x_2, \ldots, x_n)$. The set of all these vectors $\mathbf{x}$ forms the state space of the Markov chain $\{\mathbf{X}(t), t \geq 0\}$. Using set-theory notation, the state space is denoted as $\mathbf{Z} = \{0, 1, \ldots\}^n$, i.e. $\mathbf{Z}$ is the set of all those n-dimensional vectors the components of which assume nonnegative integers. Since $\mathbf{Z}$ is countably infinite, this at first glance n-dimensional Markov chain becomes one-dimensional by arranging the states as a sequence.

To determine the transition rates of $\{\mathbf{X}(t),\ t \ge 0\}$, the n-dimensional vector $\mathbf{e}_i$ is introduced. Its ith component is a 1 and the other components are zeros:

$$\mathbf{e}_i = \underset{1\ 2\ \cdots\ \ \ i\ \ \cdots\ \ n}{(0, 0, \dots, 0, 1, 0, \dots, 0)}. \tag{5.108}$$

Thus, $\mathbf{e}_i$ is the ith row of the identity matrix $\mathbf{I}$. Since the components of any state vector $\mathbf{x}$ are nonnegative integers, each $\mathbf{x}$ can be represented as a linear combination of all or some of the $\mathbf{e}_1, \mathbf{e}_2, \dots, \mathbf{e}_n$. In particular, $\mathbf{x} + \mathbf{e}_i$ $(\mathbf{x} - \mathbf{e}_i)$ is the vector which arises from $\mathbf{x}$ by increasing (decreasing) the ith component by 1. Starting from state $\mathbf{x}$, the Markov chain $\{\mathbf{X}(t),\ t \ge 0\}$ can make the following one-step transitions:

1) When a customer arrives at node i, the Markov chain makes a transition to state $\mathbf{x} + \mathbf{e}_i$.

2) When a service at node i is finished, $x_i > 0$, and the served customer leaves the network, the Markov chain makes a transition to state $\mathbf{x} - \mathbf{e}_i$.

3) When a service at node i with $x_i > 0$ is finished and the served customer leaves node i for node j, the Markov chain makes a transition to state $\mathbf{x} - \mathbf{e}_i + \mathbf{e}_j$.

Therefore, starting from state $\mathbf{x} = (x_1, x_2, \dots, x_n)$, the transition rates are

$$q_{\mathbf{x},\mathbf{x}+\mathbf{e}_i} = \lambda_i$$
$$q_{\mathbf{x},\mathbf{x}-\mathbf{e}_i} = \min(x_i, s_i)\, \mu_i\, a_i$$
$$q_{\mathbf{x},\mathbf{x}-\mathbf{e}_i+\mathbf{e}_j} = \min(x_i, s_i)\, \mu_i\, p_{ij}, \quad i \ne j$$

In view of (5.105),

$$\sum_{j, j \ne i} p_{ij} = 1 - p_{ii} - a_i .$$

Hence, the rate of leaving state $\mathbf{x}$ is

$$q_{\mathbf{x}} = \sum_{i=1}^{n} \lambda_i + \sum_{i=1}^{n} \mu_i (1 - p_{ii}) \min(x_i, s_i).$$

According to (5.28), the stationary state probabilities

$$\pi_{\mathbf{x}} = \lim_{t \to \infty} P(\mathbf{X}(t) = \mathbf{x}), \quad \mathbf{x} \in \mathbf{Z},$$

provided they exist, satisfy the system of equations

$$q_{\mathbf{x}}\, \pi_{\mathbf{x}} = \sum_{i=1}^{n} \lambda_i\, \pi_{\mathbf{x}-\mathbf{e}_i} + \sum_{i=1}^{n} a_i \mu_i \min(x_i + 1, s_i)\, \pi_{\mathbf{x}+\mathbf{e}_i}$$
$$+ \sum_{j=1}^{n} \sum_{\substack{i=1 \\ i \ne j}}^{n} a_i \mu_i \min(x_i + 1, s_i)\, p_{ij}\, \pi_{\mathbf{x}+\mathbf{e}_i-\mathbf{e}_j} . \tag{5.109}$$

In order to be able to present the solution of this system in a convenient form, recall that the stationary state probabilities of the waiting system $M/M/s_i/\infty$ with parameters α_i, μ_i and $\rho_i = \alpha_i/\mu_i$ denoting the intensity of the Poisson input, the service intensities of all servers, and the traffic intensity of the system, respectively, are given by (see formula (5.80)),

$$\varphi_i(j) = \begin{cases} \frac{1}{j!}\rho_i^j\,\varphi_i(0) & \text{for } j = 1, 2, ..., s_i - 1 \\ \frac{1}{s_i!\,s_i^{j-s_i}}\rho_i^j\,\varphi_i(0) & \text{for } j = s_i, s_i + 1, ... \end{cases}, \quad \rho_i < s_i,$$

$$\varphi_i(0) = \left[\sum_{j=0}^{s_i-1} \frac{1}{j!}\rho_i^j + \frac{\rho_i^{s_i}}{(s_i-1)!\,(s_i-\rho_i)}\right]^{-1}, \quad \rho_i < s_i.$$

(In the context queueing networks, the notation $\varphi_i(\cdot)$ for the stationary state probabilities is common practice.) The stationary state probabilities of the queueing network are simply obtained by multiplying the corresponding state probabilities of the queuing systems $M/M/s_i/\infty$; $i = 1, 2, ...n$:

> If the vector of the total input intensities $\boldsymbol{\alpha} = (\alpha_1, \alpha_2, ..., \alpha_n)$ given by (5.73) satisfies the conditions
>
> $$\alpha_i < s_i\mu_i; \quad i = 1, 2, ..., n;$$
>
> then the stationary probability of state $\mathbf{x} = (x_1, x_2, ..., x_n)$ is
>
> $$\pi_{\mathbf{x}} = \prod_{i=1}^{n} \varphi_i(x_i), \quad \mathbf{x} \in \mathbf{Z}. \tag{5.110}$$

Thus, the stationary state distribution of a Jackson queueing system is given in *product form*. This implies that each node of the network behaves like an $M/M/s_i/\infty$-system. However, the nodes need not be a queueing system of this type because the process $\{X_i(t), t \geq 0\}$ is usually not a birth- and death process. In particular, the total input into a node need not be a homogeneous Poisson process. But the product form (5.110) of the stationary state probabilities proves that the queue lengths at the nodes in the steady state are independent random variables. There is a vast amount of literature dealing with assumptions under which the stationary distribution of a queueing network has the product form (see, for instance, van Dijk [84]).

To verify that the stationary state distribution indeed has the product form (5.110), one has to substitute (5.110) into the system of equations (5.109). Using (5.105) and (5.106), one obtains an identity after some tedious algebra.

Example 5.19 The simplest Jackson queueing network arises if $n = 1$. The only difference from the queueing system $M/M/s/\infty$ is that now a positve proportion of customers who have departed from the network after having been served will return and require further service. This leads to a queueing system with *feedback*.

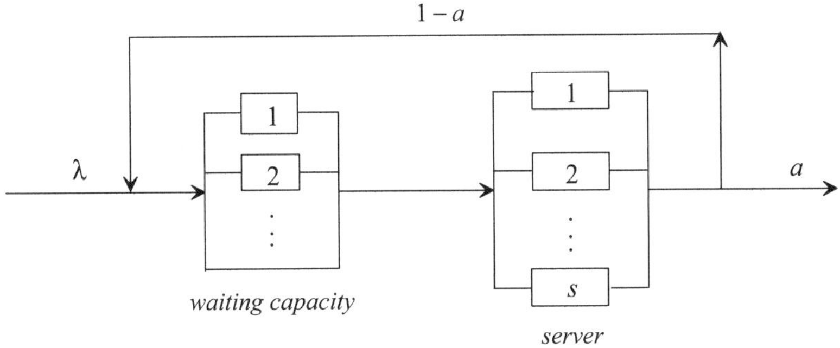

Figure 5.17 Queueing system with feedback

For instance, when servers have done a bad job, then the affected customers will soon return to exercise possible guarantee claims. Formally, these customers remain in the network. Roughly speaking, a single-node Jackson queueing network is a mixture between an open and a closed waiting system (Figure 5.17). A customer leaves the system with probability a or reenters the system with probability $p_{11} = 1 - a$. If there is an idle server, then, clearly, the service of a customer starts immediately. From (5.105) and (5.106), the total input rate α into the system satisfies

$$\alpha = \lambda + \alpha(1 - a).$$

(The index 1 is deleted from all system parameters.) Thus,

$$\alpha = \lambda/a.$$

Hence there exists a stationary distribution if

$$\lambda/a < s\mu \quad \text{or, equivalently, if} \quad \rho < as$$

with $\rho = \lambda/\mu$. In this case the stationary state probabilities are

$$\pi_j = \begin{cases} \frac{1}{j!}\left(\frac{\rho}{a}\right)^j \pi_0 & \text{for } j = 1, 2, \ldots, s-1 \\ \frac{1}{s!\, s^{j-s}}\left(\frac{\rho}{a}\right)^j \pi_0 & \text{for } j = s, s+1, \ldots \end{cases},$$

where

$$\pi_0 = \left[\sum_{j=1}^{s-1} \frac{1}{j!}\left(\frac{\rho}{a}\right)^j + \frac{\left(\frac{\rho}{a}\right)^s}{(s-1)!\left(s - \frac{\rho}{a}\right)}\right]^{-1}.$$

This is the stationary state distribution of the queueing system $M/M/s/\infty$ (without feedback), the input of which has intensity λ/a. □

Example 5.20 In technological processes, the sequence of service is usually fixed. For example, a 'customer' may be a car being manufactured on an assembly line. Therefore, queueing systems switched in series, called *sequential queueing networks* or *tandem queueing networks*, are of considerable practical interest: External customers arrive only at node 1 (arrival intensity: λ_1). They subsequently visit in this order the nodes 1, 2, ..., n and then leave the network (Figure 5.18).

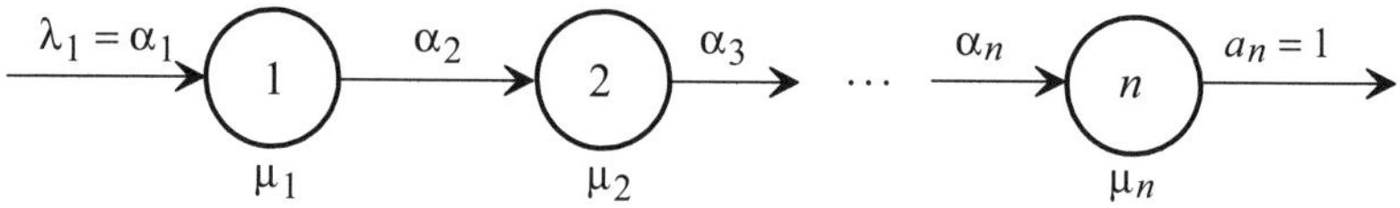

Figure 5.18 Sequential queueing network

The corresponding parameters are

$$\lambda_i = 0; \qquad i = 2, 3, \dots, n$$

$$p_{i,i+1} = 1; \quad i = 1, 2, \dots, n-1$$

$$a_1 = a_2 = \cdots = a_{n-1} = 0, \quad a_n = 1$$

According to (5.106), the (total) input intensities of all nodes in the steady state must be the same:

$$\lambda_1 = \alpha_1 = \alpha_2 = \cdots = \alpha_n.$$

Hence, for single-server nodes ($s_i = 1;\ i = 1, 2, ..., n$), a stationary state distribution exists if

$$\rho_i = \lambda_1/\mu_i < 1; \quad i = 1, 2, \dots, n$$

or, equivalently, if

$$\lambda_1 < \min(\mu_1, \mu_2, \dots, \mu_n).$$

Thus, it is the slowest server which determines the efficiency of a sequential network. The stationary probability of state $\mathbf{x} = (x_1, x_2, ..., x_n)$ is

$$\pi_{\mathbf{x}} = \prod_{i=1}^{n} \rho_i^{x_i}(1 - \rho_i); \quad \mathbf{x} \in \mathbf{Z}.$$

Of course, the sequential network can be generalized by taking feedback into account. This is left as an exercise to the reader. □

Example 5.21 Defective robots arrive at the admission's department of a maintenance workshop in accordance with a homogeneous Poisson process with intensity $\lambda = 0.2\ [h^{-1}]$. In the admissions department (denoted as (1)) a first failure diagnosis is done. Depending on the result, the robots will have to visit other departments of the workshop. These are departments for checking and repairing the mechanics (2), electronics (3), and software (4) of the robots, respectively. The failure diagnosis in

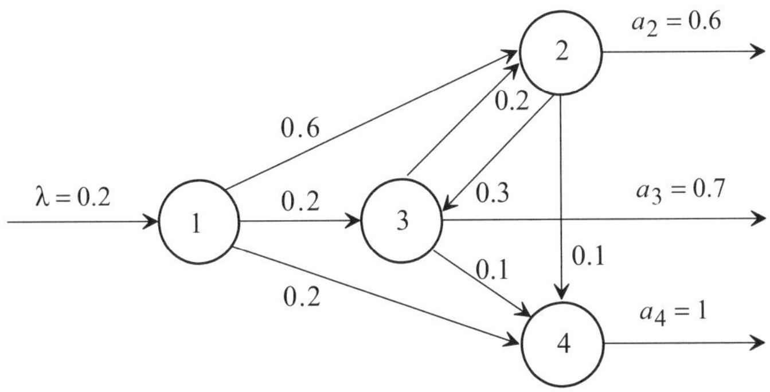

Figure 5.19 Maintenance workshop as a queueing network

the admissions department results in 60% of the arriving robots being sent to department (2) and 20% each to the departments (3) and (4). After having being maintained in department (2), 60% of the robots leave the workshop, 30% are sent to department (3), and 10% to department (4). After having being served by department (3), 70% of the robots leave the workshop, 20% are sent to department (2), and 10% are sent to department (4). After elimination of possible software failures all robots leave the workshop. Naturally, a robot can be sent several times to one and the same department.

The following transition probabilities result from the transfer of robots between the departments:

$$p_{12} = 0.6, \quad p_{13} = 0.2, \quad p_{14} = 0.2, \quad p_{23} = 0.3,$$
$$p_{24} = 0.1, \quad p_{32} = 0.2, \quad p_{34} = 0.1.$$
$$a_1 = 0, \; a_2 = 0.6, \; a_3 = 0.7, \; a_4 = 1.$$

The service intensities are assumed to be

$$\mu_1 = 1, \; \mu_2 = 0.45, \; \mu_3 = 0.4, \; \mu_4 = 0.1 \;\; [h^{-1}].$$

The graph plotted in Figure 5.19 illustrates the possible transitions between the departments. The edges of the graph are weighted by the corresponding transition probabilities. The system of equations (5.106) in the total input intensities is

$$\begin{aligned}
\alpha_1 &= 0.2 \\
\alpha_2 &= 0.6\,\alpha_1 \qquad\qquad + 0.2\,\alpha_3 \\
\alpha_3 &= 0.2\,\alpha_1 + 0.3\,\alpha_2 \\
\alpha_4 &= 0.2\,\alpha_1 + 0.1\,\alpha_2 + 0.1\,\alpha_3
\end{aligned}$$

The solution is (after rounding)

$$\alpha_1 = 0.20, \quad \alpha_2 = 0.135, \quad \alpha_3 = 0.08, \quad \alpha_4 = 0.06.$$

The corresponding traffic intensities $\rho_i = \alpha_i / \mu_i$ are

$$\rho_1 = 0.2, \quad \rho_2 = 0.3, \quad \rho_3 = 0.2, \quad \rho_4 = 0.6.$$

From (5.110), the stationary probability of state $\mathbf{x} = (x_1, x_2, x_3, x_4)$ for single-server nodes is

$$\pi_{\mathbf{x}} = \prod_{i=1}^{4} \rho^{x_i}(1 - \rho_i)$$

or

$$\pi_{\mathbf{x}} = 0.1792\,(0.2)^{x_1}\,(0.3)^{x_2}\,(0.2)^{x_3}\,(0.6)^{x_4}\,; \quad \mathbf{x} \in \mathbf{Z} = \{0, 1, \dots\}^4.$$

In particular, the stationary probability that there is no robot in the workshop is

$$\pi_{\mathbf{x}_0} = 0.1792,$$

where $\mathbf{x}_0 = (0, 0, 0, 0)$. Let X_i denote the random number of robots at node i in the steady state. Then the probability that, in the steady state, there is at least one robot in the admissions department, is

$$P(X_1 > 0) = 0.8 \sum_{i=1}^{\infty} (0.2)^i = 0.2.$$

Analogously,

$$P(X_2 > 0) = 0.3, \; P(X_3 > 0) = 0.2, \; \text{and } P(X_4 > 0) = 0.6.$$

Thus, when there is a delay in servicing defective robots, the cause is most probably department (4) in view of the comparatively high amount of time necessary for finding and removing software failures. □

5.7.6.3 Closed Queueing Networks

Analogously to the closed queueing system, customers cannot enter a *closed queueing network* 'from outside'. Customers which have been served at a node do not leave the network, but move to another node for further service. Hence, the number of customers in a closed queueing network is a constant N. Practical examples for closed queueing networks are multiprogrammed computer and communication systems.

When the service of a customer at node i is finished, then the customer moves with probability p_{ij} to node j for further service. Since the customers do not leave the network,

$$\sum_{j=1}^{n} p_{ij} = 1\,; \quad i = 1, 2, \dots, n, \tag{5.111}$$

where as usual n is the number of nodes. Provided the discrete Markov chain given by transition matrix $\mathbf{P} = ((p_{ij}))$ and state space $\mathbf{Z} = (1, 2, \dots, n\}$ is irreducible, it has a stationary state distribution $\{\pi_1, \pi_2, \dots, \pi_n\}$ which according to (4.9) is the unique solution of the system of equations

$$\pi_j = \Sigma_{i=1}^n p_{ij}\pi_i;\ \ j = 1, 2, \dots, n, \tag{5.112}$$
$$1 = \Sigma_{i=1}^n \pi_i .$$

Let $X_i(t)$ be the random number of customers at node i at time t and

$$\mathbf{X}(t) = (X_1(t), X_2(t), \dots, X_n(t)) .$$

The state space of the Markov chain $\{\mathbf{X}(t),\ t \ge 0\}$ is

$$\mathbf{Z} = \left\{\mathbf{x} = (x_1, x_2, \dots, x_n) \text{ with } \Sigma_{i=1}^n x_i = N \text{ and } 0 \le x_i \le N\right\}, \tag{5.113}$$

where the x_i are nonnegative integers. The number of elements (states) in $\mathbf{Z}$ is

$$\binom{n+N-1}{N}.$$

Let $\mu_i = \mu_i(x_i)$ be the service intensity of all servers at node i if there are x_i customers at this node, $\mu_i(0) = 0$. Then $\{\mathbf{X}(t),\ t \ge 0\}$ has the positive transition rates

$$q_{\mathbf{x},\mathbf{x}-\mathbf{e}_i+\mathbf{e}_j} = \mu_i(x_i)p_{ij};\quad x_i \ge 1,\ i \ne j,$$
$$q_{\mathbf{x}-\mathbf{e}_i+\mathbf{e}_j,\mathbf{x}} = \mu_j(x_j+1)p_{ji};\ i \ne j,\ \ \mathbf{x} - \mathbf{e}_i + \mathbf{e}_j \in \mathbf{Z},$$

where the $\mathbf{e}_i$ are given by (5.108). From (5.111), the rate of leaving state $\mathbf{x}$ is

$$q_{\mathbf{x}} = \Sigma_{i=1}^n \mu_i(x_i)(1-p_{ii}) .$$

Hence, according to (5.28), the stationary distribution $\{\pi_{\mathbf{x}},\ \mathbf{x} \in \mathbf{Z}\}$ of the Markov chain $\{\mathbf{X}(t),\ t \ge 0\}$ satisfies

$$\sum_{i=1}^n \mu_i(x_i)(1-p_{ii})\,\pi_{\mathbf{x}} = \sum_{i,j=1,i\ne j}^n \mu_j(x_j+1)p_{ji}\,\pi_{\mathbf{x}-\mathbf{e}_i+\mathbf{e}_j}, \tag{5.114}$$

where $\mathbf{x} = (x_1, x_2, \dots, x_n) \in \mathbf{Z}$. In these equations, all $\pi_{\mathbf{x}-\mathbf{e}_i+\mathbf{e}_j}$ with $\mathbf{x} - \mathbf{e}_i + \mathbf{e}_j \notin \mathbf{Z}$ are 0. Let $\varphi_i(0) = 1$ and

$$\varphi_i(j) = \prod_{k=1}^j \left(\frac{\pi_i}{\mu_i(k)}\right);\quad i = 1, 2, \dots, n;\ j = 1, 2, \dots, N.$$

Then the stationary probability of state $\mathbf{x} = (x_1, x_2, \dots, x_n) \in \mathbf{Z}$ is

$$\pi_{\mathbf{x}} = h \prod_{i=1}^n \varphi_i(x_i),\quad h = \left[\sum_{\mathbf{y}\in\mathbf{Z}} \prod_{i=1}^n \varphi_i(y_i)\right]^{-1} \tag{5.115}$$

with $\mathbf{y} = (y_1, y_2, \dots, y_n)$. By substituting (5.115) into (5.114) one readily verifies that $\{\pi_{\mathbf{x}},\ \mathbf{x} \in \mathbf{Z}\}$ is indeed a stationary distribution of the Markov chain $\{X(t),\ t \ge 0\}$.

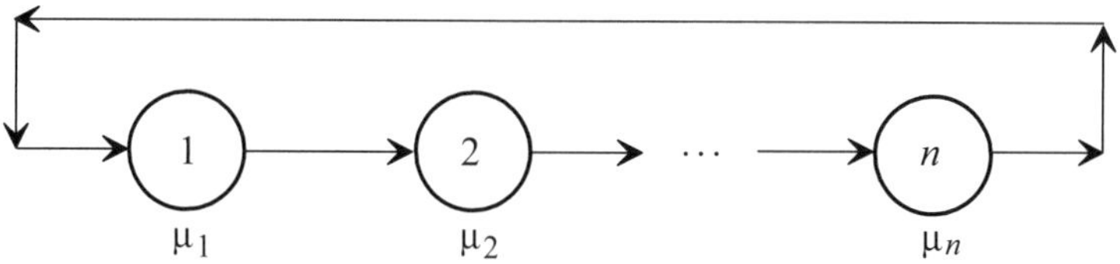

Figure 5.20 Closed sequential queueing network

Example 5.22 A closed sequential queueing network has a single server at each of its n nodes. There is only $N=1$ customer in the system. When this customer is being served at a certain node, the other nodes are empty. Hence, with vectors $\mathbf{e}_i$ as defined by (5.108), the state space of the corresponding Markov chain $\{\mathbf{X}(t), t \geq 0\}$ is $\mathbf{Z}=\{\mathbf{e}_1, \mathbf{e}_2, \ldots, \mathbf{e}_n\}$. The transition probabilities are

$$p_{i,i+1}=1;\;\; i=1,2,\ldots,n-1;\;\; p_{n,1}=1.$$

The corresponding solution of (5.114) is a uniform distribution:

$$\pi_1=\pi_2=\cdots=\pi_n=1/n.$$

Let $\mu_i=\mu_i(1)$ be the service rate at node i. Then

$$\varphi_i(0)=1 \quad \text{and} \quad \varphi_i(1)=\frac{1}{n\mu_i};\quad i=1,2,\ldots,n;$$

$$h=n\left[\sum_{i=1}^n \frac{1}{\mu_i}\right]^{-1}.$$

Hence, the stationary state probabilities are (5.115) are

$$\pi_{\mathbf{e}_i}=\frac{1/\mu_i}{\sum_{i=1}^n \frac{1}{\mu_i}};\quad i=1,2,\ldots,n.$$

In particular, if $\mu_i=\mu$; $i=1,2,\ldots,n$, then the states have a uniform distribution:

$$\pi_{\mathbf{e}_i}=1/n;\quad i=1,2,\ldots,n.$$

If there are $N \geq 1$ customers in the system and the μ_i do not depend on x_i, then the stationary state probabilities are

$$\pi_{\mathbf{x}}=\frac{(1/\mu_1)^{x_1}(1/\mu_2)^{x_2}\cdots(1/\mu_n)^{x_n}}{\sum_{\mathbf{y}\in\mathbf{Z}}\prod_{i=1}^{n}\left(\frac{1}{\mu_i}\right)^{y_i}}$$

where $\mathbf{x}=(x_1,x_2,\ldots,x_n)\in\mathbf{Z}$. Given $\mu_i=\mu$; $i=1,2,\ldots,n$; the states have again a uniform distribution:

$$\pi_{\mathbf{x}}=\frac{1}{\binom{n+N-1}{N}},\quad \mathbf{x}\in\mathbf{Z}.$$

□

Example 5.23 A computer system consists of two central processors 2 and 3, a disc drive 1, and a printer 4. A new program starts in the central processor 2. When this processor has finished its computing job, the computing phase continues in central processor 3 with probability α or the program goes to the disc drive with probability $1-\alpha$. From the disc drive the program goes to central processor 3 with probability 1. From central processor 3 it goes to the central processor 2 with probability β or to the printer with probability $1-\beta$. Here it terminates or goes back to central processor 2. When a program terminates, then another program (from outside) immediately joins the queue of central processor 2 so that there is always a fixed number of programs in the system. Hence, a program formally goes from the printer to the central processor 2 with probability 1. If N denotes the constant number of programs in the system, this situation represents a simple case of *multiprogramming* with N as the *level of multiprogramming.* The state space $\mathbf{Z}$ of this system and the matrix $\mathbf{P}$ of the transition probabilities p_{ij} are

$$\mathbf{Z} = \{\mathbf{y} = (y_1, y_2, y_3, y_4);\ y_i = 0, 1, \dots, N;\ y_1 + y_2 + y_3 + y_4 = N\}$$

and

$$\mathbf{P} = \begin{pmatrix} 0 & 0 & 1 & 0 \\ 1-\alpha & 0 & \alpha & 0 \\ 0 & \beta & 0 & 1-\beta \\ 0 & 1 & 0 & 0 \end{pmatrix}$$

(Figure 5.21). The corresponding solution of (5.114) is

$$\pi_1 = \frac{1-\alpha}{4-\alpha-\beta}, \quad \pi_2 = \pi_3 = \frac{1}{4-\alpha-\beta}, \quad \pi_4 = \frac{1-\beta}{4-\alpha-\beta}.$$

Let the service intensities of the nodes μ_1, μ_2, μ_3 and μ_4 be independent of the number of programs at the nodes. Then,

$$\varphi_i(x_i) = \left(\frac{\pi_i}{\mu_i}\right)^{x_i}, \quad i = 1, 2, \dots, n.$$

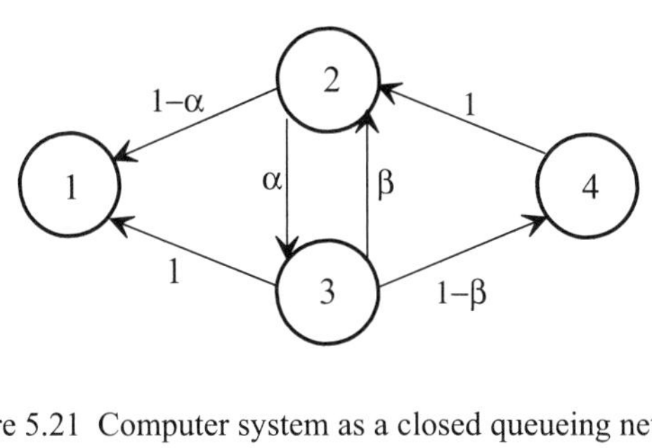

Figure 5.21 Computer system as a closed queueing network

Hence, the stationary probability of state $\mathbf{x} = (x_1, x_2, x_3, x_4)$ with

$$x_1 + x_2 + x_3 + x_4 = N$$

is given by

$$\pi_{\mathbf{x}} = \frac{h}{(4-\alpha-\beta)^N} \left(\frac{1-\alpha}{\mu_1}\right)^{x_1} \left(\frac{1}{\mu_2}\right)^{x_2} \left(\frac{1}{\mu_3}\right)^{x_3} \left(\frac{1-\beta}{\mu_4}\right)^{x_4}$$

with

$$h = \frac{(4-\alpha-\beta)^N}{\sum\limits_{\mathbf{y}\in\mathbf{Z}} \left(\frac{1-\alpha}{\mu_1}\right)^{y_1} \left(\frac{1}{\mu_2}\right)^{y_2} \left(\frac{1}{\mu 3}\right)^{y_3} \left(\frac{1-\beta}{\mu_4}\right)^{y_4}}. \qquad \square$$

Application-oriented treatments of queueing networks are, for instance, Gelenbe and Pujolle [32], Walrand [86].

5.8 SEMI-MARKOV CHAINS

Transitions between the states of a continuous-time homogeneous Markov chain are controlled by its transition probabilities. According to section 5.4, the sojourn time in a state has an exponential distribution and depends on the current state, but not on the history of the process. Since in most applications the sojourn times in system states are non-exponential random variables, an obvious generalization is to allow arbitrarily distributed sojourn times whilst retaining the transition mechanism between the states. This approach leads to the *semi-Markov chains.*

A semi-Markov chain with state space

$$\mathbf{Z} = \{0, 1, ...\}$$

evolves in the following way: Transitions between the states are governed by a discrete-time homogeneous Markov chain $\{X_0, X_1, ...\}$ with state space $\mathbf{Z}$ and matrix of transition probabilities

$$\mathbf{P} = ((p_{ij})).$$

If the process starts at time $t = 0$ in state i_0, then the subsequent state i_1 is determined according to the transition matrix $\mathbf{P}$, while the process stays in state i_0 a random time $Y_{i_0 i_1}$. After that the state i_2 following state i_1 is determined. The process stays in state i_1 a random time $Y_{i_1 i_2}$ and so on. The random variables Y_{ij} are the *conditional sojourn times* of the process in state i given that the process makes a transition from i to j. They are assumed to be independent. Hence, immediately after entering a state at a time t, the further evolvement of a semi-Markov chain depends only on its state at this time point, but not on the evolvement of the process before t. The sample paths of a semi-Markov chain are piecewise constant functions which, by convention, are continuous on the right. In contrast to homogeneous continuous-time Markov chains, for predicting the development of a semi-Markov chain from a time point t it is not only necessary to know its state $i \in \mathbf{Z}$, but also the 'age' of i at time t.

Let $T_0, T_1, ...$ denote the sequence of time points at which the semi-Markov chain makes a transition from one state to another (or to the same state). Then

$$X_n = X(T_n); \quad n = 0, 1, ..., \tag{5.116}$$

where $X_0 = X(0)$ is the initial state ($X_n = X(T_n + 0)$). Hence, the transition probabilities can be written in the following form:

$$p_{ij} = P(X(T_{n+1}) = j | X(T_n) = i); \quad n = 0, 1, ...$$

In view of (5.116), the discrete-time stochastic process $\{X_0, X_1, ...\}$ is *embedded* in the (continuous-time) semi-Markov chain $\{X(t), t \geq 0\}$ (see section 5.4).

As already pointed out, the future development of a semi-Markov chain from a *jump point* T_n is independent of the entire history of the process before T_n. Let

$$F_{ij}(t) = P(Y_{ij} \leq t); \quad i, j \in \mathbf{Z};$$

denote the distribution function of the conditional sojourn time Y_{ij} of a semi-Markov chain in state i if the subsequent state is j. By the total probability rule, the *unconditional sojourn time* Y_i of the chain in state i is

$$F_i(t) = P(Y_i \leq t) = \Sigma_{j \in \mathbf{Z}}\, p_{ij} F_{ij}(t), \quad i \in \mathbf{Z}. \tag{5.117}$$

Special cases 1) An alternating renewal process is a semi-Markov chain with state space $\mathbf{Z} = \{0, 1\}$ and transition probabilities

$$p_{00} = p_{11} = 0 \text{ and } p_{01} = p_{10} = 1.$$

The states 0 and 1 indicate that the system is under renewal or operating, respectively. In this case, $F_{01}(\cdot)$ and $F_{10}(\cdot)$ are in this order the distribution functions of the renewal time and the system lifetime.

2) A homogeneous Markov chain in continuous time with state space $\mathbf{Z} = \{0, 1, ...\}$ is a semi-Markov chain with the same state space and transition probabilities (5.34):

$$p_{ij} = \frac{q_{ij}}{q_i}, \quad i \neq j,$$

where q_{ij} (q_i) are the conditional transition rates (unconditional transition rates) of the Markov chain. By (5.31), the distribution function of the unconditional sojourn time in state i is

$$F_i(t) = 1 - e^{-q_i t}, \quad t \geq 0.$$

In what follows, semi-Markov processes are considered under the following three assumptions:

1) The embedded homogeneous Markov chain $\{X_0, X_1, ...\}$ has a unique stationary state distribution $\{\pi_0, \pi_1, ...\}$. By (4.9), this distribution is a solution of

$$\pi_j = \sum_{i \in \mathbf{Z}} p_{ij} \pi_i, \qquad \sum_{i \in \mathbf{Z}} \pi_i = 1. \tag{5.118}$$

As pointed out in section 4.3, a unique stationary state distribution exists if the Markov chain $\{X_0, X_1, ...\}$ is aperiodic, irreducible and positive recurrent.

2) The distribution functions $F_i(t) = P(Y_i \le t)$ are nonarithmetic. (As defined in section 3.3.3, a distribution function $F(t)$ is called *arithmetic* if there is a constant a with property that all points of increase of $F(t)$ have structure $t = a\,n;\ \ n = 0, 1, ...$ Otherwise, the distribution function is nonarithmetic.)

3) The mean sojourn times of the process in all states are finite:

$$\mu_i = E(Y_i) = \int_0^\infty [1 - F_i(t)]\, dt < \infty, \ \ i \in \mathbf{Z}.$$

Note μ_i denotes no longer an intensity, but a mean sojourn time.

In what follows, a transition of the semi-Markov chain into state k is called a *k-transition*. Let $N_k(t)$ be the random number of k-transitions occuring in $[0, t]$ and

$$H_k(t) = E(N_k(t)).$$

Then, for any $t > 0$,

$$\lim_{t \to \infty} [H_k(t + \tau) - H_k(t)] = \frac{\tau \pi_k}{\sum_{i \in \mathbf{Z}} \pi_i \mu_i}, \qquad k \in \mathbf{Z}. \tag{5.119}$$

This relationship implies that after a sufficiently long time period the number of k-transitions in a given time interval no longer depends on the position of this interval, but only on its length. Strictly speaking, the right-hand side of (5.119) gives the mean number of k-transitions in an interval of length τ once the process hasreached its stationary regime, or, in other words, if it is in the steady state. The following formulas and the analysis of examples is based on (5.119), but the definition and properties of stationary semi-Markov chains will not be discussed in detail.

From (5.119), when the process is in the steady state, the mean number of k-transitions per unit time is

$$U_k = \frac{\pi_k}{\sum_{i \in \mathbf{Z}} \pi_i \mu_i}.$$

Hence the portion of time the chain is in state k is

$$A_k = \frac{\pi_k \mu_k}{\sum_{i \in \mathbf{Z}} \pi_i \mu_i}. \tag{5.120}$$

Consequently, in the long run, the fraction of time the chain is in a set of states $\mathbf{Z}_0$, $\mathbf{Z}_0 \subseteq \mathbf{Z}$, is

$$A_{\mathbf{Z}_0} = \frac{\sum_{k \in \mathbf{Z}_0} \pi_k \mu_k}{\sum_{i \in \mathbf{Z}} \pi_i \mu_i} . \tag{5.121}$$

In other words, $A_{\mathbf{Z}_0}$ is the probability that a visitor, who arrives at a random time from 'outside', finds the semi-Markov chain in a state belonging to $\mathbf{Z}_0$.

Let c_k denote the cost which is caused by a k-transition of the system. Then the mean total (transition) cost per unit time is

$$C = \frac{\sum_{k \in \mathbf{Z}} \pi_k c_k}{\sum_{i \in \mathbf{Z}} \pi_i \mu_i} . \tag{5.122}$$

Note that the formulas (5.119) to (5.122) depend only on the unconditional sojourn times of a semi-Markov chain in its states. This property facilitates their application.

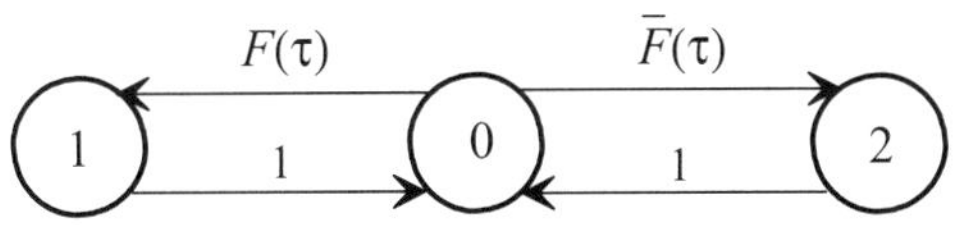

Figure 5.22 Transition graph for example 5.24

Example 5.24 ***(age renewal policy)*** The system is renewed upon failure by an *emergency renewal* or at age τ by a *preventive renewal*, whichever occurs first.

To determine the stationary system availability, system states have to be introduced:

0 operating
1 emergency renewal
2 preventive renewal

Let L be the random system lifetime, $F(t) = P(L \le t)$ its distribution function, and

$$\overline{F}(t) = 1 - F(t) = P(L > t)$$

its survival probability. Then the positive transition probabilities between the states are (Figure 5.22)

$$p_{01} = F(\tau), \;\; p_{02} = \overline{F}(\tau), \;\; p_{10} = p_{20} = 1 .$$

Let Z_e and Z_p be the random times for emergency renewals and preventive renewals, respectively. Then the conditional sojourn times of the system in the states are

$$Y_{01} = L, \;\; Y_{02} = \tau, \;\; Y_{10} = Z_e, \;\; Y_{20} = Z_p .$$

The unconditional sojourn times are

$$Y_0 = \min(L, \tau), \;\; Y_1 = Z_e, \;\; Y_2 = Z_p .$$

The system behaviour can be described by a semi-Markov chain $\{X(t), t \geq 0\}$ with state space $\mathbf{Z} = \{0, 1, 2\}$ and the transition probabilities and sojourn times given. The corresponding equations (5.118) in the stationary probabilities of the embedded Markov chain are

$$\begin{aligned} \pi_0 &= \pi_1 + \pi_2 \\ \pi_1 &= F(\tau)\,\pi_0 \\ 1 &= \pi_0 + \pi_1 + \pi_2 \end{aligned}$$

The solution is

$$\pi_0 = 1/2, \quad \pi_1 = F(\tau)/2, \quad \pi_2 = \bar{F}(\tau)/2.$$

The mean sojourn times are

$$\mu_0 = \int_0^\tau \bar{F}(t)\,dt, \quad \mu_1 = d_e, \quad \mu_2 = d_p.$$

According to (5.120), the stationary availability $A_0 = A(\tau)$ of the system is

$$A(\tau) = \frac{\mu_0 \pi_0}{\mu_0 \pi_0 + \mu_1 \pi_1 + \mu_2 \pi_2}$$

or

$$A(\tau) = \frac{\int_0^\tau \bar{F}(t)\,dt}{\int_0^\tau \bar{F}(t)\,dt + d_e F(\tau) + d_p \bar{F}(\tau)}. \tag{5.123}$$

It is important that this result does not depend on the probability distributions of the renewal times Z_e and Z_p, but only on their mean values (see also example 3.21).

If the renewal times are negligibly small, but the mean costs c_e and c_p for emergency and preventive renewals are relevant, then, from (5.122), the mean renewal cost per unit time in the steady state is

$$K(\tau) = \frac{c_e \pi_1 + c_p \pi_2}{\mu_0 \pi_0} = \frac{c_e F(\tau) + c_p \bar{F}(\tau)}{\int_0^\tau \bar{F}(t)\,dt}.$$

Analogously to the corresponding renewal times, c_e and c_p can be thought of as mean values of arbitrarily distributed renewal costs.

If $\lambda(t)$ is the failure rate of the system, a cost-optimal renewal interval $\tau = \tau^*$ satisfies the necessary condition

$$\lambda(\tau) \int_0^\tau \bar{F}(t)\,dt - F(\tau) = \frac{c}{1-c}$$

with $c = c_p/c_e$. A unique solution $\tau = \tau^*$ exists if $c < 1$ and $\lambda(t)$ strictly increases to infinity. Since $K(\tau)$ has the same functional structure as

$$1/A(\tau) - 1,$$

maximizing $A(\tau)$ and minimizing $K(\tau)$ leads to the same equation type for determining the corresponding optimal renewal intervals. □

Example 5.25 A series system consists of n subsystems $e_1, e_2, \ldots, e_n$. The lifetimes of the subsystems $L_1, L_2, \ldots, L_n$ are independent exponential random variables with parameters $\lambda_1, \lambda_2, \ldots, \lambda_n$. Let

$$G_i(t) = P(L_i \le t) = 1 - e^{-\lambda_i t}, \quad g_i(t) = \lambda_i e^{-\lambda_i t}, \quad t \ge 0;\ i = 1, 2, \ldots, n.$$

When a subsystem fails, the system interrupts its work. As soon as the renewal of the failed subsystem is finished, the system continues operating. Let μ_i be the average renewal time of subsystem e_i. As long as a subsystem is being renewed, the other subsystems cannot fail, i.e. during such a time period they are in the cold-standby mode. The following system states are introduced:

$X(t) = 0$ if the system is operating,

$X(t) = i$ if e_i is under renewal, $i = 1, 2, \ldots, n$.

Then $\{X(t), t \ge 0\}$ is a semi-Markov chain with state space $\mathbf{Z} = \{0, 1, \ldots, n\}$. The conditional sojourn times in state 0 of this semi-Markov chain are

$$Y_{0i} = L_i, \quad i = 1, 2, \ldots, n,$$

and its unconditional sojourn time in state 0 is

$$Y_0 = \min\{L_1, L_2, \ldots, L_n\}.$$

Thus, Y_0 has distribution function

$$F_0(t) = 1 - \bar{G}_1(t) \cdot \bar{G}_2(t) \cdots \bar{G}_n(t).$$

Letting $\lambda = \lambda_1 + \lambda_2 + \cdots + \lambda_n$ implies

$$F_0(t) = 1 - e^{-\lambda t}, \quad t \ge 0,$$

$$\mu_0 = E(Y_0) = 1/\lambda.$$

The system makes a transition from state 0 into state i with probability

$$p_{0i} = P(Y_0 = L_i)$$

$$= \int_0^\infty \bar{G}_1(x) \cdot \bar{G}_2(x) \cdots \bar{G}_{i-1}(x) \cdot \bar{G}_{i+1}(x) \cdots \bar{G}_n(x) g_i(x)\, dx$$

$$= \int_0^\infty e^{-(\lambda_1 + \lambda_2 + \cdots + \lambda_{i-1} + \lambda_{i+1} + \cdots + \lambda_n)x} \lambda_i e^{-\lambda_i x}\, dx$$

$$= \int_0^\infty e^{-\lambda x} \lambda_i\, dx.$$

Hence,

$$p_{0i} = \frac{\lambda_i}{\lambda}, \quad p_{i0} = 1; \quad i = 1, 2, \ldots, n.$$

Thus, the system of equations (5.118) becomes

$$\pi_0 = \pi_1 + \pi_2 + \cdots + \pi_n,$$

$$\pi_i = \frac{\lambda_i}{\lambda} \pi_0; \quad i = 1, 2, \ldots, n.$$

In view of $\pi_1 + \pi_2 + \cdots + \pi_n = 1 - \pi_0$, the solution is

$$\pi_0 = \frac{1}{2}; \quad \pi_i = \frac{\lambda_i}{2\lambda}; \quad i = 1, 2, \ldots, n.$$

With these ingredients, formula (5.20) yields the stationary system availability

$$A_0 = \frac{1}{1 + \sum_{i=1}^{n} \lambda_i \mu_i}. \qquad \square$$

Example 5.26 Consider the loss system $M/G/1/0$ on condition that the server is subjected to failures: Customers arrive according to a homogeneous Poisson process with rate λ. Hence, their interarrival times are identically distributed as an exponential random variable Y with parameter λ. The server has random lifetime L_0 when being idle, and random lifetime L_1 when being busy. L_0 is exponential with parameter λ_0 and L_1 is exponential with parameter λ_1. The service time Z has distribution function $B(t)$ with density $b(t)$. When at the time point of server failure a customer is being served, then this customer is lost, i.e. it has to leave the system. All occurring random variables are assumed to be independent (section 5.7.5.2). To describe the behaviour of this system by a semi-Markov chain, three states are introduced:

State 0 The server is idle, but available.

State 1 The server is busy.

State 2 The server is under repair (not available).

To determine the steady state probabilities of the states 0, 1 and 2, the transition probabilities p_{ij} are needed:

$$p_{00} = p_{11} = p_{22} = p_{21} = 0, \quad p_{20} = 1$$

$$p_{01} = P(L_0 > Y) = \int_0^\infty e^{-\lambda_0 t} \lambda e^{-\lambda t} dt = \frac{\lambda}{\lambda + \lambda_0}$$

$$p_{02} = 1 - p_{01} = P(L_0 \le Y) = \frac{\lambda_0}{\lambda + \lambda_0}$$

$$p_{10} = P(L_1 > Z) = \int_0^\infty e^{-\lambda_1 t}\, b(t)\, dt$$

$$p_{12} = 1 - p_{10} = P(L_1 \le Z) = \int_0^\infty [1 - e^{-\lambda_1 t}]\, b(t)\, dt.$$

With these transition probabilities, the stationary state probabilities of the embedded Markov chain $\{X_0, X_1, \ldots\}$ can be obtained from (5.118):

$$\pi_0 = \frac{\lambda + \lambda_0}{2(\lambda + \lambda_0) + \lambda p_{12}}, \quad \pi_1 = \frac{\lambda}{2(\lambda + \lambda_0) + \lambda p_{12}}, \quad \pi_2 = \frac{\lambda_0 + \lambda p_{12}}{2(\lambda + \lambda_0) + \lambda p_{12}}.$$

The sojourn times in state 0, 1 and 2 are:

$$Y_0 = \min\,(L_0, Y), \quad Y_1 = \min\,(L_1, Z), \quad Y_2 = Z.$$

Hence, the mean sojourn times are

$$\mu_0 = \frac{1}{\lambda + \lambda_0}, \quad \mu_1 = \int_0^\infty (1 - B(t))\, e^{-\lambda_1 t} dt, \quad \mu_2 = E(Z).$$

With these parameters, the stationary state probabilities of the semi-Markov process are given by (5.120). □

The time-dependent behaviour of semi-Markov chains is discussed, for instance, in Kulkarni [52].

5.9 EXERCISES

5.1) Let $\mathbf{Z} = \{0, 1\}$ be the state space and

$$\mathbf{P}(t) = \begin{pmatrix} e^{-t} & 1 - e^{-t} \\ 1 - e^{-t} & e^{-t} \end{pmatrix}$$

the transition matrix of a continuous-time stochastic process $\{X(t), t \geq 0\}$. Check whether $\{X(t), t \geq 0\}$ is a homogeneous Markov chain.

5.2) A system fails after a random lifetime L. Then it waits a random time W for renewal. A renewal takes another random time Z. The random variables L, W and Z have exponential distributions with parameters λ, ν and μ, respectively. On completion of a renewal, the system immediately resumes its work. This process continues indefinitely. All life, waiting, and renewal times are assumed to be independent. Let the system be in states 0, 1 and 2 when it is operating, waiting or being renewed..

(1) Draw the transition graph of the corresponding Markov chain $\{X(t), t \geq 0\}$.

(2) Determine the point and the stationary availability of the system on condition

$$P(X(0) = 0) = 1.$$

5.3) Consider a 1-out-of-2-system, i.e. the system is operating when at least one of its two subsystems is operating. When a subsystem fails, the other one continues to work. On its failure, the joint renewal of both subsystems begins. On its completion, both subsystems resume their work at the same time. The lifetimes of the subsystems are identically exponential with parameter λ. The joint renewal time is exponential with parameter μ. All life and renewal times are independent of each other. Let $X(t)$ be the number of subsystems operating at time t.

(1) Draw the transition graph of the corresponding Markov chain $\{X(t), t \geq 0\}$.

(2) Given $P(X(0) = 2) = 1$, determine the time-dependent state probabilities

$$p_i(t) = P(X(t) = i); \ i = 0, 1, 2.$$

(3) Determine the stationary state distribution.

Hint Consider separately the cases

$$(\lambda + \mu + \nu)^2 (=)(<)(>)\ 4(\lambda\mu + \lambda\nu + \mu\nu).$$

5.4) A launderette has 10 washing machines which are in constant use. The times between two successive failures of a washing machine have an exponential distribution with mean value 100 *hours*. There are two mechanics who repair failed machines. A defective machine is repaired by only one mechanic. During this time, the second mechanic is busy repairing another failed machine, if there is any, or this mechanic is idle. All repair times have an exponential distribution with mean value 4 *hours*. All random variables involved are independent. Consider the steady state.

1) What is the average percentage of operating machines?

2) What is the average percentage of idle mechanics?

5.5) Consider the two-unit system with standby redundancy discussed in example 5.5 a) on condition that the lifetimes of the units are exponential with respective parameters λ_1 and λ_2. The other model assumptions listed in example 5.5 remain valid.

Describe the behaviour of the system by a Markov chain and draw the transition graph.

5.6) Consider the two-unit system with parallel redundancy discussed in example 5.6 on condition that the lifetimes of the units are exponential with parameters λ_1 and λ_2, respectively. The other model assumptions listed in example 5.6 remain valid.

Describe the behaviour of the system by a Markov chain and draw the transition graph.

5.7) The system considered in example 5.7 is generalized as follows: If the system makes a direct transition from state 0 to the blocking state 2, then the subsequent renewal time is exponential with parameter μ_0. If the system makes a transition from state 1 to state 2, then the subsequent renewal time is exponential with parameter μ_1.

(1) Describe the behaviour of the system by a Markov chain and draw the transition graph.

(2) What is the stationary probability that the system is blocked?

5.8) Consider a two-unit system with standby redundancy and one mechanic. All repair times of failed units have an Erlang distribution with parameters $n = 2$ and μ. Apart from this, the other model assumptions listed in example 5.5 remain valid.

(1) Describe the behaviour of the system by a Markov chain and draw the transition graph.

(2) Determine the stationary state probabilities of the system.

(3) Sketch the stationary availability of the system as a function of

$$\rho = \lambda/\mu.$$

5.9) When being in states 0, 1, and 2 a (pure) birth process $\{X(t), t \geq 0\}$ with state space $\mathbf{Z} = \{0, 1, 2, ...\}$ has birth rates

$$\lambda_0 = 2, \ \lambda_1 = 3, \ \lambda_2 = 1.$$

Given $X(0) = 0$, determine the time-dependent state probabilities

$$p_i(t) = P(X(t) = i) \ \text{ for the states } i = 0, 1, 2.$$

5.10) Consider a linear birth process with birth rates

$$\lambda_j = j\lambda, \ j = 0, 1, ...,$$

and state space $\mathbf{Z} = \{0, 1, 2, ...\}$.

(1) Given $X(0) = 1$, determine the distribution function of the random time point T_3 at which the process enters state 3.

(2) Given $X(0) = 1$, determine the mean value of the random time point T_n at which the process enters state n, $n > 1$.

5.11) The number of physical particles of a particular type in a closed container evolves as follows: There is one particle at time $t = 0$. Its splits into two particles of the same type after an exponential random time Y with parameter λ (its lifetime). These two particles behave in the same way as the original one, i.e. after random times, which are identically distributed as Y, they split into 2 particles each, and so on. All lifetimes of the particles are assumed to be independent. Let $X(t)$ denote the number of particles in the container at time t.

Determine the absolute state probabilities

$$p_j(t) = P(X(t) = j); \ j = 1, 2, ...$$

of the stochastic process $\{X(t), t \geq 0\}$.

5.12) A death process with state space $\mathbf{Z} = \{0, 1, 2, ...\}$ has death rates

$$\mu_0 = 0, \ \mu_1 = 2, \text{ and } \mu_2 = \mu_3 = 1.$$

Given $X(0) = 3$, determine $p_j(t) = P(X(t) = j)$ for $j = 0, 1, 2, 3$.

5.13) A linear death process $\{X(t), t \geq 0\}$ has death rates

$$\mu_j = j\mu; \ j = 0, 1, ...$$

(1) Given $X(0) = 2$, determine the distribution function of the time to entering state 0 ('lifetime' of the process).

(2) Given $X(0) = n$, $n > 1$, determine the mean value of the time at which the process enters state 0.

5.14) At time $t = 0$ there are an infinite number of molecules of type a and $2n$ molecules of type b in a two-component gas mixture. After an exponential random time with parameter μ any molecule of type b combines, independently of the others, with a molecule of type a to form a molecule ab.

(1) What is the probability that at time t there are still j free molecules of type b in the container?

(2) What is the mean time till there are left only n free molecules of type b in the container?

5.15) At time $t = 0$ a cable consists of 5 identical, intact wires. The cable is subject to a constant load of 100 *kp* such that in the beginning each wire bears a load of 20 *kp*. Given a load of w *kp* per wire, the time to breakage of a wire (its lifetime) is exponential with mean value

$$\frac{1000}{w} \; [weeks].$$

When one or more wires are broken, the load of 100 *kp* is uniformly distributed over the remaining intact ones. For any fixed number of wires, their lifetimes are assumed to be independent and identically distributed.

(1) What is the probability that all wires are broken at time $t = 50\,[weeks]$?

(2) What is the mean time until the cable breaks completely?

5.16)* Let $\{X(t),\, t \ge 0\}$ be a death process with $X(0) = n$ and positive death rates

$$\mu_1, \mu_2, \ldots, \mu_n.$$

Prove: If Y is an exponential random variable with parameter λ and independent of the death process, then

$$P(X(Y) = 0) = \prod_{i=1}^{n} \frac{\mu_i}{\mu_i + \lambda}.$$

5.17) Let a birth- and death process have state space $\mathbf{Z} = \{0, 1, \ldots, n\}$ and transition rates

$$\lambda_j = (n - j)\,\lambda \;\text{ and }\; \mu_j = j\,\mu; \;\; j = 0, 1, \ldots, n.$$

Determine its stationary state probabilities.

5.18) Check whether or under what restrictions a birth- and death process with transition rates

$$\lambda_j = \frac{j+1}{j+2}\lambda \;\text{ and }\; \mu_j = \mu; \;\; j = 0, 1, \ldots,$$

has a stationary state distribution.

5.19) A birth- and death process has transition rates

$$\lambda_j = (j+1)\lambda \;\text{ and }\; \mu_j = j^2\mu; \; j = 0, 1, \ldots; \;\; 0 < \lambda < \mu.$$

Confirm that this process has a stationary state distribution and determine it.

5.20) A computer is connected to three terminals (for example, measuring devices). It can simultaneously evaluate data records from only two terminals. When the computer is processing two data records and in the meantime another data record has been produced, then this new data record has to wait in a buffer when the buffer is empty. Otherwise the new data record is lost. (The buffer can store only one data record.) The data records are processed according to the FCFS-queueing discipline. The terminals produce data records independently according to a homogeneous Poisson process with intensity λ. The processing times of data records from all terminals are independent (even if the computer is busy with two data records at the same time) and have an exponential distribution with parameter μ. They are assumed to be independent of the input. Let $X(t)$ be the number of data records in computer and buffer at time t.

(1) Verify that $\{X(t), t \geq 0\}$ is a birth- and death process, determine its transition rates and draw the transition graph.

(2) Determine the stationary loss probability, i.e. the probability that, in the steady state, a data record is lost.

5.21) Under otherwise the same assumptions as in exercise 5.20, it is assumed that a data record which has been waiting in the buffer a random *patience time*, will be deleted as being no longer up to date. The patience times of all data records are assumed to be independent, exponential random variables with parameter ν. They are also independent of all arrival and processing times of the data records.

Determine the stationary loss probability.

5.22) Under otherwise the same assumptions as in exercise 5.21, it is assumed that a data record will be deleted when its total sojourn time in the buffer <u>and</u> computer exceeds a random time Z, where Z has an exponential distribution with parameter α. Thus, the interruption of a current service of a data record is possible.

Determine the stationary loss probability.

5.23) A small filling station in a rural area provides diesel for agricultural machines. It has one diesel pump and waiting capacity for 5 machines. On average, 8 machines per hour arrive for diesel. An arriving machine immediately leaves the station without fuel if pump and all waiting places are occupied. The mean time a machine occupies the pump is 5 *minutes*. It is assumed that the station behaves like an *M/M/s/m*-queueing system.

(1) Determine the stationary loss probability.

(2) Determine the stationary probability that an arriving machine waits for diesel.

5.24) Consider a two-server loss system. Customers arrive according to a homogeneous Poisson process with intensity λ. A customer is always served by server 1 when this server is idle, i.e. an arriving customer goes only then to server 2, when server 1 is busy. The service times of both servers are idd exponential random variables with parameter μ. Let $X(t)$ be the number of customers in the system at time t.

Determine the stationary state probabilities of the stochastic process $\{X(t), t \geq 0\}$.

5.25) A 2-server loss system is subject to a homogeneous Poisson input with intensity λ. The situation considered in the previous exercise is generalized as follows: If both servers are idle, a customer goes to server 1 with probability p and to server 2 with probability $1-p$. Otherwise, a customer goes to the idle server (if there is any). The service times of the servers 1 and 2 are independent, exponential random variables with parameters μ_1 and μ_2, respectively. All arrival and service times are independent.

Describe the behaviour of the system by a suitable homogeneous Markov chain and draw the transition graph.

5.26) A single-server waiting system is subject to a homogeneous Poisson input with intensity

$$\lambda = 30\,[h^{-1}].$$

If there are not more than 3 customers in the system, the service times have an exponential distribution with mean $1/\mu = 2$ [*min*]. If there are more than 3 customers in the system, the service times are exponential with mean $1/\mu = 1$ [*min*]. All arrival and service times are independent.

(1) Show that there exists a stationary state distribution and determine it.

(2) Determine the mean length of the waiting queue in the steady state.

5.27) Taxis and customers arrive at a taxi rank in accordance with two independent homogeneous Poisson processes with intensities $\lambda_1 = 4$ an hour and $\lambda_2 = 3$ an hour, respectively. Potential customers who find two waiting customers do not wait for service, but leave the rank immediately. (Groups of customers who will use the same taxi are considered to be one customer.) On the other hand, arriving taxis who find two taxis waiting leave the rank as well.

What is the average number of customers waiting at the rank?

5.28) A transport company has 4 trucks of the same type. There are 2 maintenance teams for repairing the trucks after a failure. Each team can repair only one truck at a time and each failed truck is handled by only one team. The times between failures of a truck (lifetime) is exponential with parameter λ. The repair times are exponential with parameter μ. All life and repair times are assumed to be independent. Let

$$\rho = \lambda/\mu = 0.2.$$

What is the most efficient way of organizing the work: 1) to make both maintenance teams responsible for the maintenance of all 4 trucks so that any team which is free can repair any failed truck, or 2) to assign 2 definite trucks to each team?

5.29) Ferry boats and customers arrive at a ferry station in accordance with two independent homogeneous Poisson processes with intensities λ and μ, respectively. If there are k customers at the ferry station, when a boat arrives, then it departs with $\min(k, n)$ passengers (n is the capacity of each boat). If $k > n$, then the remaining $k - n$ customers wait for the next boat. The sojourn times of the boats at the station are assumed to be negligibly small.

Model the situation by a suitable homogeneous Markov chain $\{X(t), t \geq 0\}$ and draw the transition graph.

5.30) The life cycle of an organism is controlled by shocks (e.g. virus attacks, accidents) in the following way: A healthy organism has an exponential lifetime L with parameter λ_h. If a shock occurs, the organism falls sick and, when being in this state, its (residual) lifetime S is exponential with parameter

$$\lambda_s, \ \lambda_s > \lambda_h.$$

However, a sick organism may recover and return to the healthy state. This occurs in an exponential time R with parameter μ. If during a period of sickness another shock occurs, the organism cannot recover and will die a random time D after the occurrence of the second shock. D is assumed to be exponential with parameter

$$\lambda_d, \ \lambda_d > \lambda_s.$$

The random variables L, S, R, and D are assumed to be independent.

(1) Describe the evolvement in time of the states the organism may be in by a Markov chain.

(2) Determine the mean lifetime of the organism.

5.31) Customers arrive at a waiting system of type $M/M/1/\infty$ with intensity λ. As long as there are less than n customers in the system, the server remains idle. As soon as the nth customer arrives, the server resumes its work and stops working only then, when all customers (including newcomers) have been served. After that the server again waits until the waiting queue has reached length n and so on. Let $1/\mu$ be the mean service time of a customer and $X(t)$ be the number of customers in the system at time t.

(1) Draw the transition graph of the Markov chain $\{X(t), t \geq 0\}$.

(2) Given that $n = 2$, compute the stationary state probabilities. (Make sure that they exist.)

5.32) At time $t = 0$ a computer system consists of n operating computers. As soon as a computer fails, it is separated from the system by an automatic switching device with probability $1 - p$. If a failed computer is not separated from the system (this happens with probability p), then the entire system fails. The lifetimes of the computers are independent and have an exponential distribution with parameter λ. Thus, this distribution does not depend on the system state. Provided the switching device has operated properly when required, the system is available as long as there is at least one computer available. Let $X(t)$ be the number of computers which are available at time t. By convention, if, due to the switching device, the entire system has failed in $[0, t)$, then $X(t) = 0$.

(1) Draw the transition graph of the Markov chain $\{X(t), t \geq 0\}$.

(2) Given $n = 2$, determine the mean lifetime $E(X_s)$ of the system.

5.33) A waiting-loss system of type $M/M/1/2$ is subject to two independent Poisson inputs 1 and 2 with respective intensities λ_1 and λ_2 (type 1- and type 2-customers). An arriving type 1-customer who finds the server busy and the waiting places occupied displaces a possible type 2-customer from its waiting place (such a type 2-customer is lost), but ongoing service of a type 2-customer is not interrupted. When a type 1-customer and a type 2-customer are waiting, then the type 1-customer will always be served first, regardless of the order of their arrivals. The service times of type 1- and type 2-customers are independent and have exponential distributions with respective parameters μ_1 and μ_2.

Describe the behaviour of the system by a homogeneous Markov chain, determine the transition rates, and draw the transition graph.

5.34) A queueing network consists of two servers 1 and 2 in series. Server 1 is subject to a homogeneous Poisson input with intensity $\lambda = 5$ an hour. A customer is lost if server 1 is busy. From server 1 a customer goes to server 2 for further service. If server 2 is busy, the customer is lost. The service times of servers 1 and 2 are exponential with respective mean values

$$1/\mu_1 = 6 \text{ minutes} \text{ and } 1/\mu_2 = 12 \text{ minutes}.$$

All arrival and service times are independent.

What percentage of customers (with respect to the total input at server 1) is served by both servers?

5.35) A queueing network consists of three nodes (queueing systems) 1, 2 and 3, each of type $M/M/1/\infty$. The external inputs into the nodes have respective intensities

$$\lambda_1 = 4, \; \lambda_2 = 8, \text{ and } \lambda_3 = 12$$

customers an hour. The respective mean service times at the nodes are 4, 2 and 1 [*min*]. After having been served by node 1, a customer goes to nodes 2 or 3 with equal probabilities 0.4 or leaves the system with probability 0.2. From node 2, a customer goes to node 3 with probability 0.9 or leaves the system with probability 0.1. From node 3, a customer goes to node 1 with probability 0.2 or leaves the system with probability 0.8. The external inputs and the service times are independent.

(1) Check whether this queueing network is a Jackson network.

(2) Determine the stationary state probabilities of the network.

5.36) A closed queueing network consists of 3 nodes. Each one has 2 servers. There are 2 customers in the network. After having been served at a node, a customer goes to one of the others with equal probability. All service times are independent random variables and have an exponential distribution with parameter μ.

What is the stationary probability to find both customers at the same node?

5.37) Depending on demand, a conveyor belt operates at 3 different speed levels 1, 2, and 3. A transition from level i to level j is made with probability p_{ij} with

$$p_{12} = 0.8\,,\ p_{13} = 0.2\,,$$
$$p_{21} = p_{23} = 0.5\,,$$
$$p_{31} = 0.4\,,\ p_{32} = 0.6\,.$$

The respective mean times the conveyor belt operates at levels 1, 2, or 3 between transitions are

$$\mu_1 = 45\,,\ \mu_2 = 30\,,\ \text{and}\ \mu_3 = 12\ [hours].$$

Determine the stationary percentages of time in which the conveyor belt operates at levels 1, 2, and 3 by modeling the situation as a semi-Markov chain.

5.38) The mean lifetime of a system is 620 hours. There are two failure types: Repairing the system after a type 1-failure requires 20 hours on average and after a type 2-failure 40 hours on average. 20% of all failures are type 2-failures. There is no dependence between the system lifetime and the subsequent failure type. Upon each repair the system is 'as good as new'. The repaired system immediately resumes its work. This process is continued indefinitely. All life and repair times are independent.

(1) Describe the situation by a semi-Markov chain with 3 states and draw the transition graph of the underlying discrete-time Markov chain.

(2) Determine the stationary state probabilities of the system.

5.39) A system has two different failure types: type 1 and type 2. After a type i-failure the system is said to be in failure state i; $i = 1, 2$. The time L_i to a type i-failure has an exponential distribution with parameter

$$\lambda_i;\ \ i = 1, 2.$$

Thus, if at time $t = 0$ a new system starts working, the time to its first failure is

$$Y_0 = \min\,(L_1, L_2).$$

The random variables L_1 and L_2 are assumed to be independent. After a type 1-failure, the system is switched from failure state 1 into failure state 2. The respective mean sojourn times of the system in states 1 and 2 are μ_1 and μ_2. When in state 2, the system is being renewed. Thus, μ_1 is the mean switching time and μ_2 the mean renewal time. A renewed system immediately starts working, i.e. the system makes a transition from state 2 to state 0 with probability 1. This process continues to infinity. (For motivation, see example 5.7).

(1) Describe the system behaviour by a semi-Markov chain and draw the transition graph of the embedded discrete-time Markov chain.

(2) Determine the stationary probabilities of the system in the states 0, 1, and 2.

5.40) Under otherwise the same model assumptions as in example 5.26, determine the stationary probabilities of the states 0, 1, and 2 introduced there on condition that the service time B is a constant μ; i.e. determine the stationary state probabilities of the loss system $M/D/1/0$ with unreliable server.

CHAPTER 6

Martingales

6.1 DISCRETE-TIME MARTINGALES

6.1.1 Definition and Examples

Martingales are important tools for solving prestigious problems in probability theory and its applications. Such problems occur in areas like random walks, point processes, mathematical statistics, actuarial risk analysis, and mathematics of finance. Heuristically, martingales are stochastic models for 'fair games' in a wider sense, i.e. games in which each 'participant' has the same chance to win and to lose. In particular, *martingale* is the French word for that fair game in which a gambler doubles his bet on every loss until he wins. Martingales were introduced as a special class of stochastic processes by *J. Ville* und *P. Levy*. However, it was *J. L. Doob* (1953) who began with their systematic investigation and who recognized their large theoretical and practical potential. Martingales as stochastic processes are defined for discrete and continuous parameter spaces **T**. Analogously to Markov processes, the terminology *discrete-time martingale* and *continuous-time martingale* is adopted. The definition of a martingale relies heavily on the concept of the conditional mean value of a random variable given values of other random variables or, more generally, on the concept of the (random) conditional mean value of a random variable given other random variables (section 1.6).

Definition 6.1 A stochastic process in discrete time $\{X_0, X_1, ...\}$ with state space $\mathbf{Z}$, which satisfies $E(|X_n|) < \infty$, $n = 0, 1, 2, ...,$ is called a (*discrete-time*) *martingale* if for all vectors $(x_0, x_1, ..., x_n)$ with $x_i \in \mathbf{Z}$ and $n = 0, 1, ...$

$$E(X_{n+1} \,|\, X_n = x_n, ..., X_1 = x_1, X_0 = x_0) = x_n. \tag{6.1}$$

Under the same assumptions, $\{X_0, X_1, ...\}$ is a (*discrete-time*) *supermartingale* if

$$E(X_{n+1} \,|\, X_n = x_n, ..., X_1 = x_1, X_0 = x_0) \le x_n, \tag{6.2}$$

and a (*discrete-time*) *submartingale* if

$$E(X_{n+1} \,|\, X_n = x_n, ..., X_1 = x_1, X_0 = x_0) \ge x_n. \tag{6.3}$$

●

If, for instance, the X_n are continuous random variables, then, in view of (1.75), multiplying both sides of the (in-) equalities (6.1) to (6.3) by the joint density of the random vector $(X_0, X_1, ..., X_n)$ and integrating over its range yields:

Martingale: $E(X_{n+1}) = E(X_n);\ n = 0, 1, ...$

Supermartingale: $E(X_{n+1}) \leq E(X_n);\ n = 0, 1, ...$

Submartingale: $E(X_{n+1}) \geq E(X_n);\ n = 0, 1, ...$

Thus, the trend function of a martingale is constant:

$$m = E(X_n) = E(X_0);\ n = 0, 1, ... \tag{6.4}$$

However, despite this property, a martingale need not be a stationary process. The trend function of a supermartingale (submartingale) is nonincreasing (nondecreasing). Conditions (6.1) to (6.3) are obviously equivalent to

$$E(X_{n+1} - X_n | X_n = x_n, ..., X_1 = x_1, X_0 = x_0) = 0 \tag{6.5}$$
$$E(X_{n+1} - X_n | X_n = x_n, ..., X_1 = x_1, X_0 = x_0) \leq 0 \tag{6.6}$$
$$E(X_{n+1} - X_n | X_n = x_n, ..., X_1 = x_1, X_0 = x_0) \geq 0 \tag{6.7}$$

In particular, a stochastic process $\{X_0, X_1, ...\}$ with finite absolute first moments $E(|X_n|),\ n = 0, 1, ...$ is a martingale if and only if it satisfies condition (6.5).

If $\{X_0, X_1, ...\}$ is a martingale and X_n is interpreted as the random fortune of a gambler at time *n*, then, on condition $X_n = x_n$, the conditional mean fortune of the gambler at time $n+1$ is also x_n, and this is independent on the development in time of the fortune of the gambler before *n* (*fair game*).

Note In what follows, for notational convenience, martingales are sometimes denoted as $\{X_1, X_2, ...\}$.

Example 6.1 (*sum martingale*) Let $\{Y_0, Y_1, ...\}$ be a sequence of independent random variables with $E(Y_i) = 0$ for $n = 1, 2, ...$ Then the sequence $\{X_0, X_1, ...\}$ defined by

$$X_n = Y_0 + Y_1 + \cdots + Y_n;\quad n = 0, 1, ...$$

is a martingale. The proof is easily established:

$$\begin{aligned} & E(X_{n+1} | X_n = x_n, ..., X_1 = x_1, X_0 = x_0) \\ &= E(X_n + Y_{n+1} | X_n = x_n, ..., X_1 = x_1, X_0 = x_0) \\ &= x_n + E(Y_{n+1}) \\ &= x_n . \end{aligned}$$

The sum martingale $\{X_0, X_1, ...\}$ can be interpreted as a random walk on the real axis: X_n is the position of a particles after its *n*th jump or its position at time *n*.

The constant trend function $m = E(X_n),\ n = 0, 1, ...$ of this martingale is

$$m = E(Y_0).$$

□

Example 6.2 ***(product martingale)*** Let $\{Y_0, Y_1, ...\}$ be a sequence of independent, positive random variables with $E(Y_0) < \infty$, $\mu = E(Y_i) < \infty$ for $i = 1, 2, ...,$ and

$$X_n = Y_0\, Y_1 \cdots Y_n.$$

Then, for $n = 1, 2, ...$

$$\begin{aligned} &E(X_{n+1} \,|\, X_n = x_n, ..., X_1 = x_1, X_0 = x_0) \\ &= E(X_n\, Y_{n+1} \,|\, X_n = x_n, ..., X_1 = x_1, X_0 = x_0) \\ &= x_n\, E(Y_{n+1} \,|\, X_n = x_n, ..., X_1 = x_1, X_0 = x_0) \\ &= x_n\, E(Y_{n+1}) = x_n\, \mu . \end{aligned}$$

Thus, $\{X_0, X_1, ...\}$ is a supermartingale for $\mu \le 1$ and a submartingale for $\mu \ge 1$. For $\mu = 1$, the random sequence $\{X_0, X_1, ...\}$ is a martingale with constant trend function:

$$m = E(X_n) = E(Y_0); \quad n = 0, 1, ...$$

This martingale seems to be a realistic model for describing the development in time of share prices or derivates from these, since, from historical experience, the share price at a time point in future is usually proportional to the present share price level. With this interpretation, $Y_n - 1$ is the relative change in the share price over the interval $[n, n+1]$ with regard to X_n:

$$\frac{X_{n+1} - X_n}{X_n} = Y_n - 1; \quad n = 0, 1, ...$$

Important special cases are:

1) *Discrete Black-Scholes Model*:

$$Y_i = e^{U_i} \quad \text{with } U_i = N(\mu, \sigma^2), \quad i = 1, 2, ... \tag{6.8}$$

2) *Binomial model*:

$$Y_i = \begin{cases} r & \text{with probability } \alpha \\ 1/r & \text{with probability } 1 - \alpha \end{cases}; \quad i = 1, 2, ...; \ r > 0, \ r \ne 1.$$

In this case, with a random integer N, $|N| \le n$, the share price at time n has structure

$$X_n = Y_0\, r^N; \quad n = 0, 1, 2, ...$$

If $\alpha = 1/(r+1)$, then $E(Y_i) = 1$ so that under this condition $\{X_0, X_1, ...\}$ is a martingale. □

Specifications of the product martingale are the exponential martingale and the likelihood ratios, which are considered in the following examples.

Example 6.3 (***exponential martingale***) Let $\{Z_1, Z_2, ...\}$ be a sequence of independent, identically as Z distributed random variables and θ be a real number with

$$m(\theta) = E(e^{\theta Z}) < \infty.$$

With Y_0 given, a sequence of random variables $\{Y_0, Y_1, ...\}$ is defined by

$$Y_n = Y_0 + Z_1 + \cdots + Z_n; \quad n = 1, 2, ...$$

Then the sequence of random variables $\{X_0, X_1, ...\}$ given by

$$X_0 = e^{\theta Y_0}$$

and

$$X_n = \frac{1}{(m(\theta))^n} e^{\theta Y_n} = e^{\theta Y_0} \prod_{i=1}^{n} \left(\frac{e^{\theta Z_i}}{m(\theta)} \right); \quad n = 1, 2, ... \tag{6.9}$$

is a martingale. The proof is easily established, since, in view of the independence of the Z_i,

$$E(X_{n+1} | X_n = x_n, X_{n-1} = x_{n-1}, ..., X_0 = x_0)$$

$$= x_n E\left(\frac{e^{\theta Z_{n+1}}}{m(\theta)} \right) = x_n E\left(\frac{e^{\theta Z}}{m(\theta)} \right) = x_n \frac{m(\theta)}{m(\theta)}$$

$$= x_n .$$

In particular, if Z is a binary random variable with probability distribution

$$Z = \begin{cases} 1 & \text{with probability } p \\ -1 & \text{with probability } 1-p \end{cases},$$

then $\{Y_0, Y_1, ...\}$ can be interpreted as a random walk, which starts at Y_0, and proceeds with steps of size 1 to the right or to the left, each with probability p and $1-p$, respectively, $0 < p < 1$. In this case,

$$m(\theta) = E(e^{\theta Z}) = p\, e^{\theta} + (1-p)\, e^{-\theta}.$$

Specifically, if

$$\theta = \ln\, [(1-p)/p], \tag{6.10}$$

then $m(\theta) = 1$ so that the exponential martingale has structure

$$X_n = \left(\frac{1-p}{p} \right)^{Y_n}$$

and trend function

$$m = E(X_n) = E\left(\left(\frac{1-p}{p} \right)^{Y_0} \right); \quad n = 0, 1, ...$$

□

Example 6.4 (***likelihood ratios***) Suppose the null hypothesis has to be tested that the random variables $Y_0, Y_1, ...$ are independent and identically distributed with probability density φ via the hypothesis that these variables are independent and identically distributed with density ψ.

On condition $\{y,\ \varphi(y) > 0\} = \{y,\ \psi(y) > 0\}$, the ratio

$$r(y) = \begin{cases} \varphi(y)/\psi(y), & \psi(y) > 0 \\ 0, & \psi(y) = 0 \end{cases}$$

is introduced. Then, if ψ is the true density, the random sequence $\{X_0, X_1, ...\}$ with

$$X_n = r(Y_0)\, r(Y_1) \cdots r(Y_n)$$

is a martingale. In view of example 6.2, it is sufficient to show that the factors $r(Y_i)$ have mean value 1: For φ being a probability density,

$$E(r(Y_i)) = \int_{\{y,\, \psi(y)>0\}} \frac{\varphi(y)}{\psi(y)} \psi(y)\, dy$$

$$= \int_{\{y,\, \psi(y)>0\}} \varphi(y)\, dy = 1\,. \qquad \square$$

Example 6.5 (***branching process***) Consider a population with the property that each individual of any generation gives birth to a random number of 'children'. These numbers are independent and have mean value μ. Let X_n be the total number of children produced by the nth generation. Since each of those children will have on average μ children of its own,

$$E(X_{n+1} | X_n = x_n, ..., X_1 = x_1, X_0 = x_0) = \mu x_n\,. \tag{6.11}$$

Hence, $\{X_0, X_1, ...\}$ is a martingale if $\mu = 1$, a supermartingale if $\mu \le 1$, and a submartingale if $\mu \ge 1$. Moreover, for any positive μ, the sequence $\{Z_0, Z_1, ...\}$ with

$$Z_n = \frac{X_n}{\mu^n}$$

is a martingale. This is proved as follows:

$$E(Z_{n+1} \,\big|\, Z_n = z_n, ..., Z_1 = z_1, Z_0 = z_0)$$

$$= E\left(\frac{X_{n+1}}{\mu^{n+1}} \,\middle|\, \frac{X_n}{\mu^n} = \frac{x_n}{\mu^n}, ..., \frac{X_1}{\mu^1} = \frac{x_1}{\mu^1}, \frac{X_0}{\mu^0} = \frac{x_0}{\mu^0}\right)$$

$$= \frac{1}{\mu^{n+1}} E(X_{n+1} \,\big|\, X_n = x_n, ..., X_1 = x_1, X_0 = x_0)$$

$$= \frac{1}{\mu^{n+1}} \mu x_n = \frac{x_n}{\mu^n} = z_n\,. \qquad \square$$

6.1.2 Doob-Martingales

In this section, the concept of a (super-, sub-) martingale $\{X_0, X_1, ...\}$ as introduced in definition 6.1 is generalized by conditioning with regard to another sequence of random variables $\{Y_0, Y_1, ...\}$, which is usually related to $\{X_0, X_1, ...\}$. The following definition refers to the characterization of (super-, sub-) martingales by properties (6.5) to (6.7).

Definition 6.2 Let $\{X_0, X_1, ...\}$ and $\{Y_0, Y_1, ...\}$ be two discrete-time stochastic processes. If

$$E(|X_n|) < \infty \text{ for all } n = 0, 1, ...,$$

then the random sequence $\{X_0, X_1, ...\}$ is a *martingale with regard to* $\{Y_0, Y_1, ...\}$ or a *Doob-type martingale* if for all $(n+1)$- dimensional vectors $(y_0, y_1, ..., y_n)$ with y_i elements of the state space of $\{Y_0, Y_1, ...\}$, and for any $n = 0, 1, ...,$

$$E(X_{n+1} - X_n | Y_n = y_n, ..., Y_1 = y_1, Y_0 = y_0) = 0. \tag{6.12}$$

Under otherwise the same assumptions, $\{X_0, X_1, ...\}$ is a *supermartingale with regard to* $\{Y_0, Y_1, ...\}$ if

$$E(X_{n+1} - X_n | Y_n = y_n, ..., Y_1 = y_1, Y_0 = y_0) \le 0,$$

and a *submartingale with regard to* $\{Y_0, Y_1, ...\}$ if

$$E(X_{n+1} - X_n | Y_n = y_n, ..., Y_1 = y_1, Y_0 = y_0) \ge 0.$$

●

In what follows, under rather strong additional conditions, a criterion is derived, which ensures that a Doob-type martingale is a martingale in the sense of definition 6.1. This requires the introduction of a new concept.

Definition 6.3 Let $\{Y_0, Y_1, ...\}$ be a discrete-time Markov chain (not necessarily homogeneous) with state space $\mathbf{Z} = \{\cdots, -1, 0, +1, \cdots\}$ and transition probabilities

$$p_n(y, z) = P(Y_{n+1} = z | Y_n = y); \; y, z \in \mathbf{Z}; \; n = 0, 1, ...$$

A function $h(y, n); \, y \in \mathbf{Z}; \; n = 0, 1, ...$ is said to be *concordant* with $\{Y_0, Y_1, ...\}$ if it satisfies for all $y \in \mathbf{Z}$

$$h(y, n) = \sum_{z \in \mathbf{Z}} p_n(y, z)\, h(z, n+1). \tag{6.13}$$

●

Theorem 6.1 Let $\{Y_0, Y_1, ...\}$ be a discrete-time Markov chain with state space

$$\mathbf{Z} = \{\cdots, -1, 0, +1, \cdots\}.$$

Then, for any function $h(y, n)$ which is concordant with $\{Y_0, Y_1, ...\}$,

a) the sequence of random variables $\{X_0, X_1, ...\}$ generated by

$$X_n = h(Y_n, n)\,;\ n = 0, 1, ... \tag{6.14}$$

is a martingale with regard to $\{Y_0, Y_1, ...\}$, and

b) the sequence $\{X_0, X_1, ...\}$ is a martingale.

Proof a) By the Markov property and the concordance of h with $\{Y_0, Y_1, ...\}$,

$$E(X_{n+1} - X_n \,\big|\, Y_n = y_n, ..., Y_1 = y_1, Y_0 = y_0)$$
$$= E(X_{n+1} \,\big|\, Y_n = y_n, ..., Y_1 = y_1, Y_0 = y_0) - E(X_n \,\big|\, Y_n = y_n, ..., Y_1 = y_1, Y_0 = y_0)$$
$$= E(h(Y_{n+1}, n+1) | Y_n = y_n) - E(h(Y_n, n) | Y_n = y_n)$$
$$= \sum_{z \in \mathbf{Z}} p_n(y_n, z)\, h(z, n+1) - h(y_n, n)$$
$$= h(y_n, n) - h(y_n, n) = 0.$$

This result shows that $\{X_0, X_1, ...\}$ is a martingale with regard to $\{Y_0, Y_1, ...\}$.

b) Let, for given $x_0, x_1, ..., x_n$, the random event A be defined as the 'martingale condition'

$$A = \{X_n = x_n, ..., X_1 = x_1, X_0 = x_0\}.$$

Since the X_n are fully determined by the random variables Y_n, there exists a set $\mathbf{Y}$ of vectors $\vec{y} = (y_n, ..., y_1, y_0)$ with property that the occurrence of any of the mutually disjoint random events

$$A_{\vec{y}} = \{Y_n = y_n, ..., Y_1 = y_1, Y_0 = y_0\},\ \ \vec{y} \in \mathbf{Y},$$

implies the occurrence of event A:

$$A = \bigcup_{\vec{y} \in \mathbf{Y}} A_{\vec{y}}\,.$$

Now the martingale property of $\{X_0, X_1, ...\}$ is easily established:

$$E(X_{n+1} | X_n = x_n, ..., X_1 = x_1, X_0 = x_0)$$
$$= E(X_{n+1} | A) = \sum_{\vec{y} \in \mathbf{Y}} E\left(X_{n+1} \,\big|\, A_{\vec{y}}\right) \frac{P(A_{\vec{y}})}{P(A)}$$
$$= h(y_n, n) \sum_{\vec{y} \in \mathbf{Y}} \frac{P(A_{\vec{y}})}{P(A)}$$
$$= h(y_n, n)$$
$$= x_n\,.$$

Hence, $\{X_0, X_1, ...\}$ is a martingale according to definition 6.1. ■

Example 6.6 (***variance martingale***) Let $\{Z_1, Z_2, ...\}$ be a sequence of independent, integer-valued random variables with probability distributions

$$q_i^{(n)} = P(Z_n = i), \quad i \in \mathbf{Z} = \{\cdots, -1, 0, +1, \cdots\},$$

and numerical parameters

$$E(Z_i) = 0 \quad \text{and} \quad E(Z_i^2) = \sigma_i^2;\ i = 1, 2, ...$$

With an integer-valued constant z_0, a discrete-time Markov chain $\{Y_0, Y_1, ...\}$ with state space $\mathbf{Z} = \{\cdots, -1, 0, +1, \cdots\}$ is introduced as follows:

$$Y_n = z_0 + Z_1 + \cdots + Z_n .$$

Then,

$$E(Y_n) = z_0 \text{ for } n = 0, 1, ... \quad \text{and} \quad Var(Y_n) = \Sigma_{i=1}^n \sigma_i^2 \quad \text{for } n = 1, 2, ...$$

The function

$$h(y, n) = y^2 - \Sigma_{i=1}^n \sigma_i^2 \tag{6.15}$$

is concordant with $\{Y_0, Y_1, ...\}$. To verify this, let $p_n(y, z)$ be the transition probabilities of $\{Y_0, Y_1, ...\}$ at time n. These transition probabilities are fully determined by the probability distribution of Z_{n+1} :

$$p_n(y, z) = P(Y_{n+1} = z | Y_n = y) = P(Z_{n+1} = z - y) = q_{z-y}^{(n+1)};\ y, z \in \mathbf{Z}.$$

Therefore,

$$\begin{aligned}
\sum_{z \in \mathbf{Z}} p_n(y, z)\, h(z, n+1) &= \sum_{z \in \mathbf{Z}} q_{z-y}^{(n+1)} h(z, n+1) \\
&= \sum_{z \in \mathbf{Z}} q_{z-y}^{(n+1)} \left(z^2 - \Sigma_{i=1}^{n+1} \sigma_i^2\right) \\
&= \sum_{z \in \mathbf{Z}} q_{z-y}^{(n+1)} \left[(z - y + y)^2 - \Sigma_{i=1}^{n+1} \sigma_i^2\right] \\
&= \sum_{z \in \mathbf{Z}} q_{z-y}^{(n+1)} (z-y)^2 + 2y \sum_{z \in \mathbf{Z}} q_{z-y}^{(n+1)} (z-y) + \sum_{z \in \mathbf{Z}} q_{z-y}^{(n+1)} y^2 - \Sigma_{i=1}^{n+1} \sigma_i^2 \\
&= \sigma_{n+1}^2 + 2y \cdot 0 + 1 \cdot y^2 - \Sigma_{i=1}^{n+1} \sigma_i^2 \\
&= y^2 - \Sigma_{i=1}^n \sigma_i^2 = h(y, n).
\end{aligned}$$

Hence, the function $h(y, n)$ is concordant with $\{Y_0, Y_1, ...\}$. Thus, by theorem 6.1, the random sequence $\{X_0, X_1, ...\}$ with X_n generated by

$$X_n = Y_n^2 - Var(Y_n) \tag{6.16}$$

is a martingale. □

Example 6.7 Let Y_i be the random price of a share at time i and S_i be the random amount of share an investor holds in the interval

$$[i, i+1); \; i = 0, 1, ..., \; S_i \geq 0.$$

Thus, at time $t = 0$ the total value of the investor's shares is $X_0 = Y_0 S_0$ and in the interval $[i, i+1)$ the investor makes a 'profit' of

$$S_i (Y_{i+1} - Y_i).$$

Hence, his total profit up to time $t = n$ is

$$X_n = \Sigma_{i=0}^{n-1} S_i (Y_{i+1} - Y_i); \quad n = 1, 2, ... \tag{6.17}$$

It makes sense to assume that the investor's choice, what amount of share to hold in $[n, n+1)$ does not depend on the profit made in this and later intervals, but only on the profits made in the previous intervals. Hence, S_n is assumed to be fully determined by the $Y_0, Y_1, ..., Y_n$. Under this assumption, the sequence $\{X_1, X_2, ...\}$ is a supermartingale with regard to $\{Y_0, Y_1, ...\}$ if $\{Y_0, Y_1, ...\}$ is a supermartingale. This is proved as follows:

$$E(X_{n+1} - X_n | Y_n = y_n, ..., Y_1 = y_1, Y_0 = y_0)$$
$$= E(S_n(Y_{n+1} - Y_n) | Y_n = y_n, ..., Y_1 = y_1, Y_0 = y_0)$$
$$= S_n E(Y_{n+1} - Y_n | Y_n = y_n, ..., Y_1 = y_1, Y_0 = y_0) \leq 0.$$

The last line of this derivation makes use of the assumptions that on condition

$$'Y_n = y_n, ..., Y_1 = y_1, Y_0 = y_0'$$

the share amount S_n is constant and that $\{Y_0, Y_1, ...\}$ is a supermartingale. Hence, no matter how well-considered the investor fixes the amount of share to be held in an interval, in the long-run he cannot expect to make positive profit if the share price develops unfavourably. (A supermartingale has a decreasing trend function.) □

Example 6.8 The structure of X_n given by (6.17) includes as a special case the net profit development when applying the 'doubling strategy': A gambler bets \$ 1 on the first game. If he wins, he makes a net profit of \$ 1. But if he loses, he suffers a loss of \$ -1 and will bet \$ 2 on the next play. If he wins, he will get \$ 4 and, hence, will have made a net profit of \$ 1. But if he loses he will bet \$ 4 on the next game and so on. The following table shows the net profit development of the gambler if he loses 5 times in a row and then wins:

play	1	2	3	4	5	6
result	loss	loss	loss	loss	loss	win
bet	1	2	4	8	16	32
winnings	-1	-3	-7	-15	-31	+1

If the gambler loses the first $N-1$ games and wins the Nth game, then

$$S_i = 2^{i-1};\ i = 1, 2, ..., N,$$
$$Y_{i+1} - Y_i = -1;\ i = 0, 1, ..., N-2,$$
$$Y_N - Y_{N-1} = 1.$$

Hence, when assuming a win occurs with probability p and a loss with probability $1-p$, the $Y_1, Y_2, ...$ have structure

$$Y_i = Z_1 + Z_2 + \cdots + Z_i,\quad Y_0 = 0, \tag{6.18}$$

where the $Z_1, Z_2, ...$ are independent, identically as Z distributed binary random variables:

$$Z = \begin{cases} 1 & \text{with probability} \quad p \\ -1 & \text{with probability} \quad 1-p. \end{cases}$$

With the Y_i given by (6.18), the net winnings of the gambler at time n, $1 \le n \le N$, are given by (6.17). Now, on condition that after every win the game starts anew and the S_i are adjusted accordingly, (6.17) describes the net winning development of the gambler for all $n = 1, 2, ...$ Note, if the gambler loses at time $N+1$, his total winnings become 0. Since N is random, the S_i in (6.17) are random as well. In case the gambler makes up his mind to stop playing the doubling strategy at a fixed time point n, then, as shown in the previous example, he cannot expect to have made positive net winnings if $\{Y_1, Y_2, ...\}$ is a supermartingale. (Obviously, $\{Y_1, Y_2, ...\}$ is a supermartingale if $p \le 1/2$.) Hence, the gambler should not stop playing the doubling strategy at any time point, but at a winning time point. (If $p > 0$, then $P(N < \infty) = 1$.) However, to be able to maintain the doubling strategy in this way, the gambler must have an unlimited amount of initial capital, since each bet size 2^i; $i = 1, 2, ...$; has a positive probability to occur and the casino must allow arbitrarily large bets. Since these prerequisites are not realistic, on average no money can be made by pursuing the doubling strategy when betting on a supermartingale. □

6.1.3 Martingale Stopping Theorem and Applications

As pointed out in the beginning of this chapter, martingales are suitable stochastic models for fair games, i.e. the chances to win or to lose are equal. If one bets on a martingale, is it, nevertheless, possible to make money by finishing the game at the 'right time' ? The decision, when to finish a game can, of course, only be made on the past development of the martingale and not on its future. Hence, a proper time for finishing a game seems to be a stopping time N for $\{X_0, X_1, ...\}$, where X_n is the gambler's net profit after the nth game. According to definition 1.2, a stopping time for $\{X_0, X_1, ...\}$ is a positive, integer-valued random variable N with property that the occurrence of the random event '$N = n$' is fully determined by the random variables $X_0, X_1, ..., X_n$ and, hence, does not depend on the $X_{n+1}, X_{n+2}, ...$ However, the *martingale stopping theorem* (also called *optional stopping theorem* or *optional sampling theorem*) excludes the possibility of winning in the long-run if finishing the game is controlled by a stopping time (see also examples 6.7 and 6.8).

Theorem 6.2 (*martingale stopping theorem for discrete-time Markov chains*) Let N be a stopping time for the martingale $\{X_0, X_1, ...\}$. Then

$$E(X_N) = E(X_0) \tag{6.19}$$

if at least one of the following three conditions is fulfilled:

1) N is finite and there exists a finite constant C_1 with

$$\left|X_{\min(N,n)}\right| \le C_1 \text{ for all } n = 0, 1, ...$$

2) The stopping time N is bounded, i.e. there exists a finite constant C_2 so that, with probability 1,

$$N \le C_2.$$

3) $E(N)$ is finite and there exists a finite constant C_3 so that

$$E(\left|X_{n+1} - X_n\right| \,|\, X_1, X_2, ..., X_n) < C_3; \quad n = 0, 1, ...$$ ■

Hint When comparing formulas (6.4) and (6.19), note that in (6.19) N is a random variable.

Example 6.9 (*Wald's identity*) Theorem 6.2 implies Wald's identity (1.125) on condition that N with $E(N) < \infty$ is a stopping time for a sequence of independent, identically as Y with $E(Y) < \infty$ distributed random variables $Y_1, Y_2, ...$ To see this, let

$$X_n = \Sigma_{i=1}^{n}(Y_i - E(Y)); \quad n = 1, 2, ...$$

By example 6.1, the sequence $\{X_1, X_2, ...\}$ is a martingale. Hence, theorem 6.2 is applicable (condition 3):

$$E(X_N) = E(X_1) = 0.$$

On the other hand,

$$\begin{aligned} E(X_N) &= E\left(\Sigma_{i=1}^{N}(Y_i - E(Y))\right) \\ &= E\left(\Sigma_{i=1}^{N} Y_i - N\,E(Y)\right) \\ &= E\left(\Sigma_{i=1}^{N} Y_i\right) - E(N)\,E(Y). \end{aligned}$$

This proves Wald's identity:

$$E\left(\Sigma_{i=1}^{N} Y_i\right) = E(N)\,E(Y). \tag{6.20}$$

□

Example 6.10 (*fair game*) Let $\{Z_1, Z_2, ...\}$ be a sequence of independent, identically as Z distributed random variables:

$$Z = \begin{cases} 1 & \text{with probability } P(Z=1) \ = 1/2 \\ -1 & \text{with probability } P(Z=-1) = 1/2 \end{cases}.$$

Since $E(Z_i) = 0$, the sequence $\{Y_1, Y_2, ...\}$ defined by

$$Y_n = Z_1 + Z_2 + \cdots + Z_n; \; n = 1, 2, ...$$

is a martingale (example 6.1). Y_n is interpreted as the cumulative net profit of a gambler after the n th play if he bets one dollar on each play. The gambler finishes the game as soon he has won \$ a or lost \$ b. Thus, the game will be finished at time

$$N = \min\{n;\; Y_n = a \text{ oder } Y_n = -b\}. \tag{6.21}$$

Obviously, N is a stopping time for the martingale $\{Y_1, Y_2, ...\}$. Since $E(N)$ is finite, by theorem 6.2 (condition 3),

$$0 = E(Y_1) = E(Y_N) = a\,P(Y_N = a) + (-b)\,P(Y_N = -b).$$

Combining this relationship with

$$P(Y_N = a) + P(Y_N = -b) = 1,$$

yields the desired probabilities

$$P(Y_N = a) = \frac{b}{a+b}, \quad P(Y_N = -b) = \frac{a}{a+b}.$$

For determining the mean duration $E(N)$ of such a game, the 'variance martingale' $\{X_1, X_2, ...\}$ with

$$X_n = Y_n^2 - Var(Y_n) = Y_n^2 - n$$

is used (example 6.6). By theorem 6.2,

$$0 = E(X_1) = E(X_N) = E(Y_N^2) - E(N) = 0.$$

Therefore,

$$E(N) = E(Y_N^2) = a^2 P(Y_N = a) + b^2 P(Y_N = -b).$$

Thus, the mean duration of this fair game is

$$E(N) = a^2 \frac{b}{a+b} + b^2 \frac{a}{a+b} = a\,b.$$

□

Example 6.11 (*unfair game*) Under otherwise the same assumptions as in the previous example, let

$$Z_i = \begin{cases} 1 & \text{with probability } p \\ -1 & \text{with probability } 1-p \end{cases}, \quad p \neq 1/2. \tag{6.22}$$

Thus, the win and loss probabilities on a play are different.

The mean value of Z_i is

$$E(Z_i) = 2p - 1 .$$

Let the martingale $\{X_1, X_2, ...\}$ be defined as in example 6.9:

$$X_n = \Sigma_{i=1}^n (Z_i - E(Z_i)); \quad n = 1, 2, ...$$

By introducing $Y_n = Z_1 + Z_2 + \cdots + Z_n$, the random variable X_n can be written in the form

$$X_n = Y_n - (2p-1)\, n .$$

If this martingale is stopped at time N given by (6.21), theorem 6.2 yields

$$0 = E(X_N) = E(Y_N) - (2p-1)\, E(N), \tag{6.23}$$

or, equivalently,

$$0 = a\, P(Y_N = a) + (-b)\, P(Y_N = -b) - (2p-1)\, E(N) .$$

For establishing another equation in the three unknowns

$$P(Y_N = a), \;\; P(Y_N = -b), \text{ and } E(N),$$

the exponential martingale (example 6.3) is used. Let θ be given by

$$\theta = \ln\, [(1-p)/p] .$$

Then,

$$E\left(e^{\theta Z_i}\right) = p\, e^{\theta} + (1-p)\, e^{-\theta} = 1.$$

Hence, the sequence $\{U_1, U_2, ...\}$ with

$$U_n = \prod_{i=1}^{n} e^{\theta Z_i} = e^{\theta \Sigma_{i=1}^n Z_i} = e^{\theta Y_n}; \quad n = 1, 2, ...$$

is a martingale. Now, by applying theorem 6.2,

$$1 = E(U_1) = E(U_N) = e^{\theta a}\, P(Y_N = a) + e^{-\theta b} P(Y_N = -b) . \tag{6.24}$$

Equations (6.23) and (6.24) together with $P(Y_N = a) + P(Y_N = -b) = 1$ yield the 'hitting' probabilities

$$P(Y_N = a) = \frac{1 - \left(\frac{p}{1-p}\right)^b}{\left(\frac{1-p}{p}\right)^a - \left(\frac{p}{1-p}\right)^b}, \quad P(Y_N = -b) = \frac{\left(\frac{1-p}{p}\right)^a - 1}{\left(\frac{1-p}{p}\right)^a - \left(\frac{p}{1-p}\right)^b}$$

and the mean duration of a game

$$E(N) = \frac{a\, P(Y_N = a) - b\, P(Y_N = -b)}{2p-1}.$$

□

6.1.4 Inequalities for Discrete-Time Martingales

In what follows, some important limit properties and inequalities for discrete-time martingales $\{X_0, X_1, ...\}$ are listed.

1) Let $E(|X_n|) < C < \infty$ for $n = 0, 1, ...$ Then there exists a random variable X_∞ with property that the random sequence $X_0, X_1, ...$ converges both with probability one and in mean towards X_∞ (see section 1.9.1 for convergence criteria):

$$P(\lim_{n\to\infty} X_n = X_\infty) = 1, \quad \lim_{n\to\infty} E(|X_n - X_\infty|) = 0.$$

2) Let $\sup_n E(X_n^2) < \infty$. Then there exists a random variable X_∞ with property that the random sequence $X_0, X_1, ...$ converges in mean square towards X_∞:

$$\lim_{n\to\infty} E((X_n - X_\infty)^2) = 0.$$

3) (*Azuma's inequality*) Let $\mu = E(X_i);\ i = 1, 2, ...$ and $X_0 = \mu$. If there exist nonnegative numbers α_i and β_i with

$$-\alpha_i \le X_{i+1} - X_i \le \beta_i;\ i = 0, 1, ...;$$

then, for for all $n = 1, 2, ...$ and $\varepsilon > 0$,

$$P(X_n - \mu \ge +\varepsilon) \le \exp\left\{-2\varepsilon^2 \Big/ \Sigma_{i=1}^n \left(\alpha_i + \beta_i\right)^2\right\},$$

$$P(X_n - \mu \le -\varepsilon) \le \exp\left\{-2\varepsilon^2 \Big/ \Sigma_{i=1}^n \left(\alpha_i + \beta_i\right)^2\right\}.$$

Hence, if the increments $X_{i+1} - X_i$ of the martingale $\{X_1, X_2, ...\}$ only vary within given finite intervals, then bounds for useful probabilities can be given.

4) (*Doob's inequalities*) For all $n = 1, 2, ...,$ as well as for every $\alpha \ge 1$ and $\lambda > 0$, assuming the existence of the occurring mean values,

$$P\left(\max_{i=0,1,...,n} |X_i| \ge \lambda\right) \le \frac{E(|X_n|^\alpha)}{\lambda^\alpha}.$$

Moreover, for all $\alpha > 1$,

$$E(|X_n|^\alpha) \le E\left(\max_{i=0,1,...,n} |X_i|^\alpha\right) \le \left(\frac{\alpha}{\alpha - 1}\right)^\alpha E(|X_n|^\alpha).$$

In particular, for square-mean integrable martingales ($\alpha = 2$)

$$E\left(X_n^2\right) \le E\left(\max_{i=0,1,...,n} X_i^2\right) \le 4\,E\left(X_n^2\right).$$

6.2 CONTINUOUS-TIME MARTINGALES

This section summarizes some results on continuous-time martingales. For simplicity and with regard to applications to Brownian motion processes in the subsequent chapter, their parameter space is restricted to $\mathbf{T} = [0, \infty)$. The following definition of continuous-time martingales is based on the concept of the conditional mean value of a random variable given one or more other random variables (section 1.6.3).

Definition 6.4 A stochastic process $\{X(t), t \geq 0\}$ with $E(|X(t)|) < \infty$ for all $t \geq 0$ is called a *martingale* if for all integers $n = 0, 1, ...,$ for every sequence $t_0, t_1, ..., t_n$ with $0 \leq t_0 < t_1 < \cdots < t_n$ as well as for any t with $t > t_n$, with probability 1,

$$E(X(t)|X(t_n), \cdots, X(t_1), X(t_0)) = X(t_n). \tag{6.25}$$

●

Thus, to predict the mean value of a martingale at a time t, only the last observation point before t is relevant. The development of the process before t_n contains no additional information with respect to its mean value at time t, $t > t_n$. Hence, regardless how large the difference $t - t_n$ is, on average no increase/decrease of the process $\{X(t), t \geq 0\}$ can be expected in $[t_n, t]$. The characteristic property (6.25) of a martingale under the assumptions made is frequently written in the form

$$E(X(t)|X(y), y \leq s) = X(s), \; s < t. \tag{6.26}$$

$\{X(t), t \geq 0\}$ is a *supermartingale* (*submartingale*) if in (6.26) the sign ' = ' is replaced with ' ≤ ' (' ≥ '). If $\mathbf{Z}$ is the state space of $\{X(t), t \geq 0\}$, then, as a consequence of (6.25), a continuous-time martingale $\{X(t), t \geq 0\}$ has property

$$E(X(t)|X(t_n) = x_n, ..., X(t_1) = x_1, X(t_0) = x_0) = x_n$$

for all $(x_0, x_1, ..., x_n)$ with $x_i \in \mathbf{Z}$, and this property, under otherwise the same assumptions as in definition 6.4, can be used to define continuous-time martingales analogously to discrete-time martingales. The trend function of a continuous-time martingale is constant:

$$m(t) = E(X(t)) \equiv m(0).$$

Definition 6.5 (*stopping time*) A random variable L is a *stopping time* with regard to an (arbitrary) stochastic process $\{X(t), t \geq 0\}$ if for all $s > 0$ the occurrence of the random event '$L \leq s$' is fully determined by the evolvement of this process to time point s. Therefore, the occurrence of the random event '$L \leq s$' is independent of all $X(t)$ with $t > s$. ●

Let $I_{L>t}$ denote the indicator function for the occurrence of the event '$L > t$'.

$$I_{L>t} = \begin{cases} 1 & \text{if } L > t \text{ occurs,} \\ 0 & \text{otherwise} \end{cases}.$$

Theorem 6.3 (***martingale stopping theorem***) If $\{X(t), t \geq 0\}$ is a continuous-time martingale and L a stopping time for this martingale, then

$$E(X(L)) = E(X(0)) \tag{6.27}$$

if at least one of the following two conditions is fulfilled:

1) L is bounded.

2) $P(L < \infty) = 1, \quad E(|X(L)|) < \infty$, and $\lim_{t\to\infty} E(|X(t)| I_{L>t}) = 0.$ ■

The interpretation of this theorem is the same as in case of the martingale stopping theorem for discrete-time martingales. For proofs of theorems 6.2 and 6.3 see, for instance, Kannan [43] and Rolski et al. [67].

Example 6.12 As an application of theorem 6.3, a proof of Lundberg's inequality (3.161) in actuarial risk analysis is given: Let $\{R(t), t \geq 0\}$ be the risk process under the assumptions of section 3.4.2, i.e. $R(t) = x + \kappa t - C(t)$, where x is the initial capital of an insurance company, κ the premium rate and $\{C(t), t \geq 0\}$ the compound claim size process defined by

$$C(t) = \Sigma_{i=0}^{N(t)} M_i, \quad M_0 = 0,$$

where $\{N(t), t \geq 0\}$ is a homogeneous Poisson process with parameter $\lambda = 1/\mu$. The claim sizes $M_1, M_2, ...$ are assumed to be independent and identically as M distributed random variables with finite mean $E(M)$ and distribution function and density

$$B(t) = P(M \leq t), \; b(t) = dB(t)/dt, \quad t \geq 0.$$

Let

$$Y(t) = e^{-r R(t)} \quad \text{and} \quad h(r) = E(e^{r M}) = \int_0^\infty e^{r x} b(t) dt$$

for any positive r with property

$$h(r) < \infty. \tag{6.28}$$

Then

$$E(Y(t)) = e^{-r(x+\kappa t)} E\left(e^{+r C(t)}\right)$$

$$= e^{-r(x+\kappa t)} \sum_{i=0}^{\infty} E(e^{+r C(t)} | N(t) = n) P(N(t) = n)$$

$$= e^{-r(x+\kappa t)} \sum_{i=0}^{\infty} [h(r)]^n \frac{(\lambda t)^n}{n!} e^{-\lambda t} = e^{-r(x+\kappa t)} e^{\lambda t [h(r)-1]}.$$

Let

$$X(t) = \frac{Y(t)}{E(Y(t))} = e^{r C(t) - \lambda t [h(r)-1]}.$$

Since $\{C(t), t \geq 0\}$ has independent increments, the process $\{X(t), t \geq 0\}$ has independent increments, too. Hence, for $s < t$, since $E(X(t)) = 1$ for all $t \geq 0$,

$$E(X(t)|X(y),\ y \le s) = E(X(s) + X(t) - X(s)|X(y),\ y \le s)$$
$$= X(s) + E(X(t) - X(s)|X(y),\ y \le s)$$
$$= X(s) + E(X(t) - X(s)) = X(s) + 1 - 1 = X(s)\,.$$

Thus, $\{X(t),\ t \ge 0\}$ is a martingale. Now, let

$$L = \inf_t \{t,\ R(t) < 0\}. \tag{6.29}$$

Obviously, L is a stopping time for the martingale $\{X(t),\ t \ge 0\}$. Therefore, for any finite $z > 0$,

$$L \wedge z = \min\,(L, z)$$

is a bounded stopping time for $\{X(t), t \ge 0\}$ (exercise 6.11). Hence, theorem 6.3 is applicable with the stopping time $L \wedge z$:

$$E(X(0)) = 1 = E(X(L \wedge z))$$
$$= E(X(L \wedge z|L < z)\,P(L < z) + E(X(L \wedge z|L \ge z))\,P(L \ge z)$$
$$\ge E(X(L \wedge z|L < z)\,P(L < z)$$
$$= E(X(L\,|L < z)\,P(L < z)$$
$$= E(e^{r\,C(L) - \lambda L\,[h(r)-1]}\,|\,L < z)\,P(L < z)\,.$$

By (6.29), $x + \kappa L < C(L)$. Thus, from the first and the last line of this derivation,

$$1 > E(e^{r\,(x+\kappa L) - \lambda L\,(h(r)-1)}|L < z)\,P(L < z)\,,$$

or, equivalently,

$$1 > e^{rx} E(e^{[r\kappa - \lambda\,(h(r)-1)]\,L}|L < z)\,P(L < z)\,. \tag{6.30}$$

If the parameter r is chosen in such away that

$$r\kappa - \lambda\,[h(r) - 1] = 0\,, \tag{6.31}$$

then the inequality (6.30) simplifies to

$$P(L < z) < e^{-rx}.$$

Since this inequality holds for all finite $z > 0$, it follows that

$$P(L < \infty) \le e^{-rx}. \tag{6.32}$$

By (3.143), the probability $P(L < \infty)$ is nothing but the ruin probability $p(x)$. On the other hand, in view of $\lambda = 1/\mu$, equation (6.31) is equivalent to equation (3.202), which defines the Lundberg coefficient r. To verify this by partial integration of

$$E(e^{r\,M}) = \int_0^\infty e^{r\,x} b(t)dt,$$

note that condition (6.28) implies

$$\lim_{t \to \infty} e^{rt} \overline{B}(t) = 0.$$

Thus, (6.32) is indeed the Lundberg inequality (3.161) for the ruin probability. □

Finally, some limit properties and inequalities for continuous-time martingales $\{X(t),\, t \geq 0\}$, the samples paths of which are continuous from the right, are listed. They are quite analogous to the corresponding ones for discrete-time martingales. The existence of all occurring mean values is assumed.

1) If $\sup_t E(|X_t|) < \infty$, then there exists a random variable X_∞ with property that $X(t)$ converges both with probability one and in mean towards X_∞ as $t \to \infty$:

$$P(\lim_{t \to \infty} X_t = X_\infty) = 1, \qquad \lim_{t \to \infty} E(|X_t - X_\infty|) = 0.$$

2) If $\sup_t E(X_t^2) < \infty$, then there exists a random variable X_∞ with property that $X(t)$ converges in square mean towards X_∞ as $t \to \infty$:

$$\lim_{t \to \infty} E((X_t - X_\infty)^2) = 0.$$

3) Let $[a, b] \subseteq [0, \infty)$. Then, for any $\lambda > 0$,

$$\lambda P(\sup_{t \in [a,b]} X(t) \geq \lambda) \leq E(X(a)) + E(\max\{(0, -X(b)\}),$$

$$\lambda P(\inf_{t \in [a,b]} X(t) \leq -\lambda) \leq E(|X(b)|).$$

4) (*Doob's inequalities*) Let $[a, b] \subseteq [0, \infty)$. Then, for every $\lambda > 0$ and $\alpha \geq 1$,

$$\lambda^\alpha P(\sup_{t \in [a,b]} |X(t)| \geq \lambda) \leq E(|X(b)|^\alpha).$$

For $\alpha > 1$,

$$E(|X(b)|^\alpha) \leq E([\sup_{t \in [a,b]} X(t)]^\alpha) \leq \left(\frac{\alpha}{\alpha - 1}\right)^\alpha E(|X(b)|^\alpha).$$

In particular, for $\alpha = 2$,

$$E(X(b)^2) \leq E([\sup_{t \in [a,b]} X(t)]^2) \leq 4\, E(X(b)^2).$$

For proofs and a more prestigious treatment of martingales see, for instance, *Rolski et al.* [67] and *Williams* [88].

6.3 EXERCISES

6.1) Let $Y_0, Y_1, ...$ be a sequence of independent random variables, which are identically distributed as $N(0, 1)$. Is the discrete-time stochastic process $\{X_0, X_1, ...\}$ generated by the sums $X_n = \Sigma_{i=0}^{n} Y_i^2$; $n = 0, 1, ...$ a martingale?

6.2) Let $Y_0, Y_1, ...$ be a sequence of independent random variables with finite mean values $E(Y_i)$. Is the discrete-time stochastic process $\{X_0, X_1, ...\}$ generated by the sums $X_n = \Sigma_{i=0}^{n}(Y_i - E(Y_i))$ a martingale.

6.3) Let a discrete-time stochastic process $\{X_0, X_1, ...\}$ be defined by

$$X_n = Y_0 \cdot Y_1 \cdot \cdots \cdot Y_n,$$

where the random variables Y_i are independent and have a uniform distribution over the interval $[0, T]$. Under which condition is $\{X_0, X_1, ...\}$ (1) a martingale, (2) a submartingale, (3) a supermartingale?

6.4) Let $\{X_0, X_1, ...\}$ be the discrete Black-Scholes model defined by

$$X_n = Y_0 \cdot Y_1 \cdot \cdots \cdot Y_n,$$

where Y_0 is an arbitrary positive random variable with finite mean, and $Y_i = e^{Z_i}$ with independent $Z_i = N(\mu, \sigma^2)$; $i = 1, 2, ...$ Under which condition is $\{X_0, X_1, ...\}$ a martingale?

6.5) Starting at value 0, the profit of an investor increases per week by one unit with probability p, $p > 1/2$, or decreases per week by one unit with probability $1 - p$. The weekly increments of the investor's profit are assumed to be independent.

Let N be the random number of weeks until the investor's profit reaches for the first time a given positive integer n. By means of Wald's equation, determine $E(N)$.

6.6) Let $Z_1, Z_2, ..., Z_n$ be a sequence of independent, identically as Z distributed random variables with

$$Z = \begin{cases} 1 & \text{with probability } p \\ 0 & \text{with probability } 1-p \end{cases}, \quad 0 < p < 1,$$

$Y_n = Z_1 + Z_2 + \cdots + Z_n$ and $X_n = h(Y_n)$; $n = 1, 2, ...$; where, for any real y,

$$h(y) = [(1-p)/p]^y.$$

Prove that $\{X_1, X_2, ...\}$ is a martingale with regard to $\{Y_1, Y_2, ...\}$.

6.7) Starting at value 0, the fortune of an investor increases per week by \$ 200 with probability 3/8, remains constant with probability 3/8 and decreases by \$ 200 with probability 2/8. The weekly increments of the investor's fortune are assumed to be independent. The investor stops the 'game' as soon as he has made a total fortune of $\$ 2000$ or a loss of $\$ 1000$, whichever occurs first.

By using suitable martingales and applying the optional stopping theorem, determine

(1) the probability p_{2000} that the investor finishes the 'game' with a profit of $\$ 2000$,

(2) the probability p_{-1000} that the investor finishes the 'game' with a loss of $\$ 1000$,

(3) the mean duration $E(N)$ of the 'game'.

6.8) Let X_0 be uniformly distributed over $[0, T]$, X_1 be uniformly distributed over $[0, X_0]$, and, generally, X_{i+1} be uniformly distributed over $[0, X_i]$, $i = 0, 1, ...$

(1) Prove that the sequence $\{X_0, X_1, ...\}$ is a supermartingale.

(2) Show that $E(X_k) = \frac{T}{2^{k+1}}$; $k = 0, 1, ...$

6.9) Let $\{X_1, X_2, ...\}$ be a homogeneous discrete-time Markov chain with state space $\mathbf{Z} = \{0, 1, ..., n\}$ and transition probabilities

$$p_{ij} = P(X_{k+1} = j | X_k = i) = \binom{n}{j}\left(\frac{i}{n}\right)^j \left(\frac{n-i}{n}\right)^{n-j}; \quad i, j \in \mathbf{Z}.$$

Show that $\{X_1, X_2, ...\}$ is a martingale. (In genetics, this martingale is known as the *Wright-Fisher model without mutation.*)

6.10) Prove that every stochastic process $\{X(t), t \in \mathbf{T}\}$ with a constant trend function and independent increments which satisfies $E(|X(t)|) < \infty$, $t \in \mathbf{T}$, is a martingale.

6.11) Let L be a stopping time for a stochastic process $\{X(t), t \in \mathbf{T}\}$ in discrete or continuous time and z a positive constant. Verify that $L \wedge z = \min(L, z)$ is a stopping time for $\{X(t), t \in \mathbf{T}\}$.

6.12)* The ruin problem described in section 3.4.1 is modified in the following way: The risk reserve process $\{R(t), t \geq 0\}$ is only observed at the end of each year. The total capital of the insurance company at the end of year n is

$$R(n) = x + \kappa n - \Sigma_{i=0}^{n} M_i; \quad n = 1, 2, ...,$$

where x is the initial capital, κ is the constant premium income per year, and M_i is the total claim size the insurance company has to cover in year i, $M_0 = 0$. The random variables $M_1, M_2, ...$ are assumed to be independent and identically distributed as $M = N(\mu, \sigma^2)$ with $\kappa > \mu > 3\sigma$. Let $p(x)$ be the ruin probability of the company, i.e. the probability that there is an n with property $R(n) < 0$:

$$p(x) = P(\text{there is an } n = 1, 2, ... \text{ so that } R(n) < 0).$$

Show that

$$p(x) \leq e^{-2(\kappa-\mu)x/\sigma^2}, \quad x \geq 0.$$

CHAPTER 7

Brownian Motion

7.1 INTRODUCTION

Tiny organic and inorganic particles when immersed in fluids move randomly along zigzag paths. In 1828, the English botanist *Robert Brown* published a paper, in which he summarized his observations on this motion and tried to find its physical explanation. (Originally, he was only interested in the behaviour of pollen in liquids in order to investigate the fructification process of phanerogams.) However, at that time Brown could only speculate on the causes of this phenomenon and was at an early stage of his research even convinced that he had found an elementary form of life which is common to all particles. Other early explanations refer to attraction and repulsion forces between particles, unstable conditions in the fluids in which they are suspended, capillary actions and so on. Although the ceaseless, seemingly chaotic zigzag movement of microscopically small particles in fluids had already been detected before Brown, it is generally called *Brownian motion.*

The first approaches to mathematically modeling the Brownian motion were made by *L. Bachelier* (1900) and *A. Einstein* (1905). Both found the normal distribution to be an appropriate model for describing the Brownian motion and gave a physical explanation of the observed phenomenon: The chaotic movement of sufficiently small particles in fluids and in gases is due to the huge number of impacts with the surrounding molecules, even in small time intervals. (Assuming average physical conditions, there are about 10^{21} collisions per second between a particle and the surrounding molecules in a fluid.) More precisely, Einstein showed that water molecules could momentarily form a compact conglomerate which has sufficient energy to move a particle, when banging into it. (Note that the tiny particles are 'giants' compared with a molecule.) These bunches of molecules would hit the 'giant' particles from random directions at random times, causing its apparently irregular zigzag motion. Einstein managed to experimentially verify his theoretical findings by just a ruler and a stopwatch. As a 'byproduct', his theory of the Brownian motion and its experimental confirmation yielded another argument for the existence of atoms. Strangely, Einstein was obviously not aware of the considerable efforts, which had been made before him, to understand the phenomenon 'Brownian motion'. *N. Wiener* (1918), better known as the creator of the science of cybernetics, was the first to present a general mathematical treatment of the Brownian motion. He defined and analysed a stochastic process, which has served up till now as a stochastic model of Brownian motion. In what follows, this process is called *Brownian motion process* or, if no misunderstandings are possible, simply *Brownian motion.* Frequently, this process is also referred to as the *Wiener process.* Nowadays the enormous importance of the Brownian motion process is above all due to the fact that it is one of the most powerful tools in theory and applications of stochastic modeling, whose role can be compared with that of the normal distribution in probability theory. The Brownian motion process is an essential ingredient in stochastic calculus, plays a crucial role in mathematics of finance, is basic for defining one of the most important classes of Markov processes, the *diffusion processes*, and for solving large sample estimation problems in mathematical statistics. Brownian motion has fruitful applications in disciplines as time series analysis, operations research, communication theory (modeling signals and noise), and reliability theory (wear modeling, maintenance cost rate modeling). This chapter only deals with the one-dimensional Brownian motion.

Definition 7.1 (*Brownian motion*) A continuous-time stochastic process $\{B(t), t \geq 0\}$ with state space $\mathbf{Z} = (-\infty, +\infty)$ is called (one-dimensional) *Brownian motion process* or simply *Brownian motion* if it has the following properties:

1) $B(0) = 0$.

2) $\{B(t), t \geq 0\}$ has homogeneous and independent increments.

3) $B(t)$ has a normal distribution with

$$E(B(t)) = 0 \text{ and } Var(B(t)) = \sigma^2 t, \quad t > 0. \qquad \bullet$$

Note that condition 1, namely $B(0) = 0$, is only a normalization and as an assumption not really necessary. Actually, in what follows situations will arise in which a Brownian motion is required to start at $B(0) = u \neq 0$. In such a case, the process retains property 2, but in property 3 assumption $E(B(t)) = 0$ has to be replaced with $E(B(t)) = u$. The process $\{B_u(t), t \geq 0\}$ with $B_u(t) = u + B(t)$ is called a *shifted Brownian motion.*

In view of properties 2 and 3, the increment $B(t) - B(s)$ has a normal distribution with mean value 0 and variance $\sigma^2 |t-s|$:

$$B(t) - B(s) = N(0, \sigma^2 |t-s|), \quad s, t \geq 0. \tag{7.1}$$

In applications of the Brownian motion to finance, the parameter σ is called *volatility*. Note that

$$\sigma^2 = Var(B(1)). \tag{7.2}$$

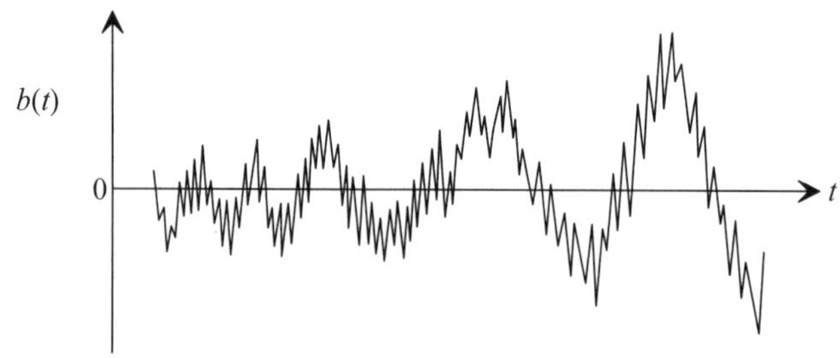

Figure 7.1 Sample path of the Brownian motion

Standard Brownian Motion If $\sigma = 1$, then $\{B(t), t \ge 0\}$ is called a *standard Brownian motion* and will be denoted as $\{S(t), t \ge 0\}$. For any Brownian motion with parameter σ,

$$B(t) = \sigma S(t). \tag{7.3}$$

Laplace Transform Since $B(t) = N(0, \sigma^2 t)$, the Laplace transform of $B(t)$ is (see example 1.12, section 1.3.2)

$$E\left(e^{-\alpha B(t)}\right) = e^{+\frac{1}{2}\alpha^2\sigma^2 t}. \tag{7.4}$$

7.2 PROPERTIES OF THE BROWNIAN MOTION

The first problem which has to be addressed is whether there exists a stochastic process having properties 1 to 3. An affirmative answer was already given by N. Wiener in 1923. In what follows, a constructive proof of the existence of the Brownian motion is given. This is done by showing that Brownian motion can be represented as the limit of a discrete-time random walk, where the size of the steps tends to 0 and the number of steps per unit time is speeded up.

Brownian Motion and Random Walk With respect to the physical background of the Brownian motion, it is not surprising that there is a close relationship between Brownian motion and the random walk of a particle along the real axis. Modifying the random walk described in example 4.1, it is now assumed that after every Δt time units the particle jumps Δx length units to the right or to the left, each with probability 1/2. Thus, if $X(t)$ is the position of the particle at time t and $X(0) = 0$,

$$X(t) = (X_1 + X_2 + \cdots + X_{[t/\Delta t]})\,\Delta x, \tag{7.5}$$

where

$$X_i = \begin{cases} +1 & \text{if the } i\text{th jump goes to the right} \\ -1 & \text{if the } i\text{th jump goes to the left} \end{cases}$$

and $[t/\Delta t]$ denotes the greatest integer less than or equal to $t/\Delta t$. The random variables X_i are independent of each other and have probability distribution

$$P(X_i = 1) = P(X_i = -1) = 1/2 \quad \text{with} \quad E(X_i) = 0, \; Var(X_i) = 1.$$

Formula (1.105) applied to (7.5) yields

$$E(X(t)) = 0, \quad Var(X(t)) = (\Delta x)^2\,[t/\Delta t].$$

With a positive constant σ, let $\Delta x = \sigma\sqrt{\Delta t}$. Then, taking the limit as $\Delta t \to 0$ in (7.5), a stochastic process in continuous time $\{X(t), t \ge 0\}$ arises which has trend and variance function

$$E(X(t)) = 0, \quad Var(X(t)) = \sigma^2 t.$$

Due to its construction, $\{X(t), t \geq 0\}$ has independent and homogeneous increments. Moreover, by the central limit theorem, $X(t)$ has a normal distribution for all $t > 0$. Therefore, the stochastic process of the 'infinitesimal random walk' $\{X(t), t \geq 0\}$ is a Brownian motion.

Even after Norbert Wiener, many amazing properties of the Brownian motion have been detected. Some of them will be considered in this chapter. The following theorem summarizes key properties of the Brownian motion.

Theorem 7.1 A Brownian motion $\{B(t),\ t \geq 0\}$ has the following properties:

a) $\{B(t),\ t \geq 0\}$ is mean-square continuous.

b) $\{B(t),\ t \geq 0\}$ is a martingale.

c) $\{B(t),\ t \geq 0\}$ is a Markov process.

d) $\{B(t),\ t \geq 0\}$ is a Gaussian process.

Proof a) From (7.1),

$$E((B(t) - B(s))^2) = Var(B(t) - B(s)) = \sigma^2 |t - s| . \tag{7.6}$$

Hence,

$$\lim_{h \to 0} E\left([B(t+h) - B(t)]^2 \right) = \lim_{h \to 0} \sigma^2 |h| = 0 .$$

Thus, the limit exists with regard to the convergence in mean-square (section 1.9.1).

b) Since a Brownian motion $\{B(t),\ t \geq 0\}$ has independent increments, for $s < t$,

$$\begin{aligned} E(B(t) | B(y),\ y \leq s)) &= E(B(s) + B(t) - B(s) |\ B(y),\ y \leq s)) \\ &= B(s) + E(B(t) - B(s) | B(y),\ y \leq s)) \\ &= B(s) + E(B(t) - B(s)) \\ &= B(s) + 0 - 0 = B(s) . \end{aligned}$$

Therefore, $\{B(t),\ t \geq 0\}$ is a martingale.

c) Any stochastic process with independent increments is a Markov process.

d) Let $t_1, t_2, ..., t_n$ be any sequence of real numbers with $0 < t_1 < t_2 < \cdots < t_n < \infty$. It has to be shown that for all $n = 1, 2, ...$ the random vector

$$(B(t_1), B(t_2), \ldots, B(t_n))$$

has an n-dimensional normal distribution. This is an immediate consequence of theorem 1.2 (section 1.6.3), since each $B(t_i)$ can be represented as a sum of independent, normally distributed random variables (increments) in the following way:

$$B(t_i) = B(t_1) + (B(t_2) - B(t_1)) + \cdots + (B(t_i) - B(t_{i-1}));\ \ i = 2, 3, \ldots, n. \qquad \blacksquare$$

Theorem 7.2 Let $\{S(t), t \geq 0\}$ be the standardized Brownian motion. Then, for any constant $\alpha \neq 0$, the stochastic processes $\{Y(t), t \geq 0\}$ defined as follows are martingales:

a) $Y(t) = e^{\alpha S(t) - \alpha^2 t/2}$ (*exponential martingale*),

b) $Y(t) = S^2(t) - t$ (*variance martingale*).

Proof a) For $s < t$,

$$E(e^{\alpha S(t)-\alpha^2 t/2} \,|\, S(y),\, y \leq s) = E(e^{\alpha[S(s)+S(t)-S(s)]-\alpha^2 t/2} \,|\, S(y),\, y \leq s)$$

$$= e^{\alpha S(s)-\alpha^2 t/2} E(e^{\alpha[S(t)-S(s)]} \,|\, S(y),\, y \leq s)$$

$$= e^{\alpha S(s)-\alpha^2 t/2} E\left(e^{\alpha[S(t)-S(s)]}\right).$$

From (7.4) with $\sigma = 1$,

$$E\left(e^{\alpha[S(t)-S(s)]}\right) = e^{+\frac{1}{2}\alpha^2 (t-s)}.$$

Hence,

$$E(e^{\alpha S(t)-\alpha^2 t/2} \,|\, S(y),\, y \leq s) = e^{\alpha S(s)-\alpha^2 s/2}. \tag{7.7}$$

b) For $s < t$, since $S(s)$ and $S(t) - S(s)$ are independent and $E(S(x)) = 0$ for all $x \geq 0$,

$$E(S^2(t) - t | S(y),\, y \leq s) = E([S(s) + S(t) - S(s)]^2 - t | S(y),\, y \leq s)$$

$$= S^2(s) + E\{2\,S(s)\,[S(t) - S(s)] + [S(t) - S(s)]^2 - t | S(y),\, y \leq s\}$$

$$= S^2(s) + 0 + E\{[S(t) - S(s)]^2\} - t$$

$$= S^2(s) + (t - s) - t$$

$$= S^2(s) - s,$$

which proves the assertion. ■

There is an obvious analogy between the exponential and the variance martingale defined in theorem 7.2 and corresponding discrete-time martingales considered in examples 6.3 and 6.6.

The relationship (7.7) can be used to generate further martingales: Differentiating (7.7) with regard to α once and twice, respectively, and letting $\alpha = 0$, 'proves' once more that $\{S(t), t \geq 0\}$ and $\{S^2(t) - t, t \geq 0\}$ are martingales. The same procedure, when differentiating (7.7) three and four times, generates the martingales

$$\{S^3(t) - 3\,t\,S(t),\ t \geq 0\} \quad \text{and} \quad \{S^4(t) - 6\,t\,S^2(t) + 3\,t^2,\ t \geq 0\}.$$

This algorithm produces martingales when differentiating $k = 2, 3, \ldots$ times.

Properties of the Sample Paths Since a Brownian motion is mean-square continuous, it is not surprising that its sample paths $b = b(t)$ are continuous functions in t. More exactly, the probability that a sample path of a Brownian motion is continuous is equal to 1. Or, equivalently, 'almost all sample paths of a Brownian motion are continuous.' In view of this, it may surprise that the sample paths of a Brownian motion are nowhere differentiable. This is here not proved either, but it is made plausible by using (7.6): For any sample path $b = b(t)$ and any sufficiently small, but positive Δt, the difference

$$\Delta b = b(t+\Delta t) - b(t)$$

is approximately equal to $\sigma\sqrt{\Delta t}$. Therefore,

$$\frac{\Delta b}{\Delta t} = \frac{b(t+\Delta t) - b(t)}{\Delta t} \approx \frac{\sigma\sqrt{\Delta t}}{\Delta t} = \frac{\sigma}{\sqrt{\Delta t}}.$$

Hence, for $\Delta t \to 0$, the difference quotient $\Delta b/\Delta t$ is likely to tend to infinity for any nonnegative t. Thus, it can be anticipated that the sample paths of a Brownian motion are nowhere differentiable. (For proofs, see e.g. Kannan [43].)

The *variation* of a sample path (as well as of any real function) $b = b(t)$ in an interval $[0, s]$ with $s > 0$ is defined as the limit

$$\lim_{n\to\infty} \sum_{k=1}^{2^n} \left| b\left(\frac{ks}{2^n}\right) - b\left(\frac{(k-1)s}{2^n}\right) \right|. \tag{7.8}$$

A consequence of the non-differentiability of the sample paths is that this limit, no matter how small s is, cannot be finite. Hence, any sample path of a Brownian motion is of *unbounded variation*. This property in its turn implies that the 'length' of a sample path over the finite interval $[0, s]$, and, hence, over any finite interval $[s, t]$, is infinite. What geometric structure is such a sample path supposed to have? The most intuitive explanation is that the sample paths of any Brownian motion are strongly dentate (in the sense of the structure of leaves), but this structure must continue to the infinitesimal. This explanation corresponds to the physical interpretation of the Brownian motion. The numerous and rapid bombardments of particles in liquids or gases by the surrounding molecules cannot lead to a smooth sample path. Unfortunately, the unbounded variation of the sample paths implies that particles move with an infinitely large velocity when dispersed in liquids or gases. Hence, the Brownian motion process cannot be a mathematically exact model for describing the movement of particles in these media. But it is definitely a good approximation. (For modeling the velocity of particles in liquids or gases the Ornstein-Uhlenbeck process has been developed, see section 7.5.2.) However, as pointed out in the introduction, nowadays the enormous theoretical and practical importance of the Brownian motion within the theory of stochastic processes and their applications goes far beyond its being a mathematical model for describing the movement of microscopically small particles in liquids or gases.

7.3 MULTIDIMENSIONAL AND CONDITIONAL DISTRIBUTIONS

Let $\{B(t), t \ge 0\}$ be a Brownian motion and $f_t(x)$ the density of $B(t), t > 0$. From property 3 of definition 7.1,

$$f_t(x) = \frac{1}{\sqrt{2\pi t}\,\sigma} e^{-\frac{x^2}{2\sigma^2 t}}, \quad t > 0. \tag{7.9}$$

Since the Brownian motion is a Gaussian process, its multidimensional distributions are multidimensional normal distributions. To determine the parameters of this distribution, next the joint density $f_{s,t}(x_1, x_2)$ of $(B(s), B(t))$ will be derived.

Because of the independence of the increments of the Brownian motion and in view of $B(t) - B(s)$ having probability density $f_{t-s}(x)$, for small Δx_1 and Δx_2,

$$f_{s,t}(x_1, x_2)\,\Delta x_1 \Delta x_2 = P(x_1 \le B(s) \le x_1 + \Delta x_1,\, x_2 \le B(t) \le x_2 + \Delta x_2)$$

$$= P(x_1 \le B(s) \le x_1 + \Delta x_1,\; x_2 - x_1 \le B(t) - B(s) \le x_2 - x_1 + \Delta x_2 - \Delta x_1)$$

$$= f_s(x_1)\, f_t(x_2 - x_1)\,\Delta x_1 \Delta x_2 .$$

Hence,

$$f_{s,t}(x_1, x_2) = f_s(x_1) f_{t-s}(x_2 - x_1). \tag{7.10}$$

(This derivation can easily be made rigorously.) Substituting (7.9) into (7.10) yields after some simple algebra

$$f_{s,t}(x_1, x_2) = \frac{1}{2\pi\sigma^2 \sqrt{s\,(t-s)}} \exp\left\{-\frac{1}{2\sigma^2 s\,(t-s)}\left(t x_1^2 - 2 s x_1 x_2 + s x_2^2\right)\right\}. \tag{7.11}$$

Comparing this density with the density of the bivariate normal distribution (1.66) shows that $\{B(s), B(t)\}$ has a joint normal distribution with correlation coefficient

$$\rho = +\sqrt{s/t}\,, \quad 0 < s < t.$$

Therefore, if $0 < s < t$, the covariance function of the Brownian motion is

$$C(s,t) = Cov(B(s), B(t)) = \sigma^2 s.$$

Since the roles of s and t can be changed,

$$C(s,t) = \sigma^2 \min(s,t). \tag{7.12}$$

However, in view of the independence of the increments of the Brownian motion, it is easier to directly determine the covariance function of $\{B(t), t \ge 0\}$: For $0 < s \le t$,

$$C(s,t) = Cov(B(s), B(t)) = Cov(B(s),\, B(s) + B(t) - B(s))$$

$$= Cov(B(s),\, B(s)) + Cov(B(s), B(t) - B(s))$$

$$= Cov(B(s),\, B(s)).$$

Hence,

$$C(s,t) = Var(B(s)) = \sigma^2 s, \quad 0 < s \le t.$$

Let $0 < s < t$. According to (1.59), the conditional density of $B(s)$ given $B(t) = b$ is

$$f_{B(s)}(x|B(t) = b) = \frac{f_{s,t}(x,b)}{f_t(b)}. \tag{7.13}$$

Substituting (7.9) and (7.11) into (7.13) yields

$$f_{X(s)}(x|B(t) = b) = \frac{1}{\sqrt{2\pi \frac{s}{t}(t-s)}\,\sigma} \exp\left\{-\frac{1}{2\sigma^2 \frac{s}{t}(t-s)}\left(x - \frac{s}{t}b\right)^2\right\}. \tag{7.14}$$

This is the density of a normally distributed random variable with parameters

$$E(B(s)|B(t) = b) = \frac{s}{t}b, \quad Var(B(s)|B(t) = b) = \sigma^2 \frac{s}{t}(t-s). \tag{7.15}$$

Obviously, the conditional variance assumes its maximum at $s = t/2$.

Let $f_{t_1,t_2,\ldots,t_n}(x_1,x_2,\ldots,x_n)$ be the n-dimensional density of the random vector

$$(B(t_1), B(t_2), \ldots, B(t_n)) \quad \text{with} \quad 0 < t_1 < t_2 < \cdots < t_n < \infty.$$

From (7.10), by induction,

$$f_{t_1,t_2,\ldots,t_n}(x_1,x_2,\ldots,x_n) = f_{t_1}(x_1)\, f_{t_2-t_1}(x_2 - x_1)\cdots f_{t_n-t_{n-1}}(x_n - x_{n-1}).$$

With $f_t(x)$ given by (7.9), the n-dimensional joint density becomes

$$f_{t_1,t_2,\ldots,t_n}(x_1,x_2,\ldots,x_n) \tag{7.16}$$

$$= \frac{\exp\left\{-\frac{1}{2\sigma^2}\left[\frac{x_1^2}{t_1} + \frac{(x_2-x_1)^2}{t_2-t_1} + \cdots + \frac{(x_n-x_{n-1})^2}{t_n-t_{n-1}}\right]\right\}}{(2\pi)^{n/2}\,\sigma^n \sqrt{t_1(t_2-t_1)\cdots(t_n-t_{n-1})}}.$$

Transforming this density analogously to the two-dimensional case shows that (7.16) has the form (1.89). This proves once more that the Brownian motion is a Gaussian process.

The Brownian motion, as any Gaussian process, is completely determined by its trend and covariance function. Actually, since the trend function of a Brownian motion is identically zero, the Brownian motion is completely characterized by its covariance function. In other words, given σ^2, there is exactly one Brownian motion process with covariance function

$$C(s,t) = \sigma^2 \min(s,t).$$

Example 7.1 (***Brownian bridge***) The *Brownian bridge* $\{\bar{B}(t), t \in [0,1]\}$ is a stochastic process, which is given by the Brownian motion in the time interval $[0,1]$ on condition that $B(1) = 0$:

$$\bar{B}(t) = B(t), \ 0 \le t \le 1; \ B(1) = 0.$$

Letting in (7.14) $b = 0$ and $t = 1$ yields the probability density of $\bar{B}(t)$:

$$f_{\bar{B}(t)}(x) = \frac{1}{\sqrt{2\pi t(1-t)}\,\sigma} \exp\left\{-\frac{x^2}{2\sigma^2 t(1-t)}\right\}, \quad 0 < t < 1.$$

Mean value and variance of $\bar{B}(t)$ are

$$E(\bar{B}(t)) = 0, \quad Var(\bar{B}(t)) = \sigma^2 t(1-t), \quad 0 \le t \le 1.$$

The two-dimensional probability density of $(\bar{B}(s), \bar{B}(t))$ can be obtained from

$$f_{t_1,t_2}(x_1, x_2) = \frac{f_{t_1,t_2,t_3}(x_1, x_2, 0)}{f_{t_3}(0)}$$

with $t_1 = s, \ t_2 = t$ and $t_3 = 1$. Hence, for $0 < s < t < 1$,

$$f_{(\bar{B}(s),\bar{B}(t))}(x_1, x_2) = \frac{\exp\left\{-\frac{1}{2\sigma^2}\left[\frac{t}{s(t-s)}x_1^2 - \frac{2}{t-s}x_1x_2 + \frac{1-s}{(t-s)(1-t)}x_2^2\right]\right\}}{2\pi\sigma^2\sqrt{s(t-s)(1-t)}}.$$

A comparison with (1.66) shows that correlation and covariance function of the Brownian bridge are

$$\rho(s,t) = \sqrt{\frac{s(1-t)}{t(1-s)}}, \qquad C(s,t) = \sigma^2 s(1-t), \qquad 0 < s < t \le 1.$$

The Brownian bridge is a Gaussian process whose trend function is identically 0. Hence, it is uniquely determined by its covariance function. □

7.4 FIRST PASSAGE TIMES

By definition, the Brownian motion $\{B(t), t \ge 0\}$ starts at $B(0) = 0$. The random time point $L(x)$ at which the process $\{B(t), t \ge 0\}$ reaches a given level x for the first time is called the *first passage time* or the *first hitting time* of $\{B(t), t \ge 0\}$ with respect to level x. Since the sample paths of the Brownian motion are continuous functions, $L(x)$ is uniquely characterized by $B(L(x)) = x$ and can, therefore, be defined as

$$L(x) = \min_t \{t, B(t) = x\}, \quad x \in (-\infty, +\infty).$$

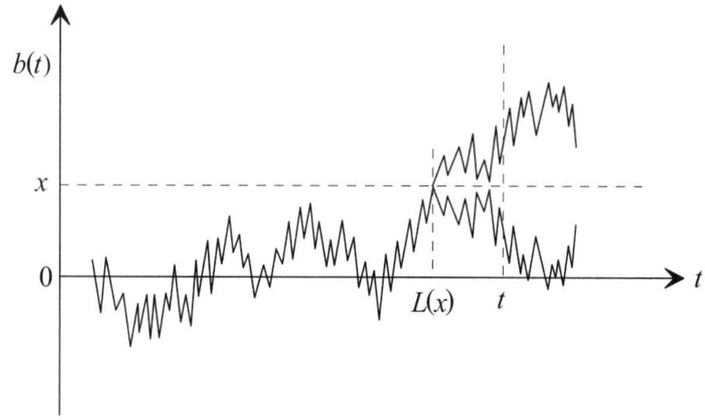

Figure 7.2 Illustration of the first passage time and the reflection principle

Next the probability distribution of $L(x)$ is derived on condition $x > 0$: Application of the the total probability rule yields

$$P(B(t) \geq x) = P(B(t) \geq x | L(x) \leq t)\, P(L(x) \leq t) \qquad (7.17)$$
$$+ P(B(t) \geq x | L(x) > t)\, P(L(x) > t).$$

The second term on the right hand side of this formula vanishes, since, by definition of the first passage time,

$$P(B(t) \geq x | L(x) > t) = 0$$

for all $t > 0$. For symmetry reasons and in view of $B(L(x)) = x$,

$$P(B(t) \geq x | L(x) \leq t) = \frac{1}{2}. \qquad (7.18)$$

This situation is illustrated in Figure 7.2: Two sample paths of the Brownian motion, which coincide up to reaching level x and which after $L(x)$ are mirror symmetric with respect to the straight line $b(t) \equiv x$, have the same chance of occurring. (The probability of this event is, nevertheless, zero.) This heuristic argument is known as the *reflection principle*. Thus, from (7.9), (7.17) and (7.18),

$$F_{L(x)}(t) = P(L(x) \leq t) = 2\, P(B(t) \geq x) = \frac{2}{\sqrt{2\pi t}\,\sigma} \int_x^{\infty} e^{-\frac{u^2}{2\sigma^2 t}}\, du\,.$$

For symmetry reasons, the probability distributions of $L(x)$ and $L(-x)$ are identical for any x. Therefore,

$$F_{L(x)}(t) = \frac{2}{\sqrt{2\pi t}\,\sigma} \int_{|x|}^{\infty} e^{-\frac{u^2}{2\sigma^2 t}}\, du, \quad t > 0.$$

The probability distribution determined by this distribution function is a special case of the *inverse Gaussian distribution* (section 1.2.3.2). Its relationship to the normal distribution (Gaussian distribution) becomes visible after substituting $u^2 = \sigma^2 t y^2$ in the integral of the distribution function of $L(x)$:

$$F_{L(x)}(t) = \frac{2}{\sqrt{2\pi}} \int_{\frac{|x|}{\sigma\sqrt{t}}}^{\infty} e^{-y^2/2}\, dy\,, \quad t > 0.$$

Hence, the distribution function of the first passage time $L(x)$ can be written as

$$F_{L(x)}(t) = 2\left[1 - \Phi\left(\frac{|x|}{\sigma\sqrt{t}}\right)\right], \quad t > 0, \tag{7.19}$$

where as usual $\Phi(\cdot)$ is the distribution function the standard normal distribution. Differentiation with respect to t yields the probability density of $L(x)$:

$$f_{L(x)}(t) = \frac{|x|}{\sqrt{2\pi}\,\sigma\, t^{3/2}} \exp\left\{-\frac{x^2}{2\sigma^2 t}\right\}, \quad t > 0. \tag{7.20}$$

The parameters $E(L(x))$ and $Var(L(x))$ do not exist, i.e. they are infinite.

Maximum Let $M(t)$ be the maximal value of the Brownian motion in $[0, t]$:

$$M(t) = \max\{B(s),\ 0 \le s \le t\}. \tag{7.21}$$

In view of (7.19), the probability distribution of $M(t)$ is obtained as follows:

$$1 - F_{M(t)}(x) = P(M(t) \ge x) = P(L(x) \le t).$$

Hence, by (7.19), distribution function and probability density of $M(t)$ are for $t > 0$,

$$F_{M(t)}(x) = 2\,\Phi\left(\frac{x}{\sigma\sqrt{t}}\right) - 1, \quad x \ge 0, \tag{7.22}$$

$$f_{M(t)}(x) = \frac{2}{\sqrt{2\pi t}\,\sigma}\, e^{-x^2/(2\sigma^2 t)}, \quad x \ge 0. \tag{7.23}$$

As a consequence from (7.22): For all finite x,

$$\lim_{t\to\infty} P(M(t) < x) = 0\,. \tag{7.24}$$

Example 7.2 A sensor for measuring high temperatures gives an unbiased indication of the true temperature during its operating time. At the start, the measurement is absolutely correct. In the course of time, its accuracy deteriorates, but on average no systematic errors occur. Let $B(t)$ be the random deviation of the temperature indicated by the sensor at time t from the true temperature. Historical observations justify the assumption that $\{B(t),\ t \ge 0\}$ is a Brownian motion with parameter

$$\sigma = Var(X(1)) = 0.1 \quad \left(\text{in}\left[{}^{o}\text{C}/\sqrt{24\,\text{h}}\,\right]\right).$$

What is the probability that within a year (365 days) $B(t)$ exceeds the critical level $x = -5\,^{o}\text{C}$, i.e. the sensor indicates at least once in a year $5\,^{o}\text{C}$ degrees less than the true temperature? The desired probability is

$$F_{L(5)}(365) = P(L(-5) < 365) = P(L(5) < 365)$$

$$= 2\left[1 - \Phi\left(\frac{5}{0.1\sqrt{365}}\right)\right] = 2\,[1 - \Phi(2.617)] = 0.009\,.$$

If the accuracy of the sensor is allowed to exceed the critical value of -5 ^{o}C with probability 0.05 during its operating time, then the sensor has to be exchanged by a new one after a time $\tau_{0.05}$ given by

$$P(L(-5) \le \tau_{0.05}) = 0.05\,.$$

According to (7.19), $\tau_{0.05}$ satisfies equation

$$2\left[1 - \Phi\left(\frac{5}{0.1\sqrt{\tau_{0.05}}}\right)\right] = 0.05$$

or, equivalently,

$$\frac{5}{0.1\sqrt{\tau_{0.05}}} = \Phi^{-1}(0.975) = 1.96\,.$$

Thus, $\tau_{0.05} = 651$ [days]. □

The following example presents another, more prestigious application of the probability distribution of $M(t)$.

Example 7.3 Let $p_{(1,d]}$ be the probability that the Brownian motion $\{B(t), t \ge 0\}$ crosses the t-axis at least once in the interval $(1, d]$, $1 < d$. To determine $p_{(1,d]}$ note that for symmetry reasons and in view of (7.22), for any $b > 0$,

$$P(B(t) = 0 \text{ for a } t \text{ with } 1 < t \le d \,|\, B(1) = b)$$

$$= P(B(t) = 0 \text{ for a } t \text{ with } 1 < t \le d \,|\, B(1) = -b)$$

$$= P(B(t) \le -b \text{ for a } t \text{ with } 0 < t \le d - 1)$$

$$= P(B(t) \ge b \text{ for a } t \text{ with } 0 < t \le d - 1)$$

$$= P(M(d-1) \ge b)$$

$$= \frac{2}{\sqrt{2\pi(d-1)}\,\sigma} \int_b^\infty e^{-\frac{u^2}{2\sigma^2(d-1)}}\,du\,, \tag{7.25}$$

where $M(d-1)$ is the maximum of the Brownian motion in the interval $[0, d-1]$. Since b is a value the random variable $B(1)$ can assume, the mean value of the random probability

$$P(B(t)=0 \text{ for a } t \text{ with } 1<t\le d|B(1))$$

is the desired probability $p_{(1,d]}$. Taking into account negative values of $B(1)$ by the factor 2, (7.25) and (7.9) yield

$$p_{(1,d]} = 2\int_0^\infty P(B(t)=0 \text{ for a } t \text{ with } 1<t\le d \,|\, B(1)=b) f_{B(1)}(b)\,db$$

$$= \frac{2}{\pi\sqrt{d-1}\,\sigma^2}\int_0^\infty \int_b^\infty e^{-\frac{u^2}{2\sigma^2(d-1)}}\,du\; e^{-\frac{b^2}{2\sigma^2}}\,db.$$

By substituting

$$u = x\sigma\sqrt{d-1} \text{ and } y = b/\sigma$$

in the inner and outer integral, respectively,

$$p_{(1,d]} = \frac{2}{\pi}\int_0^\infty \int_{\frac{y}{\sqrt{d-1}}}^\infty e^{-\frac{x^2+y^2}{2}}\,dx\,dy.$$

The integration can be simplified by transition to polar coordinates (r,φ). Then the domain of the (x,y)-integration has to be transformed as follows:

$$\left\{0<y<\infty,\ \frac{y}{\sqrt{d-1}}<x<\infty\right\} \to \left\{0<r<\infty,\ \arctan\frac{1}{\sqrt{d-1}}<\varphi<\frac{\pi}{2}\right\}.$$

Since

$$\int_0^\infty r e^{-r^2/2}\,dr = 1,$$

the desired probability becomes

$$p_{(1,d]} = \frac{2}{\pi}\int_0^\infty \int_{\arctan\frac{1}{\sqrt{d-1}}}^{\pi/2} e^{-r^2/2}\,r\,d\varphi\,dr$$

$$= \frac{2}{\pi}\left[\frac{\pi}{2} - \arctan\frac{1}{\sqrt{d-1}}\right]\int_0^\infty r e^{-r^2/2}\,dr$$

$$= 1 - \frac{2}{\pi}\arctan\frac{1}{\sqrt{d-1}}$$

$$= \frac{2}{\pi}\arccos\frac{1}{\sqrt{d}}.$$

By introducing the time unit c, $0 < c < d$, i.e. replacing d with d/c, this formula yields the probability $p_{(c,d]}$ that the Brownian motion crosses the t-axis at least once in the interval $(c, d]$:

$$p_{(c,d]} = \frac{2}{\pi} \arccos \sqrt{\frac{c}{d}}.$$

Hence, the probability that the Brownian motion does not cross the x-axis in $(c, d]$ is

$$1 - p_{(c,d]} = \frac{2}{\pi} \arcsin \sqrt{\frac{c}{d}}. \tag{7.26}$$

Now, let

$$\tau = \max_t \{t,\ t \le d,\ B(t) = 0\},$$

i.e. τ is the largest time point less than or equal to d with property $B(\tau) = 0$. Then the random event '$\tau \le c$' with $c < d$ occurs if and only if there is no time point t in $(c, d]$ satisfying $B(t) = 0$. Hence, as a corollary from (7.26), for $0 < c < d$,

$$P(\tau \le c) = \frac{2}{\pi} \arcsin \sqrt{\frac{c}{d}}.$$ □

The next example considers a first passage time problem with regard to the Brownian motion leaving an interval.

Example 7.4 Let $L(a, b)$ be the random time at which $\{B(t),\ t \ge 0\}$ for the first time hits either value a or value b :

$$L(a, b) = \min_t \{t,\ B(t) = a \text{ or } B(t) = b\};\ \ b < 0 < a.$$

Then the probability $p_{a,b}$ that $\{B(t),\ t \ge 0\}$ assumes value a before value b is

$$p_{a,b} = P(L(a) < L(b)) = P(L(a, b) = L(a))$$

(Figure 7.3) or

$$p_{a,b} = P(B(L(a, b)) = a).$$

To determine $p_{a,b}$, note that $L(a, b)$ is a stopping time for $\{B(t),\ t \ge 0\}$. In view of formula (7.24), $E(L(a, b))$ is finite. Hence, theorem 6.3 is applicable and yields

$$0 = E(B(L(a, b))) = a\,p_{a,b} + b\,(1 - p_{a,b}).$$

Therefore, the probability that the Brownian motion hits value a before value b is

$$p_{a,b} = \frac{|b|}{a + |b|}. \tag{7.27}$$

For determining the mean value of $L(a, b)$, the martingale $\{Y(t),\ t \ge 0\}$ with

$$Y(t) = \frac{1}{\sigma^2} B^2(t) - t$$

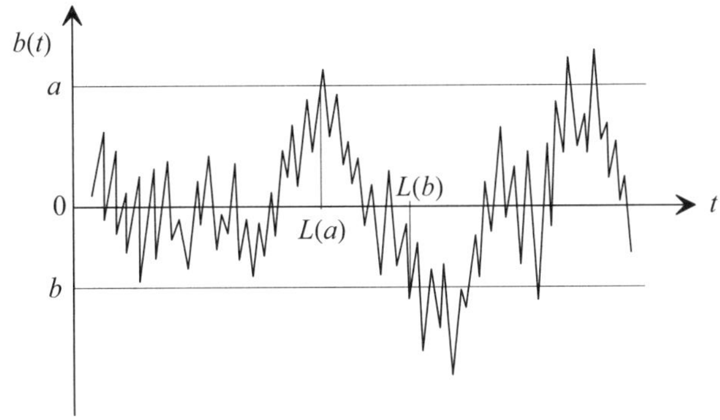

Figure 7.3 First-passage times with regard to an interval

is used (theorem 7.2 b). In this case, theorem 6.3 yields

$$0 = E\left(\frac{1}{\sigma^2} B^2(L(a,b))\right) - E(L(a,b)).$$

Hence,

$$E(L(a,b)) = E\left(\frac{1}{\sigma^2} B^2(L(a,b))\right)$$

$$= \frac{1}{\sigma^2}\left[p_{a,b}\, a^2 + (1 - p_{a,b})\, b^2\right].$$

Thus, by (7.27),

$$E(L) = \frac{1}{\sigma^2}\, a\, |b|. \tag{7.28}$$

As an application of the situation considered in this example, assume that the total profit which a speculator makes with a certain investment develops according to a Brownian motion process $\{B(t), t \geq 0\}$, i.e. $B(t)$ is the cumulative 'profit', the speculator has achieved at time t (possibly negative). If the speculator stops investing after having achieved a profit of a or after having suffered a loss of b, then $p_{a,b}$ is the probability that he finishes with a profit of a. With reference to example 7.2: The probability that the sensor reads 8 ^{o}C high before it reads 4°C low is equal to

$$4/(8+4) = 1/3.$$

Or, if in the same example the tolerance region for $B(t)$ is

$$[-5\,^o\text{C},\ 5\,^o\text{C}],$$

then $B(t)$ on average leaves this region for the first time after

$$E(L) = 25/0.01 = 2500 \text{ days}.$$

7.5 TRANSFORMATIONS OF THE BROWNIAN MOTION

7.5.1 Identical Transformations

Transforming the Brownian motion leads to stochastic processes which are important in their own right, both from the theoretical and practical point of view. Some transformations again lead to the Brownian motion. Theorem 7.3 compiles three transformations of this type.

Theorem 7.3 If $\{S(t), t \geq 0\}$ is the standard Brownian motion, then each of the following stochastic processes is also the standard Brownian motion:

(1) $\{X(t), t \geq 0\}$ with $X(t) = c\,S(t/c^2), \;\; c > 0,$

(2) $\{Y(t), t \geq 0\}$ with $Y(t) = S(t+h) - S(h), \;\; h > 0,$

(3) $\{Z(t), t \geq 0\}$ with $Z(t) = \begin{cases} t\,S(1/t) & \text{for } t > 0 \\ 0 & \text{for } t = 0 \end{cases}.$

Proof The theorem is proved by verifying properties 1) to 3) in definition 7.1. Obviously, the processes (1) to (3) start at the origin: $X(0) = Y(0) = Z(0) = 0$. Since the Brownian motion has independent, normally distributed increments, the processes (1) to (3) have the same property. Their trend functions are identically zero. Therefore, it remains to show that the increments of the processes (1) to (3) are homogeneous. In view of (7.1), it suffices to prove that the variances of the increments of the processes (1) to (3) in any interval $[s, t]$ with $s < t$ are equal to $t - s$. The following derivations make use of $E(S^2(t)) = t$ and (7.12).

(1)
$$Var(X(t) - X(s)) = E([X(t) - X(s)]^2)$$
$$= E(X^2(t)) - 2Cov(X(s), X(t)) + E(X^2(s))$$
$$= c^2\left[E(S^2(t/c^2)) - 2Cov(S(s/c^2), S^2(t/c^2)) + E(S^2(s/c^2))\right]$$
$$= c^2\left[\frac{t}{c^2} - 2\frac{s}{c^2} + \frac{s}{c^2}\right] = t - s.$$

(2)
$$Var(Y(t) - Y(s)) = E([S(t+h) - S(s+h)]^2)$$
$$= E(S^2(t+h)) - 2\,Cov(S(s+h)\,S(t+h)) + E(S^2(s+h))$$
$$= (t+h) - 2(s+h) + (s+h) = t - s.$$

(3)
$$Var(Z(t) - Z(s)) = E([t\,S(1/t) - s\,S(1/s)]^2)$$
$$= t^2 E(S^2(1/t)) - 2\,s\,t\,Cov(S(1/s)\,S(1/t)) + s^2 E(S^2(1/s))$$
$$= t^2 \cdot \frac{1}{t} - 2\,s\,t \cdot \frac{1}{t} + s^2 \cdot \frac{1}{s} = t - s.$$

Thus, the theorem is proved. ■

For any Brownian motion $\{B(t),\ t \geq 0\}$, with probability 1,

$$\lim_{t\to\infty} \frac{1}{t} B(t) = 0. \tag{7.29}$$

(For a proof, see, for example, Lawler [54].) If t is replaced with $1/t$, then taking the limit as $t \to \infty$, is equivalent to taking the limit as $t \to 0$. Hence, with probability 1,

$$\lim_{t\to 0} t B(1/t) = 0. \tag{7.30}$$

A consequence of (7.29) is that any Brownian motion $\{B(t),\ t \geq 0\}$ crosses the t-axis with probability 1 at least once in the interval $[s, \infty)$, $s > 0$, and, therefore, even countably infinite times. Since

$$\{t B(1/t),\ t \geq 0\}$$

is also a Brownian motion, it must have the same property. Therefore, for any $s > 0$, no matter how small s is, a Brownian motion $\{B(t), t \geq 0\}$ crosses the t-axis in $(0, s]$ countably infinite times with probability 1.

7.5.2 Reflected Brownian Motion

A stochastic process $\{X(t),\ t \geq 0\}$ defined by $X(t) = |B(t)|$ is called a *reflected Brownian motion* (reflected at the t-axis). Its trend and variance function are

$$m(t) = E(X(t)) = \frac{2}{\sqrt{2\pi t}\ \sigma} \int_0^\infty x\, e^{-\frac{x^2}{2\sigma^2 t}} dx = \sigma \sqrt{\frac{2t}{\pi}}\ , \quad t \geq 0,$$

$$Var(X(t)) = E(X^2(t)) - [E(X(t))]^2 = \sigma^2 t - \sigma^2 \frac{2t}{\pi} = (1 - 2/\pi)\, \sigma^2 t.$$

The reflected Brownian motion is a homogeneous Markov process with state space $\mathbf{Z} = [0, \infty)$. This can be seen as follows: For

$$0 \leq t_1 < t_2 < \cdots < t_n < \infty, \quad x_i \in \mathbf{Z},$$

taking into account the Markov property of the Brownian motion and its symmetric stochastic evolvement with regard to the t-axis,

$$P(X(t) \leq y \big| X(t_1) = x_1, X(t_2) = x_2, ..., X(t_n) = x_n)$$

$$= P(-y \leq B(t) \leq +y \big| B(t_1) = \pm x_1, B(t_2) = \pm x_2, ..., B(t_n) = \pm x_n)$$

$$= P(-y \leq B(t) \leq +y | B(t_n) = \pm x_n)$$

$$= P(-y \leq B(t) \leq +y | B(t_n) = x_n).$$

Hence, for $0 \leq s < t$, the transition probabilities

$$P(X(t) \leq y | X(s) = x)$$

of the reflected Brownian motion are determined by the increment of the Brownian motion in $[s, t]$ if it starts at time s at state x. Because such an increment has an $N(x, \sigma^2\tau)$-distribution with $\tau = t - s$,

$$P(X(t) \le y | X(s) = x) = \frac{1}{\sqrt{2\pi\tau}\,\sigma} \int_{-y}^{y} e^{-\frac{(u-x)^2}{2\sigma^2\tau}}\, du.$$

Equivalently,

$$P(X(t) \le y | X(s) = x) = \Phi\left(\frac{y-x}{\sigma\sqrt{\tau}}\right) - \Phi\left(-\frac{y+x}{\sigma\sqrt{\tau}}\right); \quad x, y \ge 0; \;\; \tau = t - s.$$

Since the transition probabilities depend on s and t only via $\tau = t - s$, the reflected Brownian motion is a homogeneous Markov process.

7.5.3 Geometric Brownian Motion

A stochastic process $\{X(t), t \ge 0\}$ with

$$X(t) = e^{B(t)} \tag{7.31}$$

is called *geometric Brownian motion.*

Unlike the Brownian motion, the sample paths of a geometric Brownian motion cannot become negative. Therefore and for analytical convenience, the geometric Brownian motion is a favourite tool in mathematics of finance for modeling share prices, interest rates and so on.

According to (7.4), the Laplace transform of $B(t)$ is

$$\hat{B}(\alpha) = E(e^{-\alpha B(t)}) = e^{+\frac{1}{2}\alpha^2\sigma^2 t}. \tag{7.32}$$

Substituting in (7.32) the parameter α with an integer n yields all the moments of $X(t)$:

$$E(X^n(t)) = e^{+\frac{1}{2}n^2\sigma^2 t}; \quad n = 1, 2, ... \tag{7.33}$$

In particular, mean value and second moment of $X(t)$ are

$$E(X(t)) = e^{+\frac{1}{2}\sigma^2 t}, \quad E(X^2(t)) = e^{+2\sigma^2 t}. \tag{7.34}$$

From (7.34) and (1.19):

$$Var(X(t)) = e^{t\sigma^2}(e^{t\sigma^2} - 1).$$

Although the trend function of the Brownian motion is constant, the trend function of the geometric Brownian motion is increasing:

$$m(t) = e^{\sigma^2 t/2}, \quad t \ge 0. \tag{7.35}$$

7.5.4 Ornstein-Uhlenbeck Process

As mentioned before, if the Brownian motion process would be the absolutely correct model for describing the movements of particles in liquids or gases, the particles had to move with an infinitely large velocity. To overcome this unrealistic situation, *Ornstein* and *Uhlenbeck* developed a stochastic process for modeling the velocity of tiny particles in liquids and gases.

Definition 7.2 Let $\{B(t), t \geq 0\}$ be a Brownian motion with parameter σ. Then the stochastic process $\{U(t), t \in (-\infty, +\infty)\}$ defined by

$$U(t) = e^{-\alpha t} B(e^{2\alpha t}) \tag{7.36}$$

is said to be an *Ornstein-Uhlenbeck process* with parameters σ and α, $\alpha > 0$. ●

The density of $U(t)$ is easily derived from (7.9):

$$f_{U(t)}(x) = \frac{1}{\sqrt{2\pi}\,\sigma} e^{-x^2/(2\sigma^2)}, \quad -\infty < x < \infty.$$

Thus, $U(t)$ has a normal distribution with parameters

$$E(U(t)) = 0, \quad Var(U(t)) = \sigma^2.$$

In particular, the trend function of the Ornstein-Uhlenbeck process is identically 0 and $U(t)$ is standard normal if $\{B(t), t \geq 0\}$ is the standard Brownian motion.

Since $\{B(t), t \geq 0\}$ is a Gaussian process, the Ornstein-Uhlenbeck process has the same property. (This is a corollary from theorem 1.2.) Hence, the multidimensional distributions of the Ornstein-Uhlenbeck process are multidimensional normal distributions. Moreover, there is a unique correspondence between the sample paths of the Brownian motion and the sample paths of the corresponding Ornstein-Uhlenbeck process. Thus, the Ornstein-Uhlenbeck process, like the Brownian motion, is a Markov process. Its covariance function is

$$C(s,t) = \sigma^2 e^{-\alpha(t-s)}, \quad s \leq t. \tag{7.37}$$

This is proved as follows: For $s \leq t$, in view of (7.12),

$$\begin{aligned} C(s,t) &= Cov(U(s), U(t)) = E(U(s)U(t)) \\ &= e^{-\alpha(s+t)} E(B(e^{2\alpha s}) B(e^{2\alpha t})) \\ &= e^{-\alpha(s+t)} Cov(B(e^{2\alpha s}), B(e^{2\alpha t})) \\ &= e^{-\alpha(s+t)} \sigma^2 e^{2\alpha s} = \sigma^2 e^{-\alpha(t-s)}. \end{aligned}$$

Corollary The Ornstein-Uhlenbeck process is weakly stationary. Therefore, as a Gaussian process, it is also strongly stationary.

The stationary Ornstein-Uhlenbeck process arises from the nonstationary Brownian motion by time transformation and standardization. In contrast to the Brownian motion, the Ornstein-Uhlenbeck process has the following properties:

1) The increments of the Ornstein-Uhlenbeck process are not independent.

2) The Ornstein-Uhlenbeck process is mean-square differentiable.

7.5.5 Brownian Motion with Drift

7.5.5.1 Definitions and First Passage Times

Definition 7.3 A stochastic process $\{D(t),\ t \ge 0\}$ is called *Brownian motion with drift* if it has the following properties:

1) $D(0) = 0$.

2) $\{D(t),\ t \ge 0\}$ has homogeneous, independent increments.

3) Every increment $D(t) - D(s)$ has a normal distribution with mean value $\mu\,(t-s)$ and variance $\sigma^2 |t-s|$. ●

An equivalent definition of the Brownian motion with drift is: $\{D(t),\ t \ge 0\}$ is a Brownian motion with drift if and only if $D(t)$ has structure

$$D(t) = \mu\, t + B(t), \tag{7.38}$$

where $\{B(t),\ t \ge 0\}$ is the Brownian motion. The constant μ is called *drift parameter* or simply *drift*. Thus, a Brownian motion with drift arises by superimposing a Brownian motion on a deterministic function. This deterministic function is a straight line and coincides with the trend function of the Brownian motion with drift:

$$m(t) = E(D(t)) = \mu\, t.$$

If properties 2) and 3) are fulfilled, but the process starts at time $t = 0$ at level u, $u \ne 0$, then the resulting stochastic process $\{D_u(t),\ t \ge 0\}$ is called a *shifted Brownian motion with drift.* $D_u(t)$ has structure

$$D_u(t) = u + D(t).$$

The one-dimensional density functions of the Brownian motion with drift are

$$f_{D(t)}(x) = \frac{1}{\sqrt{2\pi t}\ \sigma}\, e^{-\frac{(x-\mu t)^2}{2\sigma^2 t}}; \quad -\infty < x < \infty, \quad t > 0. \tag{7.39}$$

Brownian motion processes with drift are, amongst other applications, used for modeling wear parameters, maintenance cost rates, productivity criteria and capital increments over given time periods as well as for modeling physical noise. Generally speaking, Brownian motion with drift can successfully be applied to modeling situations in which causally linear processes are permanently disturbed by random influences.

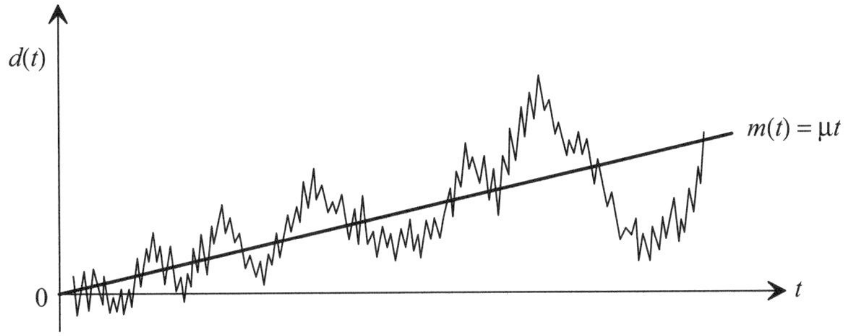

Figure 7.4 Sample path of a Brownian motion with positive drift

Let $L(x)$ be the first passage time of $\{D(t), t \geq 0\}$ with regard to level x. Then,

$$L(x) = \min_t \{t, D(t) = x\}, \quad x \in (-\infty, +\infty).$$

Since every Brownian motion with drift has independent increments and is a Gaussian process, the following relationship between the probability densities of $D(t)$ and $L(x)$ holds:

$$f_{L(x)}(t) = \frac{x}{t} f_{D(t)}(x), \quad x > 0, \ \mu > 0.$$

(For more general assumptions guaranteeing the validity of this formula, see Franz [30].) Hence, the probability density of $L(x)$ is

$$f_{L(x)}(t) = \frac{x}{\sqrt{2\pi}\,\sigma\, t^{3/2}} \exp\left\{-\frac{(x-\mu t)^2}{2\sigma^2 t}\right\}, \quad t > 0. \tag{7.40}$$

(See Scheike [71] for a direct proof of this result.) For symmetry reasons, the probability density of the first passage time $L(x)$ of a Brownian motion with drift starting at u can be obtained from (7.40) by replacing there x with $x - u$.

The probability distribution given by the density (7.40) is the *inverse Gaussian distribution* with parameters μ, σ^2 and x (section 1.2.3.2). Contrary to the first passage time of the Brownian motion, now mean value and variance of $L(x)$ exist:

$$E(L(x)) = \frac{x}{\mu}, \qquad Var(L(x)) = \frac{x\sigma^2}{\mu^3}. \tag{7.41}$$

For $\mu = 0$, the density (7.40) simplifies to the first passage time density (7.20) of the Brownian motion. If $x < 0$ and $\mu < 0$, formula (7.40) yields the density of the corresponding first passage time $L(x)$ by substituting $|x|$ and $|\mu|$ for x and μ, respectively.

Let

$$F_{L(x)}(t) = P(L(x) \leq t) \ \text{ and } \ \overline{F}_{L(x)}(t) = 1 - F_{L(x)}(t), \quad t \geq 0.$$

Assuming $x > 0$ and $\mu > 0$, integration of (7.40) yields

$$\bar{F}_{L(x)}(t) = \Phi\left(\frac{x-\mu t}{\sqrt{t}\,\sigma}\right) - e^{-2x\mu}\,\Phi\left(-\frac{x+\mu t}{\sqrt{t}\,\sigma}\right), \quad t > 0. \tag{7.42}$$

If the second term on the right-hand side of (7.42) is sufficiently small, then one obtains an interesting result: The Birnbaum-Saunders distribution (3.179) as a limit distribution of first passage times of compound renewal processes (theorem 3.20) approximately coincides with the inverse Gaussian distribution.

After some tedious algebra, the Laplace transform of $f_{L(x)}(t)$ is seen to be

$$E\left(e^{-sL(x)}\right) = \int_0^\infty e^{-st} f_{L(x)}(t)\,dt = \exp\left\{-\frac{x}{\sigma^2}\left(\sqrt{2\sigma^2 s + \mu^2} - \mu\right)\right\}. \tag{7.43}$$

Theorem 7.4 Let M be the absolute maximum of the Brownian motion with drift on the positive semiaxis $(0, \infty)$:

$$M = \max_{t \in (0,\infty)} D(t).$$

Then,

$$P(M > x) = \begin{cases} 1 & \text{for } x > 0 \text{ and } \mu > 0 \\ e^{-2|\mu|x/\sigma^2} & \text{for } x > 0 \text{ and } \mu < 0 \end{cases}. \tag{7.44}$$

Proof In view of (7.24), it is sufficient to prove (7.44) for $\mu < 0$. The exponential martingale $\{Y(t), t \geq 0\}$ with

$$Y(t) = e^{\alpha S(t) - \alpha^2 t/2}$$

(theorem 7.2) is stopped at time $L(x)$. Since

$$D(L(x)) = \mu L(x) + \sigma S(L(x)) = x,$$

$Y(L(x))$ can be represented as

$$Y(L(x)) = \exp\left\{\tfrac{\alpha}{\sigma}[x - \mu L(x)] - \alpha^2 L(x)/2\right\} = \exp\left\{\tfrac{\alpha}{\sigma}x - \left[\tfrac{\alpha\mu}{\sigma} + \alpha^2/2\right]L(x)\right\}.$$

Hence,

$$E(Y(L(x))) = e^{\frac{\alpha}{\sigma}x} E\left(\exp\left\{\frac{\alpha|\mu|}{\sigma} - \alpha^2/2\right\}L(x) \,\middle|\, L(x) < \infty\right) P(L(x) < \infty)$$

$$+ e^{\frac{\alpha}{\sigma}x} E\left(\exp\left\{\frac{\alpha|\mu|}{\sigma} - \alpha^2/2\right\}L(x) \,\middle|\, L(x) = \infty\right) P(L(x) = \infty).$$

Assume $\alpha > 2|\mu|/\sigma$. Then the second term disappears and theorem 6.3 yields

$$1 = e^{\frac{\alpha}{\sigma}x} E\left(\exp\left\{\frac{\alpha|\mu|}{\sigma} - \alpha^2/2\right\}L(x) \,\middle|\, L(x) < \infty\right) P(L(x) < \infty).$$

Since $P(M > x) = P(L(x) < \infty)$, letting $\alpha \downarrow 2|\mu|/\sigma$ yields the desired result. ■

Corollary The maximal value a Brownian motion with negative drift assumes on the positive semiaxis $(0, +\infty)$ has an exponential distribution with parameter

$$\lambda = \frac{2\,|\mu|}{\sigma^2}. \tag{7.45}$$

Example 7.5 (*Leaving an interval*) Analogously to example 7.4, let $L(a, b)$ denote the first time point at which the Brownian motion with drift $\{D(t), t \geq 0\}$ hits either value a or value b, $b < 0 < a$, $\mu \neq 0$, and

$$p_{a,b} = P(L(a) < L(b)) = P(L(a, b) = a).$$

Thus, $p_{a,b}$ is the probability that $\{D(t), t \geq 0\}$ hits level a before level b. For establishing an equation in $p_{a,b}$, the exponential martingale in theorem 7.2 with

$$S(t) = \frac{D(t) - \mu t}{\sigma}$$

is stopped at time $L(a, b)$. From theorem 6.3,

$$1 = E\left(\exp\left\{\tfrac{\alpha}{\sigma}\,(D(L(a, b)) - \mu\, L(a, b)) - \frac{\alpha^2 L(a,b)}{2}\right\}\right).$$

Equivalently,

$$1 = E\left(\exp\left\{\tfrac{\alpha}{\sigma}\,(D(L(a, b)) - \left[\frac{\alpha\mu}{\sigma} + \frac{\alpha^2}{2}\right] L(a, b)\right\}\right).$$

Let $\alpha = -2\mu/\sigma$. Then,

$$1 = E\left(e^{\frac{\alpha}{\sigma}(D(L(a,b))}\right) = p_{a,b}e^{-2\mu\, a/\sigma^2} + (1 - p_{a,b})\, e^{-2\mu\, b/\sigma^2}.$$

Solving this equation for $p_{a,b}$ yields

$$p_{a,b} = \frac{1 - e^{-2\mu b/\sigma^2}}{e^{-2\mu a/\sigma^2} - e^{-2\mu b/\sigma^2}}. \tag{7.46}$$

If $\mu < 0$ and b tends to $-\infty$ in (7.46), then the probability $p_{a,b}$ becomes $P(L(a) < \infty)$, which proves once more formula (7.44) with $x = a$.

Generally, for a shifted Brownian motion with drift $\{D_u(t), t \geq 0\}$,

$$D_u(t) = u + D(t), \quad b < u < a, \quad \mu \neq 0,$$

formula (7.46) yields the corresponding probability $p_{a,b}$ by replacing a and b with $a - u$ and $b - u$, respectively (u can be negative):

$$p_{a,b} = P(L(a) < L(b) | D_u(0)) = \frac{e^{-2\mu u/\sigma^2} - e^{-2\mu b/\sigma^2}}{e^{-2\mu a/\sigma^2} - e^{-2\mu b/\sigma^2}}. \qquad \square$$

Geometric Brownian Motion with Drift Let $\{D(t), t \geq 0\}$ be a Brownian motion with drift. Then the stochastic process $\{X(t),\ t \geq 0\}$ with

$$X(t) = e^{D(t)} \tag{7.47}$$

is called *geometric Brownian motion with drift.* If the drift μ is 0, then $\{X(t),\ t \geq 0\}$ is simply the *geometric Brownian motion* as defined by (7.31).

The Laplace transform of $D(t)$ is obtained by multiplying (7.4) by $e^{-t\mu\alpha}$:

$$E(e^{-\alpha D(t)}) = e^{-t\mu\alpha + \frac{1}{2}\sigma^2 t \alpha^2}. \tag{7.48}$$

Letting $\alpha = -1$ and $\alpha = -2$ yields the first two moments of $X(t)$:

$$E(X(t)) = e^{t(\mu+\sigma^2/2)}, \quad E(X^2(t)) = e^{2t\mu + 2\sigma^2 t}. \tag{7.49}$$

Therefore, by (1.19),

$$Var(X(t)) = e^{t(2\mu+\sigma^2)}(e^{t\sigma^2} - 1).$$

Since the inequalities

$$e^{D(t)} \geq x \quad \text{and} \quad D(t) \geq \ln x$$

are equivalent, the first passage time results obtained for the Brownian motion with drift can immediately be used for characterizing the first passage time behavior of the geometric Brownian motion with drift with $\ln x$ instead of x, $x > 0$.

7.5.5.2 Application to Option Pricing

In finance, Brownian motion and its transformations are used to model the evolvement in time of prices of risky securities and combinations of them. The concept of a *risky security* comprises all risky assets, e.g. shares and precious metals. An *option* is a contract, which entitles (but not obliges) its *holder* (*owner*) to either buy or sell a risky security at a fixed, predetermined price, called *strike price* or *exercise price.* A *call* (*put*) *option* gives its holder the right to buy (to sell). An option has a finite or an infinite *expiration* or *maturity date.* An *American option* can be exercised at any time point to its expiration, a *European option* can only be exercised at the time point of its expiration.

A basic problem in option trading is: What amount of money should a speculator pay to the *writer* (seller) of an option at the time of signing the contract? Common sense tells that the writer will fix the option price at a level which is somewhat higher than the mean payoff (profit) the speculator will achieve by acquiring this option. Hence, the following examples focus on determining the mean (expected) payoff of a holder. For instance, if a European call option has the finite expiration date τ, a strike price x_s and the random price (value) of the underlying risky security at time τ is $X(\tau)$, then the holder will achieve a positive random payoff of $X(\tau) - x_s$ if $X(\tau) > x_s$. If $X(\tau) \leq x_s$, then the owner will not exercise because this would make no financial sense. In case of a European put option, the owner will exercise at time τ if $X(\tau) < x_s$ and make a random profit of $x_s - X(\tau)$. Thus, owners of European call or put options will achieve the respective random payoffs (notation: $z_+ = \max(z, 0)$)

$$(X(\tau) - x_s)_+ \text{ and } (x_s - X(\tau))_+ .$$

Another basic aspect in finance is *discounting*. Due to interest and inflation rates, the value which a certain amount of money has today, will not be the value which the same amount of money has tomorrow. In financial calculations, in particular in option pricing, this phenomenon is taken into account by a *discount factor*.

The following examples deal with option pricing under rather simplistic assumptions. For detailed and more general expositions, see, e.g. Bouchaud and Potters [12], Shafer and Vovk [74].

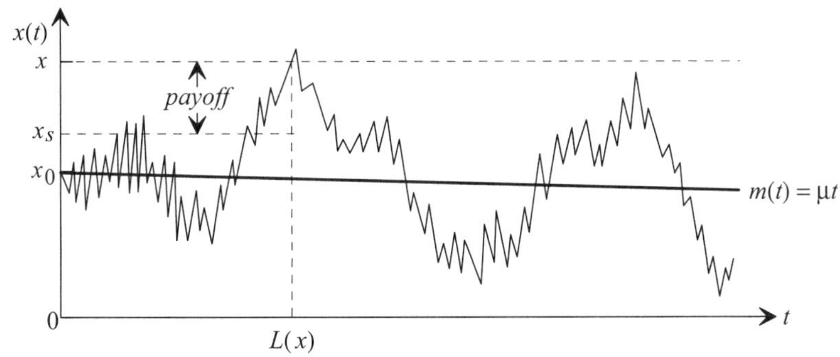

Figure 7.5 Payoff from random share price fluctuations

Example 7.6 The price of a share at time t is given by a shifted Brownian motion $\{X(t) = D_{x_0}(t), t \geq 0\}$ with negative drift μ and volatility $\sigma^2 = Var(B(1))$:

$$X(t) = x_0 + D(t) = x_0 + \mu t + B(t). \tag{7.50}$$

Thus, x_0 is the initial price of the share: $x_0 = X(0)$. Based on this share, a speculator holds an American call option with strike price x_s, $x_s \geq x_0$. The option has no finite expiry date. Although the price of the share is on average decreasing, the speculator hopes to profit from random share price fluctuations. He makes up his mind to exercise the option at that time point, when the share price for the first time reaches value x with $x > x_s$. Thus, if the holder exercises, his payoff will be $x - x_s$ (Figure 7.5). By following this policy, the holder's mean payoff (gain) is

$$G(x) = (x - x_s)p(x) + 0 \cdot (1 - p(x)) = (x - x_s)p(x),$$

where $p(x)$ is the probability that the share price will ever reach level x. Equivalently, $p(x)$ is the probability that the Brownian motion with drift $\{D(t), t \geq 0\}$ will ever reach level $x - x_0$. Since the option has no finite expiration date, this probability is given by (7.44) if there x is replaced with $x - x_0$. Hence the holder's mean payoff is

$$G(x) = (x - x_s)\, e^{-\lambda(x - x_0)} \quad \text{with } \lambda = 2|\mu|/\sigma^2. \tag{7.51}$$

The condition $dG(x)/dx = 0$ yields the optimal value of x: The holder will exercise as soon as the share price hits level

$$x^* = x_s + 1/\lambda. \tag{7.52}$$

The corresponding maximal mean payoff is

$$G(x^*) = \frac{1}{\lambda\, e^{\lambda(x_s - x_0) + 1}}. \tag{7.53}$$

Discounted Payoff Let the constant (risk free) discount rate α be positive. The discounted payoff from exercising the option at time t on condition that the share has at time t price x with $x > x_s$ is $e^{-\alpha t}(x - x_s)$. Since under the policy considered the holder exercises the option at the random time $L_D(x - x_0)$ (= first passage time of $\{D(t), t \geq 0\}$ with respect to level $x - x_0$), his random discounted payoff is

$$e^{-\alpha L_D(x - x_0)}\,(x - x_s).$$

Hence, the holder's mean discounted payoff is

$$G_\alpha(x) = (x - x_s) \int_0^\infty e^{-\alpha t} f_{L_D(x - x_0)}(t)\, dt, \tag{7.54}$$

where the density $f_{L_D(x-x_0)}(t)$ is given by (7.40) with x replaced by $x - x_0$. The integral in (7.54) is equal to the Laplace transform of $f_{L_D(x-x_0)}(t)$ with parameter $s = \alpha$. Thus, from (7.43),

$$G_\alpha(x) = (x - x_s)\, \exp\left\{ -\frac{x - x_0}{\sigma^2} \left(\sqrt{2\sigma^2\alpha + \mu^2} - \mu \right) \right\}. \tag{7.55}$$

The functional structures of the mean undiscounted payoff and the mean discounted payoff as given by (7.51) and (7.55), respectively, are identical. Hence the optimal parameters with respect to $G_\alpha(x)$ are again given by (7.52) and (7.53) with λ replaced by

$$\gamma = \frac{1}{\sigma^2} \left(\sqrt{2\sigma^2\alpha + \mu^2} - \mu \right). \tag{7.56}$$

Note that minimizing $G_\alpha(x)$ also makes sense for a positive drift parameter μ. □

Example 7.7 Since for a negative drift parameter μ the sample paths of a stochastic process $\{X(t), t \geq 0\}$ of structure (7.50) eventually become negative with probability one, the share price model (7.50) is only limitedly applicable, in particular in case of finite expiration dates. In such a situation it seems to be more realistic to model the share price development, apart from a constant factor, by a geometric Brownian motion with drift:

$$X(t) = x_0\, e^{D(t)}, \quad t \geq 0.$$

The other assumptions as well as the formulation of the problem and the notation introduced in the previous example remain valid. In particular, the price of the share at time $t = 0$ is again equal to x_0.

The random event '$X(t) \geq x$' with $x > x_0$ is equivalent to

$$D(t) \geq \ln(x/x_0).$$

Therefore, by (7.44), the probability that the share price will ever reach level x is

$$p(x) = e^{-\lambda \ln(x/x_0)} = \left(\frac{x_0}{x}\right)^{\lambda}.$$

If the holder exercises the option as soon as the share price is x, his mean payoff is

$$G(x) = (x - x_s)\left(\frac{x_0}{x}\right)^{\lambda}. \tag{7.57}$$

The optimal level $x = x^*$ is

$$x^* = \frac{\lambda}{\lambda - 1} x_s. \tag{7.58}$$

To ensure that $x^* > x_s > 0$, an additional assumption has to be made:

$$\lambda = 2|\mu|/\sigma^2 > 1.$$

The corresponding maximal mean payoff is

$$G(x^*) = \left(\frac{\lambda - 1}{x_s}\right)^{\lambda-1} \left(\frac{x_0}{\lambda}\right)^{\lambda}. \tag{7.59}$$

Discounted Payoff The undiscounted payoff $x - x_s$ is made when $\{D(t), t \geq 0\}$ hits level $\ln(x/x_0)$ for the first time, i.e. at time $L_D(\ln(x/x_0))$. Using this and processing as in the previous example, the mean discounted payoff is seen to be

$$G_\alpha(x) = (x - x_s)\left(\frac{x_0}{x}\right)^{\gamma} \tag{7.60}$$

with γ given by (7.56). The functional forms of the mean undiscounted payoff (7.57) and the mean discounted payoff (7.60) are identical. Hence, the corresponding optimal values x^* and $G_\alpha(x^*)$ are given by (7.58) and (7.59) if in these formulas λ is replaced with γ. Note that condition $\gamma > 1$ is equivalent to

$$2(\alpha - \mu) > \sigma^2.$$

As in the previous example, a positive drift parameter μ need not be excluded. □

Example 7.8 (*Formula of Black-Scholes-Merton*) A European call option is considered with strike price x_s and expiration date τ. The option is based on a risky security the price of which, apart from a constant factor x_0, develops according to a geometric Brownian motion with drift $\{X(t), t \geq 0\}$:

$$X(t) = x_0\, e^{D(t)}, \quad t \geq 0.$$

The holder will buy if $X(\tau) > x_s$. Given a constant discount factor α, his random discounted payoff is

$$[e^{-\alpha\tau}(X(\tau)-x_s)]_+ = \max\,[e^{-\alpha\tau}(X(\tau)-x_s),\,0]\,.$$

The holder's mean discounted profit is denoted as

$$G_\alpha(\tau,\mu,\sigma) = E([e^{-\alpha\tau}(X(\tau)-x_s)]_+). \tag{7.61}$$

In view of $D(\tau) = N(\mu\tau, \sigma^2\tau)$,

$$G_\alpha(\tau;\mu,\sigma) = e^{-\alpha\tau}\int_{\ln(x_s/x_0)}^{\infty}(x_0 e^y - x_s)\frac{1}{\sqrt{2\pi\sigma^2\tau}}\exp\left\{-\frac{1}{2\tau}\left(\frac{y-\mu\tau}{\sigma}\right)^2\right\}dy$$

Substituting $u = \dfrac{y-\mu\tau}{\sigma\sqrt{\tau}}$ and letting $c = \dfrac{[\ln(x_s/x_0)-\mu\tau]}{\sigma\sqrt{\tau}}$ yields

$$G_\alpha(\tau;\mu,\sigma) = x_0\, e^{(\mu-\alpha)\tau}\frac{1}{\sqrt{2\pi}}\int_c^\infty e^{u\sigma\sqrt{\tau}}\, e^{-u^2/2}du - x_s\, e^{-\alpha\tau}\frac{1}{\sqrt{2\pi}}\int_c^\infty e^{-u^2/2}du\,.$$

By substituting in the first integral $u = y + \sigma\sqrt{\tau}$,

$$\int_c^\infty e^{u\sigma\sqrt{\tau}}\, e^{-u^2/2}du = e^{\frac{1}{2}\sigma^2\tau}\int_{c-\sigma\sqrt{\tau}}^{\infty} e^{-y^2/2}dy\,.$$

Hence,

$$\begin{aligned}G_\alpha(\tau;\mu,\sigma) &= x_0\, e^{(\mu-\alpha+\sigma^2/2)\tau}\frac{1}{\sqrt{2\pi}}\int_{c-\sigma\sqrt{\tau}}^{\infty} e^{-y^2/2}dy - x_s\, e^{-\alpha\tau}\frac{1}{\sqrt{2\pi}}\int_c^\infty e^{-u^2/2}du\\ &= x_0\, e^{(\mu-\alpha+\sigma^2/2)\tau}\,\Phi(\sigma\sqrt{\tau}-c) - x_s\, e^{-\alpha\tau}(\Phi(-c)).\end{aligned}$$

At time t, the discounted price of the risky security is

$$X_\alpha(t) = e^{-\alpha t}X(t) = x_0\, e^{-(\alpha-\mu)\,t+\sigma S(t)},$$

where $\{S(t),\ t \geq 0\}$ is the standard Brownian motion. In view of theorem 7.2, the stochastic process $\{X_\alpha(t),\ t \geq 0\}$ is a martingale (exponential martingale) if and only if

$$\alpha - \mu = \sigma^2/2.$$

Under this condition, the mean discounted payoff of the holder is given by the *Formula of Black-Scholes-Merton*

$$\tilde{G}_\alpha(\tau,\sigma) = x_0\,\Phi(\sigma\sqrt{\tau}-c) - x_s\, e^{-\alpha\tau}\,\Phi(-c). \tag{7.62}$$

(Black and Scholes [10], Merton [61]). In this formula, the influence of the drift parameter μ on the price development is eliminated by the assumption that the discounted price of the risky security develops according to a martingale. The formula of

Black-Scholes-Merton gives the *fair price* of the option. This is motivated by the fact that a martingale has a constant trend function and that, on average, holder and writer of this option will neither lose nor win. Of course, this statement is only theory, since the price development of the underlying risky security will never strictly follow a geometric Brownian motion with drift. Hence, other stochastic models have been proposed for the price development of risky securities [12, 50, 64, 74]. □

7.5.5.3 Application to Maintenance

In examples 7.9 and 7.10, functionals of the Brownian motion will be used to model the cumulative repair cost arising over a time period and to model the cumulative repair cost rate. It is a formal disadvantage of this model assumption that cumulative repair costs modeled in this way do not have nondecreasing sample paths. However, the problem to be analyzed is not directly based on sample paths generated by the process, but on its trend function and its mean first passage times. Both have 'reasonable' properties with respect to the application considered. Moreover, the results obtained are relevant for all those stochastic maintenance cost developments, where the pair 'trend function and mean first passage time' approximately exhibit the same behaviour as the corresponding pair resulting from the Brownian motion model.

In all examples, the following basic situation is considered: A system starts working at time $t = 0$. The random repair cost accumulating over the time interval $[0, t]$ is denoted as $X(t)$. The sample paths of the stochastic process $\{X(t),\ t \geq 0\}$ are assumed to be continuous and its trend function $m(t) = E(X(t)),\ t \geq 0$, to be progressively (faster than linear) increasing. The cost of each replacement is c, a replacement takes negligibly small time, and after a replacement a system is 'as good as new'. With regard to cost and length, all replacement cycles are independent of each other. In each case, the optimal scheduling of replacements is based on the long-run total maintenance cost per unit time, in what follows referred to as *maintenance cost rate.*

In this section, replacement policies based on limiting the cumulative repair cost $X(t)$ and the cumulative repair cost per unit time (in what follows called *repair cost rate*) $R(t) = X(t)/t$ are considered. These replacement policies need the same basic input as the already classic 'economic lifetime policy', which is introduced next for serving as standard of comparison. The repair-replacement process continues to infinity.

Policy 1 The system is replaced by a new one after reaching its economic lifetime.

Let $K_1(\tau)$ be the maintenance cost rate if the system is always replaced after τ time units. Then, by (3.79),

$$K_1(\tau) = \frac{m(\tau) + c}{\tau}. \tag{7.63}$$

That value of τ minimizing $K_1(\tau)$ is called the *economic lifetime* of the underlying system and denoted as τ^*. If τ^* exists, then

$$K_1(\tau^*) = m'(\tau^*).$$

Policy 2 The system is replaced by a new one as soon as the cumulative repair cost $X(t)$ reaches a given positive level x.

When scheduling replacements according to this policy, a 'typical replacement cycle' has random length $L_X(x)$, where $L_X(x)$ is the first passage time of $\{X(t), t \geq 0\}$ with regard to level x. Under policy 2, the maintenance cost rate has structure

$$K_2(x) = \frac{x + c}{E(L_X(x))}. \tag{7.64}$$

Policy 3 The system is replaced by a new one as soon as the repair cost rate

$$R(t) = X(t)/t$$

reaches a given positive level r.

Under policy 3, the maintenance cost rate has structure

$$K_3(r) = r + \frac{c}{L_R(r)}, \tag{7.65}$$

where $L_R(r)$ is the first passage time of the stochastic process $\{R(t), t \geq 0\}$ with regard to level r. Formulas (7.65) and (7.64) follow from the strong law of the large numbers (theorem 1.8).

Example 7.9 The cumulative repair cost $X(t)$ is assumed to have structure

$$X(t) = x_0\left[e^{D(t)} - 1\right], \tag{7.66}$$

where $\{D(t), t \geq 0\}$ is a Brownian motion with positive drift μ and variance parameter σ^2. Since for a level x with $0 < x_0 < x_0$,

$$X(t) = x \text{ if and only if } D(t) = \ln\left(\frac{x + x_0}{x_0}\right),$$

by (7.41), the mean value of $L_X(x)$ is

$$E(L_X(x)) = \frac{1}{\mu}\ln\left(\frac{x + x_0}{x_0}\right).$$

Therefore, under policy 2,

$$K_2(x) = \frac{x + c}{\ln\left(\frac{x + x_0}{x_0}\right)}\mu.$$

A limit x being optimal with respect to $K_2(x)$ satisfies the condition $dK_2(x)/dx = 0$:

$$\ln\left(\frac{x + x_0}{x_0}\right) = \frac{x + c}{x + x_0}.$$

A unique solution $x = x^*$ exists and the corresponding maintenance cost rate is

$$K_2(x^*) = (x^* + x_0)\mu.$$

Comparison to policy 1 Making use of (7.49) yields

$$m(t) = E(X(t)) = x_0\left(e^{(\mu+\sigma^2/2)\,t} - 1\right), \quad t \ge 0.$$

Hence, the corresponding maintenance cost rate (7.63) is

$$K_1(\tau) = \frac{x_0\left\lfloor e^{(\mu+\sigma^2/2)\tau} - 1\right\rfloor + c}{\tau}\,. \tag{7.67}$$

There exists a unique $\tau = \tau^*$ mimimizing $K_1(\tau)$. By introducing the notation

$$K_1(\tau,\sigma),\ m(\tau,\sigma) \text{ and } \tau^*(\sigma) \text{ for } K_1(\tau),\ m(\tau), \text{ and } \tau^*,$$

$K_1(\tau)$ on condition $\sigma = 0$ is

$$K_1(\tau, 0) = \frac{x_0\,[e^{\mu\tau} - 1] + c}{\tau}\,.$$

Since $m(\tau,\sigma) \ge m(\tau, 0)$ for all σ, there holds

$$K_1(\tau,\sigma) \ge K_1(\tau, 0).$$

One readily verifies that minimizing $K_2(x)$ with respect to $x = x_0\,(e^{\mu\tau} - 1)$ and minimizing $K_1(\tau, 0)$ with respect to τ are equivalent problems. Hence,

$$K_1(\tau^*(\sigma),\sigma) \ge K_1(\tau^*(0), 0) = K_2(x^*)\,.$$

Therefore, applying the economic lifetime on condition that the cumulative repair cost evolves deterministically according to function $m(t, 0)$ is equivalent to applying the optimal total repair cost limit x^*. Thus, policy 2 equalizes the cost-increasing influence of random fluctuations of individual repair costs, which are ignored under policy 1. As a consequence, under the assumptions stated, applying policy 2 leads to a lower maintenance cost rate than applying the economic lifetime. Moreover, the efficiency of policy 2 relative to policy 1 increases with increasing σ. □

Example 7.10 Let the repair cost rate $R(t) = X(t)/t$ be given by

$$R(t) = r_0\,B^4(t); \quad r_0 > 0,\ t \ge 0\,,$$

where $\{B(t),\ t \ge 0\}$ is the Brownian motion with parameter σ. For $r > r_0$,

$$R(t) = r \ \text{ if and only if } \ B(t) = \pm\left(\frac{r}{r_0}\right)^{1/4}.$$

Hence, the mean value of the first passage time of the stochastic process $\{R(t),\ t \ge 0\}$ with regard to level r is given by (7.28) with $a = |b| = (r/r_0)^{1/4}$:

$$E(L_R(r)) = \frac{1}{\sigma^2}\sqrt{\frac{r}{r_0}}\,.$$

Thus, when applying policy 3, the corresponding maintenance cost rate (7.65) is

$$K_3(r) = r + \frac{c\sqrt{r_0}\,\sigma^2}{\sqrt{r}}. \tag{7.68}$$

The necessary condition $dK_3(r)/dr = 0$ yields the optimal repair cost rate limit and the corresponding maintenance cost rate:

$$r^* = \left(\tfrac{1}{4}c^2 r_0 \sigma^4\right)^{1/3}, \quad K_3(r^*) = 1.89\left(c^2 r_0 \sigma^4\right)^{1/3}. \tag{7.69}$$

Comparison to policy 1 Since $B(t) = N(0, \sigma^2 t)$, the trend function of the cumulative repair cost process $\{X(t), t \geq 0\}$ with $X(t) = r_0\, t B^4(t)$ is

$$m(t) = r_0 t E(B^4(t)) = 3\, r_0 \sigma^4 t^3, \quad t \geq 0.$$

The corresponding maintenance cost rate (7.63) is

$$K_1(\tau) = 3\, r_0\, \sigma^4 \tau^2 + \tfrac{c}{\tau}. \tag{7.70}$$

Minimizing (7.70) with regard to τ gives

$$\tau^* = \left(\frac{c}{6\, r_0 \sigma^4}\right)^{1/3}, \quad K_1(\tau^*) = 2.73\left(c^2 r_0 \sigma^4\right). \tag{7.71}$$

With $K_3(r^*)$ given by (7.69) and $K_1(\tau^*)$ given by (7.71),

$$\frac{K_3(r^*)}{K_1(\tau^*)} = 0.69.$$

Hence, applying the optimal repair cost rate limit r^* instead of the economic lifetime τ^* reduces the total maintenance cost on average by 31%. □

The next example illustrates that optimizing replacement intervals on the basis of limits on the cumulative repair cost (rate) does not need full information on the underlying stochastic process $\{X(t), t \geq 0\}$ of the cumulative repair cost development if making use of the fact that this process actually has nondecreasing sample paths.

Example 7.11 It is assumed that the sample paths of the cumulative repair cost process $\{X(t), t \geq 0\}$ are nondecreasing. Then,

$$P(X(t) \leq x) = P(L_X(x) \geq t).$$

Thus, if the one-dimensional probability distribution of $\{X(t), t \geq 0\}$ is given by

$$F_t(x) = P(X(t) \leq x) \quad \text{for all } t \geq 0,$$

trend function and mean first passage time with respect to level x of the cumulative repair cost process $\{X(t), t \geq 0\}$ are

$$m(t) = \int_0^\infty (1 - F_t(x))\, dx,$$

$$E(L_X(x)) = \int_0^\infty F_t(x)\,dt\,.$$

In what follows, policy 2 is applied on condition that $X(t)$ has a Rayleigh distribution with probability density

$$f_t(x) = \frac{2x}{\lambda^2 t^{2y}} \exp\left\{-\left(\frac{x}{\lambda t^y}\right)^2\right\}; \quad x \ge 0,\ y > 1,\ \lambda > 0\,.$$

Then,

$$E(L_X(x)) = \int_0^\infty \int_0^x f_t(u)\,du\,dt = \int_0^x \int_0^\infty f_t(u)\,dt\,du\,.$$

Integration yields

$$E(L_X(x)) = \left(\frac{1}{\lambda}\right)^{1/y} \Gamma\left(1 - \frac{1}{2y}\right) x^{1/y} = k_1 x^{1/y}.$$

Minimizing the corresponding long-run total maintenance cost rate (7.64) yields the optimal limit x^* and the corresponding maintenance cost rate $K_2(x^*)$:

$$x^* = \frac{c}{y-1}, \quad K_2(x^*) = \frac{y}{k_1}\left(\frac{c}{y-1}\right)^{(y-1)/y}.$$

Comparison to policy 1 The trend function of $\{X(t),\ t \ge 0\}$ is

$$m(t) = \frac{\sqrt{\pi}}{2}\lambda t^y = k_2 t^y, \quad t \ge 0.$$

Minimizing the corresponding maintenance cost rate (7.63) yields

$$\tau^* = \left(\frac{c}{k_2\,(y-1)}\right)^{1/y}, \quad K_1(\tau^*) = y\,k_2^{1/y}\left(\frac{c}{y-1}\right)^{(y-1)/y}.$$

For all $y > 1$, the inequality $K_2(x^*) < K_1(\tau^*)$ is equivalent to

$$\frac{2}{\sqrt{\pi}} < U(x), \quad 0.5 \le x < 1, \tag{7.72}$$

where

$$U(x) = [\Gamma(x)]^{\frac{1}{2(1-x)}}.$$

The function $U(x)$ is decreasing in $[0.5 \le x < 1]$ with

$$U(0.5) = \sqrt{\pi} > 2/\sqrt{\pi} \quad \text{and} \quad \lim_{x\to 1} U(x) = e^{E/2} > 2/\sqrt{\pi}\,,$$

where $E \approx 0.5772$ is the *Euler number*. Hence, inequality (7.72) holds for all $y > 1$ so that, as in example 7.9, policy 2 is superior to policy 1. In particular, if $1.1 \le y \le 5$, then average cost savings between 25 and 9% are achieved by applying the optimal cumulative repair cost limit x^* instead of the economic lifetime τ^*. □

The examples analyzed indicate that policies 2 and 3 belong to the most cost efficient replacement policies. Moreover, in spite of the terminology applied, in practice $X(t)$ will not only include pure repair costs, but also costs due to monitoring, servicing, stockkeeping as well as personnel costs. A great advantage to the 'repair cost limit replacement policy' considered in section 3.2.6.4 is that knowledge on the lifetime distribution of the system is not required. Hence, from the modeling point of view and with regard to their applicability, policies 2 and 3 are superior to the 'repair cost limit replacement policy'. Finally it should be mentioned that the maintenance cost rate criterion can be readily replaced with a discounted cost criterion.

7.5.5.4 Point Estimation for the Brownian Motion with Drift

The parameters of a probability distribution are generally estimated from samples taken from this distribution. But if a random variable X is the first passage time of a Brownian motion process with drift, then X has an inverse Gaussian distribution and the parameters of this distribution can also be estimated on the basis of samples generated by scanning sample paths of the underlying process. Therefore, the maximum-likelihood estimators $\hat{\mu}$ and $\hat{\sigma}^2$ for the parameters μ and σ^2 of a Brownian motion with drift, which will be constructed in what follows, are also point estimates of the parameters μ and σ^2 of the corresponding inverse Gaussian distribution.

Let $\{D_u(t), t \geq 0\}$ be a shifted Brownian motion with drift which starts at value

$$D_u(0) = u$$

and let

$$d_i = d_i(t); \quad i = 1, 2, ..., n$$

be n of its sample paths, which have been observed in n independent random experiments. The sample path $d_i = d_i(t)$ is scanned at time points

$$t_{i1}, t_{i2}, ..., t_{im_i} \text{ with } 0 < t_{i1} < t_{i2} < \cdots < t_{m_i} \text{ and } m_i \geq 2, \; i = 1, 2, ..., n.$$

The outcomes are

$$d_{ij} = d(t_{ij}); \; j = 1, 2, ..., m_i; \; i = 1, 2, ..., n.$$

The total number m of observations is

$$m = \sum_{i=1}^{m} m_i.$$

Further, let

$$\Delta d_{ij} = d_{ij} - d_{ij-1}; \;\; \Delta t_{ij} = t_{ij} - t_{ij-1}$$

with $j = 2, 3, ..., m_i; \; i = 1, 2, ..., n.$ If the initial value u is a constant, then

$$u = d_i(0); \;\; i = 1, 2, ..., n.$$

In this case, the maximum-likelihood estimators of μ and σ^2 are

$$\hat{\mu} = \frac{\sum_{i=1}^{n} d_{i m_i} - n u}{\sum_{i=1}^{n} t_{i m_i}} ,$$

$$\hat{\sigma}^2 = \frac{1}{m}\left\{ \sum_{i=1}^{n} \frac{(d_{i1} - \hat{\mu} t_{i1} - u)^2}{t_{i1}} + \sum_{i=1}^{n} \sum_{j=2}^{m_i} \frac{(\Delta d_{ij} - \hat{\mu}\,\Delta t_{ij})^2}{\Delta t_{ij}} \right\}.$$

Unfortunately, these estimators are biased. The structure of the estimator $\hat{\mu}$ confirms the intuitively obvious fact that for estimating μ only the initial value u and the last tuples (t_{im_i}, d_{im_i}) of each sample path are relevant.

If u is random, then the maximum-likelihood estimator of its mean value is

$$\hat{u} = \frac{\sum_{i=1}^{n} d_{i1}\, t_{i1}^{-1} - n \sum_{i=1}^{n} d_{i m_i} \left(\sum_{i=1}^{n} t_{i m_i}\right)^{-1}}{\sum_{i=1}^{n} t_{i1}^{-1} - n^2 \left(\sum_{i=1}^{n} t_{i m_i}\right)^{-1}} . \tag{7.73}$$

The following maximum-likelihood estimators were derived on condition that u is a random variable.

Special case $n = 1$ In this case only one sample path is available for estimating. Let the time points at which the sample path is scanned and the corresponding outcomes be $t_1, t_2, ..., t_m$ and $d_1, d_2, ..., d_m$, respectively. With the notation

$$\Delta d_j = d_j - d_{j-1}, \quad \Delta t_j = t_j - t_{j-1},$$

the bias-corrected maximum-likelihood estimators of μ and σ^2 are

$$\hat{\mu} = \frac{d_m - \hat{u}}{t_m} ,$$

$$\hat{\sigma}^2 = \frac{1}{m-2}\left\{ \frac{(d_1 - \hat{\mu} t_1 - \hat{u})^2}{t_1} + \sum_{j=2}^{m} \frac{(\Delta d_j - \hat{\mu}\,\Delta t_j)^2}{\Delta t_j} \right\}.$$

Special case $m_i = 1$; $i = 1, 2, ..., n$ In this case the estimation is based on n sample paths, but each sample path is only scanned at one time point. This requires to drop the assumption $m_i \geq 2$ stated above. Hence, $m = n$. The bias-corrected maximum-likelihood estimators of μ and σ^2 are

$$\hat{\mu} = \frac{\sum_{i=1}^{m} d_i - m\hat{u}}{\sum_{i=1}^{m} t_i} , \qquad \hat{\sigma}^2 = \frac{1}{m-2} \sum_{i=1}^{m} \frac{(d_i - \hat{\mu} t_i - \hat{u})^2}{t_i} . \tag{7.74}$$

Example 7.12 *Pieper* (1988) measured the mechanical wear of 35 identical items (cylinder running bushes) used in diesel engines of ships over a time span of 11,355 hours, each at one time point. He assumed that the stochastic wear process develops according to a Brownian motion with drift starting at u:

$$D_u(t) = u + D(t), \quad t \geq 0.$$

The point estimates of u, μ and σ^2 obtained from (7.73) and (7.74) are

$$\hat{u} = 36.145 \; [\mu m], \quad \hat{\mu} = 0.0029 \, [\mu m/h], \quad \hat{\sigma}^2 = 0.137 \; [\mu m^2/h].$$

The point estimate of the wear at time t can be written in the form

$$D_u(t) = 36.145 + 0.0029\, t + 0.137\, S(t), \tag{7.75}$$

where $\{S(t),\, t \geq 0\}$ is the standard Brownian motion.

Hint If the model (7.75) is correct, then the test function

$$T(t) = \frac{D_u(t) - 0.0029\, t - 36.145}{\sqrt{0.137}\; t}$$

has a standard normal distribution for all t (according to property 3 of definition 7.1). In particular, this must hold for all measurement points t_i. Hence, model (7.75) can be supported or rejected by a chi-square goodness of fit test.

Let $w = 1000$ $[\mu m]$ be an upper critical wear level with property that a cylinder running bush will experience a drift failure when the wear reaches this level. Then the lifetime of such a wear part is the first passage time $L_{D_u}(w)$ of the stochastic process $\{D_u(t), t \geq 0\}$ with regard to level $w = 1000$. By (7.41), estimates for mean value, variance and standard deviation of the first passage time $L_{D_u} = L_{D_u}(1000)$ are

$$E(L_{D_u}) \approx \frac{1000 - 36.145}{0.0029} = 332,364 \; [\text{h}],$$

$$Var(L_{D_u}) \approx \frac{(1000 - 36.145) \cdot 0.137}{(0.0029)^3} = 5.41425 \cdot 10^9 \left[\text{h}^2\right],$$

$$\sqrt{Var(L_{D_u})} \approx 73,581 \; [\text{h}].$$

Let $t = \tau_\varepsilon$ be that time point at which a wear part must be preventively replaced in order to avoid drift failures with a given probability ε. With the survival function given by (7.42), a point estimate of $\hat{\tau}_\varepsilon$ of τ_ε satisfies

$$\overline{F}(\hat{\tau}_\varepsilon) = \varepsilon. \tag{7.76}$$

Since

$$e^{-2(w-\hat{u})\,\hat{\mu}} \approx e^{-5.6},$$

the second term in (7.42) can be neglected. Therefore, equation (7.76) becomes

$$\Phi\left(\frac{w-\hat{u}-\hat{\mu}\,\hat{\tau}_\varepsilon}{\hat{\sigma}\sqrt{\hat{\tau}_\varepsilon}}\right)=\varepsilon \quad \text{or} \quad \frac{w-\hat{u}-\hat{\mu}\,\hat{\tau}_\varepsilon}{\hat{\sigma}\sqrt{\hat{\tau}_\varepsilon}}=z_\varepsilon\,, \tag{7.77}$$

where z_ε is the ε–percentile of the standard normal distribution. The relevant solution of (7.77) is

$$\hat{\tau}_\varepsilon=\frac{w-\hat{u}}{\hat{\mu}}+\frac{1}{2}\left(\frac{z_\varepsilon\hat{\sigma}}{\hat{\mu}}\right)^2-\frac{z_\varepsilon\hat{\sigma}}{\hat{\mu}}\sqrt{\frac{w-\hat{u}}{\hat{\mu}}+\left(\frac{z_\varepsilon\hat{\sigma}}{2\hat{\mu}}\right)^2}\,.$$

In particular, if $\varepsilon=0.95$, then $z_{0.95}=1.65$ so that $\hat{\tau}_{0.95}=231,121$ $[h]$. Thus, with probability 0.95, the wear remains below the critical level of $1000\,\mu\text{m}$ within an operating time of 231,121 hours . □

The Brownian motion with drift was firstly investigated by Schrödinger [72] and Smoluchowski [75]. Both found the first passage time distribution of this process. Folks and Chhikara [18] give a survey of the theory and discuss numerous applications: distribution of the water level of dams, duration of strikes, length of employment times of people in a company, wind velocity, and cost caused by system breakdowns. Moreover, they were the first to publish tables of the percentiles of the inverse Gaussian distribution. As a distribution of first passage times, the inverse Gaussian distribution naturally plays a significant role as a statistical model for lifetimes of systems which are subject to drift failures, see Kahle and Lehmann [42]. Seshadri [73] presents an up to date and comprehensive treatment of the inverse Gaussian distribution.

7.5.6 Integral Transformations

7.5.6.1 Integrated Brownian Motion

If $\{B(t),\, t\geq 0\}$ is a Brownian motion, then its sample paths $b=b(t)$ are continuous. Hence, the integrals

$$b(t)=\int_0^t b(y)\,dy\,.$$

exist for all sample paths. They are realizations of the *random integral*

$$U(t)=\int_0^t B(y)\,dy\,. \tag{7.78}$$

The stochastic process $\{U(t),\, t\geq 0\}$ is called *integrated Brownian motion*. This process can be a suitable model for situations in which the observed sample paths seem to be 'smoother' than those of the Brownian motion. Analogously to the definition of the Riemann integral, for any n-dimensional vector $(t_1, t_2, ..., t_n)$ with

$$0 = t_0 < t_1 < \cdots < t_n = t \text{ and } \Delta t_i = t_{i+1} - t_i; \quad i = 0, 1, 2, \ldots, n-1,$$

the random integral $U(t)$ is defined as the limit

$$U(t) = \lim_{\substack{n\to\infty \\ \Delta t_i \to 0}} \left\{ \sum_{i=0}^{n-1} [B(t_i + \Delta t_i) - B(t_i)]\, \Delta t_i \right\}. \tag{7.79}$$

(Note that passing to the limit refers here and in what follows to mean-square convergence.) The random variable $U(t)$, being the limit of a sum of independent, normally distributed random variables, is itself normally distributed. More generally, by theorem 1.2, the integrated Brownian motion is a Gaussian process. Therefore, the integrated Brownian motion is uniquely characterized by its trend and covariance function. In view of

$$E\left(\int_0^t B(y)\, dy\right) = \int_0^t E(B(y))\, dy = \int_0^t 0\; dy \equiv 0,$$

the trend function of the integrated Brownian motion $\{U(t), t \geq 0\}$ is 0:

$$m(t) = E(U(t)) \equiv 0.$$

Its covariance function of $\{U(t), t \geq 0\}$,

$$C(s,t) = Cov(U(s), U(t)) = E(U(s)U(t)), \quad s \leq t,$$

is obtained as follows:

$$C(s,t) = E\left\{ \int_0^s B(y)\, dy \int_0^t B(z)\, dz \right\}$$

$$= E\left\{ \int_0^t \int_0^s B(y)\, B(z)\, dy\, dz \right\}$$

$$= \int_0^t \int_0^s E(B(y)\, B(z))\, dy\, dz.$$

Since

$$E(B(y), B(z)) = Cov(B(y), B(z)) = \sigma^2 \min\,(y,z),$$

it follows that

$$C(s,t) = \sigma^2 \int_0^t \int_0^s \min(y,z)\, dy\, dz$$

$$= \sigma^2 \int_0^s \int_0^s \min(y,z)\, dy\, dz + \sigma^2 \int_s^t \int_0^s \min(y,z)\, dy\, dz$$

$$= \sigma^2 \int_0^s \left[\int_0^z y\, dy + \int_z^s z\; dy \right] dz + \sigma^2 \int_s^t \int_0^s y\, dy\, dz$$

$$= \sigma^2 \frac{s^3}{3} + \sigma^2 \frac{s^2}{2}(t-s).$$

Thus,

$$C(s,t) = \frac{\sigma^2}{6}(3t - s)s^2, \quad s \leq t.$$

Letting $s = t$ yields

$$Var(U(t)) = \frac{\sigma^2}{3} t^3.$$

The structure of the covariance function implies that the integrated Brownian motion is nonstationary. But it can be shown that for any τ the process $\{V(t), t \geq 0\}$ with

$$V(t) = U(t+\tau) - U(t)$$

is stationary. (Recall that for a Gaussian process strong and weak stationarity are equivalent.)

7.5.6.2 White Noise

Since the sample paths of a Brownian motion are nowhere differentiable with probability 1, a stochastic process of the form $\{X(t), t \geq 0\}$ with

$$X(t) = \frac{dB(t)}{dt} = B'(t) \quad \text{or} \quad dB(t) = X(t)\, dt$$

cannot be introduced by taking the limit in a difference quotient. However, a definition via an integral is possible. To establish an approach to this definition, let $g(t)$ be any function with a continuous derivative $g'(t)$ in the interval $[a, b]$ and $t_0, t_1, ..., t_n$ any sequence of numbers satisfying

$$a = t_0 < t_1 < \cdots < t_n = b \quad \text{and} \quad \Delta t_i = t_{i+1} - t_i; \quad i = 0, 1, 2, ..., n-1.$$

Then the *stochastic integral* $\int_a^b g(t)\, dB(t)$ is defined as the limit

$$\int_a^b g(t)\, dB(t) = \lim_{\substack{n\to\infty \\ \max_{i=1,2,...,n} \Delta t_i \to 0}} \left\{ \sum_{i=0}^{n-1} g(t_i) \left[B(t_i + \Delta t_i) - B(t_i) \right] \right\}. \tag{7.80}$$

The sum in (7.80) can be written as follows:

$$\sum_{i=0}^{n-1} g(t_i) \left(B(t_i + \Delta t_i) - B(t_i) \right)$$

$$= g(b)\, B(b) - g(a)\, B(a) - \sum_{i=0}^{n-1} B(t_{i+1}) \frac{g(t_i + \Delta t_i) - g(t_i)}{\Delta t_i} \Delta t_i.$$

Taking the limit on both sides as in (7.80) yields

$$\int_a^b g(t)\, dB(t) = g(b)\, B(b) - g(a)\, B(a) - \int_a^b B(t)\, g'(t)\, dt. \tag{7.81}$$

This explanation of the stochastic integral is usually preferred to (7.80). As a limit of a sum of normally distributed random variables, the stochastic integral also has a normal distribution. From (7.81),

$$E\left(\int_a^b g(t)\,dB(t)\right) = 0. \tag{7.82}$$

By making use of

$$Var(B(t) - B(s)) = \sigma^2 |t - s|,$$

the variance of the following sum is seen to have a simple structure:

$$\begin{aligned} & Var\left(\sum_{i=0}^{n-1} g(t_i)\,[B(t_i + \Delta t_i) - B(t_i)]\right) \\ & = \sum_{i=0}^{n-1} g^2(t_i)\, Var\,(B(t_i + \Delta t_i) - B(t_i)) \\ & = \sigma^2 \sum_{i=0}^{n-1} g^2(t_i)\,\Delta t_i\,. \end{aligned}$$

Passing in this equation to the limit as in (7.80) yields the variance of the stochastic integral:

$$Var\left(\int_a^b g(t)\,dB(t)\right) = \sigma^2 \int_a^b g^2(t)\,dt. \tag{7.83}$$

The relationship (7.81) motivates the following definition.

Definition 7.3 (*White noise*) Let $\{B(t),\, t \geq 0\}$ be the Brownian motion. A stochastic process $\{X(t),\, t \geq 0\}$ is called *white noise* if it satisfies for any function $g(t)$ with a continuous derivative $g'(t)$ in $[a, b],\ a < b,$ the relationship

$$\int_a^b g(t)\,X(t)\,dt = g(b)\,B(b) - g(a)\,B(a) - \int_a^b B(t)\,g'(t)dt. \tag{7.84}$$

●

If $B(t)$ had a first derivative, then $X(t) = dB(t)/dt$ would satisfy (7.84) anyway. Thus, $X(t)$ as introduced in definition 7.3 can be interpreted as a 'generalized derivative' of $B(t)$, because it exists although the differential quotient does not exist. However, this interpretation of the white noise does not facilitate its intuitive understanding. To get an idea of the nature of the white noise process $\{X(t),\, t \geq 0\}$, a heuristic argument is presented by 'deriving' the covariance function of $\{X(t),\, t \geq 0\}$: Assuming that the order of 'generalized differentiation' and integration can be exchanged, one obtains for all s and t with $s \neq t$,

$$\begin{aligned} C(s,t) = Cov(X(s), X(t)) &= Cov\left(\frac{\partial B(s)}{\partial s}, \frac{\partial B(t)}{\partial t}\right) \\ &= \frac{\partial}{\partial s}\frac{\partial}{\partial t} Cov(B(s), B(t)) \\ &= \frac{\partial}{\partial s}\frac{\partial}{\partial t} \min(s,t)\,. \end{aligned}$$

Thus, if $s < t$, then

$$C(s,t) = \frac{\partial}{\partial s}\frac{\partial}{\partial t}\, s = \frac{\partial}{\partial s}\, 0 = 0\,.$$

If $s > t$, then

$$C(s,t) = \frac{\partial}{\partial s}\frac{\partial}{\partial t}\, t = \frac{\partial}{\partial s}\, 1 = 0\,.$$

Hence,

$$C(s,t) = 0 \quad \text{for } s \neq t\,. \tag{7.85}$$

Therefore, for $s \neq t$, there is no correlation between $X(s)$ and $X(t)$, no matter how small the absolute difference $|s-t|$ may be. Thus, white noise can be interpreted as the 'most random stochastic process', and this property explains its favourite role as a process for modeling random noise, which is superimposed on a useful signal. However, in view of its key property (7.85), white noise cannot exist in the real world. Nevertheless, the white noise process is of great importance for approximately modeling various phenomena in electronics, electrical engineering, communication, econometrics, time series analysis et alia. Its role can be compared with the concept of the 'point mass' in mechanics, which also only exists in theory.

Intuitively, white noise can be thought of as a sequence of extremely sharp pulses, which occur after extremely short time intervals, and which have independent, identically distributed amplitudes. The times in which the pulses rise and fall are so short that they cannot be registered by measuring instruments. Moreover, the response times of the measuring instruments are so large that during any response time a huge number of pulses occur which cannot be registered.

In practice, a weakly stationary stochastic process $\{X(t), t \geq 0\}$ can approximately be considered white noise if the covariance between $X(t)$ and $X(t+\tau)$ tends extremely fast to 0 with increasing τ. For example, if $X(t)$ denotes the absolute value of the force which particles in a liquid are subjected to at time t (causing their Brownian motion), then this force arises from the about 10^{21} collisions per second between the particles and the surrounding molecules. The process $\{X(t), t \geq 0\}$ is known to be weakly stationary with a covariance function of type

$$C(\tau) = e^{-b\,|\tau|},$$

where

$$b \geq 10^{19}\,\text{sec}^{-1}.$$

Hence, $X(t)$ and $X(t+\tau)$ are practically uncorrelated if

$$|\tau| \geq 10^{-18}.$$

A similar fast drop of the covariance function can be observed if $\{X(t), t \geq 0\}$ describes the electromotive force in a conductor, which is caused by the thermal movement of electrons.

7.6 EXERCISES

Note In all exercises, $\{B(t),\ t \geq 0\}$ is the Brownian motion with $Var(B(1)) = \sigma^2$.

7.1) Verify that the probability density $f_t(x)$ of $B(t)$,

$$f_t(x) = \frac{1}{\sqrt{2\pi t}\,\sigma}\, e^{-x^2/(2\sigma^2 t)}, \quad t > 0,$$

satisfies the *thermal conduction equation*

$$\frac{\partial f_t(x)}{\partial t} = c\,\frac{\partial^2 f_t(x)}{\partial x^2}.$$

7.2) Determine the conditional probability density of $B(t)$ given $B(s) = y,\ \ 0 \leq s < t.$

7.3)* Prove that the stochastic process $\{\overline{B}(t),\ 0 \leq t \leq 1\}$ given by

$$\overline{B}(t) = B(t) - t\,B(1)$$

is the Brownian bridge.

7.4) Let $\{\overline{B}(t),\ 0 \leq t \leq 1\}$ be the Brownian bridge. Prove that the stochastic process $\{S(t), t \geq 0\}$ defined by

$$S(t) = (t+1)\,\overline{B}\left(\tfrac{t}{t+1}\right)$$

is the standard Brownian motion.

7.5) Determine the probability density of $B(s) + B(t)$.

7.6) Let n be any positive integer. Determine mean value and variance of

$$X(n) = B(1) + B(2) + \cdots + B(n).$$

Hint Make use of formula (1.100).

7.7) Prove that for any positve h the stochastic process $\{V(t),\ t \geq 0\}$ defined by

$$V(t) = B(t+h) - B(t)$$

is weakly stationary.

7.8) Prove that the stochastic process $\{X(t),\ t \geq 0\}$ with $X(t) = S^3(t) - 3t\,S(t)$ is a continuous-time martingale, i.e show that

$$E(X(t)\,|\,X(y),\ y \leq s) = X(s), \quad s < t.$$

7.9) Prove that the increments of the Ornstein-Uhlenbeck process are not independent.

7.10)* Starting from $x = 0$, a particle makes independent jumps of length

$$\Delta x = \sigma \sqrt{\Delta t}$$

to the right or to the left every Δt time units. The respective probabilities of jumps to the right and to the left are

$$p = \frac{1}{2}\left(1 + \frac{\mu}{\sigma}\sqrt{\Delta t}\right) \text{ and } 1 - p,$$

where, for $\sigma > 0$,

$$\sqrt{\Delta t} \leq \left|\frac{\sigma}{\mu}\right|.$$

Show that as $\Delta t \to 0$ the position of the particle at time t is governed by a Brownian motion with drift with parameters μ and σ.

7.11) Let $\{D(t), t \geq 0\}$ be a Brownian motion with drift with parameters μ and σ. Determine

$$E\left(\int_0^t (D(s))^2\, ds\right).$$

7.12) Show that for $c > 0$ and $d > 0$

$$P(B(t) \leq c\,t + d \text{ for all } t \geq 0) = 1 - e^{-2\,c\,d/\sigma^2}.$$

Hint Make use of formula (7.29).

7.13) (1) What is the mean value of the first passage time of the reflected Brownian motion $\{|B(t)|, t \geq 0\}$ with regard to a positive level x?

(2) Determine the distribution function of $|B(t)|$.

7.14) At time $t = 0$ a speculator acquires an American call option with infinite expiration time and strike price x_s. The price $X(t)$ of the underlying risky security at time t is given by

$$X(t) = x_0 e^{B(t)}.$$

The speculator makes up his mind to exercise this option at that time point, when the price of the risky security hits a level x with

$$x > x_s \geq x_0$$

for the first time,.

1) What is the speculator's mean discounted payoff $G_\alpha(x)$ under a constant discount rate α?

2) What is the speculator's payoff $G(x)$ without discounting?

In both cases, cost of acquiring the option is not included in the speculator's payoff.

7.15) The price $X(t)$ of a risky security at time t is

$$X(t) = x_0\, e^{\mu t + B(t) + a|B(t)|}, \quad t \geq 0, \quad 0 < a \leq 1,$$

with a negative drift parameter μ. At time $t = 0$ a speculator acquires an American call option with strike price x_s on this risky security. The option has no finite expiration date. The speculator makes up his mind to exercise this option at that time point, when the price of the risky security hits a level x with $x > x_s \geq x_0$ for the first time. Otherwise, i.e. if the price of the risky security never reaches level x, the speculator will never exercise.

Determine the level $x = x*$ at which the speculator should schedule to exercise this option to achieve

1) maximal mean payoff without discounting and

2) maximal mean discounted payoff (constant discount rate α).

7.16) The value of a share at time t is

$$X(t) = x_0 + D(t),$$

where $x_0 > 0$ and $\{D(t),\ t \geq 0\}$ is a Brownian motion with positive drift parameter μ and variance parameter σ^2. At time point $t = 0$ a speculator acquires an American call option on this share with finite expiry date τ. Assume that

$$x_0 + \mu t > 3\sigma\sqrt{t}, \quad 0 \leq t \leq \tau.$$

(1) Why does the assumption make sense?

(2) When should the speculator exercise to make maximal mean undiscounted profit?

7.17) At time $t = 0$, a speculator acquires a European call option with strike price x_s and finite expiration time τ. Thus, the option can only be exercised at time τ at price x_s, independently of its market value at time τ. The price $X(t)$ of the underlying risky security at time t is

$$X(t) = x_0 + D(t),$$

where $\{D(t),\ t \geq 0\}$ is the Brownian motion with positive drift parameter μ and volatility σ^2. If $X(\tau) > x_s$, the speculator will exercise the option. Otherwise, he will not. As in example 7.16, assume that

$$x_0 + \mu t > 3\sigma\sqrt{t}, \quad 0 \leq t \leq \tau.$$

1) What will be the mean undiscounted payoff of the speculator (cost of acquiring the option not included)?

2) Under otherwise the same assumptions, what is the investor's mean undiscounted profit if

$$X(t) = x_0 + B(t) \text{ and } x_0 = x_s\,?$$

7.18) Let $X(t)$ be the cumulative repair cost of a system arising in the interval $(0, t]$ (excluding replacement costs) and

$$R(t) = X(t)/t$$

the corresponding cumulative repair cost rate. Assume

$$R(t) = r_0 B^2(t), \quad r_0 > 0.$$

The system is replaced by an equivalent new one as soon as $R(t)$ reaches level r.

(1) Given a constant replacement cost c, determine a level $r = r^*$ which is optimal with respect to the long-run total maintenance cost per unit time $K(r)$. (Make sure that an optimal level r^* exists.)

(2) Compare $K(r^*)$ to the minimal long-run total maintenance cost per unit time $K(\tau^*)$ which arises by applying the corresponding economic lifetime τ^*.

7.19)* Let $\{S(t), t \geq 0\}$ be the standard Brownian motion and

$$X(t) = \int_0^t S(s)\, ds.$$

(1) Determine the covariance between $S(t)$ and $X(t)$.

(2) Verify that

$$E(X(t)|S(t) = x) = \frac{tx}{2} \quad \text{and} \quad Var(X(t)|S(t) = x) = \frac{t^3}{12}.$$

Hint Make use of the fact that the random vector $(S(t), X(t))$ has a two-dimensional normal distribution.

7.20) Show that for any constant α

$$E(e^{\alpha X(t)}) = e^{\alpha^2 t^3/6},$$

where $X(t)$ is defined as in exercise 7.19.

Hint Make use of the moment generating function of the normal distribution.

ANSWERS TO SELECTED EXERCISES

Chapter 1

1.2) (2) Let 1 and 0 indicate that a person has gene g or not, respectively. Then the sample space **M** consists of all the $2^3 = 8$ vectors (z_1, z_2, z_3) with

$$z_i = \begin{cases} 1 & \text{if person } i \text{ has gene } g \\ 0 & \text{otherwise} \end{cases} \quad ; \; i = 1, 2, 3.$$

$$A = \{(0,0,0)\}, \;\; B = \{(1,0,0), (0,1,0), (0,0,1)\}, \;\; A \cap B = \emptyset$$

$$C = \{(1,1,0), \;(1,0,1), \;(0,1,1), \;(1,1,1)\}.$$

$$B \cup C = \mathbf{M} \backslash A, \quad (A \cup B) \cap \overline{C} = A \cup B$$

1.3) 0.6, 0.3, 0.8

1.4) (1) 0.925, 0.89, 0.85. 0.965, 0.15 (2) no

1.5) 0.59, 0.61, 0.52, 0.68, 0.852, 0.881, 0.179, 0.8205

1.6) 0.0902

1.7) 13

1.8) (1) and (2): don't check (3) check

1.9) (1) 0.6475 (2) 0.9979

1.10) (1) 0.023 (2) 0.2978

1.11) (1) $2p^2(1 + p + p^3) - 5p^4$

1.12) (1) 0.7800 (2) 0.9744

1.13) (1) $n = 132$. Probability distribution of X:
$\{p_i = P(X = x_i) = n_i/n;\; i = 1, 2, \ldots, 10\}$. (2) 0.8182, 0.5

1.15) 45.18, 5.3421

1.16) 15.22, 0.0151

1.17) 0.9535

1.18) 0.1329

1.19) 0.01185

1.20) (1) 0.8701 (2) 0.0411

1.21) 0.191

1.22) 0.920

1.23) 0.4493

1.24) (1) $c = 3/64$ (2) $c = 1/6$ (3) $c = 1$

1.25) 0.6931, 0.6931, 0.0513

1.26) 0.4

1.27) 0.54

1.28) (1) 3, 0.6 (2) 11/9, 23/81 (3) 1, 1

1.29) (1) a) 0.1009 b) 0.7364, (2) 0.9963.

1.30) (1) $F(x) = (x-2)^3\left[10 - 15(x-2) + 6(x-2)^2\right], \; 2 \le x \le 3,$

(2) 0.0579 (3) 2.5

1.31) 5, -1.56, 6.45, such an x does not exist, 15.3

1.32) (1) 0.0475 (2) 0.2975

1.33) (1) 0.7725 (2) 6.68%

1.34) (1) 0.1524 (2) 125.6 hours

1.35) $f(p) = \begin{cases} 1, & 0 \le p \le 1 \\ 0, & \text{otherwise} \end{cases}$

1.36) $G(x) = \int_0^\infty e^{-\alpha/x} \lambda e^{-\lambda\alpha} d\alpha = \frac{\lambda x}{1+\lambda x}, \; x \ge 0$ (*Pareto distribution*)

1.37) 0.7165

1.38) $[0, \infty)$

1.41) (1) $\{p_0 = 0.2,\ p_1 = 0.5,\ p_2 = 0.3\}$, $\{q_0 = 0.2,\ q_1 = 0.6,\ q_2 = 0.2\}$, (2) no

(3) $E(X|Y=1) = 7/6, \; E(Y|X=0) = 1/2.$

1.42) (1) no (2) $f(z) = 2(1-z), \; 0 \le z \le 1.$

1.43) (1) yes (2) $f(z) = 6z(1-z), \; 0 \le z \le 1.$

1.44) (1) $E(Z) = 200{,}000, \; \sqrt{Var(Z)} = 12{,}296$ (2) 0.792

1.45) (1) 0.032, 0.406 (2) 726 kg

1.46) $n_{\min} = 43.$

1.47) 0.293

1.49) (1) $M_X(z) = \frac{pz}{1-(1-p)z}$, (2) $M_Z(z) = \left(\frac{pz}{1-(1-p)z}\right)^2.$

1.50) $p_1 = p_2 = \cdots = p_k.$

1.51) (1) yes, (2) $f(z) = \begin{cases} z/T^2, & 0 \le z \le T \\ (2T-z)/T^2, & T < z \le 2T \\ 0, & \text{otherwise} \end{cases}$

Simpson- or *triangle- distribution.*

1.52) $\hat{f}(s) = \frac{\lambda^2}{\lambda^2 - s^2} e^{-\mu s}$

1.53) (1) $n_0 = 11{,}280$ (2) $n_0 = 2167.$

Chapter 2

2.1) not stationary

2.2) $m(t) = \mu t,\ \ Var(X(t)) = \sigma^2 t$

2.3) (1) $m(t) \equiv 0,\ C(\tau) = \frac{1}{2} E(A^2)\cos\omega\tau,\ \ \rho(\tau) = \cos\omega\tau$. (2) weakly stationary

2.5) $C(\tau) = \frac{1}{2}\sum_{i=1}^{n} a_i^2 \cos\omega\tau\,,\ \ \rho(\tau) = \cos\omega\tau$

2.6) (2) $C(s,t) = \begin{cases} 1, & n \le s, t \le (n + 1/2),\ n = 0, \pm 1, \dots \\ 0, & \text{elsewhere} \end{cases}$, (3) no

2.7) The trend function of the second order stochastic process $\{Z(t),\, t \ge 0\}$ is identically 0 and its covariance function is $C_Z(\tau) = C(\tau)\cos\omega\tau$.

2.8) Note that the one-dimensional distributions of $\{X(t),\, t \ge 0\}$ depend on t.

2.9) $C_U(s,t) = C_V(s,t) = C_X(s,t) + C_Y(s,t)$

Chapter 3

3.1) (1) 0.4422 (2) 0.4422

3.2) $Cov(N(s), N(t)) = Cov(N(s), N(s)) + Cov(N(s), N(t) - N(s))$

3.3) 0.2739

3.4) (1) 0.9084 (2) $E(Y) = 1/4$ min, $Var(Y) = (1/4)^2$

3.5) 0.1341

3.6) λ_2/λ_1

3.8) $C(\tau) = \begin{cases} \frac{\lambda}{2}(\pi - |\tau|)\cos|\tau| + \sin(\pi - |\tau|) & 0 \le |\tau| \le \pi \\ 0, & \text{elsewhere} \end{cases}$

3.10) $E(K) = E(C)\frac{\lambda}{\alpha}(1 - e^{-\alpha t})$

3.11) (1) 64 (2) 0.89

3.13) $\tau^* = \theta\left[\frac{c_p}{(\beta-1)c_m}\right]^{1/\beta}$

3.14) (1) $P(N_L(t) = n) = \frac{1}{t}\left[1 - e^{-t}\sum_{k=0}^{n} t^k/k!\right];\ \ k = 0, 1, \dots$

(2) $E(N_L(t)) = t/2,\ \ Var(N_L(t)) = t/2 + t^2/12$

(3) $\alpha = 3,\ \ \beta = 6$

3.17) (1) $K(c) = \dfrac{\frac{1}{\overline{R}(c)}\int_0^c \overline{R}(x)dx + c_r - c}{\int_0^\infty [\overline{F}(t)]^{\overline{R}(c)} dt}$. (2) $c^* = \left[1 - \sqrt{\frac{\beta-1}{\beta+1}}\right]c_r,\ \ \beta > 1$

3.18) (1) $n^* = 86$ (2) $n^* = 88$

3.19) (1) $H(t) = \frac{1}{4}\left(e^{-2\lambda t} + 2\lambda t - 1\right)$

3.22) 0.2642

3.25) (2) $H(n) = \dfrac{n+p}{1-p}$

3.26) $\mu = \sqrt{\pi}/2,\ \mu_2 = 1,\ \sigma^2 = 1 - \pi/4$

(1) $\lim\limits_{t\to\infty} \int_0^t (t-x+1)^{-2}\, dH(x) = \dfrac{2}{\sqrt{\pi}}$, (2) $\lim\limits_{t\to\infty} (H(t) - t/\mu) = \frac{2}{\pi} - 1$

3.27) $\frac{1}{\mu} \int_0^t \bar{F}(x)\, dx$

3.28) (1) $P(A(t) > y - t \mid B(t) = x) = \dfrac{\bar{F}(t-x+y)}{\bar{F}(t-x)}$,

(2) $P(A(t) \le y \mid B(t) = x) = \dfrac{F(x+y) - F(x)}{\bar{F}(x)}$

3.30) $\frac{1}{3}(\lambda x + 2)\, e^{-\lambda x}$

3.31) (1) 0.9841 (2) 0.9970

3.33) (1) $K(\tau) = \dfrac{c_e F(\tau) + c_p \bar{F}(\tau)}{\int_0^\tau \bar{F}(t)\, dt}$

(2) $\lambda(\tau) \int_0^\tau \bar{F}(t)\, dt - F(\tau) = c/(1-c)$ with $0 < c = c_p/c_e < 1 - \dfrac{1}{\mu\, \lambda(\infty)}$

(3) $\tau^* = \frac{z}{1-c}\left[\sqrt{c(2-c)} - c\right]$ with $0 < c = c_p/c_e < 1$

3.34) (1) $K(\tau) = \dfrac{c_p + c_e H(\tau)}{\tau}$

(2) $(1 + 3\lambda\tau)\, e^{-3\lambda\tau} = 1 - 9c/2$ with $0 < c = c_p/c_e < 2/9$

3.35) (1) 0.2163 (2) $p(x) \le e^{-\frac{x}{13{,}600}}$

3.36) (1) 0.1315 (2) 0.1342

3.38) (1) 860.5 [$/h] (2) ≈ 0

Chapter 4

4.1) (1) 0.5, 0.2 (2) 0.25, 0.25 (3) 0.5, 0.072

4.2) (1) $\mathbf{P}^{(2)} = \begin{pmatrix} 0.58 & 0.12 & 0.3 \\ 0.32 & 0.28 & 0.4 \\ 0.36 & 0.18 & 0.46 \end{pmatrix}$ (2) 0.42, 0

4.3) (1) 0.2864 (3) $\pi_0 = 0.4,\ \pi_1 = \pi_2 = 0.3$

4.4) yes

4.5) (2) $\pi_i = 0.25;\ i = 0, 1, 2, 3$

Note that **P** is a doubly stochastic matrix. See exercise 4.6.

4.6) (2) no

4.8) (2) $\mathbf{P} = \begin{pmatrix} 0.8 & 0.2 & 0 & 0 \\ 0 & 0 & 0.4 & 0.6 \\ 0.6 & 0.4 & 0 & 0 \\ 0 & 0 & 0.2 & 0.8 \end{pmatrix}, \quad \pi_1 = 3/8,\ \pi_2 = \pi_3 = 1/8,\ \pi_4 = 3/8$

4.11) (1) minimal closed sets: {1, 2, 3}, {3, 4} (2) There are no inessential states.

4.12) essential: {0, 1, 2}, inessential: {3}

4.13) (3) $\pi_0 = 50/150,\ \pi_1 = 10/150,\ \pi_2 = 40/150,\ \pi_3 = 13/150,\ \pi_4 = 37/150$

4.14) (1) essential: {0, 1}, inessential: {2, 3, 4}
(2) recurrent: {0, 1}, transient: {2, 3, 4}

4.15) $\pi_i = p(1-p)^i;\ i = 0, 1, \ldots$

4.18) (1) positive recurrent (2) transient

4.19) $E(N_i) = \frac{p_{ii}}{1-p_{ii}}, \quad Var(N_i) = \frac{p_{ii}}{(1-p_{ii})^2}$

Hint N_i has a geometric distribution:

$$P(N_i = n) = (1 - p_{ii})p_{ii}^{n-1};\ n = 1, 2, \ldots$$

Chapter 5

5.1) no

5.3) $\pi_0 = \frac{2\lambda}{2\lambda + 3\mu}, \quad \pi_1 = \frac{2\mu}{2\lambda + 3\mu}, \quad \pi_2 = \frac{\mu}{2\lambda + 3\mu}$

5.4) (1) 96% (2) 81%

5.5) state (i, j): i, j respective states of unit 1 and 2: 0 down, 1 operating

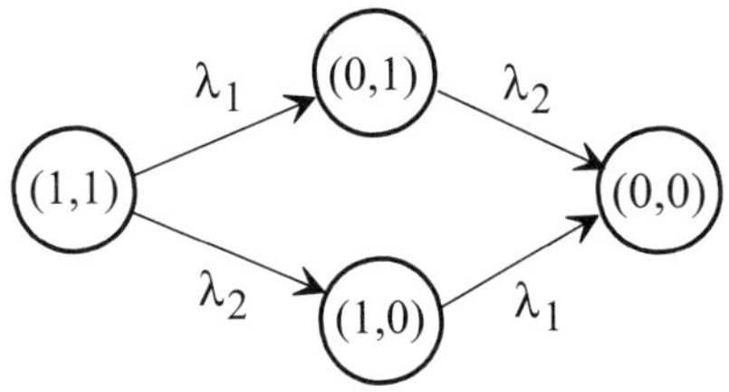

5.7) states: 0 system operating, 1 dangerous state, 2 system blocked, 3 system blocked after dangerous failure

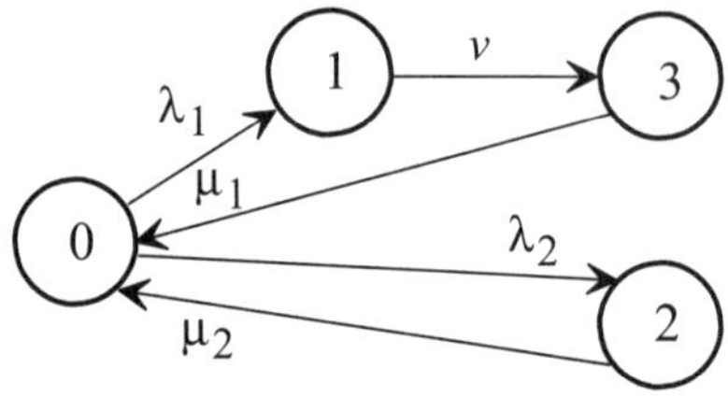

$$\pi_1 = \frac{\lambda_1}{\nu}\pi_0, \quad \pi_2 = \frac{\lambda_2}{\mu_2}\pi_0, \quad \pi_3 = \frac{\lambda_1}{\mu_1}\pi_0, \quad \pi_0 = \frac{1}{1+\frac{\lambda_1}{\nu}+\frac{\lambda_1}{\mu_1}+\frac{\lambda_2}{\mu_2}}$$

$P(\text{system blocked}) = \pi_2 + \pi_3$

5.9) $p_0(t) = e^{-2t}, \; p_1(t) = 2\left(e^{-2t} - e^{-3t}\right), \; p_2(t) = 3e^{-t}(1-e^{-t})^2$

5.10) (1) $(1-e^{-\lambda t})^2$ (2) $\frac{1}{\lambda}\left(1+\frac{1}{2}+\cdots+\frac{1}{n-1}\right)$

5.11) $p_j(t) = e^{-\lambda t}(1-e^{-\lambda t})^{j-1}; \quad j = 1, 2, ...$

5.14) (1) $\binom{2n}{j} e^{-j\mu t}(1-e^{-\mu t})^{2n-j}$, (2) $\frac{1}{\mu}\left(\frac{1}{2n}+\frac{1}{2n-1}+\cdots+\frac{1}{n+1}\right), \; n \geq 1$

5.15) (1) 0.56 (2) 50 weeks

(*Hint*: $p_0(t) = P(\text{cable completely broken at time } t)$ is given by an Erlang distribution with parameters $n = 5$ and $\lambda = 0.1$.)

5.17) see example 5.14

5.18) $\lambda < \mu$

5.20)

(1)

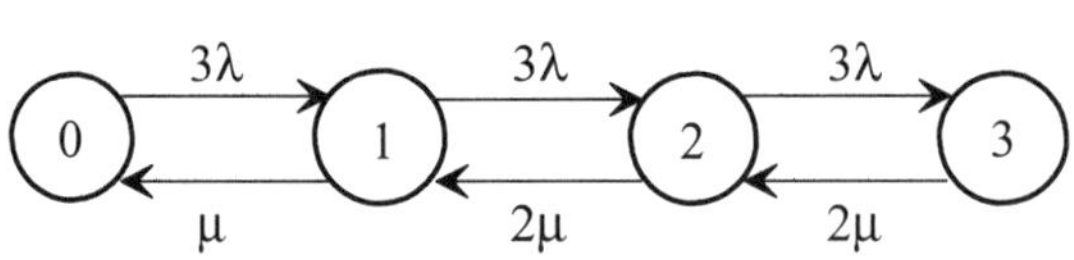

(2) $$\pi_{loss} = \pi_3 = \frac{6.75\rho^3}{1+3\rho+4.5\rho^2+6.75\rho^3}, \quad \rho = \lambda/\mu$$

5.21) $$\pi_{loss} = \pi_3 = \frac{13.5\rho^3/(2+\nu/\mu)}{1+3\rho+4.5\rho^2+13.5\rho^3/(2+\nu/\mu)}$$

5.23) $\pi_{loss} = 0.0311, \; \pi_{wait} = 0.6149$

5.24) state (i, j) : i, j customers at server 1, 2; $i, j = 0, 1$

$$\pi_{(1,0)} = \rho\,\pi_{(0,0)}, \quad \pi_{(1,1)} = \frac{\rho^2}{2}\pi_{(0,0)}, \quad \pi_{(0,1)} = \frac{\rho^2}{2(\rho+1)}\pi_{(0,0)}$$

5.26) (1) $\pi_0 = \pi_1 = \pi_2 = \pi_3 = 1/4$ (2) 1.2273

5.28) see example 5.14

5.32) (1)

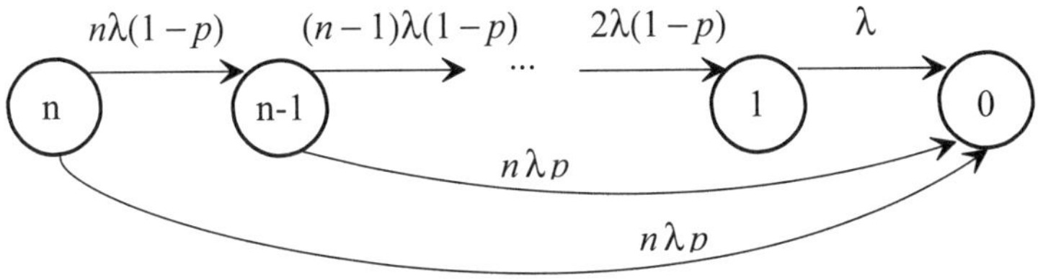

(2) $1 - F_s(t) = P(X_s > t) = p_1(t) + p_2(t), \quad E(X_s) = \frac{1}{\lambda}[1.5 - p]$

5.34) $66\frac{2}{3}\%$

5.37) 0.3153, 0.4144, 0.2703

5.38) (2) states: 0 working, 1 repair after type 2 failure, 2 repair after type 1 failure

$$P(X=0) = 360/372, \quad P(X=1) = 4/372, \quad P(X=2) = 8/372$$

5.39) (1)

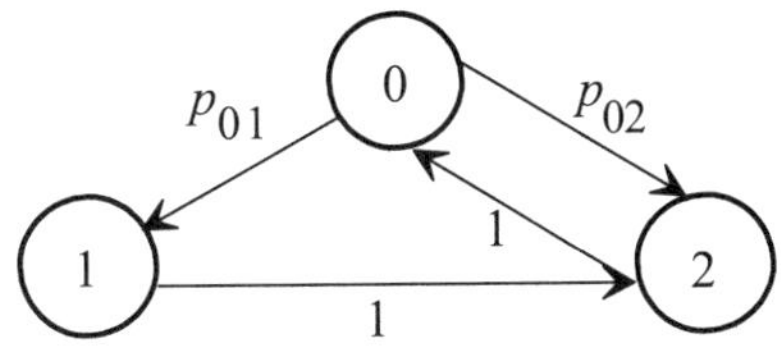

(2) $A_0 = \frac{1}{1 + \lambda_1\mu_1 + (\lambda_1 + \lambda_2)\mu_2}$ (*stationary availability*)

$$A_1 = \frac{\lambda_1\mu_1}{1 + \lambda_1\mu_1 + (\lambda_1 + \lambda_2)\mu_2}, \quad A_2 = \frac{(\lambda_1 + \lambda_2)\mu_2}{1 + \lambda_1\mu_1 + (\lambda_1 + \lambda_2)\mu_2}$$

5.40) The stationary state probabilities of the embedded Markov chain are:

$$\pi_0 = \frac{\lambda + \lambda_0}{2\lambda_0 + \lambda(3 - e^{-\lambda_1\mu})}, \quad \pi_1 = \frac{\lambda}{2\lambda_0 + \lambda(3 - e^{-\lambda_1\mu})},$$

$$\pi_2 = \frac{\lambda_0 + \lambda(1 - e^{-\lambda_1\mu})}{2\lambda_0 + \lambda(3 - e^{-\lambda_1\mu})}.$$

The mean sojourn times μ_0 and μ_2 are the same as in example 5.26, whereas μ_1 is

$$\mu_1 = \frac{1 - e^{-\lambda_1\mu}}{\lambda_1}.$$

Chapter 6

6.1) no, since $E(Y_i^2) > 0$ (see example 6.1).

6.2) Example 6.1 is applicable with $X_i = Y_i - E(Y_i)$ since $E(X_i) = 0$.

6.3) (1) $T = 2$, (2) $T > 2$, (3) $T < 2$.

6.4) $\sigma^2 = -2\mu$ (condition $\mu < 0$ is necessary)

6.5) $E(N) = \dfrac{n}{2p-1}$

6.7) (1) $p_{2000} = 0.8703$, (2) $p_{-1000} = 0.1297$, (3) $E(N) = 64.4$.

6.9) Note that the transition from i to j is governed by a binomial distribution with parameters n and $p = i/n$.

6.10) See example 6.12 or proof of theorem 7.1 b)

Chapter 7

7.2) $f_t(x|B(s) = y) = \dfrac{1}{\sqrt{2\pi(t-s)}\,\sigma}\exp\left(-\dfrac{(x-y)^2}{2(t-s)\sigma^2}\right), \quad 0 \le s < t$

7.5) $f_{B(s)+B(t)}(x) = \dfrac{1}{\sqrt{2\pi(t+3s)}\,\sigma}\exp\left\{-\dfrac{1}{2}\dfrac{x^2}{(t+3s)\sigma^2}\right\}, \quad -\infty < x < +\infty$

7.6) $E(X(n)) = 0, \quad Var(X(n)) = \dfrac{n(n+1)(2n+1)}{6}\sigma^2$

7.11) $\dfrac{t^2}{6}\left(2\mu^2 t + 3\sigma^2\right)$

7.13) (1) $\dfrac{1}{\sigma^2}x^2$ (see example 7.4) (3) $P(|B(t)| \le x) = 2\Phi(x/\sigma\sqrt{t}\,) - 1$

7.14) 1) $G_a(x) = (x - x_s)\left(\dfrac{x_0}{x_s}\right)^{\gamma}$ with $\gamma = \dfrac{\sqrt{2\alpha}}{\sigma}$ 2) $G(x) = x - x_s$

7.15) (1) Optimal level $x*$ given by formula (7.42) with $\lambda = \dfrac{2|\mu|}{(1+a)^2\sigma^2}$

(2) Optimal level $x*$ given by formula (7.42) with λ replaced by

$$\gamma = \frac{1}{(1+a)^2\sigma^2}\left(\sqrt{2(1+a)^2\sigma^2\alpha + \mu^2} - \mu\right)$$

Hint Note that $\{X(t), t \ge 0\}$ hits a positive level x with $x > x_0$ at the same time point as the geometric Brownian motion with drift $\{x_0 e^{\mu t + (1+a)B(t)}, t \ge 0\}$.

7.16) at time τ

7.17) 1) $G = \sigma\sqrt{\tau}\; c\,\Phi(c) + \sigma\sqrt{\dfrac{\tau}{2\pi}}\; e^{-\frac{1}{2}c^2}$, where $c = \dfrac{x_0 + \mu\tau - x_s}{\sigma\sqrt{\tau}}$ 2) $G = \sigma\sqrt{\dfrac{\tau}{2\pi}}$

REFERENCES

[1] Asmussen, S., *Ruin Probabilities*, World Scientific. Singapore, London, 2000.

[2] Bachelier, L., Théorie de la spéculation,
Ann. Scient. de l' École Normale Supér., 17, 21, 1900.

[3] Barlow, R.E. and Proschan, F., *Mathematical Theory of Reliability,*
Wiley & Sons, New York, 1965.

[4] Barlow, R.E. and Proschan, F., *Statistical Theory of Reliability and Life Testing,*
Holt, Rinehart & Winston, New York, 1975.

[5] Beichelt, F., A general preventive maintenance policy,
Mathem. Operationsforschung und Statistik, 7, 927, 1976.

[6] Beichelt, F., *Stochastische Prozesse für Ingenieure*,
B.G. Teubner, Stuttgart, 1997.

[7] Beichelt, F. and Fatti, P., *Stochastic Processes and their Applications*,
Taylor and Francis, London, New York, 2002.

[8] Beichelt, F. and Montgomery, D.C., Eds., *Teubner-Taschenbuch der Stochastik*,
B.G. Teubner, Stuttgart (2003).

[9] Bhat, U.N. and Miller, G.K., *Elements of Applied Stochastic Processes*, 3rd ed.,
Wiley, New York, 2002.

[10] Black, F. and Scholes, M., The pricing of options and corporate liabilities.
Journal of Political Economy 81, 637, 1973.

[11] Borovkov, K., *Elements of Stochastic Modeling*,
World Scientific, Singapore, 2003.

[12] Bouchaud, J-P. and Potters, M., *Theory of Financial Risks*,
Cambridge University Press, 2000.

[13] Brandt, A., Franken, P., and Lisek, B., *Stationary Stochastic Models,*
Wiley, New York, 1990.

[14] Brown, M., Bounds, inequalities and monotonicity properties for some specialized renewal processes,
Annals Probab., 22, 93, 1980.

[15] Brown, R., A brief account of microscopial observations made in the months of June, July, and August, 1827, on particles contained in the pollen of plants; and on the general existence of active molecules in organic and inorganic bodies,
Phil. Mag., Series 2, 161, 1828.

[16] Brzeźniak, Z. and Zastawniak, T., *Basic Stochastic Processes*,
Springer, New York-Berlin, 1999.

[17] Capasso, V. and Bakstein, D., *An Introduction to Continuous-Time Stochastic Processes*, Birkhäuser, 2005.

[18] Chhikara, R.S. and Folks, J.L., *The Inverse Gaussian Distribution: Theory, Methodology and Applications*, Marcel Dekker, New York, 1989.

[19] Chung, K. L., *Markov Chains with Stationary Transition Probabilities*, Springer-Verlag, Berlin, 1960.

[20] Chung, K.L. and AitSahlia, F., *Elementary Probability*, Springer, New York, Berlin, Heidelberg, 2003.

[21] Cox, D.R., Some statistical methods connected with series of events. *J. Roy. Statist. Soc. B,* 17, 129, 1955.

[22] Cramér, H. and Leadbetter, M.R., *Stationary and Related Stochastic Processes,* Wiley, New York, 1967.

[23] Doob, J.L., *Stochastic Processes*, Wiley & Sons, New York, 1953.

[24] Dubourdieu, J., Remarques relatives a la théorie mathématique de l' assurance accidents, *Bull. Trim. de l'Inst. des Actuaires Français,* 49, 76, 1938.

[25] Durrett, R., *Essentials of Stochastic Processes,* Springer, New York, Berlin, 1999.

[26] Einstein, A., Über die von der molekularkinetischen Theorie der Wärme geforderte Bewegung von in ruhenden Flüssigkeiten suspendierten Teilchen. Annalen der Physik 17, 549, 1905.

[27] Feller, W., *An Introduction to Probability Theory and its Applications*, Vol. I, 3rd ed., Wiley & Sons, New York, 1968.

[28] Feller, W., *An Introduction to Probability Theory and its Applications*, Vol. II, Wiley & Sons, New York, 1971.

[29] Franken, P. et al. (authors are Franken, König, Arndt, and Schmidt) *Queues and Point Processes,* Akademie-Verlag, Berlin, 1981.

[30] Franz, J., Niveaudurchgangszeiten zur Charakterisierung sequentieller Schätzverfahren, *Mathem. Operationsforsch.* Series Statistics, 8, 499, 1977.

[31] Gardner, W. A., *Introduction to Random Processes with Applications to Signals and Systems*, Mc Graw-Hill Publishing Company, New York, 1989.

[32] Gelenbe, E. and Pujolle, G., *Introduction to Queueing Networks,* Wiley & Sons, New York, 1987.

[33] Gnedenko, B.W. and König, D., *Handbuch der Bedienungstheorie* I, II, Akademie-Verlag, Berlin 1983, 1984.

[34] Grandell, J., *Aspects of Risk Theory*, Springer-Verlag, New York, Berlin, 1991.

[35] Grandell J., *Mixed Poisson Processes*, Chapman & Hall, London, 1997.

[36] Grimmett, G.R. and Stirzaker, D.R., *Probability and Random Processes*, 3rd ed., Oxford University Press, Oxford, 2001.

[37] Guo, R., Ascher, H., and Love, E., Generalized models of repairable systems -a survey via stochastic processes formalism, *South African Journ. Oper. Res.* (*ORION*) 16, 2, 87, 2000.

[38] Gut, A., Cumulative shock models, *Adv. Appl. Prob.*, 22, 504, 1990.

[39] Hellstrom, C. W., *Probability and Stochastic Processes for Engineers*, Macmillan Publishing Company, New York, London, 1984.

[40] Hunt, P.J. and Kennedy, J.E., *Financial Derivatives in Theory and Practice*, Wiley & and Sons, New York, 2000.

[41] Kaas, R. et al. (authors are Kaas, Goovaerts, Dhaene, and Denuit), *Modern Actuarial Risk Theory*, Springer, New York, 2004.

[42] Kahle, W. and Lehmann, A., Parameter Estimation in Damage Processes, in: *Advances in Stochastic Models for Reliability, Quality and Safety*, Kahle, W., von Collani, E., Franz, J., and Jensen, U., Eds., Birkhäuser, Boston, Basel, 1998.

[43] Kannan, D., *An Introduction to Stochastic Processes,* North Holland; New York, 1979.

[44] Kapur, P.K., Garg, R.B., and Kumar, S., *Contributions to Hardware and Software Reliability*, World Scientific, Singapore, London, 1999.

[45] Karlin, S. and Taylor, H. M., *A Second Course to Stochastic Processes,* Academic Press, New York, 1981.

[46] Karlin, S. and Taylor, H. M., *An Introduction to Stochastic Modeling*, Academic Press, New York, 1994.

[47] Kendall, D.G., On the generalized 'birth-and-death' process, *Ann. Math. Statistics*, 19, 1, 1948.

[48] Kijima, M., Morimura, H., and Suzuki, Y., Periodical Replacement Policy without assuming Minimal Repair. *Europ. Journ. Oper. Research* 37, 194, 1988.

[49] Kijima, M., *Markov Processes for Stochastic Modeling*, Chapman & Hall, London, New York, 1996.

[50] Kijima, M., *Stochastic Processes with Applications to Finance*, Chapman & Hall, Boca Raton, Fla, 2003.

[51] König, D. and Schmidt, V., *Zufällige Punktprozesse. Eine Einführung mit Anwendungsbeispielen*, Teubner, Stuttgart, 1991.

[52] Kulkarni, V.G., *Modeling and Analysis of Stochastic Systems*, Chapman & Hall, New York, London, 1995.

[53] Kulkarni, V.G., *Modeling, Analysis, Design, and Control of Stochastic Systems*, Springer, New York, 1999.

[54] Lawler, G., *Introduction to Stochastic Processes*, Chapman & Hall, London, 1995.

[55] Lorden, G., On excess over the boundary, *Ann. Math. Statistics*, 41, 520, 1970.

[56] Lundberg, O., *On Random Processes and their Application to Sickness and Accident Statistics*, Almqvist och Wiksell, Uppsala, 1964.

[57] Makabe, H. and Morimura, H., A new policy for preventive maintenance. *J. Oper. Res. Soc. Japan* 5, 110, 1963.

[58] Mann, H.B. and Whitney, D.R., On a test whether one of two random variables is stochastically larger than the other, *Ann. Math. Statistics*, 18, 50, 1947.

[59] Marshall, K.T., Linear bounds on the renewal function, *SIAM J. Appl. Math.*, 24, 245, 1973.

[60] Matthes, K., Kerstan, J., and Mecke, J., *Unbegrenzt teilbare Punktprozesse*, Akademie-Verlag, Berlin, 1974,
English edition: *Infinitely Divisible Point Processes*, Wiley, New York, 1978.

[61] Merton, R.C., Theory of rational option pricing, *Bell Journal of Economics and Management Science* 4, 141, 1973.

[62] Müller, A. and Stoyan, D., *Comparison Methods for Stochastic Models and Risks*, Wiley & Sons, 2002.

[63] Ochi, M. K., *Applied Probability and Stochastic Processes in Engineering and Physical Sciences*, Wiley, New York, 1990.

[64] Paul, W. and Baschnagel, J., *Stochastic Processes. From Physics to Finance*, Springer, New York, 2000.

[65] Pieper, V., *Zuverlässigkeitsuntersuchungen auf der Grundlage von Niveauüberschreitungsuntersuchungen*,
Dissertation, Technical University 'Otto von Guericke', Magdeburg, 1988.

[66] Resnick, S.I., *Adventures in Stochastic Processes*, Birkhäuser, Basel, 1992.

[67] Rolski, T. et al (authors are: Rolski, Schmidli, Schmidt, Teugels), *Stochastic Processes for Insurance and Finance*, Wiley, NewYork, 1999.

[68] Rosin, E. and Rammler, E., The laws governing the fineness of powdered coal, *J. Inst. Fuel* 7, 29, 1933.

[69] Ross, S. M., *Stochastic Processes*, 2nd ed., Wiley & Sons, New York, 1996.

[70] Ross, S. M., *Introduction to Probability Models*, 8th ed., Academic Press, San Diego, 2003.

[71] Scheike, T. H., A boundary-crossing result for Brownian motion. *J. Appl. Prob.,* 29, 448, 1992.

[72] Schrödinger, E., Zur Theorie der Fall- und Steigversuche an Teilchen mit Brownscher Bewegung, *Physikalische Zeitschrift,* 16, 289, 1915.

[73] Seshradi, V., *The Inverse Gaussian Distribution,* Springer, New York, Berlin, 1999

[74] Shafer, G. and Vovk, V., *Probability and Finance. It's Only a Game,* Wiley & Sons, New York, 2001.

[75] Smoluchowski, M., Notiz über die Berechnung der Brownschen Molekularbewegung bei der Ehrenhaft-Millikanschen Versuchsanordnung. *Physikalische Zeitschrift,* 16, 318, 1915.

[76] Snyder, D.L., *Random Point Processes,* Wiley, New York, London, 1975.

[77] Solov'ev, A.D., *Rascet i ocenka characteristic nadežnosti,* Izd. Znanie, 1978.

[78] Stigman, K., *Stationary Marked Point Processes,* Chapman & Hall, New York, 1995.

[79] Stoyan, D., *Qualitative Eigenschaften und Abschätzungen Stochastischer Modelle,* Akademie-Verlag, Berlin, 1977.

[80] Stoyan, D., *Comparison Methods for Queues and other Stochastic Models,* Chichester, Wiley & Sons, 1983.

[81] Tijms, H. C., *Stochastic Models-An Algorithmic Approach.* Wiley & Sons, New York, 1994.

[82] Tijms, H.C., *A First Course in Stochastic Models,* Wiley & Sons, New York, 2003.

[83] Uematsu, K. and Nishida, T., One unit system with a failure rate depending upon the degree of repair, *Mathematica Japonica* 32, 139, 1987.

[84] van Dijk, N. *Queueing Networks and Product Forms,* Wiley, New York, 1993.

[85] Vinogradov, O.P., O primenenijach odnoj formuly obraščenija preobrasovanija laplasa, *Teorija verojatnost. i primenen.,* 21, 857, 1976.

[86] Walrand, J., *An Introduction to Queueing Networks,* Prentice Hall, Englewood Cliffs, 1988.

[87] Wiener, N., Differential space *J. Math. and Phys.,* 2, 131, 1923.

[88] Williams, D., *Probability with Martingales,* Cambridge University Press, Cambridge, 1992.

[89] Willmot, G.E. and Lin, X.S., *Lundberg Approximations for Compound Distributions with Insurance Applications*,
Springer, New York, 2001.

[90] Yates, R.D. and Goodman, D.J., *Probability and Stochastic Processes*, Wiley & Sons, New York, 2005.

INDEX

M

N

O

P